Biomass for Renewable Energy, Fuels, and Chemicals

Biomass for Renewable Energy, Fuels, and Chemicals

Donald L. Klass

Entech International, Inc.

Academic Press

San Diego   London   Boston   New York   Sydney   Tokyo   Toronto

This book is printed on acid-free paper. ⊗

Academic Press
a division of Harcourt Brace & Company
525 B Street, Suite 1900, San Diego, California 92101-4495, USA
http://www.apnet.com

Academic Press Limited
24-28 Oval Road, London NW1 7DX, UK
http://www.hbuk.co.uk/ap/

Library of Congress Catalog Card Number: 98-84422

International Standard Book Number: 0-12-410950-0

PRINTED IN THE UNITED STATES OF AMERICA
98 99 00 01 02 03 MM 9 8 7 6 5 4 3 2 1

CONTENTS

13 Organic Commodity Chemicals from Biomass

14 Integrated Biomass Production–Conversion Systems and Net Energy Production

Epilogue

ABOUT THE AUTHOR

Dr. Donald L. Klass is Director of Research for Entech International, Inc., an energy and environmental consulting firm headquartered in Barrington, Illinois. He also serves as President of the Biomass Energy Research Association, a membership association in Washington, D.C. founded in 1982 by industry and university researchers throughout North America. He is a member of several research advisory boards and business development committees, and a consultant to industry and government, both U.S. and foreign. Formerly, Dr. Klass managed biomass, natural gas, and petroleum research and educational programs for the Institute of Gas Technology and the petroleum industry. Over a 40-year period, his R&D and commercialization experience has been concentrated on the conversion of virgin and waste biomass to gaseous and liquid fuels and chemicals, the development of petrochemical, refinery, and gas separation processes, the microbial production of substitute natural gas from biomass and single cell protein from methane and methanol, and the development of electroviscous fluids. Dr. Klass is the author or co-author of over 300 papers and patents in these fields and the editor of 27 books on energy and fuels. He received his B.S. in chemistry from the University of Illinois, and his A.M. and Ph.D. in organic chemistry from Harvard University.

PREFACE

The need for energy and fuels is one of the common threads throughout history and is related to almost everything that man does or wishes to do. Energy, in its many useful forms, is a basic element that influences and limits our standard of living and technological progress. It is clearly an essential support system for all of us. In the twentieth century, the subject did not receive much attention until well into the middle of the century, that is, the fossil fuel era, and then usually only in crisis situations of one kind or another. Until we were confronted with energy and fuel shortages that affected our daily lives, most of us assumed that the petroleum, natural gas, and electric power industries would exist forever. A bountiful supply of energy in whatever forms needed was taken for granted.

An energy corollary to the economic law of supply and demand gradually evolved. In the early 1970s, the law's first derivative might legitimately have been called the law of energy availability and cost. The oil marketing policies of the Organization of Petroleum Exporting Countries initiated the so-called First Oil Shock in 1973–1974 and changed, probably forever, the international oil markets and the energy policies of most industrialized nations. Oil prices increased dramatically, seemingly overnight. Markets were disrupted and shortages developed. Crash programs to develop alternatives to petroleum-based fuels began in earnest in many parts of the world. Many of these programs continue today.

Intensive research programs were started to develop renewable energy resources such as active and passive solar energy, photovoltaic, wind, and ocean power systems, and biomass—the only indigenous renewable energy resource capable of displacing large amounts of solid, liquid, and gaseous fossil fuels. As a widely dispersed, naturally occurring carbon resource, biomass was a logical choice as a raw material for the production of a broad range of fossil fuel substitutes. Environmental issues such as air quality and global climate change that many believe are related to fossil fuel consumption also began to come to the fore. The world appeared ready to resurrect biomass as a major indigenous energy resource for industrialized nations, as it had been up to the end of the nineteenth century. It now appears that biomass energy will displace increasingly larger amounts of fossil fuels as time passes.

This book addresses biomass energy technologies and the development of virgin and waste biomass as renewable, indigenous, energy resources for the production of heat, steam, and electric power, as well as solid, liquid, and gaseous fuels that are suitable as substitutes for fossil fuels and their refined products. Biomass is defined as nonfossil, energy-containing forms of carbon and includes all land- and water-based vegetation and such materials as municipal solid wastes, forestry and agricultural residues, municipal biosolids, and some industrial wastes. In other words, biomass is all nonfossil organic materials that have an intrinsic chemical energy content. The history, status, and future expectations of biomass research, development, and deployment efforts are examined from the standpoint of the role of biomass in our global and national energy economy, the impact of biomass energy use on the environment, its potential to replace fossil fuels, and the commercial systems already in place. The development of advanced technology and improved biomass growth and conversion processes and environmental issues are also discussed. One chapter is also devoted to organic commodity chemicals from biomass.

Because of the special organization of most chapters, this book should serve as an introduction to the subject for the student and professional who wish to become knowledgeable about the production and consumption of biomass energy and its potential long-range impact. This book is also useful for energy professionals interested in some of the technical details of and references for specific biomass energy applications. One special feature of the book that will become apparent to the reader is that it is multidisciplinary in content and treatment of the subject matter, because many scientific and engineering disciplines are directly or indirectly involved in the development of biomass energy. For example, the biological gasification of biomass is described in terms of its microbiology and biochemistry, but the practical use of this information for the design and operation of combined waste disposal–methane production processes for feedstocks such as municipal solid waste is also discussed. Another example of the multidisciplinary nature of the book is the treatment given to biomass production. The study and selection of special strains of hybrid trees for use as biomass resources, as well as the advanced agricultural practices used for growth and harvesting in short-rotation biomass plantations, are discussed. An important feature of this book is the effort to discuss barriers that hinder biomass energy utilization and what must be done to overcome them. An example of one of the barriers is net energy production. Given that the prime objective of a biomass energy system is to replace fossil fuels in a specific application, the system cannot effectively attain this objective if the net energy output available for market is less than the total of nonrenewable energy inputs required to operate the system.

Throughout this book, the International System of Units, or le Système International d'Unités (SI), is used. SI is a modern metrication system of measurement consisting of coherent base and derived units and is used by many scientists, engineers, and energy specialists. Most major technical associations and publishers require that SI be used in their publications. Because of the United States' position as the world's largest energy-consuming country, commonly used U.S. energy units, several of which are somewhat unusual, such as quads of energy (1×10^{15} Btu/quad) and barrels of oil equivalent (BOE), and their conversion factors are presented in Appendix A along with the definitions and conversion factors for SI units. This makes it possible for the reader who is familiar only with U.S. units to readily convert them to SI units and to convert the SI units used in the text to common U.S. units. In the text, common U.S. units are sometimes cited in parentheses after the SI units for clarification.

ACKNOWLEDGMENTS

I would like to take this opportunity to thank several groups and individuals who helped formulate my thinking on the subject of this book. It may seem unusual to some, but the Institute of Gas Technology, an education and research institute that specializes in the fossil fuel natural gas, is where I first became interested in biomass after spending several years in the petroleum industry. IGT's policy, which was very close to one of academic freedom, and my association with colleagues in both research and education were invaluable in encouraging and stimulating me to structure and sustain a biomass research program. I also had the opportunity to develop the conference "Energy from Biomass and Wastes" that was started almost simultaneously with the renewal of interest in biomass R&D in North America in the 1970s. The conference was presented annually until I retired from IGT in 1992. Literally hundreds of researchers and project developers presented the results of their efforts at this conference. I learned much about biomass energy research and commercialization from these meetings. The exchange of new ideas and information always inspired in me fresh approaches to new projects. My association with the directors and many of the members of the Biomass Energy Research Association (BERA) and direct contact with the Washington scene as a result of this affiliation since the early 1980s had the same stimulatory effect. I thank all of my colleagues, many of whom are still involved in biomass energy development, for sharing their thoughts and expertise with me. Without their contributions to my "data bank" over a period of three decades, it would have been impossible for me to prepare this book. Finally, I want to extend a special thank you to Dr. Don J. Stevens, a director of BERA and consultant with Cascade Research, Inc. I invited him to review the manuscript of this book. He accepted and performed a superb job of providing me with an objective assessment and numerous suggestions.

CHAPTER 1

Energy Consumption, Reserves, Depletion, and Environmental Issues

I. INTRODUCTION

It is well known that developed or industrialized nations consume more energy per capita than developing or Third World countries, and that there is a correlation between a country's living standards and energy consumption. In general, the higher the per-capita energy consumption, the higher the living standard. However, the rapid worldwide increase in the consumption of fossil fuels in the twentieth century to meet energy demand, mostly by industrialized nations, suggests that the time is not too distant before depletion begins to adversely affect petroleum and natural gas reserves. This is expected to result in increased usage of alternative biomass energy resources.

The potentially damaging environmental effect of continued fossil fuel usage is another factor that will affect biomass energy usage. It has not been established with certainty that on a global basis, there is a specific relationship between fossil fuel consumption and environmental quality. There is also considerable disagreement as to whether increased fossil fuel consumption is the primary cause of global climate change. But most energy and environmental specialists agree that there is a strong relationship between localized and regional air quality in terms of pollutant concentrations and fossil fuel consumption. The greater the consumption of fossil fuels, especially by motor vehicles and power plants, the greater the levels of air pollution in a given region.

These issues are briefly examined in this chapter to provide a starting point and a foundation for development of the primary subject of this book—energy and fuels from virgin and waste biomass. Special emphasis is given to the United States because it utilizes about one quarter of the energy consumed in the world.

II. HISTORICAL ENERGY CONSUMPTION PATTERNS

It was not too many years ago that humans' basic survival depended in whole or in part on the availability of biomass as a source of foodstuffs for human and animal consumption, of building materials, *and* of energy for heating and cooking. Not much has changed in this regard in the Third World countries since preindustrial times. But industrial societies have modified and added to this list of necessities, particularly to the energy category. Biomass is now a minor source of energy and fuels in industrialized countries. It has been replaced by coal, petroleum crude oil, and natural gas, which have become the raw materials of choice for the manufacture and production of a host of derived products and energy as heat, steam, and electric power, as well as solid, liquid, and gaseous fuels. The fossil fuel era has indeed had a large impact on civilization and industrial development. But since the reserves of fossil fuels are depleted as they are consumed, and environmental issues, mainly those concerned with air quality problems, are perceived by many scientists to be directly related to fossil fuel consumption, biomass is expected to exhibit increasing usage as an energy resource and feedstock for the production of organic fuels and commodity chemicals. Biomass is one of the few renewable, indigenous, widely dispersed, natural resources that can be utilized to reduce both the amount of fossil fuels burned and several greenhouse gases emitted by or formed during fossil fuel combustion processes. Carbon dioxide, for example, is one of the primary products of fossil fuel combustion and is a greenhouse gas that is widely believed to be associated with global warming. It is removed from the atmosphere via carbon fixation by photosynthesis of the fixed carbon in biomass.

A. ENERGY CONSUMPTION IN THE UNITED STATES

The gradual change in the energy consumption pattern of the United States from 1860 to 1990 is illustrated in Fig. 1.1. In the mid-1800s, biomass, princi-

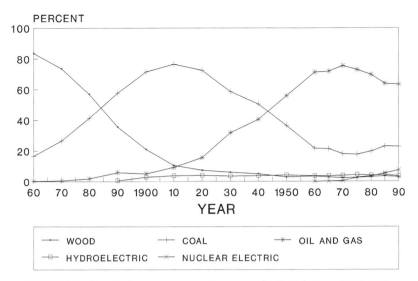

FIGURE 1.1   Historical energy consumption pattern for United States, 1860–1990.

pally woody biomass, supplied over 90% of U.S. energy and fuel needs, after which biomass consumption began to decrease as fossil fuels became the preferred energy resources. For many years, a safe illuminant had been sought as a less expensive substitute for whale oils. By the mid-1800s, distillation of coal oils yielded naphthas, coal oil kerosines, lubricants, and waxes, while liquid fuels were manufactured by the distillation of petroleum, asphalt, and bituminous shales. Coal slowly displaced biomass and became the primary energy resource until natural gas and oil began to displace coal. In 1816, the first gas company was established in Baltimore, and by 1859, more than 300 U.S. cities were lighted by gas. Natural gas was no longer a curiosity, but illuminating gas manufactured from coal by thermal gasification processes still ruled the burgeoning gas industry. Natural gas did not come to the fore until manufactured gas was widely adopted for cooking, space heating, water heating, and industrial uses. Installation of a nationwide pipeline grid system after World War II for transmission of natural gas eventually made it available in most urban areas.

After the first oil well was drilled in 1859 in Titus, Pennsylvania, for the specific purpose of bringing liquid petroleum to the surface in quantity, producing oil wells were drilled in many states. The installation of long-distance pipelines for transport of oil from the producing regions to the refineries and the natural gas pipeline grid signaled the end of coal's dominance as an energy resource in the United States. As shown in Fig. 1.1, the percentage contributions to total primary U.S. energy demand in the 1990s were about 70% for petroleum and natural gas and 20% for coal. Biomass, hydroelectric power, and nuclear power made up the balance. It is noteworthy that since the advent of nuclear power, its overall contribution to U.S. energy demand has remained relatively small.

Over the period 1860 to 1990, U.S. fossil fuel consumption correlated well with the growth in population (Fig. 1.2), but more revealing is the trend over the same period, in annual and per-capita U.S. energy consumption (Fig. 1.3). As technology advanced, the efficiency of energy utilization increased. Less energy per capita was consumed even though living standards were dramatically improved. Large reductions in per-capita energy consumption occurred from over 600 GJ/capita-year (102 BOE/capita-year) in 1860 to a level of about 200 GJ/capita-year in 1900. Per-capita energy consumption then remained relatively stable until the 1940s when it began to increase again. In the 1970s, energy consumption stabilized again at about 350 GJ/capita-year (59 BOE/capita-year). This is undoubtedly due to the emphasis that has been given to energy conservation and the more efficient utilization of energy and because of improvements in energy-consuming processes and hardware.

Because of the increasing efficiency of energy utilization, the energy consumed per U.S. gross national product dollar exhibited substantial reductions also over the period 1930 to the early 1990s (Fig. 1.4). The U.S. gross national

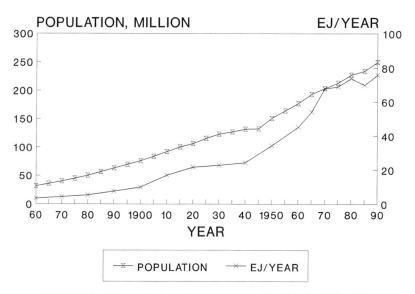

FIGURE 1.2   U.S. population and consumption of fossil fuels, 1860–1990.

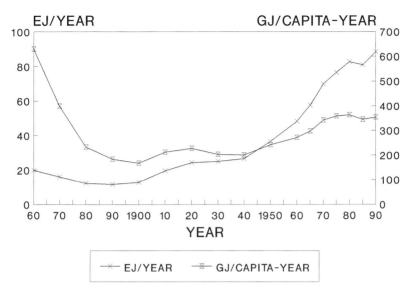

FIGURE 1.3 Annual and per-capita energy consumption for United States, 1860–1990.

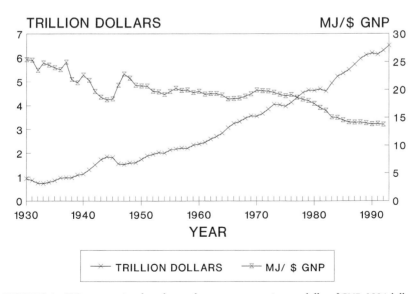

FIGURE 1.4 U.S. gross national product and energy consumption per dollar of GNP, 1994 dollars.

product increased more than sixfold in 1994 dollars over this period, while energy consumption per dollar of gross national product decreased from about 26 MJ/$ GNP to 14 MJ/$ GNP.

B. GLOBAL ENERGY CONSUMPTION

The relationship of gross national product per capita to energy consumption per capita for most countries of the world correlates very well with the status of economic and technological development. The World Bank defines developing countries as low- and middle-income countries for which the annual gross national product is $5,999 or less per capita (World Bank, 1989; U.S. Congress, 1991). With the exceptions of Brunei, Bahrain, Japan, Kuwait, Qatar, Saudi Arabia, Singapore, and the United Arab Emirates, it includes all countries in Africa, Asia, Latin America, and the Middle East, and Bulgaria, Greece, Hungary, Papua New Guinea, Poland, Turkey, and the former Yugoslavia. All of the developing countries that have annual gross national products of less than $5,999 per capita also consume less than 25 BOE/capita-year (3,300 kg of oil equivalent/capita-year). In fact, there is a good correlation between the magnitude of annual energy consumption per capita and the corresponding gross national product per capita for both the developing and developed countries (Fig. 1.5).

Annual global energy consumption statistics by region show that although fossil fuels supply the vast majority of energy demand, the developing areas of the world consume more biomass energy than the developed or more industrialized regions (Tables 1.1 and 1.2). More than one-third of the energy consumed in Africa, for example, is supplied by biomass. But examination of the energy consumption and population statistics in modern times of the world's 10 highest energy-consuming countries reveals some interesting trends that may not generally be intuitively realized. Excluding biomass energy consumption, these countries consumed about 65% of the world's primary energy demand in 1992 and contained about one-half of the world's population (Table 1.3). The industrialized countries and some of the more populated countries of the world are responsible for most of the world's primary energy consumption (65%) and for most of the fossil fuel consumption. One extreme, however, is represented by the United States, which has only about 5% of the world's population, and yet consumes about one quarter of the total global primary energy demand. Coal, oil, and natural gas contributed 23, 41, and 25%, respectively, to total U.S. energy demand in 1992, about 80% of which was produced within the United States. Oil has been the single largest source of energy for many years. The U.S. per-capita energy consumption in 1992, 56.3 BOE/capita, was second only to that of Canada, 69.8 BOE/capita, in this group of countries.

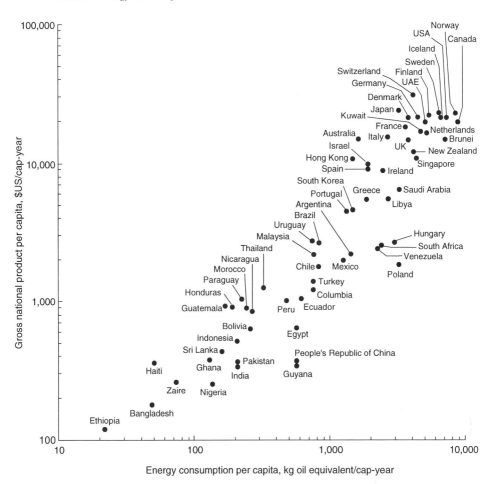

FIGURE 1.5   Gross national product vs energy consumption of selected countries, 1990.

Another extreme is represented by China and India, which rank first and second in population. Their respective per-capita energy consumptions were 4.4 and 1.7 barrels of oil equivalent in 1992, the smallest in this group of countries. Of the three fossil fuels—coal, oil, and natural gas—coal contributed 78 and 60% to energy demand in China and India, while natural gas contributed only 2 and 6%, respectively. This suggests that the indigenous reserves of coal are large and those of natural gas are small in these countries.

Globally, total energy consumption exhibited an almost exponential increase from 1860 to 1990. Total consumption increased from 16 to 403 EJ, or by a

TABLE 1.1   Global Energy Consumption by Region and Energy Source in 1990[a]

Region[b]	Fossil fuel[c] (EJ)			Electricity[d] (EJ)	Biomass[e] (EJ)	Total (EJ)
	Solids	Liquids	Gases			
Africa	2.96	3.36	1.55	0.18	4.68	12.73
America, N.	21.55	38.48	22.13	4.69	3.75[f]	90.60
America, S.	0.68	4.66	2.09	1.29	2.71	11.43
Asia	35.52	27.58	8.38	2.57	8.89	82.94
Europe	35.18	40.90	37.16	6.25	1.29	120.85
Oceania	1.64	1.70	0.85	0.14	0.19	4.53
World	97.52	116.68	72.18	15.13	21.51	323.02

[a]Adapted from United Nations (1992). The sums of individual figures may not equal the totals because of rounding.
[b]Europe includes the former U.S.S.R.
[c]Solids are hard coal, lignite, peat, and oil shale. Liquids are crude petroleum and natural gas liquids. Gases are natural gas.
[d]Electricity includes hydro, nuclear, and geothermal sources, but not fossil fuel-based electricity, which is included in fossil fuels.
[e]Biomass includes fuelwood, charcoal, bagasse, and animal, crop, pulp, paper, and municipal solid wastes, but does not include derived biofuels.
[f]Estimated by the author: 2.95 EJ for the U.S.A., 0.5 EJ for Canada, and 0.3 EJ for Mexico. More details are presented in Chapter 2.

factor of about 25 (Klass, 1992) (Fig. 1.6). The world's population exhibited about a fivefold increase to 5.3 billion people over this same period. From 1860 to the mid-1930s, the world's population, total fossil fuel consumption,

TABLE 1.2   Global Energy Consumption in Percent by Region and Energy Source in 1990[a]

Area	Fossil fuels (%)	Nonfossil electricity (%)	Biomass[b] (%)
Africa	61.8	1.4	36.8
America, N.	90.7	5.2	4.1
America, S.	65.0	11.3	23.7
Asia	86.2	3.1	10.7
Europe	93.7	5.2	1.1
Oceania	92.7	3.1	4.2
World	88.6	4.7	6.7

[a]Derived from Table 1.1.
[b]Does not include derived biofuels such as ethanol or methane.

TABLE 1.3 Primary and Per-Capita Energy Consumption in 1992 for World's 10 Highest Energy-Consuming Countries[a]

Country	% of world total	Total (EJ)	Oil (EJ)	Natural gas (EJ)	Coal (EJ)	Net nuclear (EJ)	Net hydro (EJ)	Population (1,000s)	World rank	BOE per capita
United States	23.9	86.71	35.36	21.46	19.92	7.01	2.96	260,714	3	56.3
Russia	9.5	34.52	9.46	15.86	6.00	1.36	1.84	149,609	6	39.1
China	8.5	30.83	5.86	0.55	22.98	—	1.44	1,190,431	1	4.4
Japan	5.5	20.05	11.68	2.21	2.86	2.24	1.06	125,107	9	27.1
Germany	4.1	14.89	6.18	2.52	4.26	1.64	0.28	81,088	12	31.1
Canada	3.2	11.58	3.46	2.75	1.32	0.93	3.12	28,114	35	69.8
France	2.8	10.24	4.18	1.36	0.90	3.08	0.71	57,840	22	30.0
United Kingdom	2.8	10.21	3.87	2.31	2.87	1.04	0.12	58,135	21	29.7
Ukraine	2.5	9.24	1.76	3.87	2.73	0.81	—	51,847	23	30.2
India	2.5	8.98	2.75	0.51	4.90	0.08	0.74	919,903	2	1.7
Total		237.25	84.56	53.40	68.74	18.19	12.27	2,922,788		51.8
% of world total		65.4	58.8	68.1	73.3	80.3	50.8	51.8		

[a]Energy consumption data adapted from U.S. Department of Energy (1994). Population data are for mid-1994 (U.S. Bureau of the Census, 1994). Sums of individual figures may not equal totals because of rounding.

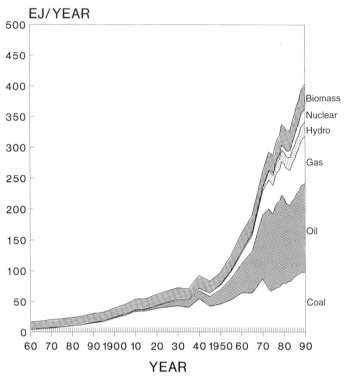

FIGURE 1.6   World energy consumption by resource, 1860–1990.

and per-capita fossil fuel consumption gradually increased, but then increased much more rapidly after the beginning of World War II (Figs. 1.7 and 1.8).

Since the 1940s, fossil energy resources have clearly become the world's largest source of energy. Interestingly, the average overall per-capita fossil fuel consumption by the world's population started to level off in the range of 60 GJ/capita-year (10 BOE/capita-year) in 1970 (Fig. 1.8). Meanwhile, the contribution of biomass energy, which was over 70% of the world's total energy demand in 1860, decreased to about 7% of total demand in the early 1990s.

III. FOSSIL FUEL RESERVES AND DEPLETION

In 1955, Farrington Daniels, professor of chemistry at the University of Wisconsin from 1920 to 1959 and a pioneer in solar energy applications, stated (Daniels and Duffie, 1955):

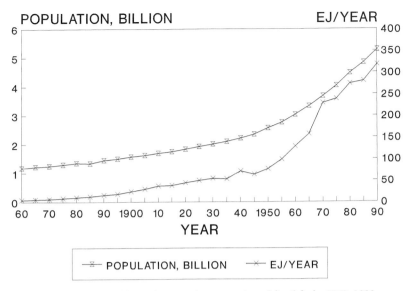

FIGURE 1.7   World population and consumption of fossil fuels, 1860–1990.

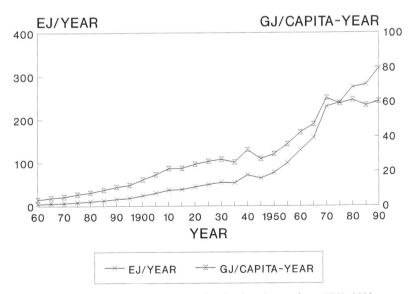

FIGURE 1.8   World consumption of coal, oil, and natural gas, 1860–1990.

> . . . our fuels were produced millions of years ago and through geological accident preserved for us in the form of coal, oil, and gas. These are essentially irreplaceable, yet we are using them up at a rapid rate. Although exhaustion of our fossil fuels is not imminent, it is inevitable.

Few people paid any attention to such remarks at that time. Many regarded them as the usual gloom-and-doom commentary of the day.

Between 1860 and 1990, the world's population and the consumption of fossil fuels per capita sequentially doubled almost three times and four times, but over the same period of years, global consumption of fossil fuels passed through six sequential doubling cycles. The doubling times for global fossil fuel consumption, population, and fossil fuel consumption per capita in the mid-1990s were approximately 25, 35, and 50 years, respectively (Table 1.4). These trends suggest several features of a society whose gradual and then rapid industrialization has depended on the availability of energy and fuels, namely that fossil fuel consumption is disproportionately increasing as more and more of the world's population is industrialized despite the large improvements in the efficiency of energy utilization over the past 50 years. Human activity and interactions at all levels require the acquisition and consumption of energy and fuels, no matter what the living standards are. It is simply a matter of increasing population and the apparent preference for energy-rich, high-quality fossil fuels. Questions of where recoverable fossil fuel deposits are located and the size of these deposits are obvious. How long will it be, for example, before the world's supplies of petroleum crude oils begin to permanently fall short of demand?

Energy specialists and reservoir engineers in the United States and several other countries use "proved reserves" to predict the amounts of coal, oil, and natural gas that can be produced and marketed. Proved reserves are defined

TABLE 1.4 Approximate Times in Years for Sequential Doubling of World Population, Fossil Fuel Consumption, and Fossil Fuel Consumption Per Capita from 1860 to 1990

Doubling sequence	Population		Fossil fuel consumption		Fossil fuel consumption/capita	
	Period	Time	Period	Time	Period	Time
First	1860–1945	85	1860–1875	15	1860–1880	20
Second	1945–1980	35	1875–1895	20	1880–1900	20
Third	1980–2015	35 est.	1895–1910	15	1900–1940	40
Fourth			1910–1940	30	1940–1990	50
Fifth			1940–1965	25		
Sixth			1965–1990	25		

as the estimated portion of a natural fossil fuel deposit that is projected from analysis of geological and engineering data with a reasonably high degree of certainty, usually a combination of experimental field data, modeling, and experience, to be economically recoverable in future years under existing economic and operating conditions. Unfortunately, there are no international standards for estimating or defining reserves, and there are many problems associated with development of accurate proved reserves figures. They are, however, the best running accounting method available today to project fossil energy supplies.

Examination of the world's proved reserves of coal, crude oil, and natural gas and their regional locations shows that well over half of the world's crude oil and natural gas supplies are located in the Middle East and the former Soviet Union, while North America, the Far East, and the former Soviet Union have over 70% of the coal reserves (Table 1.5, Fig. 1.9).

Intuitively, these data suggest that countries in those regions having large amounts of specific proved fossil fuel reserves would tend, because of proximity to these resources, to consume more of the indigenous fossil fuels than those

TABLE 1.5 Global Proved Coal, Oil, and Natural Gas Reserves by Region[a]

Region	Coal (10^6 ton)	(EJ)	Oil (10^9 bbl)	(EJ)	Natural gas (10^{12} ft³)	(EJ)
Africa	68,420	1716	75	441	326	344
America, N.	276,285	5382	81	476	329	347
America, S. and Central	10,703	224	74	439	189	199
Eastern Europe and former U.S.S.R.	329,457	6444	189	1113	2049	2160
Far East and Oceania	334,947	6928	54	319	343	361
Middle East	213	5	596	3520	1366	1440
Western Europe	129,904	2185	24	142	216	227
Total:	1,145,002	22,884	1092	6449	4817	5078

[a]The coal data are for the end of 1990 (World Energy Council, 1992). The oil and natural gas data are for January 1, 1993 (Gulf Publishing Company, 1993). The reserves data for coal, oil, and natural gas are indicated in tons, barrels, and cubic feet, respectively, as published and were not converted to SI units. The world average heating values for subbituminous, bituminous, and anthracite coals; lignite; oil; and natural gas are assumed to be 27.9 GJ/t (24 million Btu/ton), 16.3 GJ/t (14 million Btu/ton), 5.9 GJ/bbl (5.6 million Btu/bbl), and 39.3 MJ/m³(n) (1,000 Btu/ft³), respectively. The result of multiplying the amount of reserves by the world average heating value may not equal the EJs in this table because of the variation in fuel value of specific reserves within a given fuel type. The sums of individual figures may not equal the totals because of rounding.

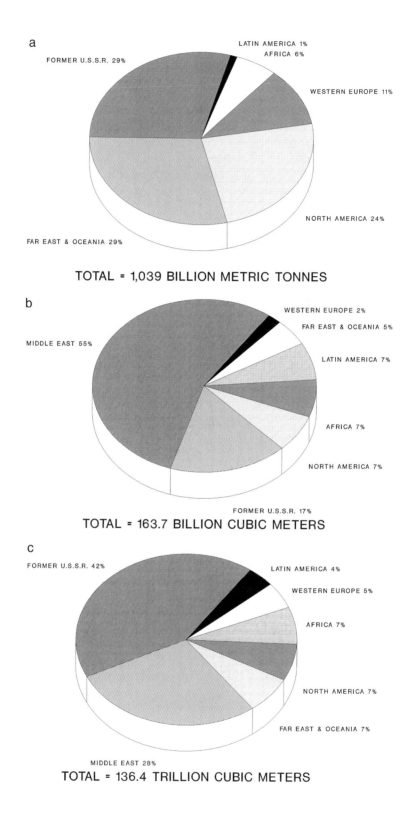

a

LATIN AMERICA 1%
AFRICA 6%
FORMER U.S.S.R. 29%
WESTERN EUROPE 11%
NORTH AMERICA 24%
FAR EAST & OCEANIA 29%

TOTAL = 1,039 BILLION METRIC TONNES

b

WESTERN EUROPE 2%
FAR EAST & OCEANIA 5%
MIDDLE EAST 55%
LATIN AMERICA 7%
AFRICA 7%
NORTH AMERICA 7%
FORMER U.S.S.R. 17%

TOTAL = 163.7 BILLION CUBIC METERS

c

FORMER U.S.S.R. 42%
LATIN AMERICA 4%
WESTERN EUROPE 5%
AFRICA 7%
NORTH AMERICA 7%
FAR EAST & OCEANIA 7%
MIDDLE EAST 28%

TOTAL = 136.4 TRILLION CUBIC METERS

that are not within their confines. This is often the case, as illustrated by some of the data in Table 1.3 for the world's 10 highest energy-consuming countries. There are many exceptions. The proved reserves-to-annual consumption ratios calculated from the proved reserves and annual consumption data for coal, crude oil, and natural gas for a few selected countries illustrate some of these exceptions (Table 1.6). In theory, these ratios indicate the number of years until the proved reserves of a particular resource are exhausted, assuming no imports of fossil fuels, a constant rate of fuel consumption, and no further discoveries of economically recoverable coal, oil, or natural gas. According to these data, a 258-year supply of coal, the world's largest energy resource of the three conventional fossil fuels, is available in the United States, whereas oil and natural gas have much shorter depletion times. Nevertheless, coal currently contributes less to energy demand than either oil or natural gas. In contrast, other countries such as China, Germany, and India have large proved reserves of coal and consume relatively large amounts, while Saudi Arabia has essentially no proved coal reserves and consumes none. Worldwide, coal consumption grew at an annual rate of 1.4% between 1980 and 1993 and accounted for about 25% of the world's total energy use in 1993, so it continues to be an important energy resource.

Oil is clearly a much smaller fossil energy resource than coal. Because of its intrinsic properties such as high energy density, ease of transport, storage, and conversion to storable liquid fuels, and an existing infrastructure that facilitates worldwide distribution of refined products to the consumer, it is the fossil fuel of choice for the manufacture of motor fuels. Some countries, such as Japan, that have little or no proved reserves of oil consume relatively large quantities and are therefore strongly dependent on imports to meet demand. Some countries, such as Saudi Arabia, have an abundance of proved oil reserves and supply their own demands as well as a large fraction of the world's markets. Global consumption of oil increased by 18.4 EJ between 1983 and 1992 at an annual rate of growth of 1.5% (U.S. Dept. of Energy, 1994). Motor fuels from oil are expected to remain the dominant international transportation fuel for the foreseeable future. Other projections indicate that global consumption of oil will exhibit a growth rate of nearly 2% per year up to 2015 (U.S. Dept. of Energy, 1996). While natural gas and renewables are making inroads into the energy markets of OECD (Organization for Economic Cooperation and Development) nations, leading to a decline in oil's share in those

FIGURE 1.9 (a) World coal reserves by region, December 31, 1990. (b) World oil reserves by region, January 1, 1993. (c) World natural gas reserves by region, January 1, 1993.

TABLE 1.6   Proved Reserves-to-Annual Consumption Ratios for Fossil Fuels for Selected Countries and World[a]

Country	Proved reserves (EJ)	Annual consumption (EJ)	Ratio
United States			
Coal	5144	20	258
Oil	140	35	4
Natural gas	174	21	8
China			
Coal	2586	23	113
Oil	175	5.9	30
Natural gas	47	0.6	86
Japan			
Coal	23	2.9	8
Oil	0	12	0
Natural gas	0	2.2	0
Germany			
Coal	1581	4.3	371
Oil	1.2	6.2	0.2
Natural gas	8.2	2.5	3
India			
Coal	1773	4.9	362
Oil	35	2.8	13
Natural gas	25	0.5	49
Saudi Arabia			
Coal	0	0	0
Oil	1541	2.4	647
Natural gas	195	1.3	147
World			
Coal	22,886	94	244
Oil	6449	144	45
Natural gas	5078	78	65

[a]Data adapted from U.S. Department of Energy (1994).

markets, its share is rising in the developing nations as transportation, industrial, and other uses for oil expand.

Natural gas is somewhat similar to oil in that it is a relatively clean-burning fuel compared to coal. Long-distance pipelines have been built in many developed and developing countries to deliver gas from the producing areas to large urban markets where it is delivered to the consumer via local gas distribution networks. In modern combined-cycle, cogeneration systems, it is generally the fossil fuel of choice for electric power production and stationary applications. Again, a correlation does not necessarily exist between the location of indigenous proved reserves in a given country and energy consumption in that country. Japan is an example of a country that has no natural gas reserves, yet consumes considerable natural gas that is transported to Japan from producing countries as liquefied natural gas (LNG) in large cryogenic tankers. Another example is the utilization of the large reserves of natural gas in Eastern Europe. Consumption is high in Eastern Europe, but high-pressure pipelines are used to transport natural gas from producing regions in Eastern Europe to Western Europe where proved reserves are small. Natural gas is the fastest-growing fossil fuel in the world's energy mix. Its annual rate of growth in production was 3.7% from 1983 to 1992, and it contributed 22% to world energy demand in 1993.

A somewhat more quantitative estimate of depletion times for fossil fuels can be calculated under specific conditions using a simple model that accounts for proved reserves and growth rates in consumption (Appendix B). Application of this model to the consumption of global proved reserves of petroleum crude oils is presented here. Calculation of global depletion times eliminates the problem of accounting for imports and exports. The conditions assumed for these calculations are those for 1992. The world's proved reserves are 6,448 EJ, the annual consumption is 144 EJ, and the average annual growth rate in consumption of petroleum products is assumed to be a conservative 1.2%, which is projected by the U.S. Department of Energy to hold until 2010. Under these assumed conditions, the depletion time of the proved reserves of petroleum is 35 years, or the year 2027.

Current estimates of proved reserves do not represent the ultimate recoverable reserves because of ongoing oil exploration activities and new discoveries, which have generally been able to sustain proved reserves for several decades. For this reason, and because changing economic conditions and technical improvements affect the assessment of proved reserves and the economic recoverability of oil from lower-grade reserves and unconventional reserves of tar sands and oil shales, calculation of the depletion time for several multiples of the proved reserves is also of interest. The depletion time for five times the proved reserves (32,240 EJ) at the same consumption rate is 108 years, or the year 2100. The ultimate recoverable reserves are believed to be closer to two times the world's proved reserves of oil and syncrudes (12,896 EJ) from unconventional sources (Institute of Gas Technology, 1989). Note that the depletion time of 108 years for five times the proved reserves is not a factor of five greater than that calculated for proved reserves of 6,448 EJ because of the compounding effect of the growth rate in consumption of 1.2% per year; it is about three times greater. The changes in remaining reserves with time from these calculations are illustrated in Fig. 1.10.

Despite the facts that world trade in the international oil and natural gas markets is flourishing and there is little sign of a significant reduction in energy

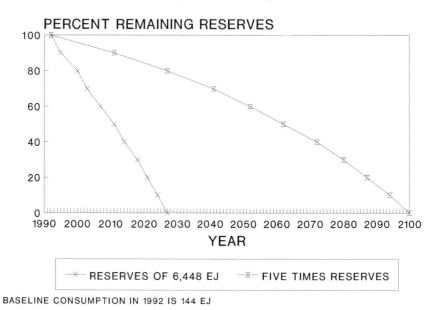

BASELINE CONSUMPTION IN 1992 IS 144 EJ

FIGURE 1.10   Global depletion of petroleum reserves at annual consumption growth rate of 1.2%.

consumption, the limited data and simplified analysis presented here suggest that gradual depletion of oil and natural gas reserves can be expected to become a major problem by the middle of the twenty-first century. Without preparation and long-range planning to develop alternative fuels, particularly nonpolluting liquid motor fuels for large-scale worldwide distribution and clean-burning fuels for power production in stationary applications, energy and fuel shortages could become severe. The disruptions in energy and fuel supply and availability that occurred in the 1970s illustrate the potential impact on society. The oil marketing policies of the Organization of Petroleum Exporting Countries (OPEC) and the resulting First Oil Shock in 1973–74 had a lasting impact on the international oil markets and the energy policies of most industrialized nations. In 1973, Mideast light crude oil spot market prices rose to about $13 per barrel from a low of about $2 per barrel. The Second Oil Shock began in 1979 as a result of OPEC's curtailment of production until spot Mideast oil prices peaked in early 1980 at $38.63 per barrel. Major policy changes and legislative actions occurred in many industrialized countries to try to counteract these conditions. The First Oil Shock resulted in a flurry of legislative activities and executive orders by the executive and legislative branches of the U.S. Government, for example, that affected literally all energy-related sectors. This was actually the beginning of national policies in many countries to develop new indigenous energy supplies. In the United States, the federal laws that have been enacted since the First and Second Oil Shocks have had a profound and continuing impact on all U.S. energy production and utilization. When it was realized that oil prices and availability could be manipulated or controlled to a significant extent by outside forces and how important these factors and their impact are for the U.S. economy, massive programs were undertaken to make the United States less dependent on imported oil. Other nations have taken similar actions. Many of these programs continue today.

A few words of caution are warranted in dealing with depletion times and the proved reserves of fossil fuels, that is, the possibility of new discoveries, the variability of depletion time, the effects of new technologies, and the uncertainty of predictions. Detailed assessment of the proved reserves-to-consumption ratios for oil and natural gas over the past several decades shows that although there has been a slight decline in the values of specific proved reserves reported by some sources, new additions to proved reserves have been able to sustain market demands over many years while the calculations indicated that depletion should have occurred in just a few years. The estimated depletion times calculated in the mid-1970s showed, for example, that the global reserves of natural gas should have been depleted by about 1995. Discovery of large new reserves capable of economic production, the development of significantly improved gas producing and processing methods, higher gas utilization efficiencies by end-use equipment, and lower actual annual growth rates in consumption than those predicted have all contributed to prolong depletion and the time of depletion. Basically, the estimates of the world's total remaining recoverable reserves of oil and natural gas have been sustained and continue to keep pace with consumption. But given the extensive periods of time required to replenish finite supplies of fossil fuels, the earth is not an infinite source of these materials when considered in terms of world energy demand and population growth. Presuming Professor Daniels' prediction that depletion of coal, oil, and natural gas is truly inevitable, it is still prudent to use these natural resources wisely. This will help conserve our valuable fossil fuels and extend the time when depletion and the unavoidable rise in energy prices and shortages occur and become a fact of life. The coupling of fossil fuel usage and environmental problems may eventually result in the equivalent of mandated conservation of fossil fuels.

IV. ENVIRONMENTAL ISSUES

A. The Greenhouse Effect

Since the early 1960s, climate change and air quality have become major and often controversial issues in many countries and among groups from governments to various scientific communities. Prominent among these issues is the greenhouse effect, in which the gradually increasing tropospheric concentrations of carbon dioxide (CO_2), methane (CH_4), and nitrous oxide (N_2O) are believed to trap an excessive amount of solar radiation reflected from the earth. The trapped radiation is predicted to cause significant ambient temperature increases. Other issues include ozone (O_3) formation over populated areas due to photochemical interactions of hydrocarbon, carbon monoxide (CO), and nitrogen oxide (NO_x) emissions, primarily from motor vehicles; natural ozone layer destruction in the stratosphere by photochemical reactions of organic chlorofluorocarbon compounds (CFCs) resulting in increased penetration to the earth's surface of shorter-wavelength ultraviolet light that can cause skin cancers; and acid rain, which has harmful effects on buildings and the growth of biomass and is caused by sulfur oxide (SO_x) emissions from the combustion of sulfur-containing fossil fuels. The predictions of some of the resulting environmental effects are quite dramatic. In the U.S. National Research Council's first assessment of the greenhouse effect in 1979, one of the primary conclusions was that if the CO_2 content of the atmosphere is doubled and thermal equilibrium is achieved, a global surface warming of between 2 and 3.5°C can occur, with greater increases occurring at higher latitudes (National Research Council, 1979). Some of the earlier predictions indicated that this increase is sufficient to cause warming of the upper layers of the oceans and a substantial rise in sea level, a pronounced shift of the agricultural zones, and major but unknown changes in the polar ice caps.

There has by no means been universal acceptance among the experts of many of the predictions that have been made, and there are many who have opposing views of the causes of some of the phenomena that have been observed and experimentally measured. However, several detailed reports were issued in the 1990s in which the consensus of large groups of experts is that human activities, largely the burning of fossil fuels, are affecting global climate. At any one location, annual variations can be large, but analyses of meteorological and other data over decades for large areas provide evidence of important systematic changes.

One of the first comprehensive estimates of global mean, near-surface temperature over the earth's lands and oceans was reported in 1986 (Jones *et al.*, 1986). The data showed a long-timescale warming trend. The three warmest years were 1980, 1981, and 1983, and five of the nine warmest years in the entire 124-year record up to 1984 were found to have occurred after 1978. It was apparent from this study that over this period, annual mean temperature increased by about 0.6 to 0.7°C, and that about 40 to 50% of this increase occurred since about 1975. According to many analysts, the warmest year on record up to 1995 is 1995, and recent years have been the warmest since 1860 despite the cooling effect of the volcanic eruption of Mt. Pinatubo in 1991 (*cf.* Intergovernmental Panel on Climate Change, 1991 and 1995). Nighttime temperatures over land have generally increased more than daytime temperatures, and regional changes are also evident. Warming has been the greatest over the mid-latitude continents in winter and spring, with a few areas of cooling such as the North Atlantic Ocean. Precipitation has increased over land in the high latitudes of the Northern Hemisphere, especially during the cold season. Global mean surface temperature has increased by between 0.3 and 0.6°C since the late nineteenth century and average global surface temperature increases of 1 to 3.5°C, somewhat lower than originally predicted, are expected to occur by the middle of the twenty-first century. Global sea level has risen by between 10 and 25 cm over the past 100 years, and much of the rise may be related to the increase in global mean temperature.

Since preindustrial times, ambient concentrations of the greenhouse gases have exhibited substantial increases, *inter alia* CO_2 by 30% to about 360 parts per million (ppm), CH_4 by 145% to more than 1,700 parts per billion (ppb), and N_2O by 15% to more than 300 ppb. The growth rates in the concentrations of these gases in the early 1990s were lower than predicted, while subsequent data indicate that the growth rates are comparable to those averaged over the 1980s. If CO_2 emissions were maintained near mid-1990 levels, analysts have predicted that this would lead to a nearly constant increase in atmospheric concentrations for at least two centuries, reaching about 500 ppm by the end of the twenty-first century, and that stabilization of atmospheric CO_2 concentrations at 450 ppm could only be achieved if global anthropogenic emissions drop to 1990 levels by about 2035, and subsequently drop substantially below 1990 levels (Intergovernmental Panel on Climate Change, 1995). It is estimated that the corresponding atmospheric lifetimes of CO_2, CH_4, and N_2O are about 50 to 200, 12, and 120 years, respectively, and that together with increasing emissions to the atmosphere, they account for the steadily rising ambient concentrations of the greenhouse gases.

These gases are called greenhouse gases because they selectively allow more of the shorter wavelengths of solar radiation to reach the earth's surface, but absorb more of the reflected longer wavelength infrared radiation than that allowed to leave the atmosphere. The result is the greenhouse effect on reradiation of the absorbed energy. An example of the change in atmospheric concentration of CO_2 at one measuring site is shown in Fig. 1.11 (Whorf, 1996). These data were accumulated from 1958 to 1995 by experimental measurement at Mauna Loa, Hawaii and show how the concentration increased from about 315 to 360 ppm over the measurement period and how it varies during the biomass growing season. The data show an approximate proportionality between the rising atmospheric concentrations and industrial CO_2 emissions (Keeling *et al.*, 1995). The distribution and a few properties of selected atmospheric gases that have infrared absorption in the atmospheric window (7 to

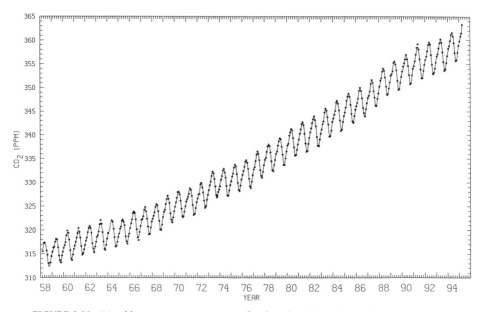

FIGURE 1.11   Monthly average concentration of carbon dioxide in dry air observed at Mauna Loa Observatory, Hawaii from March 1958 to April 1995. Note: The measurements were made with a continuously recording non-dispersive infrared gas analyzer. The smooth curve fit is a fit of the data to a four harmonic annual cycle which increases linearly with time plus a spline fit of the interannual component of the variation. From Whorf (1996).

13 μm) are listed in Table 1.7. Carbon dioxide is by far the most abundant and is indicated in this table as the relative infrared standard. The gas-to-carbon dioxide infrared absorption ratios in the atmospheric window of CH_4, N_2O, and the CFCs are much greater than 1.0. The effect of doubling the concentration of N_2O, CO, CH_4, and CO_2 on the earth's surface temperature is estimated to be 0.25, 0.6 to 0.9, 0.95, and 2 to 3°C, respectively.

Methane is present at much lower concentrations than CO_2, but is estimated to increase the surface temperature by almost 1°C on doubling of its concentration. This is predicted to occur because the methane-to-carbon dioxide infrared ratio in the atmospheric infrared window is about 25, and hence CH_4 is a much stronger absorber of infrared radiation than CO_2. Presuming the current rates of increase in ambient concentrations of the greenhouse gases continue, the doubling times can be estimated at which the surface temperature effects in Table 1.7 can be expected. For CO_2, various studies indicate that its concentration will double by the latter part of the twenty-first century. Although there is disagreement as to the exact time of doubling, there is virtually no

TABLE 1.7 Distribution and Selected Properties of Some Atmospheric Polluting Gases Having Absorption in the Radiative Window[a]

Gas	Atmospheric mass (Gt)	Concentration Preindustrial (ppm)	Concentration Current (ppm)	IR absorption (ratio gas/CO_2)	ΔT on doubling concentration[b] (°C)
CO_2	2640	275	345	1	2–3
CH_4	4.9	0.75	1.65	25	0.95
N_2O	2.5	0.25	0.35	250	0.25
CO	0.6		0.12		0.6–0.9
CCl_2F_2[c]	0.006	0	0.00038	20,000	0.02–0.05
CCl_3F[d]	0.004	0	0.00023	17,500	0.008–0.016

[a]Adapted from Chamberlain et al. (1982), Ramanathan (1988), International Energy Agency (1989), and Intergovernmental Panel on Climate Change (1995).
[b]Change in earth's surface temperature resulting from doubling of concentration of indicated gas as estimated in Chamberlain et al. (1982).
[c]Freon 12.
[d]Freon 11.

dispute among scientists that the concentrations of atmospheric CO_2 have increased about 30% since 1850.

A note of caution is necessary regarding the predictions that have been made regarding global temperature increases. The predictions made in the mid-1990s by the Intergovernmental Panel on Climate Change rely heavily on the use of computerized climate models. There is much uncertainty inherent in this technique because few models can reliably simulate even the present climate without "flux adjustments" (cf. Kerr, 1997). Consequently, there is considerable disagreement about the specific effects on global temperature of the greenhouse gases, and even clouds and pollutant hazes, and whether global warming can be correlated with human activities or is a natural phenomenon. Application of improved computer models that do not use flux adjustments indicates that global warming is occurring at the lower end of the many predictions that have been made.

B. SOURCES OF GREENHOUSE GASES

It is of interest to examine potential sources of atmospheric CO_2 by analysis of the global distribution of carbon in all its forms. The data presented in Table 1.8 show that atmospheric carbon, which can be assumed to be essentially all in the form of CO_2 (i.e., 700 Gt carbon equals 2,570 Gt of CO_2) comprises

TABLE 1.8   Global Carbon Distribution[a]

Location	Mass (Gt)	Percent of world total	
		With lithosphere	Without lithosphere
Terrestrial			
Sediments (lithosphere)	20,000,000	99.780	
Live biomass	450	0.002	1.02
Dead biomass	700	0.003	1.59
Subtotal terrestrial:	20,001,150	99.785	2.61
Atmosphere	700	0.003	1.59
Oceans			
Deep sea (inorganic)	34,500	0.172	78.43
Sea surface layers (dissolved)	500	0.002	1.14
Live biomass	10	0.00005	0.02
Dead biomass	3000	0.015	6.82
Subtotal oceans:	38,010	0.189	86.41
Fossil fuel deposits			
Coal	3510	0.018	7.98
Oil	230	0.001	0.52
Gas	140	0.0007	0.32
Others	250	0.001	0.57
Subtotal fossil fuel:	4130	0.021	9.39
Total carbon deposits	20,043,990		

[a]Adapted from Watts (1982) and Klass (1983).

only about 1.6% of total global carbon, excluding lithospheric carbon. Obvious sources of direct or indirect additions of CO_2 to the atmosphere are therefore fossil fuel deposits, since portions of them are combusted each year as fuels, and terrestrial biomass. Biomass, the photosynthetic sink for removal of CO_2 from the atmosphere, is important because any changes that modify natural biomass growth can affect ambient CO_2 concentration. Reducing the size of the photosynthetic sink by such practices as slash-and-burn agriculture, large-scale wood burning, and rain-forest destruction cause an overall reduction in the amount of natural photosynthesis.

To develop more quantitative information regarding atmospheric CO_2, the emissions on combustion of coal, oil, and natural gas per energy input unit (Appendix C) were used to calculate the CO_2 generated from fossil fuel combustion for the world's regions and each of the top 10 energy-consuming countries (Table 1.9). Oil is the largest CO_2 source, followed by coal and natural gas. It is obvious that the largest energy-consuming regions of the world generate relatively more fossil-based CO_2, and that the world's 10 top energy-consuming

TABLE 1.9   Carbon Dioxide Generated from Fossil Fuel Combustion by World Region and the 10 Highest Energy-Consuming Countries[a]

Region	% of world total	Total (Gt)	Oil (Gt)	Natural gas (Gt)	Coal (Gt)
Africa	3.5	0.79	0.35	0.09	0.35
America, N.	27.6	6.17	3.04	1.28	1.86
America, S. and Central	3.4	0.76	0.56	0.12	0.08
Eastern Europe and Former U.S.S.R.	17.7	3.94	1.19	1.34	1.42
Western Europe	17.7	3.94	2.13	0.58	1.23
Middle East	3.7	0.82	0.58	0.22	0.02
Far East and Oceania	26.4	5.90	2.36	0.35	3.20
Total		22.34	10.21	3.98	8.15
% of world total			45.7	17.8	36.5
Country					
United States	23.9	5.33	2.51	1.09	1.73
Russia	8.9	2.00	0.67	0.81	0.52
China	10.9	2.44	0.42	0.03	2.00
Japan	5.3	1.19	0.83	0.11	0.25
Germany	4.2	0.94	0.44	0.13	0.37
Canada	2.2	0.50	0.25	0.14	0.11
France	2.0	0.44	0.30	0.07	0.08
United Kingdom	2.9	0.64	0.27	0.12	0.25
Ukraine	2.5	0.56	0.12	0.20	0.24
India	2.9	0.65	0.20	0.03	0.43
Total		14.69	6.00	2.71	5.97
% of world total		65.7	26.9	12.1	26.7

[a] Adapted from Klass (1992). The energy consumption data for the countries in Table 1.3 and for the world's regions (U.S. Department of Energy, 1994) were used for the calculations. The factors for converting energy consumption in EJ to carbon dioxide emissions for oil, natural gas, and coal are 0.07098, 0.05076, and 0.08690, respectively, and were derived from the data in Appendix C. The sums of individual figures may not equal the totals because of rounding.

countries generate almost two-thirds of the world's total CO_2 emitted on combustion of fossil fuels. This kind of information has led to several national plans and international agreements to attempt to lower or at least maintain atmospheric CO_2 by reducing fossil fuel consumption through such mechanisms as fossil carbon consumption taxes and higher-efficiency hardware. A variety of technologies for removal of CO_2 from the environment have also been proposed.

Although the position has been supported with limited and sometimes questionable data, it has come to be accepted as fact by many if not most climate change specialists that fossil fuel consumption is the major cause of atmospheric CO_2 buildup. The CO_2 in the atmosphere is estimated to have a mass of about 2,640 Gt (Table 1.7). Uncertainty is a factor because it is only by inference that the mass is calculated. But many direct analyses of atmospheric CO_2 have been made at different locations throughout the world. Analysis of air trapped in ancient ice cores shows that about 160,000 years ago, atmospheric CO_2 concentration was about 200 ppm and then peaked at about 300 ppm 130,000 and 10,000 years ago. The concentration then began to increase from an apparent equilibrium value of about 280 ppm in the eighteenth century to its present level of about 360 ppm, the highest concentration in the past 160,000 years. Atmospheric CO_2 concentration has increased at least 50 ppm since 1860 and is currently increasing at an annual rate of about 1.5 ppm according to analyses carried out continuously over the last several decades. Presuming the atmospheric mass of 2,640 Gt is correct, this corresponds to an annual increase of about 11.3 Gt/year.

Compared to other carbon flows, CO_2 emissions from fossil fuel consumption by country are perhaps the most accurate, large-scale carbon flux calculations that can be performed. The reason for this is that detailed data on fossil fuel production and consumption are compiled and reported worldwide. Since the mid-1800s, fossil fuel usage has increased significantly, notably since World War II as discussed earlier, to over 300 EJ/year (Fig. 1.6). Global CO_2 emissions from fossil fuel combustion have been calculated and reported to four significant figures for many years; the annual average from 1978 to 1987 was 18.91 Gt/year (Klass, 1993) and is in the 22-Gt/year range in the 1990s (Table 1.9). So fossil fuel emissions are about twice the annual atmospheric CO_2 buildup. This type of "factual data" comprises the essence of the argument that fossil fuel consumption is the primary cause of CO_2 buildup in the atmosphere, and sic climate change. Much of the additional evidence is qualitative and uncertain because the study of global CO_2 buildup is inextricably related to global carbon cycles and reservoirs and the myriad of processes that take place over time on a living planet. The problem from an investigative standpoint is extremely difficult to elaborate. Few direct measurements can be made with precision and then be reproduced. Broad use is made of modeling, and real-world confirmation of the conclusions is often anecdotal. As will be shown later (Chapter 2), biomass has a very important role in atmospheric CO_2 fluxes and may affect ambient concentrations much more than fossil fuel consumption alone. Because of the environmental trends today, it appears that international agreements to limit fossil fuel consumption will be implemented sometime in the twenty-first century. This will require much greater usage of alternative fuels, especially renewable biomass energy and biofuels manufactured from biomass.

REFERENCES

Chamberlain, J. C., Foley, H. M., MacDonald, G. J., and Ruderman, M. A. (1982). *In* "Carbon Dioxide Review 1982," (W. C. Clark, ed.), p. 255. Oxford University Press, New York.

Daniels, F., and Duffie, J. A. (1955). *In* "Solar Energy Research," (F. Daniels and J. A. Duffie, eds.), p. 3. University of Wisconsin Press, Madison, WI.

Gulf Publishing Company (August 1993). *World Oil* **214** (8).

Institute of Gas Technology (1989). "IGT World Reserves Survey," (H. Feldkirchner, ed.). Institute of Gas Technology, Chicago.

Intergovernmental Panel on Climate Change (August 1991). "Estimation of Greenhouse Gas Emissions and Sinks, Final Report from the OECD Experts Meeting, 18–21 February 1991 [Paris]." ECD/OCDE, United Nations.

Intergovernmental Panel on Climate Change (December 1995). "IPCC Working Group I 1995 Summary for Policymakers," approved at 5th WGI Session, Madrid, Spain, 27–29 November 1995, and associated detailed report.

International Energy Agency (1989). "Energy and Environment: Policy Overview." OECD/IEA, Paris.

Jones, P. D., Wigley, T. M. L., and Wright, P. B. (1986). *Nature* **322**, 430.

Keeling, C. D., Whorf, T. P., Wahle, M., and van der Plicht, J. (1995). *Nature* **375**, 666.

Kerr, R. A. (1997). *Science* **276**(5315), 1040.

Klass, D. L. (1983). *In* "Handbook of Energy Technology and Economics," (R. A. Meyers, ed.), p. 712. John Wiley, New York.

Klass, D. L. (1992). *Energy & Environment* **3** (2), 109.

Klass, D. L. (1993). *Energy Policy* **21** (11), 1076.

National Research Council (1979). "Carbon Dioxide and Climate: A Scientific Assessment," Report of an Ad Hoc Study Group on Carbon Dioxide and Climate (J. G. Charney, Chairman). National Academy of Sciences, Washington, D.C.

Ramanathan, V. (1988). *Science* **240** (4850), 293.

United Nations (1992). "1990 Energy Statistics Yearbook," Department for Economic and Social Development, New York.

U.S. Bureau of the Census (1994). "Statistical Abstract of the United States 1994," 114th Ed. Washington, D.C.

U.S. Congress (January 1991). "Energy in Developing Countries," OTA-E-486. Office of Technology Assessment, Washington, D.C.

U.S. Dept. of Energy (January 1994). "International Energy Annual 1992," DOE/EIA-0219(92). Energy Information Administration, Washington, D.C.

U.S. Dept. of Energy (May 1996). "International Energy Annual 1996," DOE/EIA-0484(96). Energy Information Administration, Washington, D.C.

Watts, J. A. (1982). *In* "Carbon Dioxide Review 1982," (W. C. Clark, ed.), p. 432. Oxford University Press, New York.

Whorf, T. P. (1996). Scripps Institution of Oceanography, University of California, San Diego, personal communication.

World Bank (1989). *In* "World Development Report," p. 164. Oxford University Press, New York.

World Bank (1991). "Social Indicators of Development 1990." The Johns Hopkins University Press, Baltimore.

World Energy Council (1992). "1992 Survey of Energy Resources." World Energy Conference.

CHAPTER 2

Biomass as an Energy Resource: Concept and Markets

I. INTRODUCTION

As late as the mid 1800s, biomass supplied the vast majority of the world's energy and fuel needs and only started to be phased out in industrialized countries as the fossil fuel era began, slowly at first and then at a rapid rate. But with the onset of the First Oil Shock in the late 1970s, biomass was again realized by many governments and policy makers to be a viable, domestic, energy resource that has the potential of reducing oil consumption and imports and improving the balance of payments and deficit problems caused by dependency on imported oil. For example, the contribution of biomass energy to U.S. energy consumption in the late 1970s was over 850,000 BOE/day, or more than 2% of total energy consumption at that time. By 1990, it had increased to about 1.4 million BOE/day, or 3.3% of energy consumption, and conservative projections indicate that by the year 2000, biomass energy consumption is expected to increase to 2.0 million BOE/day (Klass, 1994). Other industrialized countries have also increased biomass energy consumption. Canada, for example, consumed about 134,000 BOE/day of biomass energy, or 3% of its total energy demand in the late 1970s, and by 1992 had increased consumption to 250,000 BOE/day, or 4.4% of total energy demand. Although biomass energy has continued to be utilized in Third World countries as a source of fuels and energy for many years, it has become a renewable carbon resource for energy and fuels once again for industrialized countries and is expected to exhibit substantial growth in the twenty-first century. In this chapter, the concept of virgin and waste biomass as an alternative source of supply for energy and fuels is examined and the energy potential of biomass energy and its market penetration are evaluated.

II. BASIC CONCEPT

The terminology "renewable carbon resource" for virgin and waste biomass is actually a misnomer because the earth's carbon is in a perpetual state of flux. Carbon is not consumed in the sense that it is no longer available in any form. Many reversible and irreversible chemical reactions occur in such a manner that the carbon cycle makes all forms of carbon, including fossil carbon resources, renewable. It is simply a matter of time that makes one form of carbon more renewable than another. If society could wait several million years so that natural processes could replenish depleted petroleum or natural gas deposits, presuming that replacement occurs, there would never be a shortage of organic fuels as they are distributed and accepted in the world's energy markets. Unfortunately, this cannot be done, so fixed carbon-containing materials that renew themselves over a time span short enough to make them continuously available in large quantities are needed to maintain and supplement energy supplies. Biomass is a major source of carbon that meets these requirements.

The capture of solar energy as fixed carbon in biomass via photosynthesis, during which carbon dioxide (CO_2) is converted to organic compounds, is the key initial step in the growth of biomass and is depicted by the equation

$$CO_2 + H_2O + \text{light} + \text{chlorophyll} \rightarrow (CH_2O) + O_2.$$

Carbohydrate, represented by the building block (CH_2O), is the primary organic product. For each gram mole of carbon fixed, about 470 kJ (112 kcal) is absorbed. Oxygen liberated in the process comes exclusively from the water, according to radioactive tracer experiments. Although there are still many unanswered questions regarding the detailed molecular mechanisms of photosynthesis, the prerequisites for virgin biomass growth are well established; CO_2, light in the visible region of the electromagnetic spectrum, the sensitizing catalyst chlorophyll, and a living plant are essential. The upper limit of the capture efficiency of the incident solar radiation in biomass has been variously estimated to range from about 8% to as high as 15%, but in most actual situations, it is generally in the 1% range or less (Klass, 1974).

The main features of how biomass is used as a source of energy and fuels are schematically illustrated in Fig. 2.1. Conventionally, biomass is harvested for feed, food, fiber, and materials of construction or is left in the growth areas where natural decomposition occurs. The decomposing biomass or the waste products from the harvesting and processing of biomass, if disposed of on or in land, can in theory be partially recovered after a long period of time as fossil fuels. This is indicated by the dashed lines in Fig. 2.1. Alternatively, biomass and any wastes that result from its processing or consumption could be converted directly into synthetic organic fuels if suitable conversion processes were available. The energy content of biomass could be diverted instead to direct heating applications by combustion. Another route to energy products is to grow certain species of biomass such as the rubber tree (*Hevea braziliensis*) in which high-energy hydrocarbons are formed within the species by natural biochemical mechanisms. In this case, biomass serves the dual role of a carbon-fixing apparatus and a continuous source of hydrocarbons without being consumed in the process. Other biomass species, such as the guayule bush, produce hydrocarbons also, but must be harvested to recover them. Conceptually, it can be seen from Fig. 2.1 that there are several different pathways by which energy products and synthetic fuels might be manufactured.

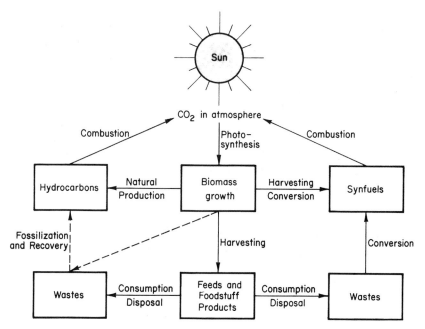

FIGURE 2.1   Main features of biomass energy technology.

Another approach to the development of fixed carbon supplies from renewable carbon sources is to convert CO_2 outside the biomass species into synthetic fuels and organic intermediates. The ambient air, which contains about 360 ppm of CO_2, the dissolved CO_2 and carbonates in the oceans, and the earth's large terrestrial carbonate deposits could serve as renewable carbon sources. But since CO_2 is the final oxidation state of fixed carbon, it contains no chemical energy. Energy must be supplied in a reduction step. A convenient method of supplying the required energy and of simultaneously reducing the oxidation state is to reduce CO_2 with elemental hydrogen. The end product, for example, can be methane, the dominant component of natural gas:

$$CO_2 + 4H_2 \rightarrow CH_4 + 2H_2O$$

With all components in the ideal gas state, the standard enthalpy of the process is exothermic by -165 kJ (-39.4 kcal) per gram mole of methane formed. Biomass feedstocks could also serve as the original source of hydrogen via partial oxidation or steam reforming to an intermediate hydrogen-containing product gas. Hydrogen would then effectively act as an energy carrier from the biomass to the CO_2 to yield substitute or synthetic natural gas (SNG). The production of other synthetic organic fuels can be conceptualized in a similar manner.

The basic concept then of biomass as a renewable energy resource comprises the capture of solar energy and carbon from ambient CO_2 in growing biomass, which is converted to other fuels (biofuels, synfuels) or is used directly as a source of thermal energy or hydrogen. One cycle is completed when the biomass or derived fuel is combusted. This is equivalent to releasing the captured solar energy and returning the carbon fixed during photosynthesis to the atmosphere as CO_2. Hydrocarbons identical to those in petroleum or natural gas can be manufactured from biomass feedstocks. This means that essentially all of the products manufactured from petroleum and natural gas can be produced from biomass feedstocks. Alternatively, biomass feedstocks can be converted to organic fuels that are not found in petroleum or natural gas. The practical uses of biomass feedstocks and the applications of biomass energy and derived fuels, however, are limited by several factors.

III. DISTRIBUTION OF RENEWABLE CARBON RESOURCES AND BIOMASS ABUNDANCE

A. BIOSPHERIC CARBON FLUXES

Most global studies of the transport and distribution of the earth's carbon eventually lead many analysts to conclude that the continuous exchange of carbon with the atmosphere and the assumptions and extrapolations that must be employed make it next to impossible to eliminate large errors in the results and uncertainty in the conclusions. Only a very small fraction of the immense mass of carbon at or near the earth's surface is in relatively rapid circulation in the earth's biosphere, which includes the upper portions of the earth's crust, the hydrosphere, and biomass. There is a continuous flow of carbon between the various sources and sinks. The atmosphere is the conduit for most of this flux, which occurs primarily as CO_2.

Some of the difficulties encountered in analyzing this flux are illustrated by estimating the CO_2 exchanges with the atmosphere (Table 2.1). Despite the possibilities for errors in this tabulation, especially regarding absolute values, several important trends and observations are apparent and should be valid for many years. The first observation is that fossil fuel combustion and industrial operations such as cement manufacture emit much smaller amounts of CO_2 to the atmosphere than biomass respiration and decay, and the physical exchanges between the oceans and the atmosphere. The total amount of CO_2 emissions from coal, oil, and natural gas combustion is also less than 3% of that emitted by all sources. This is perhaps unexpected because most of the climate change literature indicates that the largest source of CO_2 emissions is fossil fuel combustion. Note that human and animal respiration are projected to emit more than five times the CO_2 emissions of all industry exclusive of energy-related emissions. Note also that biomass burning appears to emit almost as much CO_2 as oil and natural gas consumption together.

One of the CO_2 sources not listed in Table 2.1 that can result in significant net CO_2 fluxes to the atmosphere is land cover changes such as those that result from urbanization, highway construction, and the clear-cutting of forestland for agricultural purposes. It has been estimated that the net flux of CO_2 to the atmosphere in 1980, for example, was 5.13 Gt, or 1.40 Gt of carbon, because of land cover changes (Houghton and Hackler, 1995). Land cover changes are usually permanent, so the loss in atmospheric carbon-fixing capacity and annual biomass growth are essentially permanent also. It has been estimated from the world's biomass production data that losses of only 1% of standing forest biomass and annual forest biomass productivity correspond to the ultimate return of approximately 27 Gt of CO_2 to the atmosphere, and an annual loss of about 1.22 Gt in atmospheric CO_2 removal capacity (*cf.* Klass, 1993).

Overall, the importance of the two primary sinks for atmospheric CO_2—terrestrial biota and the oceans—is obvious. No other large sinks have been identified. It is evident that only small changes in the estimated CO_2 uptake and release rates of these sinks determine whether there is a net positive or negative exchange of CO_2 with the atmosphere. A small change in either or both carbon fixation in biomass by photosynthesis or biomass respiration estimates tends to cause a large percentage change in the arithmetic difference

TABLE 2.1 Estimated Annual Global Carbon Dioxide and Carbon Exchanges with the Atmosphere[a]

Source and/or sink	Carbon dioxide		Carbon equivalent	
	To atmosphere (Gt/year)	From atmosphere (Gt/year)	To atmosphere (Gt/year)	From atmosphere (Gt/year)
Terrestrial:				
Cement production	0.51		0.14	
Other industrial processes	0.47		0.13	
Human respiration	1.67		0.46	
Animal respiration	3.34		0.91	
Methane emissions equivalents	1.69		0.46	
Natural gas consumption	3.98		1.09	
Oil consumption	10.21		2.79	
Coal consumption	8.15		2.22	
Biomass burning	14.3		3.90	
Gross biomass photosynthesis		388		106
Biomass respiration	194		53	
Soil respiration and decay	194		53	
Total terrestrial:	432	388	118	106
Oceans:				
Gross biomass photosynthesis		180		49
Biomass respiration	90		25	
Physical exchange	275	202	75	55
Total oceans:	365	382	100	104
Total terrestrial and oceans:	797	770	218	210

[a]The fossil fuel, human, and animal emissions were estimated by the author (Appendix C). Most of the other exchanges are derived from exchanges that have been reported in the literature (cf. Boden, Marlund, and Andres, 1995) or they are based on assumptions that have generally been used by climatologists. It was assumed that 50% of the terrestrial biomass carbon fixed by photosynthesis is respired and that an equal amount is emitted by the soil. The total uptake and emission of carbon dioxide by the oceans were assumed to be 104 and 100 Gt C/year (Houghton and Woodwell, 1989), and biomass respiration was assumed to emit 50% of the carbon fixed by photosynthesis. The carbon dioxide emissions from cement production and other industrial processes are process emissions that exclude energy-related emissions; they are included in the fossil fuel consumption figures.

between them. And the impact of the assumptions is very large. The assumption that live biomass respires about 50% per year of the total carbon that is photochemically fixed results in a substantial calculated addition of CO_2 to the atmosphere, far more than that from fossil fuel combustion. The other assumption incorporated in most biospheric carbon budgets concerns the annual emission of CO_2 from soils by microbial action and the oxidation of dead biomass, namely that the emission of CO_2 occurs at an annual rate approximately equal to 50% of the gross annual photosynthetic carbon uptake. This assumption has little experimental support. The end result of the use of these assumptions with respect to terrestrial biomass, the soils, and the oceans is that they are almost neutral factors in the scenarios generally published on carbon exchanges with the atmosphere and the buildup of atmospheric CO_2; that is, about the same amount of CO_2 is emitted as is taken up each year, as shown in the tabulation. This conclusion can be subject to major error when attempting to quantify carbon exchanges with the atmosphere. The largest reservoir of biomass carbon resides in live forest biomass, as will be shown later, and unless this biomass is removed or killed, it fixes atmospheric CO_2 with the passage of time during most of its life cycle. To sustain the environmental benefits of biomass growth as a sink for the removal of CO_2 from the atmosphere, it is evident that biomass growth should be sustained and expanded. The large-scale use of virgin biomass for energy will not adversely affect these benefits if it is replaced at the same or a greater rate than the rate of consumption.

B. Global Biomass Carbon Distribution

Detailed estimation of the amounts of biomass carbon on the earth's surface is the ultimate problem in global statistical analysis. Yet what appear to be reasonable projections have been made using available data, maps, and surveys. The validity of the conclusions in their entirety is difficult to support with hard data because of the nature of the problem. But such analyses must be performed to assess the practical feasibility of biomass energy systems and the gross types of biomass that might be available for energy applications.

The results of one such study are summarized in Table 2.2. Ignoring the changes in agricultural practice and the deforestation that has taken place over the last few decades, this is perhaps one of the better attempts to conduct an analysis of the earth's biomass carbon distribution (Whittaker and Likens, 1975). Each ecosystem on the earth is considered in terms of area, mean net carbon production per year, and standing biomass carbon. Standing biomass carbon is that contained in biomass on the earth's surface and does not include the carbon stored in biomass underground. A condensation of this data (Table 2.3) facilitates interpretation. Of the total net carbon fixed on the earth each year, forest biomass, which is produced on only 9.5% of the earth's surface, contributes more than any other source. Marine sources of net fixed carbon are also high, as might be expected because of the large area of the earth occupied by water. But the high turnover rates of carbon in a marine environment result in relatively small steady-state quantities of standing carbon. In contrast, the low turnover rates of forest biomass make it the largest contributor

TABLE 2.2   Estimated Net Photosynthetic Production of Dry Biomass Carbon for World Biosphere[a]

Ecosystem	Area (10^6 km^2)	Mean net biomass carbon production		Standing biomass carbon	
		(t/ha-year)	(Gt/year)	(t/ha)	(Gt)
Tropical rain forest	17.0	9.90	16.83	202.5	344
Boreal forest	12.0	3.60	4.32	90.0	108
Tropical season forest	7.5	7.20	5.40	157.5	118
Temperate deciduous forest	7.0	5.40	3.78	135.0	95
Temperate evergreen forest	5.0	5.85	2.93	157.5	79
Total	48.5		32.26		744
Extreme desert-rock, sand, ice	24.0	0.01	0.02	0.1	0.2
Desert and semidesert scrub	18.0	0.41	0.74	3.2	5.8
Savanna	15.0	4.05	6.08	18.0	27.0
Cultivated land	14.0	2.93	4.10	4.5	6.3
Temperate grassland	9.0	2.70	2.43	7.2	6.5
Woodland and shrubland	8.5	3.15	2.68	27.0	23.0
Tundra and alpine	8.0	0.63	0.50	2.7	2.2
Swamp and marsh	2.0	13.50	2.70	67.5	14.0
Lake and stream	2.0	1.80	0.36	0.1	0.02
Total	100.5		19.61		85
Total continental	149.0		52.87		829
Open ocean	332.0	0.56	18.59	0.1	3.3
Continental shelf	36.6	1.62	4.31	0.004	0.1
Estuaries excluding marsh	1.4	6.75	0.95	4.5	0.6
Algae beds and reefs	0.6	11.25	0.68	9.0	0.5
Upwelling zones	0.4	2.25	0.09	0.9	0.04
Total marine	361.0		24.62		4.5
Grand total	510.0		77.49		833.5

[a]Adapted from Whittaker and Likens (1975).

to standing carbon reserves. According to this assessment, the forests produce about 43% of the net carbon fixed each year and contain over 89% of the standing biomass carbon of the earth. Tropical forests are the largest sources of these carbon reserves. Temperate deciduous and evergreen forests are also major sources of biomass carbon. Next in order of biomass carbon supply would probably be the savanna and grasslands. Note that cultivated land is one of the smaller producers of fixed carbon and is only about 9% of the total terrestrial area of the earth.

TABLE 2.3   Estimated Distribution of World's Biomass Carbon[a]

	Forests	Savanna and grasslands	Swamp and marsh	Remaining terrestrial	Marine
Area (10^6 km^2)	48.5	24.0	2.0	74.5	361
Percent	9.5	4.7	0.4	14.6	70.8
Net C production (Gt/year)	33.26	8.51	2.70	8.40	24.62
Percent	42.9	11.0	3.5	10.8	31.8
Standing C (Gt)	744	33.5	14.0	37.5	4.5
Percent	89.3	4.0	1.7	4.5	0.5

[a]Adapted from Table 2.2.

It is necessary to emphasize that anthropological activities and the increasing population, particularly in developing and Third World countries, continue to make it more difficult to sustain the world's biomass growth areas. It has been estimated that tropical forests are disappearing at a rate of tens of thousands of square miles per year. Satellite imaging and field surveys show that Brazil alone has a deforestation rate of about 8×10^6 ha/year (19.8×10^6 ac/year; 30,888 mi.^{2}/year) (Repetto, 1990). At mean net biomass carbon yields of 9.90 t/ha-year for tropical rain forests (Table 2.2), this rate of deforestation corresponds to a loss of 79.2×10^6 t/year of net biomass carbon productivity.

The remaining carbon transport mechanisms on earth are primarily physical mechanisms, such as the solution of carbonate sediments in the sea and the release of dissolved CO_2 to the atmosphere by the hydrosphere. Because of the relatively short lifetimes of live biomass (phytoplankton and zooplankton) in the oceans compared to those of land biomass, there is a much larger amount of carbon in viable land biomass at any given time. The great bulk of carbon, however, is contained in the lithosphere as carbonates in rock. The carbon deposits that contain little or no stored chemical energy, although some high-temperature deposits can provide considerable thermal energy, consist of lithospheric sediments and atmospheric and hydrospheric CO_2. Together, these carbon sources comprise 99.96% of the total carbon estimated to exist on the earth (Table 2.4). The carbon in fossil fuel deposits is only about 0.02% of the total, and live and dead biomass carbon makes up the remainder, about 0.02%. Biomass carbon is thus a very small fraction of the total carbon inventory of the earth, but it is an extremely important fraction. It helps to maintain the delicate balance among the atmosphere, hydrosphere, and biosphere necessary to support all life forms, and is essential to maintain the diversity of species that inhabit the earth and to sustain their gene pools. Any large-scale utilization of biomass carbon, especially virgin material, therefore requires that it be replaced, preferably as it is consumed so that the biomass reservoirs are not

TABLE 2.4   Estimated Carbon Distribution on Earth[a]

Carbon type	Mass (Gt)	Percent of total
Lithospheric sediments	20,000,000	99.78
Deep sea	34,500	0.172
Fossil deposits	4130	0.021
Dead organic matter in sea	3000	0.015
Dead organic matter on land	700	0.0035
Atmosphere	700	0.0035
Sea surface layers (dissolved)	500	0.0025
Live terrestrial biomass	450	0.0022
Live phytoplankton	5	0.00002
Live zooplankton	5	0.00002
Total	20,043,990	

[a]Adapted from Table 1.8.

reduced. Indeed, enlargement of these reservoirs may become necessary as the world's population expands and climate changes occur.

IV. ENERGY POTENTIAL OF BIOMASS

It is important to examine the potential amounts of energy and biofuels that might be produced from biomass carbon resources and to compare these amounts with fossil fuel demands. This would make it possible to estimate the percentage of energy demand that might be satisfied by particular biomass types.

A. VIRGIN BIOMASS

Consider first the incident solar radiation, or insolation, that strikes the earth's surface. At an average daily insolation worldwide of about 220 W/m^2 (1676 Btu/ft^2), the annual insolation on about 0.01% of the earth's surface is approximately equal to all the primary energy consumed by humans each year. For the United States alone, the insolation on about 0.1 to 0.2% of its total surface is equivalent to its total annual energy consumption.

The most widespread and practical process for capture of this energy as organic fuels is the growth of virgin biomass. As already discussed, extremely large quantities of carbon are fixed each year in the form of terrestrial and aquatic biomass. Using the figures in Table 2.2, the energy content of standing biomass carbon; that is, the renewable, above-ground biomass reservoir that in theory could be harvested and used as an energy resource, is about 100 times the world's annual energy consumption. At a nominal biomass heating value of 18.6 GJ/dry t (16 $\times$ 10^6 Btu/dry ton) and assuming that the world's total annual coal, oil, and natural gas consumption is about 315 EJ (1993), the solar energy trapped in 16.9 Gt of dry biomass, or about 7.6 Gt of biomass carbon, would be equivalent to the world's consumption of these fossil fuels. Since it is estimated that about 77 Gt of carbon, or 171 Gt of dry virgin biomass equivalent, most of which is wild and not controlled by humans, is fixed on the earth each year, it is certainly in order to consider biomass as a raw material for direct use as fuel or for conversion to large supplies of substitute fossil fuels. Under controlled conditions, dedicated biomass species might be grown specifically as energy crops or for multiple uses including energy. Relatively rapid replacement of the biomass utilized can take place through regrowth.

A more realistic assessment of biomass as an energy resource can be made by calculating the average surface areas needed to produce sufficient biomass at different annual yields to meet certain percentages of fuel demand for a particular country, and then to compare these areas with those that might be made available. Such an assessment for the United States could, for example, address the potential of biomass for conversion to SNG as shown in Table 2.5. For this analysis, the annual U.S. demand for natural gas is projected to reach 26.5 EJ (25.1 quad) by 2010 at an annual growth rate in consumption of 1.2% (U.S. Dept. of Energy, 1994). It is assumed that biomass, whether it be trees, plants, grasses, algae, or water plants, has a heating value of 18.6 GJ/dry t, is grown under controlled conditions in "methane plantations" at yields of 20 and 50 dry t/ha-year, and is converted in integrated biomass planting, harvesting, and conversion systems to SNG at an overall thermal efficiency of 50%. These conditions of biomass production and conversion either are within

TABLE 2.5 Potential Substitute Natural Gas Production in United States from Virgin Biomass Feedstocks at Different Biomass Yields

Percent of natural gas demand supplied	Average area required at indicated biomass yield (10^6 ha)	
	20 dry t/ha-year	50 dry t/ha-year
1.42	2.02	0.81
10	14.3	5.7
50	71.2	28.5
100	142.5	57.0

the range of present technology and agricultural practice, or are believed to be attainable in the near future. The average total plantation areas were then calculated to meet 1.42, 10, 50, and 100% of the projected U.S. demand in 2010 for natural gas. A percentage of 1.42 is equal to a daily production of 26.9×10^6 m^3 at normal conditions (1×10^9 SCF) of dry SNG. The range of areas required at the low yield level is between 2,023,000 and 142,500,000 ha, or 0.2 and 14.9% of the 50-state U.S. area. At the high yield level, the areas are between 809,000 and 57,000,000 ha, or 0.08 and 6.0% of the 50-state U.S. area. To put this analysis in the proper perspective, the results are shown in graphical form in Fig. 2.2 at the two yield levels together with the percentage area of the United States needed at any selected gas demand supplied by SNG from biomass. Relatively large areas are required, but not so much as to make the use of land or freshwater biomass for energy applications impractical. When compared with the area distribution pattern of the United States (Table 2.6) (USDA Forest Service, 1989), it is seen that selected areas or combinations of areas might be utilized for biomass energy. Areas that are not used for productive purposes might be suitable, or possibly biomass for both energy and foodstuffs or energy and forest products applications can be grown simultaneously or sequentially in ways that would benefit both. Also, relatively small portions of the bordering oceans might supply the needed biomass growth areas, in which case, marine plants would be grown and harvested.

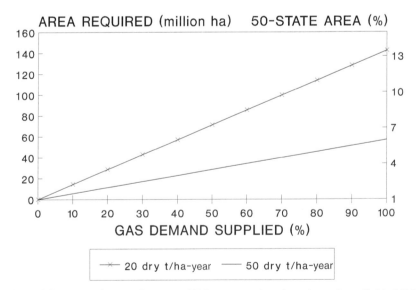

FIGURE 2.2   Required area and percent of U.S. area vs projected gas demand supplied in 2010.

TABLE 2.6 Land and Water Areas of United States[a]

Area classification	10^6 ha	Percent
Nonfederal land		
Forest	179.41	18.8
Rangeland	178.66	18.7
Other land	279.09	29.2
Transition land	14.41	1.5
Total	651.57	68.2
Federal land		
Forest	102.14	10.7
Rangeland	133.10	13.9
Other land	25.70	2.7
Total	260.94	27.3
Water		
Inland water	24.75	2.6
Other water	19.28	2.0
Total	44.03	4.6
Grand total land and water	956.54	

[a]Adapted from USDA Forest Service (1989). The data for forest, rangeland, and other land are for 1982. The data for inland water are for 1990. The data for other water are for 1970. Forest areas are at least 10% stocked by trees of any size, or formerly having such tree cover and not currently developed for nonforest use. Transition land is forest land that carries grasses or forage plants used for grazing as the predominant vegetation. Climax vegetation on rangelands is predominantly grasses, grass-like plants, forbs, and shrubs suitable for grazing and browsing. Other land areas include crop and pasture land and farmsteads, strip mines, permanent ice and snow, and land that does not fit any other land cover.

This approach to the preliminary assessment of the potential of biomass energy presumes that suitable conversion processes are available for conversion of biomass to SNG. Other processes could be used to manufacture other synfuels such as synthesis gas, alcohols, esters, and hydrocarbons. The direct route, alluded to in Fig. 2.1 as natural production of hydrocarbons, can possibly bypass the harvesting-conversion routes. As already mentioned, some biomass species produce hydrocarbons as metabolic products. Natural rubber, glycerides, and terpenes from selected biomass species, for example, as well as other reduced compounds could be extracted and refined to yield conventional or substitute fossil fuels.

A second source of renewable carbon is the deposits and reservoirs of essentially non-energy carbon forms—ambient CO_2 and the lithospheric carbonates. The availability of such raw materials cannot be questioned, although low-cost separation and energy-efficient recovery of very small concentrations of CO_2 from the atmosphere present technological challenges. Another basic problem resides in the fact that all of the energy must be supplied by a second raw material, such as elemental hydrogen. Hydrogen would have to be made available in large quantities from a nonfossil source or the purpose of the synfuel system to produce renewable fuels would be defeated. Conceptually, there is no difficulty in developing such hydrogen sources. Hydrogen can be produced by water electrolysis and thermochemical and photolytic splitting of water. Electrical power and thermal energy can be supplied by nonfossil-powered nuclear reactors, and by means of hydroelectric and wind systems, ocean thermal gradients, wave action, and solar-actuated devices. Hydrogen can also be manufactured from biomass and by direct action of solar energy on certain catalytic surfaces.

As already pointed out, about 16.9 Gt of dry biomass, or about 7.7 Gt of biomass carbon, would have approximately the same energy content as the total global consumption of coal, oil, and natural gas (in 1993). This amount of carbon corresponds to less than 1.0% of the total standing biomass carbon of the earth. Under present conditions of controlled and natural production of fixed carbon supplies, the utilization of some of this carbon for energy applications seems to be a logical end use of a renewable raw material. Forest biomass is especially interesting for these applications because of its abundance. The expansion of controlled production of virgin biomass in dedicated energy crop systems should also be considered because this would result in new additions to natural biomass carbon supplies. For example, the biomass carbon supplies in marine ecosystems might conceivably be increased under controlled conditions over the current low levels by means of marine biomass energy plantations in areas of the ocean that are dedicated to this objective. Unused croplands and federal lands might also be used for the production of herbaceous or woody biomass energy crops.

B. Waste Biomass

Another large source of renewable carbon supplies is waste biomass. It consists of a wide range of materials and includes municipal solid wastes (MSW), municipal biosolids (sewage), industrial wastes, animal manures, agricultural crop and forestry residues, landscaping and tree clippings and trash, and dead biomass that results from nature's life cycles. Several of these wastes can cause serious health or environmental problems if they are not disposed of properly. Some wastes such as MSW can be considered to be a source of recyclables such as metals and glass in addition to energy. Thus, waste biomass is a potential energy resource in the same manner as virgin biomass.

To assess the potential impact of energy from waste biomass on supplying energy demand, it is necessary to consider the amounts of the different types of wastes generated, their energy contents, and their availabilities. Every person in the United States, for example, discards about 2.3 kg (5 lb) of MSW per day. From an energy standpoint, one short ton of MSW has an as-received energy content of about 9.5 GJ (9.0×10^6 million Btu), so about 2.2 EJ/year (2.1 quad/year) of energy potential resides in the MSW generated in the United States.

As for the amount of energy that can actually be recovered from a given waste and utilized, much depends on the waste type. The amount of available MSW, for example, is larger than the total amounts of available agricultural wastes even though much larger quantities of agricultural wastes are generated. This is caused by the fact that a larger fraction of MSW is collected for centralized disposal than the corresponding amounts of agricultural wastes, most of which are left in the fields where generated. The collection costs are prohibitive for most of these wastes. Note that municipal biosolids on a dry solids basis is generated in the smallest quantity of all wastes. Its disposal, however, is among the most costly and difficult of all waste treatment operations.

Many studies have been carried out to estimate the potential of available virgin and waste biomass as energy resources. One is presented in Table 2.7 for the United States for the year 2000 (Klass, 1990). The estimated energy potential of the recoverable materials is about 25% of the theoretical maximum. Wood and wood wastes are about 70% of the total recoverable energy potential and 50% of the estimated maximum energy potential. These estimates of virgin and waste biomass energy potential are based on existing, sustainable biomass production and do not include new, dedicated biomass energy plantations that might be developed and placed in commercial operation.

An assessment of the energy potential of waste biomass that is more localized can often provide better leads for the development of biomass energy supplies. The results of one such preliminary study performed for the state of Indiana are summarized in Table 2.8 (Klass, 1981). Indiana is a farm state. More than 60% of the state area was devoted to cropland at the time of the study and about 52% of state area was under active cultivation. The major agricultural crop and farm animal wastes as well as forestry and municipal wastes were therefore selected for the assessment of waste biomass energy potential. The waste biomass generated in the state each year was first inventoried, and each waste was then converted to gross energy content using generic conversion factors as a first approximation of energy potential. Comparison of the results with annual commercial energy utilization in the form of petroleum motor

TABLE 2.7   Potential Biomass Energy Available in United States in 2000[a]

Energy source	Estimated recoverable (EJ)	Theoretical maximum (EJ)
Wood and wood wastes	11.0	26.4
Municipal solid wastes		
Incineration	1.9	2.1
Methane from landfills	0.2	1.1
Herbaceous biomass and agricultural residues	1.1	15.8
Aquatic biomass	0.8	8.1
Industrial solid wastes	0.2	2.2
Methane from municipal biosolids	0.1	0.2
Methane from farm animal manures	0.05	0.9
Miscellaneous wastes	0.05	1.1
Total	15.4	57.9

[a]Klass (1990). The energy values are the higher heating values of the indicated biomass or derived methane. The conversion of biomass or methane to another biofuel or to steam, heat, or electric power requires that the process efficiency be used to reduce the potential energy available. These figures do not include additional biomass that could be grown as a dedicated energy crop.

fuels indicated that grain crop residues, particularly corn and soybean residues, and cattle manures have the largest potential as feedstocks for conversion to substitute motor fuels. Most of the other wastes are generated in insufficient quantities to make a large contribution. This simple assessment provided direction to the initiation of programs to develop systems using waste biomass feedstocks generated in the state of Indiana.

An example of a different type of assessment of waste and virgin biomass energy potential is one performed for the state of Wisconsin, another farm state in the Corn Belt of the United States. This assessment evaluated the economic impacts of shifting a portion of Wisconsin's future energy investment from imported fossil fuels toward renewable energy resources. It assumed a 75% increase in the state's renewable energy use by 2010—775 MW of new electric generating capacity to supply electricity to 500,000 Wisconsin homes, and 100 million gallons per year of new ethanol production to supply gasohol (blends of 10 vol % ethanol and 90 vol % gasoline) to 45% of Wisconsin's automobiles (Clemmer and Wichert, 1994). This scenario generated about three times more jobs, earnings, and output (sales) in Wisconsin than the same level of imported fossil fuel usage and investment, and was equivalent to 63,234 more job-years of net employment, $1.2 billion in higher wages, and $4.6 billion in additional output. Over the operating life of the technologies

TABLE 2.8 Energy Potential of Waste Biomass in Indiana[a]

Source and type	Estimated residue (dry Mt/year)	Estimated energy content (PJ/year)	Percent of petroleum motor fuel consumption
Grain crops			
Corn	14.27	249	51.4
Soybeans	2.92	50.9	10.5
Wheat	1.27	22.1	
Oats	0.33	5.7	
Rye	0.03	0.47	
Sorghum	0.04	0.63	
Barley	0.01	0.16	
Total:		329	68.0
Farm animal manures			
Cattle	3.22	56.2	11.6
Hogs	0.73	12.7	
Sheep	0.01	0.2	
Chickens	0.29	5.1	
Total:		74.2	15.3
Forest residues			
Hardwoods	0.395	7.89	1.6
Softwoods	0.002	0.04	
Total:		7.93	1.6
Sawmill residues			
Slabs and edgings	0.149	2.97	0.6
Sawdust	0.132	2.63	
Bark	0.087	1.74	
Total:		7.34	1.5
Municipal wastes			
MSW	2.40	27.9	5.8
Industrial	0.36	3.2	
Biosolids	0.18	3.0	
Total:		34.1	7.0
Grand total:		453	94

[a]Klass (1981).

analyzed, about $2 billion in avoided payments for imported fossil fuels would remain in Wisconsin to pay for the state-supplied renewable resources, labor, and technologies. Wood, corn, and waste biomass contributed 47% of the increase in net employment.

This review of the concept of utilizing biomass energy shows that when sufficient supplies of renewable carbon are available, virgin and waste biomass have the potential of becoming basic energy resources. Presuming that suitable conversion processes are available, and that the demand for energy and established organic fuels and intermediates continues, an industry based on renewable biomass fuels and feedstocks that can supply a significant portion of this demand is, at the very least, a technically feasible concept.

V. MARKET PENETRATION

A. U.S. MARKETS

As mentioned in the introduction to this chapter, biomass energy is already a substantial contributor to commercial primary energy demand. Market penetration is significant and is expected to increase. A comparison of U.S. consumption of biomass energy in 1990 with projections for 2000 (Table 2.9) (Klass, 1994) shows that consumption in 2000 is expected to be about 50% greater. This assessment is based on the following assumptions: Noncrisis conditions prevail; the U.S. tax incentives in place continue and are not changed; no

TABLE 2.9   Consumption of Biomass Energy in United States in 1990 and Projected for 2000[a]

	1990		2000	
Resource	EJ	BOE/day	EJ	BOE/day
Wood and wood wastes				
Industrial sector	1.646	763,900	2.2	1,021,000
Residential sector	0.828	384,300	1.1	510,500
Commercial sector	0.023	10,700	0.04	18,600
Utilities	0.013	6000	0.01	4600
Total:	2.510	1,164,900	3.35	1,554,700
Municipal solid wastes	0.304	141,100	0.63	292,400
Agricultural and industrial wastes	0.040	18,600	0.08	37,100
Methane				
Landfill gas	0.033	15,300	0.100	46,400
Biological gasification	0.003	1400	0.004	1900
Thermal gasification	0.001	500	0.002	900
Total:	0.037	17,200	0.106	49,200
Transportation fuels				
Ethanol	0.063	29,200	0.1	46,400
Other biofuels	0	0	0.1	46,400
Total	0.063	29,200	0.2	92,800
Grand total:	2.954	1,371,000	4.37[b]	2,026,200
Percent of primary energy consumption	3.3		4.8	

[a]Klass (1994) and U.S. Department of Energy (1990) for 1990; Klass (1990, 1994) for 2000.
[b]Other estimates range from 3.5 to 5.8 EJ/year in 2000 (cf. Hohenstein and Wright, 1994).

legislative mandates to embark on an off-oil campaign via fossil carbon consumption taxes or related disincentives to use fossil fuels, such as those in place in certain parts of Europe, are enacted; and total energy consumption in 2000 is 92 EJ (87 quad).

In 1990, industrial and residential utilization of biomass energy as wood and wood wastes was responsible for almost 84% of total biomass energy consumption, while MSW contributed about 10%. When these figures are compared with the estimated recoverable amounts of biomass energy available in the United States in 2000 (Table 2.7), it is evident that biomass energy consumption can be substantially increased. The development of large-scale biomass energy plantations in which system designs incorporate total replacement of virgin biomass resources as utilized could provide much larger increases in biomass energy consumption beyond these estimates. At an average U.S. wellhead price of petroleum of $20/bbl in 1990, total biomass consumption in 1990 was equivalent to about $27.4 million per day retained in the country and not expended on fossil fuels. There are clearly strong beneficial economic impacts of biomass energy consumption on U.S. trade deficits, a good portion of which is caused by oil imports.

A few comments are in order regarding the utilization of fuel ethanol, most of which is manufactured from corn in the United States. Fuel ethanol is used in motor gasoline blends as an octane enhancer and as an oxygenate to reduce emissions. The Clean Air Act Amendments of 1990 (U.S. Public Law 101-549) mandate the use of oxygenates in reformulated gasolines, and the market for ethanol from biomass is therefore expected to exhibit substantial growth as time passes, provided the tax incentives in place for fuel ethanol from biomass continue or fossil fuel consumption taxes are implemented to attempt to reduce atmospheric pollution. As will be shown in later chapters, advanced technologies may eventually make it possible for fuel ethanol to be manufactured from low-grade cellulosic biomass feedstocks and to be economically competitive with motor gasolines without the need for tax incentives. In the mid-1990s, the production capacity for fuel ethanol from biomass was about 4.2 billion L/year, or 0.088 EJ. Total U.S. production of fuel ethanol has increased by more than an order of magnitude since it was first marketed in modern times in the United States as a gasoline extender and octane enhancer in 1979. Fuel ethanol is a major biomass energy commodity, the production of which is expected to increase by another 2.3 billion L/year as the Clean Air Act Amendments are fully implemented. But note that the U.S. motor gasoline market in the mid-1990s was more than 379 billion L/year (100 billion gal/yr), so fuel ethanol only displaced about 1% by volume of petroleum gasolines.

The estimate of U.S. biomass energy usage in 2000 (Table 2.9) indicates that the largest contributions are still expected to come from wood and wood wastes in the industrial and residential sectors, or about three-quarters of total estimated U.S. biomass energy consumption. Because of the technical and economic problems associated with solid waste disposal, the increasing amounts of MSW generated by increasing urban populations, and the phase-out of sanitary landfilling as a preferred method of MSW disposal, the contribution of MSW to biomass energy usage is expected to double by 2000.

A projection of biomass energy consumption for the United States is shown for the years 2000, 2010, 2020, and 2030 by end-use sector in Table 2.10 (U.S. Dept. of Energy, 1990). This particular analysis is based on a national premiums scenario which assumes specific market incentives are applied to all new renewable energy technology deployment and continue to 2030. The premiums are 2¢/kWh on electricity generation from fossil fuels, $1.90/GJ ($2.00/10^6 Btu) on direct coal and petroleum consumption, and $0.95/GJ ($1.00/10^6 Btu) on direct natural gas consumption. This scenario depends on the enactment of federal legislation that is equivalent to a fossil fuel consumption tax. Any incentives over and above those assumed for the assessment in Table 2.9 can be a strong stimulus to increase biomass energy consumption.

The market penetration of synthetic fuels from virgin and waste biomass in the United States depends on several basic factors such as demand, price, performance, competitive feedstock uses, government incentives, whether an established fuel is replaced by a chemically identical fuel or a different fuel, and the cost and availability of other fuels such as oil and natural gas. Many detailed analyses have been performed to predict the market penetration of biomass energy over the next 10 to 50 years. There seems to be a range from about 4 to 20 quads per year that characterize the growth of biomass energy consumption. All of these projections of future market penetrations for biomass energy in the United States should be viewed in the proper perspective.

TABLE 2.10 Projected Biomass Energy Contribution in the United States under a National Premiums Scenario from 2000 to 2030[a]

End-use sector[b]	2000 (EJ)	2010 (EJ)	2020 (EJ)	2030 (EJ)
Industry	2.85	3.53	4.00	4.48
Electricity	3.18	4.41	4.95	5.48
Buildings	1.05	1.53	1.90	2.28
Liquid fuels	0.33	1.00	1.58	2.95
Total:	7.41	10.47	12.43	15.19

[a]U.S. Department of Energy (1990).
[b]Industrial end uses: combustion of wood and wood wastes. Electric end uses: electric power derived from 1992 technology via the combustion of wood and wood wastes, MSW, agricultural wastes, landfill and digester gas, and advanced digestion and turbine technology. Buildings end uses: biomass combustion in wood stoves. Liquid fuels are ethanol from grains, and ethanol, methanol, and gasoline from energy crops.

The potential of biomass energy is easily demonstrated as shown in this chapter, but the necessary infrastructure does not exist to realize this potential without large investments by industry. Government incentives will probably be necessary, too. U.S. capacity for producing virtually all biofuels manufactured by biological or thermal conversion of biomass would have to be dramatically increased to approach the potential contributions of virgin and waste biomass. For example, an incremental quad per year of methane from biomass feedstocks in the United States requires about 200 times the biological methane production capacity in place, and an incremental quad per year of fuel ethanol requires about 13 to 14 times the existing plant capacity to manufacture fermentation ethanol. Given the long lead times necessary to design and construct large biomass conversion plants, it is unrealistic to assume that sufficient capacity and the associated infrastructure could be placed on-line in the near term to satisfy quad-blocks of energy demand. This is not to say that plant capacities cannot be rapidly increased if a concerted effort is made by the private sector to do so.

Conversely, the upside of any assessment of virgin biomass feedstocks is that energy and fuel markets are very large and expand with the population, so there should be no shortage of demand for economically competitive energy supplies in the foreseeable future. Systems that offer improved waste disposal together with efficient energy recovery are also expected to fare quite well.

Projections of market penetrations and contributions to primary energy demand by biomass can contain significant errors. It is important, therefore, to keep in mind that even though some of these projections may turn out to be incorrect, they are still necessary to assess the future role and impact of renewable energy resources. They are also of great help in deciding whether a potential renewable energy resource should be developed and commercialized.

B. GLOBAL MARKETS

The United Nations estimate of global biomass energy consumption was about 6.7% of the world's energy consumption in 1990 (Table 1.2). Biomass energy continues to be a major source of energy and fuels in the developing regions of the world—Africa, South America, and Asia. The markets for biomass energy and biofuels as replacements and substitutes for fossil fuels are obviously large, but have only been developed to a limited extent.

There are still major barriers that must be overcome to permit biomass energy to have a truly large role in displacing fossil fuels. Among these are developing large-scale biomass energy plantations that can supply sustainable amounts of low-cost feedstocks; developing integrated biomass production–conversion systems that are capable of producing quad blocks of energy at competitive prices; developing nationwide biomass energy distribution systems that simplify consumer access and ease of use; and increasing the availability of capital for financing biomass projects in the private sector. Niche markets for biomass energy will continue to expand, and as fossil fuels either are phased out because of environmental issues or become less available and uneconomical because of depletion, biomass energy is expected to acquire an increasingly larger share of the global energy market.

REFERENCES

Boden, T. A., Marland, G., and Andres, R. J. (1995). "Estimates of Global, Regional, and National CO_2 Emissions from Fossil-Fuel Burning, Hydraulic Cement Production, and Gas Flaring: 1950-1992," ORNL/CDIAC-90, NDP-030/R6. Oak Ridge National Laboratory, Oak Ridge, TN, December.

Clemmer, S., and Wichert, D. (April 1994). "The Economic Impacts of Renewable Energy Use in Wisconsin." Wisconsin Energy Bureau, Division of Energy and Intergovernmental Relations, Madison, WI.

Hohenstein, W. G., and Wright, L. L . (1994). *Biomass and Bioenergy* 6(3), 161.

Houghton, R. A., and Woodwell, G. M. (1989). *Sci. Am.* **260** (4), 36.

Houghton, R. A., and Hackler, J. L. (1995). "Continental Scale Estimates of the Biotic Carbon Flux from Land Cover Change: 1850 to 1980," (R. C. Daniels, ed.), ORNL/CDIAC-79, NDP-050. Oak Ridge National Laboratory, Oak Ridge, TN.

Klass, D. L. (1974). *Chemtech 4* (3), 161.

Klass, D. L. (1981). *In* "Fuels from Biomass and Wastes," (D. L. Klass and G. H. Emert, eds.), p. 519. Ann Arbor Science Publishers, Inc., The Butterworth Group, Ann Arbor, MI.

Klass, D. L. (1990). *Chemtech* **20** (12), 720.

Klass, D. L. (1993). *Energy Policy* **21** (11), 1076.

Klass, D. L. (1994). *In* "Kirk-Othmer Encyclopedia of Chemical Technology," 4th Ed., Vol. 12, p. 16. John Wiley, New York.

Repetto, R. (1990). *Sci. Am. 262* (4), 36.

USDA Forest Service (October 1989). "RPA Assessment of the Forest and Rangeland Situation in the U.S.," No. 26. Washington, D.C.

U.S. Dept. of Energy (March 1990). "The Potential of Renewable Energy, An Interlaboratory White Paper," SERI/TP-260-3674, DE90000322. Office of Policy, Planning and Analysis, Washington, D.C.

U.S. Dept. of Energy (October 1991). "Estimates of U.S. Biofuels Consumption 1990," DOE/EIA-0548. Energy Information Administration, Washington, D.C.

U.S. Dept. of Energy (December 1994). *In* "Annual Energy Outlook 1995 with Projections to 2010," DOE/EIA-0383(95). Energy Information Administration, Washington, D.C.

Whittaker, R. H., and Likens, G. E. (1975). "Primary Productivity of the Biosphere," (H. Leith and R. H. Whittaker, eds.). Springer Verlag, New York.

CHAPTER 3

Photosynthesis of Biomass and Its Conversion-Related Properties

I. INTRODUCTION

Although many questions remain to be answered regarding the complex chemistry of biomass growth, the reactions that occur when carbon dioxide (CO_2) is fixed in live green biomass are photochemical and biochemical conversions that involve the uptake of CO_2, water, and the solar energy absorbed by plant pigments. Carbon dioxide is reduced in the process and water is oxidized. The overall process is called photosynthesis and is expressed by a simple equation that affords monosaccharides as the initial organic products. Light energy is converted by photosynthesis into the chemical energy contained in the biomass components:

$$6CO_2 + 6H_2O + light \rightarrow C_6H_{12}O_6 + 6O_2.$$
$$\text{Hexose}$$

The inorganic materials, CO_2 and water, are converted to organic chemicals, and oxygen is released. The initial products of a large group of biochemical reactions that occur in the photosynthetic assimilation of ambient CO_2 are sugars. Secondary products derived from key intermediates include polysaccharides, lipids, proteins and a wide range of organic compounds, which may or may not be produced in a given biomass species, such as simple low-molecular-weight organic chemicals (e.g., acids, alcohols, aldehydes, ethers, and esters), and complex alkaloids, nucleic acids, pyrroles, steroids, terpenes, waxes, and high-molecular-weight polymers such as the polyisoprenes. When biomass is burned, the process is reversed and the energy absorbed in photosynthesis is liberated along with the initial reactants.

The fundamentals of photosynthesis are examined in this chapter, with emphasis on how they relate to biomass production and its limitations. The compositions of different biomass species and the chemical structures of the major components are also examined in the context of biomass as an energy resource and feedstock.

II. PHOTOSYNTHESIS

A. FUNDAMENTALS

The historical development of our understanding of the photosynthesis of biomass began in 1772 when the English scientist Joseph Priestley discovered that green plants expire a life-sustaining substance (oxygen) to the atmosphere, while a live mouse or a burning candle removes this same substance from the atmosphere. A variety of suggestions were offered by the scientific community during the ensuing 30 years to explain these observations until in 1804, the Swiss scientist Nicolas Théodore de Sausseure showed that the amount of CO_2 absorbed by green plants is the molecular equivalent of the oxygen expired. From that point on, the stoichiometry of the process was developed and major advancements were made to detail the chemistry of photosynthesis and how the assimilation of CO_2 takes place. Much of this work paralleled the development of research done to understand the biochemical pathways of the cellular metabolism of foodstuffs. Indeed, there is much overlap in the chemistry of both processes.

About 75% of the energy in solar radiation, after passage through the atmosphere where much of the shorter wavelength, high-energy radiation is filtered out, is contained in light of wavelengths between the visible and near-infrared portions of the electromagnetic spectrum, 400 to 1100 nm. The light-absorbing pigments effective in photosynthesis have absorption bands in this range. Chlorophyll *a* and chlorophyll *b*, which strongly absorb wavelengths in the red and blue regions of the spectrum, and accessory carotenoid and phycobilin pigments participate in the process. Numerous investigations have established many of the parameters in the complex photosynthetic reactions occurring in biomass membrane systems which contain the necessary pigments and electron carriers. Tracer studies to establish the chemical structures of the intermediates in photosynthesis were initiated in the 1940s by the U.S. investigators Melvin Calvin, J. A. Bassham, and Andrew A. Benson. These studies showed that the oxygen evolved in photosynthesis comes exclusively from water and not CO_2. With green algae (*Chlorella*), reducing power is accumulated during illumination in the absence of CO_2 and can later be used for the reduction of CO_2 in the absence of light (Calvin and Benson, 1947). After a short 30- to 90-second exposure to light, the main portion of the newly reduced, labeled CO_2 was found to be distributed in a dozen or more organic compounds. By progressively shortening the light exposure to 2 seconds before killing the cells, almost all of the ^{14}C in the labeled CO_2 was found to be incorporated in 3-phosphoglyceric acid, a compound that occurs in practically all plant and animal cells. Such experiments led to elaboration of the biochemical pathways and the essential compounds required for photosynthesis. It was found that the pentose ribulose-1,5-diphosphate is a key intermediate in the process. It reacts with CO_2 to yield 3-phosphoglyceric acid. Equimolar amounts of ribulose-1,5-diphosphate and CO_2 react to form 2 mol of 3-phosphoglyceric acid, and in the process, inorganic carbon is transformed into organic carbon.

$$
\begin{array}{ll}
\begin{array}{l}
CH_2OPO_3H_2 \\
| \\
C=O \\
| \\
H-C-OH \\
| \\
H-C-OH \\
| \\
CH_2OPO_3H_2
\end{array}
\quad + CO_2 \longrightarrow 2 &
\begin{array}{l}
CO_2H \\
| \\
H-C-OH \\
| \\
CH_2OPO_3H_2
\end{array}
\end{array}
$$

Ribulose–1,5–diphosphate   3–Phosphoglyceric acid

These reactions of course occur in the presence of the proper enzymic catalysts and cofactors. Glucose is the primary photosynthetic product. As will be shown later in the discussion of the structures of the organic intermediates in the various pathways, the dark reactions take place in such a manner that ribulose-1,5-diphosphate is regenerated.

B. LIGHT REACTIONS FOR CARBON DIOXIDE ASSIMILATION

In addition to a suitable environment, appropriate pigments, whose cumulative light-absorbing properties determine the range of wavelengths over which photosynthesis occurs, a reaction center where the excited pigments emit

electrons, and an electron transfer chain that generates the high-energy phos-
phorylating agent adenosine triphosphate (ATP) by photophosphorylation are
necessary for ambient CO_2 reduction. The pigment chlorophyll absorbs light
and is oxidized by ejection of an electron. The electron is accepted by ferredoxin
(Fd), a nonheme iron protein, to form reduced ferredoxin (Fd^{+2}), which
through other electron carriers generates ATP and the original oxidized ferredoxin (Fd^{+3}). Chlorophyll functions as both a light absorber and a source of
electrons in the excited state, and as the site of the initial photochemical
reaction. Accessory pigments function to absorb and transfer light energy
to chlorophyll.

Two photochemical systems are involved in these "light reactions": photosystems II (PS II) and I (PS I). PS II consists of the first series of reactions
that occur in the light phase of photosynthesis during which the excited
pigment participates in the photolysis of water to liberate free oxygen, protons,
and electrons. PS I is the second series of reactions that occur in the light
phase of photosynthesis; they result in the transfer of reducing power to
nicotinamide adenine dinucleotide phosphate (NADP) for ultimate utilization
by CO_2. The light reactions yield ATP, and the reduced form of nicotinamide
adenine dinucleotide phosphate ($NADPH_2$), both of which facilitate the dark
reactions that yield sugars. Hydrogen is transferred by $NADPH_2$. The low-
energy adenosine diphosphate (ADP) and NADP produced in the dark reactions
are reconverted to ATP and $NADPH_2$ in the light reactions.

The chemistry of the light reactions was elucidated in 1954 by the U.S.
biochemist Daniel Arnon and co-workers (Arnon, Allen, and Whatley, 1954).
Light energy is absorbed by the chlorophyll pigments in plant chloroplasts
and transferred to the high-energy bonds in ATP, which is produced in noncyclic and cyclic photophosphorylation reactions. Noncyclic photophosphorylation occurs in the presence of light, requires Fd catalyst, and yields ATP and
oxygen. NADP is then reduced by Fd^{+2} in the absence of light:

$$4Fd^{+3} + 2ADP + 2P_i + 2H_2O \rightarrow 4Fd^{+2} + 2ATP + O_2 + 4H^+$$
$$4Fd^{+2} + 2NADP + 4H^+ \rightarrow 4Fd^{+3} + 2NADPH_2.$$

For each molecule of oxygen evolved, two molecules each of ATP and $NADPH_2$
are formed. Cyclic photophosphorylation requires Fd catalyst and produces
ATP only:

$$ADP + P_i \rightarrow ATP.$$

This process provides the additional molecule of ATP needed for assimilation
of one molecule of CO_2.

PS I appears to promote cyclic photophosphorylation and proceeds best in
light of wavelengths greater than 700 nm, whereas PS II promotes noncyclic
photophosphorylation and proceeds best in light of wavelengths shorter than

700 nm. PS II and PS I operate in series in the chloroplast membranes and transfer reducing power from water to Fd and NADP by an electron chain that includes plastoquinone and three proteins in chloroplasts—cytochrome b, cytochrome f, and the copper protein plastocyanin. To overcome the potential difference between the carbon dioxide-glucose couple and the water-oxygen couple, two photons are absorbed, one by PS II and one by PS I. In the traditional "Z scheme" concept of photosynthesis first proposed in 1960 (Hill and Bendall, 1960), and which is now well accepted, the strong oxidizing agent, oxygen, is at the bottom, and the strong reducing agent, reduced ferredoxin, is at the top (Fig. 3.1) (Bassham, 1976). By transfer of electrons from water to ferredoxin, the chemical potential for reduction of CO_2 is created. PS II is depicted as absorbing one photon to raise an electron to an intermediate energy level, after which the electron falls to an intermediate lower energy level while generating ATP. PS I then absorbs the second photon and raises the electron to a still higher intermediate energy level, and subsequently generates $NADPH_2$ via reduced ferredoxin, after which CO_2 is reduced to yield sugars. In effect, the Z scheme transfers electrons from a low chemical potential in water to a higher chemical potential in $NADPH_2$, which is necessary to reduce CO_2.

It has been discovered, however, that both photosystems do not seem to be necessary as depicted in the Z scheme to reduce CO_2. PS II seems to be adequate alone to generate the chemical potential for reduction, at least in

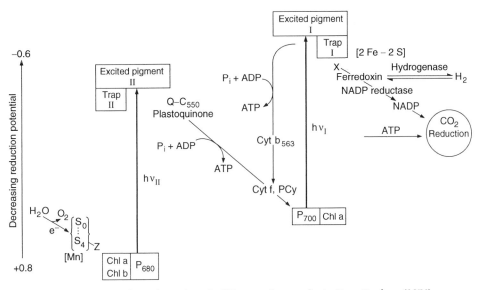

FIGURE 3.1 Traditional "Z scheme" of biomass photosynthesis. From Bassham (1976).

one biomass species under certain conditions. PS I is absent in a mutant of the alga *Chlamydomonas reinhardtii*, but the organism has been found to be capable of photoautrophic assimilation of CO_2 and the simultaneous evolution of oxygen and hydrogen (Greenbaum *et al.*, 1995). The investigators interpreted their results to mean that a single-photon light reaction has the potential of increasing the efficiency of photosynthesis by overcoming the thermodynamic limitations of converting light energy into chemical energy. This will be referred to later in the discussion of photosynthesis efficiency and biomass yield. In any case, this observation tends to verify Arnon's argument that water oxidation and NADP reduction can be driven by PS II alone (Barber, 1995).

In summary, ambient CO_2 fixation by photosynthesis involves the photochemical decomposition of water to form oxygen, protons, and electrons; the transport of these electrons to a higher energy level via PS II and I and several electron transfer agents; the concomitant generation of $NADPH_2$ and ATP; and reductive assimilation of CO_2 to monosaccharides. The initial process is the absorption of light by chlorophyll, which promotes the decomposition of water. The ejected electrons are accepted by the oxidized form of Fd. The reduced Fd then starts a series of electron transfers to generate ATP from ADP and inorganic phosphate, and $NADPH_2$ in the light reactions. The stoichiometry, including the reduction in the dark reactions of 1 mol CO_2 to carbohydrate, represented by the building block (CH_2O), is illustrated in simplified form as follows:

$$2H_2O(l) \rightarrow 4H^+ + 4e + O_2(g)$$
$$4Fd^{+3} + 4e \rightarrow 4Fd^{+2}$$
$$3ADP + 3P_i \rightarrow 3ATP$$
$$4Fd^{+2} \rightarrow 4Fd^{+3} + 4e$$
$$2NADP + 4H^+ + 4e \rightarrow 2NADPH_2$$
$$CO_2(g) + 3ATP + 2NADPH_2 \rightarrow (CH_2O) + 3ADP + 2NADP + 3P_i + H_2O(l)$$
$$\text{Overall: } CO_2(g) + H_2O(l) \rightarrow (CH_2O) + O_2(g).$$

For each of the two light reactions, one photon is required to transfer each electron; a total of eight photons is thus required to fix one molecule of CO_2. Assuming CO_2 is in the gaseous phase and the initial product is glucose, the standard Gibbs free energy change at 25°C is +0.48 MJ (+114 kcal) per mole of CO_2 assimilated and the corresponding enthalpy change is +0.47 MJ (+112 kcal).

C. DARK REACTIONS FOR CARBON DIOXIDE ASSIMILATION

The discussion of photosynthesis to this point has concentrated more on the light reactions that occur in photosynthesis. The organic components in biomass are formed during the dark reactions. Some discussion of the biochemical pathways and organic intermediates involved in the reduction of CO_2 to sugars is beneficial because they play a significant role in our understanding of the molecular events of biomass growth, and in differentiating between the various kinds of biomass.

Before discussion of these pathways, it is important to note that the photosynthetic pathways also involve several dark reactions that occur in the glycolysis of glucose. The metabolic pathways provide energy for cellular maintenance and growth, form 3-phosphoglyceric acid and 3-phosphoglyceraldehyde, from which almost all other cellular organic components are synthesized, and are virtually identical in a large variety of living organisms, for example, in corn, wheat, oats, legumes, algae, many bacteria, the muscle, brain, liver, and other organs of humans and animals, and many birds, insects, and reptiles. Several reactions in the dark reactions on CO_2 uptake are the same as those that occur in the metabolism of foodstuffs by these organisms. Seven reactions of photosynthesis are common to the Embden-Meyerhof metabolic pathway, and three reactions are common to the pentose phosphate metabolic pathway. Each pathway converts glucose to pyruvic acid.

In photosynthesis, CO_2 generally enters the leaves or stems of biomass through the stoma, the small intercellular openings in the epidermis. These openings provide the main route for both photosynthetic gas exchange and for water vapor loss in transpiration. At least three different biochemical pathways can occur during CO_2 reduction to sugars (Rabinovitch, 1956; Loomis et al., 1971; Osmond, 1978).

One pathway is called the Calvin or Calvin–Benson cycle and involves the three-carbon intermediate 3-phosphoglyceric acid. This cycle, which is sometimes referred to as the reductive pentose phosphate cycle, is used by autotrophic photochemolithotrophic bacteria, algae, and green plants. As shown in Fig. 3.2, ribulose-1,5-diphosphate (I) and CO_2 react to form 3-phosphoglyceric acid (II), which in turn is converted via 1,3-diphosphoglyceric acid (III) and 3-phosphoglyceraldehyde (IV) to glucose (V) and ribulose-5-phosphate (VI), from which I is regenerated. For every 6 molecules of CO_2 converted to 1 molecule of glucose in a dark reaction, 18 ATP, 12 NADPH, and 24 Fd molecules are required. Twelve molecules of II are formed in the chloroplasts from 6 molecules each of CO_2 and I. After these carboxylation reactions, a reductive phase occurs in which 12 molecules of II are successively transformed into 12 molecules of III and 12 molecules of IV, a triose phosphate. Ten molecules of the triose phosphate are then used to regenerate 6 molecules of I, which initiates the cycle again. The other 2 triose phosphate molecules are used to generate glucose.

Plant biomass species that use the Calvin–Benson cycle are called C_3 plants. The cycle is common in many fruits, legumes, grains, and vegetables. C_3 plants usually exhibit low rates of photosynthesis at light saturation, low light

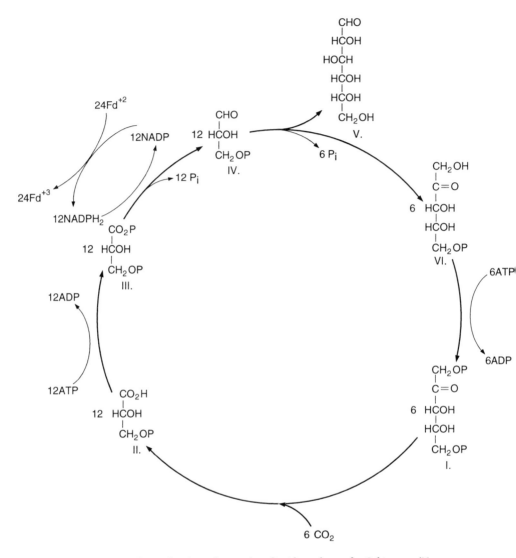

FIGURE 3.2 Biochemical pathway from carbon dioxide to glucose for C_3 biomass. (Net process: 3 ATP, 2 $NADPH_2$, 4 Fd^{+2}/CO_2 assimilated.)

saturation points, sensitivity to oxygen concentration, rapid photorespiration, and high CO_2 compensation points (about 50 ppm). The CO_2 compensation point is the CO_2 concentration in the surrounding environment below which more CO_2 is respired by the plant than is photosynthetically fixed. Typical C_3

biomass species are alfalfa, barley, *Chlorella,* cotton, Eucalyptus, *Euphorbia lathyris,* oats, peas, potato, rice, soybean, spinach, sugar beet, sunflower, tall fescue, tobacco, and wheat.

The second pathway is called the C_4 cycle because CO_2 is initially converted to the four-carbon dicarboxylic acids, malic or aspartic acids (Fig. 3.3). Phosphoenolpyruvic acid (I) reacts with one molecule of CO_2 to form oxaloacetic acid (II) in the mesophyll of the biomass, and then malic or aspartic acid (III) is formed. The C_4 acid is transported to the bundle sheath cells, where decarboxylation occurs to regenerate pyruvic acid (IV), which is returned to

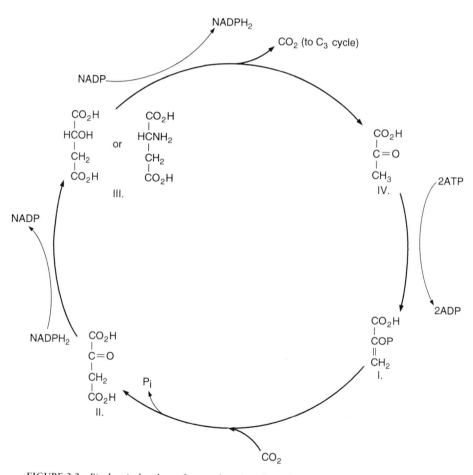

FIGURE 3.3   Biochemical pathway from carbon dioxide to glucose for C_4 biomass. (Net process: 5 ATP, 2 $NADPH_2$, 4 Fd^{+2}/CO_2 assimilated.)

the mesophyll cells to initiate another cycle. The CO_2 liberated in the bundle sheath cells enters the C_3 cycle in the usual manner. Thus, no net CO_2 is fixed in the portion of the C_4 cycle shown in Fig. 3.3, and it is the combination with the C_3 cycle which ultimately results in CO_2 fixation.

The subtle differences between the C_4 and C_3 cycles are believed responsible for the wide variations in biomass properties. In contrast to C_3 biomass, C_4 biomass is usually produced at higher yields and has higher rates of photosynthesis, high light saturation points, insensitivity to atmospheric oxygen concentrations below 21 mol %, low levels of respiration, low CO_2 compensation points, and greater efficiency of water usage. C_4 biomass often occurs in areas of high insolation, hot daytime temperatures, and seasonal dry periods. Typical C_4 biomass includes important crops such as corn, sugarcane, and sorghum, and forage species and tropical grasses such as Bermuda grass. Even crabgrass is a C_4 biomass. At least 100 genera in 10 plant families are known to exhibit the C_4 cycle.

The third pathway is called crassulacean acid metabolism, or CAM. CAM refers to the capacity of chloroplast-containing biomass tissues to fix CO_2 via phosphoenolpyruvate carboxylase in dark reactions leading to the synthesis of free malic acid. The mechanism involves the β-carboxylation of phosphoenolpyruvic acid by this enzyme and the subsequent reduction of oxaloacetic acid by malate dehydrogenase. CAM has been documented in at least 18 families, including the family Crassulaceae, and at least 109 genera of the *Angeospermae*. Biomass species in the CAM category are typically adapted to arid environments, have low photosynthesis rates, and have high water usage efficiencies. Examples are cactus plants and the succulents, such as pineapple. The information developed to date on CAM biomass indicates that CAM has evolved so that the initial CO_2 fixation can take place in the dark with much less water loss than the C_3 and C_4 pathways. CAM biomass also conserves carbon by recycling endogenously formed CO_2. Several CAM species show temperature optima in the range 12 to 17°C for CO_2 fixation in the dark. The stomates in CAM plants open at night to allow entry of CO_2 and then close by day to minimize water loss. The carboxylic acids formed in the dark are converted to sugars when the radiant energy is available during the day. Relatively few CAM plants have been exploited commercially.

D. OTHER BIOCHEMICAL PATHWAYS

A short discussion is also in order regarding the biochemical pathways to the polysaccharides (celluloses and hemicelluloses), which are the dominant organic components in most biomass, and the lignins, proteins (polypeptides), and triglycerides (lipids or fats) that are found in biomass. Most biomass on a dry basis contains about 50 wt % celluloses. The other components are present in lower concentrations. It is evident that since the simple sugars are the initial products of photosynthesis, they are the primary precursors of all the organic components in biomass. More details on the composition of biomass and the chemical structures of the major components are presented later in Section III.

Celluloses and Hemicelluloses

The pathways to the high-molecular-weight polysaccharides involve successive condensations of the monosaccharides, mainly the C_6 hexoses to yield celluloses and starches, and mainly the C_5 pentoses to yield hemicelluloses. Celluloses are composed of β-glucosidic units in the polymer chain, and starches are composed of α-glucosidic units. Glucose is the dominant immediate precursor of the celluloses. Similarly, the dominant repeating unit in the hemicelluloses is the pentoses, which are intermediates in the phototosynthetic pathways to glucose of C_3 and C_4 plants. Since the celluloses always occur in terrestrial biomass together with the hemicelluloses, it is likely that some of the C_5 intermediates are shunted from the glucose pathway to form the hemicelluloses.

Lignins

This group of biomass components is unique in that it occurs mainly in woody biomass and cellulosic, terrestrial biomass as aromatic polymers containing phenyl propane units in which the benzene rings are substituted by methoxyl and hydroxyl groups. The linkages in the polymers occur directly between the rings, between the propane units, and through ether linkages via the hydroxyl groups. About 25 wt % of dry wood consists of lignins; slightly more of which is usually contained in softwoods than hardwoods. The precursors of the lignins appear to be C_9 compounds such as p-hydroxyphenylpyruvic acid, which can be derived through a series of condensation reactions starting with glucose. Amino acids seem to play a role in this biochemical pathway. One pathway proposed for the formation of spruce lignin involves the biochemical conversion of glucose, which is transformed into the lignins by a series of reactions via shikimic acid (3,4,6-trihydroxy-cyclohexene-1-carboxylic acid) and coniferin, a glycoside of coniferyl alcohol (3-(4-hydroxy-3-methoxy-phenyl)-2-propene-1-ol) (Freudenberg, 1957). Note that because of the chemical complexity of the lignins and the variety of specific lignin structures formed by different biomass, numerous biochemical pathways have been proposed to explain how lignins are formed from sugars. Coniferyl alcohol is a precursor in many of these pathways.

Proteins

Proteins are polymers composed of natural amino acids that are bonded together through peptide linkages. They are formed via condensation of the acids through the amino and carboxyl groups by abstraction of water to form polyamides and are widely distributed in biomass as well as animals. Indeed, although they are present in some systems at concentrations approaching 0 wt %, they are found in all living systems because the enzymic catalysts that promote the various biochemical reactions are proteins. The apparent precursors of the proteins are the amino acids in which an amino group, or an imino group in a few cases, is bonded to the carbon atom adjacent to the carboxyl group. Many amino acids have been isolated from natural sources, but only about 20 of them are used for protein biosynthesis. This does not mean that all 20 of these amino acids appear in each polypeptide molecule. The number of amino acids used and the possible sequences in the polymeric chains correspond to an infinite number of potential polypeptide structures. Natural selection controls these parameters.

Regarding the precursors of the amino acids, several biochemical intermediates and various nitrogen sources are utilized. The amino acids have been divided into five families: glutamate (glutamine, arginine, proline), aspartate (asparagine, methionine, threonine, isoleucine, lysine), aromatic (tryptophan, phenylalanine, tyrosine), serine (glycine, cysteine), and pyruvate (alanine, valine, and leucine) (Stanier *et al.*, 1986). The corresponding precursors for these families are α-ketoglutarate, oxalacetate, phosphoenolpyruvate and erythrose-4-phosphate, 3-phosphoglycerate, and pyruvate. The biosynthesis of histidine uses an isolated pathway and requires a phosphoribosylpyrophosphate intermediate. All of these precursors are either biochemical intermediates or are derived from them. The nitrogen source is ambient nitrogen for many biomass species and ammonia, urea, or ammonium salts for most cash crops.

Triglycerides

Lipids are esters of the triol, glycerol, and long chain fatty acids. The fatty acids are any of a variety of monobasic fatty acids such as palmitic and oleic acids. The esters are formed in a large variety of oilseed crops, green plants, and some microalgae. Examples are soybean, cottonseeed, and corn oils. One pathway to the lipids produces glycerol and the other produces fatty acids, which can then combine to afford the triglycerides. In the chloroplasts, the Calvin–Benson cycle produces phosphoglyceric acid, which undergoes successive noncyclic photophosphorylation and isomerization to yield 3-phosphoglyceraldehyde and dihydroxyacetone phosphate. Three molecules of the aldehyde and one molecule of the phosphate are translocated out of the chloroplasts and combine to form fructose-1,6-diphosphate, which then successively undergoes a series of hydrolysis, isomerization, and condensation reactions to yield the disaccharide sucrose from glucose and fructose intermediates. This pathway to glycerol involves reduction of dihydroxyacetone phosphate to glycerol-1-phosphate. The fatty acids are derived from pyruvic acid formed on glycolyis of glucose. The pathway involves decarboxylation of pyruvic acid, the formation of acetyl coenzyme A, which is involved in the synthesis (and breakdown) of fatty acids, and the buildup of fatty acid chains by insertion of two-carbon units into the growing chain.

E. TERPENES

Some biomass species are able to reduce CO_2 via the initial sugars produced in photosynthesis to higher energy hydrocarbons, most of which have terpene structures. Because the energy values of terpene hydrocarbons can be as high or higher in some cases than conventional motor fuel components, and because some of the terpenes have been used as motor fuels and chemical feedstocks, a somewhat more detailed discussion of the biochemical pathways of hydrocarbon production in biomass is worthwhile. The terpenes are isoprene adducts having the generic formula $(C_5H_8)_n$, where n is 2 or more. A large number of terpene derivatives in various states of oxidation and unsaturation (terpenoids) may also be formed. Perhaps the best-known example of natural hydrocarbon production is high-molecular-weight polyisoprene rubber having very high stereospecificity from the cambium of the hevea rubber tree (*Hevea braziliensis*), a member of the Euphorbiaceae family that grows in Brazil and in other tropical climates. In contrast, the Brazilian tree, *Copaifera multijuga*, a member of the Caesalpiniaceae family, produces relatively pure liquid sesquiterpene hydrocarbons ($C_{15}H_{24}$), not in the cambium but in the heartwood from pores that run vertically throughout the tree trunk (Calvin, 1983). In some members of the Euphorbiaceae family such as *E. lathyris*, the biosynthetic pathway leads mainly to the acyclic dihydrotriterpene squalene, $C_{30}H_{50}$, which then undergoes internal cyclization to form C_{30} terpenoid alcohols and sterols. In still other biomass species, lower molecular weight acyclic and alicyclic isoprene adducts are formed as monoterpenes ($C_{10}H_{16}$) and diterpenes ($C_{20}H_{48}$). Certain aquatic, unicellular biomass such as the green microalgae *Botryococcus braunii* are reported to accumulate terpene-type hydrocarbon liquids within the cells, sometimes in large amounts depending on the growth conditions.

The major steps in the mechanisms of terpene and polyisoprene formation in plants and trees are known, and this knowledge should help improve the natural production of terpene hydrocarbons (Fig. 3.4). Mevalonic acid (I), a key intermediate derived from plant sugars via acetylcoenzyme A, is succes-

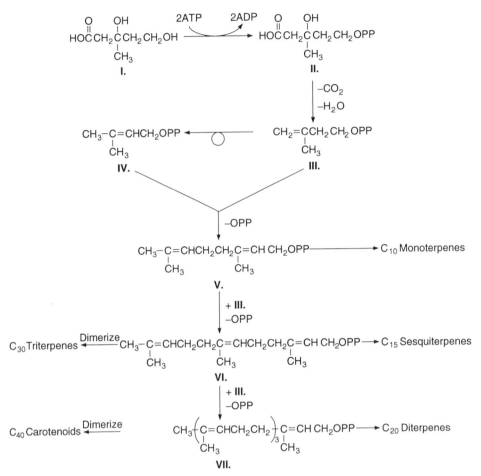

FIGURE 3.4　Biochemical pathways to terpenes.

sively transformed into 5-diphosphomevalonic acid (II) and the five-carbon intermediate isopentenylpyrophosphate (III) via phosphorylation, dehydration, and decarboxylation. Isomerization of a portion of III then occurs to form dimethylallylpyrophosphate (IV). The isomers III and IV combine by head-to-tail condensation to form another allylic pyrophosphate containing 10 carbon atoms (V), which can be converted to monoterpenes. Inorganic phosphate is released in the process. Continuation of this process leads to all other terpenes. The chain can successively build up by five-carbon units to yield sesquiterpenes and diterpenes containing 15 and 20 carbon atoms via VI and VII by additional head-to-tail condensations. The same structural unit is inserted in each step. A single enzyme, farnesyl pyrophosphate cyclase, is reported to be involved in the cyclization of the farnesyl units in *C. multijuga* to yield many sesquiterpenes (Calvin, 1983). Alternatively, tail-to-tail condensation of two C_{15} farnesyl units can yield the 30-carbon compound squalene followed by a large number of cyclizations and rearrangements to yield an array of natural triterpenoids, as already mentioned. Similar condensation of two C_{20} units yields phytoene, a precursor of carotenoids. This information is expected to help in the development of genetic engineering methods to control the hydrocarbon structures and yields.

F. PHOTOSYNTHESIS EFFICIENCY AND BIOMASS YIELD

Major differences in net photosynthetic assimilation of CO_2 are apparent between C_3, C_4, and CAM biomass species. Biomass species that fix CO_2 by the C_4 pathway usually exhibit higher rates of photosynthesis at warm temperatures. At cool temperatures, C_3 biomass species usually have higher rates of photosynthesis. Plants that grow well in the early spring such as forage grasses and wheat, are all C_3 species, whereas many desert plants, tropical species, and species originating in the tropics such as sugarcane are C_4 plants. One of the major reasons for the generally lower yields of C_3 biomass is its higher rate of photorespiration. If the photorespiration rate could be reduced, the net yield of biomass would increase. Considerable research has been done to achieve this rate reduction by chemical and genetic methods, but only limited yield improvements have been made. Such an achievement if broadly applicable to C_3 biomass would be expected to be very beneficial for both the production of foodstuffs and biomass energy. Another advancement that will probably evolve from research concerns increasing the yields of the secondary derivatives such as the liquid terpene hydrocarbons and triglycerides produced by certain biomass species to make direct fuel production from biomass practical and economically competitive. Detailed study and manipulation of the biochemical pathways involved will undoubtedly be neccessary to achieve some of these improvements, particularly since most of the advancements that have been made by controlling growth conditions and trying to select improved strains of biomass have not been very successful in increasing the natural production of liquid fuels.

The CO_2-fixing pathways used by a specific biomass species will affect the efficiency of photosynthesis, so from a biomass energy standpoint, it is desirable to choose species that exhibit high photosynthesis rates to maximize the yields of biomass in the shortest possible time. Obviously, however, there are numerous factors that affect the efficiency of photosynthesis other than the carbon dioxide-fixing pathway. Insolation; the amounts of available water and macronutrients and micronutrients; the CO_2 in the surrounding environment; the atmospheric concentration of CO_2, which is normally about 0.03 mol %; the temperature; and the transmission, reflection, and biochemical energy losses within or near the plant affect the efficiency of photosynthesis. For lower plants such as the green algae, many of these parameters can be controlled, but for conventional biomass growth that is subjected to the natural elements, it is not feasible to control all of them.

The maximum efficiency with which photosynthesis can occur has been estimated by several methods. The upper limit has been projected to range from about 8 to 15%, depending on the assumptions made, that is, the maximum amount of solar energy trapped as chemical energy in the biomass is 8 to 15% of the energy content of the incident solar radiation. It is worthwhile to examine the rationale in support of this efficiency limitation because it will help to point out some aspects of biomass production as they relate to energy applications.

The relationship of the energy and frequency of a photon is given by

$$e = (hc)/\lambda,$$

where e = energy content of one photon, J; h = Planck's constant, 6.626×10^{-34} J · s; c = velocity of light, 3.00×10^8 m/s; and λ = wavelength of light, nm. Assume that the wavelength of the light absorbed is 575 nm and is equivalent to the light absorbed between the blue (400 nm) and red (700 nm) ends of the visible spectrum. This assumption has been made for green plants by several investigators to calculate the upper limit of photosynthesis efficiency. The energy absorbed in the fixation of 1 mol CO_2, which requires 8 photons per molecule, is then given by

$$\text{Energy absorbed} = \frac{(6.626 \times 10^{-34})(3.00 \times 10^8)(575 \times 10^{-9})^{-1}(8)}{(6.023 \times 10^{23})}$$
$$= 1.67 \text{ MJ (399 kcal).}$$

Since 0.47 MJ of solar energy is trapped as chemical energy in this process, the maximum efficiency for total white light absorption is 28.1%. Further adjustments are usually made to account for the percentages of photosynthetically active radiation in white light that can actually be absorbed, and respiration. The fraction of photosynthetically active radiation in solar radiation that reaches the earth is estimated to be about 43%. The fraction of the incident light absorbed is a function of many factors such as leaf size, canopy shape, and reflectance of the plant; it is estimated to have an upper limit of 80%. This effectively corresponds to the utilization of 8 photons out of every 10 in the active incident radiation. The third factor results from biomass respiration. A portion of the stored energy is used by the plant, the amount of which depends on the properties of the particular biomass species and the environment. For purposes of calculation, assume that about 25% of the solar energy trapped as chemical energy is used by the plant, thereby resulting in an upper limit for retention of the nonrespired energy of 75%. The upper limit for the efficiency of photosynthetic fixation of biomass can then be estimated to be 7.2% ($0.281 \times 0.43 \times 0.80 \times 0.75$). For the case where little or no energy is lost by respiration, the upper limit is estimated to be 9.7% ($0.281 \times 0.43 \times 0.80$). The low-efficiency limit might correspond to terrestrial biomass, while the higher efficiency limit might be closer to the efficiency of aquatic biomass such as unicellular algae. These figures can be converted to dry biomass yields by assuming that all of the CO_2 fixed is contained in the biomass as cellulose, $-(C_6H_{10}O_5)_x-$, from the equation

$$Y = (CIE)/F,$$

where Y = yield of dry biomass, t/ha-year; C = constant, 3.1536; I = average insolation, W/m^2; E = solar energy capture efficiency, %; and F = energy content of dry biomass, MJ/kg.

Thus, for high-cellulose dry biomass, an average isolation of 184 W/m^2 (1404 Btu/ft^2-day), which is the average insolation for the continental United States, a solar energy capture efficiency of 7.2%, and a higher heat of combustion of 17.51 MJ/kg for cellulose, the yield of dry biomass is 239 t/ha-year (107 ton/ac-year). The corresponding value for an energy capture efficiency of 9.7% is 321 t/ha-year (143 ton/ac-year). These yields of organic matter can be viewed as an approximation of the theoretical upper limits for land- and water-based biomass. Some estimates of maximum yield reported by others are higher and some are lower than these figures, depending on the values used for I, E, and F, but they serve as a guideline to indicate the highest theoretical yields of a biomass production system. Unfortunately, real biomass yields rarely approach these limits. Sugarcane, for example, which is one of the high-yielding biomass species, typically produces total dry plant matter at yields of about 80 t/ha-year (36 ton/ac-year).

Yield is plotted against solar energy capture efficiency in Fig. 3.5 for insolation values of 150 and 250 W/m^2 (1142 and 1904 Btu/ft^2-day), which span the range commonly encountered in the United States, and for dry biomass energy values of 12 and 19 MJ/kg (5160 and 8170 Btu/lb). The higher the efficiency of photosynthesis, the higher the biomass yield. But it is interesting to note that for a given solar energy capture efficiency and incident solar radiation, the yield is projected to be lower at the higher biomass energy values (curves A and C, curves B and D). From a gross energy production standpoint, this simply means that a higher-energy-content biomass could be harvested at lower yield levels and still compete with higher-yielding but lower-energy-content biomass species. It is also apparent that for a given solar energy capture

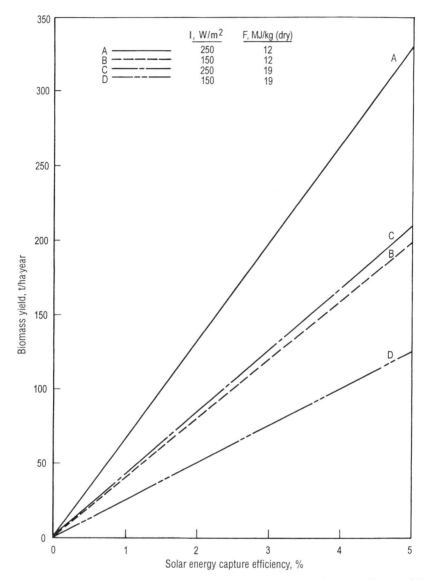

FIGURE 3.5 Effect of solar energy capture efficiency on biomass yield. (I is insolation and F is biomass energy content.)

efficiency, yields similar to those obtained with higher-energy-content species should be possible with a lower-energy-content species even when it is grown at lower insolation (curves B and C). Finally, at the solar energy capture efficiency usually encountered in the field, the spread in yields is much less than at the higher-energy-capture efficiencies. It is important to emphasize that this interpretation of biomass yield as functions of insolation, energy content, and energy capture efficiency, although based on sound principles, is still a theoretical analysis of living systems that can exhibit unexpected behavior.

Because of the many uncontrollable factors, such as climatic changes and the fact that the atmosphere only contains 0.03 mol% CO_2, biomass production outdoors generally corresponds to photosynthesis efficiencies in the 0.1 to 1.0% range. Significant departures from the norm can be obtained, however, with certain plants such as sugarcane, napier grass, algae, maize, and water hyacinth (Tables 3.1 and 3.2). The average insolation values at the locations corresponding to the biomass growth areas listed in Table 3.1 were used to calculate solar energy capture efficiency at the reported annual dry yields. Other than insolation, all environmental factors and the nonfuel components in the biomass were ignored for the solar energy capture efficiency estimates listed in Table 3.1 and it was assumed that all dry matter is organic and has an energy content of 18.6 MJ/dry kg (16 million Btu/dry ton). There is still a reasonably good correlation between dry biomass yield and solar energy capture efficiency in Table 3.1. The estimates of the efficiencies are only approximations and most are probably higher than the actual values. They indicate, however, that C_4 biomass species are usually better photosynthesizers than C_3 biomass species and that high insolation alone does not necessarily correlate with high biomass yield and solar energy capture efficiency. The biomass production data shown in Table 3.2 are some of the high daily rates of biomass photosynthesis reported for the indicated species. It has been estimated that water hyacinth could be produced at rates up to about 150 t/ha-year (67 ton/ac-year) if the plant were grown in a good climate, the young plants always predominated, and the water surface was always completely covered (Westlake, 1963). Some evidence has been obtained to support these estimates (McGarry, 1971; Yount and Grossman, 1970). Unicellular algae, such as the species *Chlorella* and *Scenedesmus*, have been produced by continuous processes in outdoor light at high photosynthesis efficiencies (Burlew, 1953; Enebo, 1969; Kok and Van Oorschot, 1954; Oswald, 1969). Growth rates as high as 1.10 dry t/ha-day have been reported for *Chlorella* (Retovsky, 1966). In tropical climates, this rate might be sustainable over most of the year, in which case the annual yield might be expected to approach 401 dry t/ha-year. This yield range is beyond the theoretical upper limit estimated here

TABLE 3.1   Examples of Biomass Productivity and Estimated Solar Energy Capture Efficiency[a]

Location	Biomass community	Annual yield dry matter (t/ha-year)	Average insolation (W/m²)	Solar energy capture efficiency (%)
Alabama	Johnsongrass	5.9	186	0.19
Sweden	Enthrophic lake angiosperm	7.2	106	0.38
Denmark	Phytoplankton	8.6	133	0.36
Minnesota	Willow and hybrid poplar	8–11	159	0.30–0.41
Mississippi	Water hyacinth	11.0–33.0	194	0.31–0.94
California	Euphorbia lathyris	16.3–19.3	212	0.45–0.54
Texas	Switchgrass	8–20	212	0.22–0.56
Alabama	Switchgrass	8.2	186	0.26
Texas	Sweet sorghum	22.2–40.0	239	0.55–0.99
Minnesota	Maize	24.0	169	0.79
New Zealand	Temperate grassland	29.1	159	1.02
West Indies	Tropical marine angiosperm	30.3	212	0.79
Nova Scotia	Sublittoral seaweed	32.1	133	1.34
Georgia	Subtropical saltmarsh	32.1	194	0.92
England	Coniferous forest, 0–21 years	34.1	106	1.79
Israel	Maize	34.1	239	0.79
New South Wales	Rice	35.0	186	1.04
Congo	Tree plantation	36.1	212	0.95
Holland	Maize, rye, two harvests	37.0	106	1.94
Marshall Islands	Green algae	39.0	212	1.02
Germany	Temperate reedswamp	46.0	133	1.92
Puerto Rico	Panicum maximum	48.9	212	1.28
California	Algae, sewage pond	49.3–74.2	218	1.26–1.89
Colombia	Pangola grass	50.2	186	1.50
West Indies	Tropical forest, mixed ages	59.0	212	1.55
Hawaii	Sugarcane	74.9	186	2.24
Puerto Rico	Pennisetum purpurcum	84.5	212	2.21
Java	Sugarcane	86.8	186	2.59
Puerto Rico	Napier grass	106	212	2.78
Thailand	Green algae	164	186	4.90

[a]Dry matter yield data from Berguson et al. (1990), Bransby and Sladden (1991), Burlew (1953), Cooper (1970), Loomis et al. (1963, 1971), Lipinsky (1978), Rodin and Brazilevich (1967), Sachs et al. (1981), Sanderson et al. (1995), Schneider (1973), Westlake (1963).

TABLE 3.2 High Daily Production Data for Selected Biomass[a]

Location	Biomass community	Daily yield (t/ha-day)
United Kingdom	Kale	0.20
United Kingdom	Barley	0.22
Netherlands	Potato	0.22
Trinidad	Pangola grass	0.27
California	Algae	0.27
United Kingdom	Beet	0.31
Hawaii	*Saccharum officinale*	0.38
Thailand	Algae	0.45
California	Maize	0.52
California	Sorghum	0.52
Florida	Water hyacinth	0.54

[a]Adapted from Burlew (1953) and Schneider (1973).

from the basic chemistry of photosynthesis. It will be shown in later chapters that there are many species of biomass that can be grown at sufficiently high yields in moderate climates to make them promising candidates as biomass energy crops.

As indicated in the discussion of the chemistry of photosynthesis, PS II alone instead of both PS II and I is sufficient for photosynthesis of a green algal mutant under certain conditions (Greenbaum *et al.*, 1995). The investigators suggested that the maximum thermodynamic conversion efficiency of light energy into chemical energy can be potentially doubled because a single photon rather than two is required to span the potential difference between water oxidation/oxygen evolution and proton reduction/hydrogen evolution. Comparison of the experimental results from the wild strain that contained both PS II and I and the mutant indicated this did not occur because the quantum efficiencies are similar. Also, the phenomenon was not observed under aerobic conditions. But this research still suggests that yields can be improved if the single-photon system can be incorporated into other biomass that grows under atmospheric conditions. For example, such biomass might exhibit higher productivity not only because of more efficient usage of the solar energy that is absorbed, but also because more CO_2 could be fixed at lower insolation values due to longer equivalent growth times during the growing season. However, it has been suggested that since both PS II and PS I are required for photosynthesis under normal aerobic conditions, the validity of the Z-scheme remains secure (Barber, 1995).

III. BIOMASS COMPOSITION AND ENERGY CONTENT

A. PROXIMATE AND ULTIMATE ANALYSES AND HEATING VALUES

Typical proximate analyses and higher heating values (product water in liquid state, HHV) of representative biomass types and species illustrate the wide range of some parameters such as moisture and ash contents and the relatively limited range of heating values (Table 3.3). The physical moisture contents of biomass are generally detemined experimentally by drying a sample at 100 to 105°C at atmospheric pressure or at lower temperature and reduced pressure. In a few cases, some organic compounds may be lost by these procedures because of volatilization and/or steam distillation, but generally the results are suitable for biomass characterization. The moisture contents listed in Table 3.3 range from a low of 2 to 3 wt % for the biomass derivatives char and paper to a high of 98 wt % for primary biosolids (primary sewage sludge). Green wood in the field before drying usually contains about 50 wt % moisture, whereas primary biosolids contain only a few percent suspended and dissolved solids in water. Similarly, the marine biomass giant brown kelp (*Macrocystis pyrifera*) and most other aquatic biomass contain only a few percent organic matter when first harvested; the main component is intracellular water.

Total organic matter is estimated by difference between 100 and the ash percentage that is experimentally determined by ashing the biomass samples at elevated temperature using standard methods (*cf.* annual volumes of *ASTM Standards*, American Society for Testing and Materials; *Methods for the Examination of Water and Wastewater*, American Public Health Association). The chemical reactions that occur during ashing result in the uptake of oxygen and the formation of metal oxides, so the experimental ash content is not identical to the inorganic mineral matter in the original sample. Ideally, all carbon in the original sample is eliminated on ashing, the metals are not oxidized, and none of the metals is lost. Such is not the case for some ashing procedures, particularly when samples contain high alkali metal concentrations. The loss of material due to the volatility of some alkali metal oxides at the ashing temperature causes errors in the analysis. Adjustments are sometimes made to the experimental ash determinations so that they correspond more closely to the inorganic matter present in the unashed samples. Nevertheless, subtraction of the experimental ash values in percent dry weight of the biomass from 100 to obtain the percent organic matter is adequate for most purposes.

Detailed chemical analyses of the components in the ash from two woody and one herbaceous biomass samples (Table 3.4) show that many metal oxides

TABLE 3.3 Typical Proximate Analyses and High Heating Values of Representative Biomass, Coal, and Peat[a]

Category	Name	Type	Moisture range (wt %)	Organic matter (dry wt %)	Ash (dry wt %)	High heating value (MJ/dry kg)
Wastes	Cattle manure	Feedlot	20–70	76.5	23.5	13.4
	Activated biosolids	Sewage	90–97	76.5	23.5	18.3
	Primary biosolids	Sewage	90–98	73.5	26.5	19.9
	Refuse-derived fuel (RDF)	Municipal	15–30	86.1	13.9	12.7
	Sawdust	Woody	15–60	99.0	1.0	20.5
Herbaceous	Cassava	Tropical	20–60	96.1	3.9	17.5
	Euphorbia lathyris	Warm season	20–60	92.7	7.3	19.0
	Kentucky bluegrass	Cool season	10–70	86.5	13.5	18.7
	Sweet sorghum	Warm season	20–70	91.0	9.0	17.6
	Switchgrass	Warm season	30–70	89.9	10.1	18.0
Aquatic	Giant brown kelp	Marine	85–97	54.2	45.8	10.3
	Water hyacinth	Fresh water	85–97	77.3	22.7	16.0
Woody	Black alder	Hardwood	30–60	99.0	1.0	20.1
	Cottonwood	Hardwood	30–60	98.9	1.1	19.5
	Eucalyptus	Hardwood	30–60	97.6	2.4	18.7
	Hybrid poplar	Hardwood	30–60	99.0	1.0	19.5
	Loblolly pine	Softwood	30–60	99.5	0.5	20.3
	Redwood	Hardwood	30–60	99.8	0.2	21.0
	Sycamore	Hardwood	30–60	98.9	1.1	19.4
Derivatives	Paper		3–13	94.0	6.0	17.6
	Pine bark	Softwood	5–30	97.1	2.9	20.4
	Rice straw		5–15	80.8	19.2	15.2
	Redwood char		2–6	95.9	4.1	30.5
Coal	Illinois bituminous	Soft	5–10	91.3	8.7	28.3
	North Dakota lignite	Soft	5–15	89.6	10.4	14.0
Peat	Reed sedge	Young coal	70–90	92.3	7.7	20.8

[a]The ash, organic matter, and heating values were obtained from Boley and Landers (1969), Bowerman (1969), Chow et al. (1995), Hodgman (1949), Jerger et al. (1982), Klass (1980, 1984), Monk et al. (1984), Paisley et al. (1993), Pober and Bauer (1977), Tillman (1978), Wen et al. (1974). The moisture content ranges of biomass in the field were measured, estimated, or obtained from various literature sources.

TABLE 3.4   Analysis of Ash from Hybrid Poplar, Pine, and Switchgrass[a]

Component	Hybrid poplar (dry wt %)	Pine (dry wt %)	Switchgrass (dry wt %)
CaO	47.20	49.20	4.80
K_2O	20.00	2.55	15.00
P_2O_5	5.00	0.31	2.60
MgO	4.40	0.44	2.60
SiO_2	2.59	32.46	69.92
Al_2O_3	0.94	4.50	0.45
BaO	0.70		0.22
Fe_2O_3	0.50	3.53	0.45
TiO_2	0.26	0.40	0.12
Na_2O	0.18	0.44	0.10
Mn_2O_4	0.14		0.15
SrO	0.13		0.04
CO_2[b]	14.00		
SO_3[b]	2.74	2.47	1.90
Total:	98.78	96.30	98.35

[a]Paisley et al. (1993).
[b]The reason for the presence of carbon and sulfur in the ash is that the ashing procedure was probably performed at an insufficient temperature and/or for an insufficient time to volatilize all nonmineral components.

are present, but that the distribution of the metallic elements is quite different in each sample analyzed. The oxides of calcium and potassium are dominant in hybrid poplar ash; the oxides of calcium and silica are dominant in pine ash; and the oxides of potassium and silica are dominant in switchgrass ash. As will be shown in later chapters, the distribution of the metals in biomass and the compositions of the ash are important in the development of certain types of biomass conversion processes because they can affect process performance. Also, some biomass species that have an unusually large amount of a specific metal have been harvested and used as a commercial source of that material during times of shortages.

It is evident from the data in Table 3.3 that the organic matter content and the HHV are affected by the ash, which in almost all cases has no energy value. The higher the ash value, the lower the organic matter and the HHV, as expected.

Intuitively, it might also be expected that the composition of biomass would vary over a broad range because there are so many different types and species.

The elemental compositions summarized in Table 3.5 support this hypothesis. In this table, typical ultimate and proximate analyses and the HHVs of land- and water-based biomass (pine wood, Kentucky bluegrass, giant brown kelp) and waste biomass (cattle feedlot manure, municipal solid waste, primary biosolids) are compared with those of cellulose, peat, and bituminous coal. On a dry basis, the ash values for these particular samples range from 0.5 wt % for pine wood to about 39 wt % for giant brown kelp. Also, on a dry basis, the total organic matter and the elemental analyses for carbon and hydrogen do not vary quite as much as the moisture and ash contents. Pure cellulose, a representative primary photosynthetic product, has a carbon content of 44.4%. Most of the renewable carbon sources listed in Table 3.5 have carbon contents near this value. When adjusted for moisture and ash contents, it is seen that with the exception of the biosolids sample, the carbon contents are slightly higher than that of cellulose, but span a relatively narrow range. It is also evident from the data in Table 3.5 that the HHVs per unit mass of carbon are quite close. Even those for reed sedge peat and Illinois bituminous coal are close to those calculated per unit mass of biomass carbon. As will be shown below, a reasonably good correlation exists between the carbon content of biomass and its energy content.

One of the analyses not included in the compositional information presented here on biomass is the percentage of so-called fixed carbon. This subject will be discussed in Chapter 8 under pyrolysis because there is no fixed carbon as such in biomass.

B. CORRELATION OF CARBON AND ENERGY CONTENTS

The energy content of biomass is obviously a very important parameter from the standpoint of conversion of biomass to energy and synfuels. The different components in biomass would be expected to have different heats of combustion because of the different chemical structures and carbon content. This is illustrated by the HHVs listed in Table 3.6 for each of the main classes of organic compounds in biomass. The more reduced the state of carbon in each class, the higher the energy content. Monosaccharides have the lowest carbon content, highest degree of oxygenation, and lowest heating value. As the carbon content increases and the degree of oxygenation is reduced, the structures become more hydrocarbon-like and the heating value increases. The terpene hydrocarbon components thus have the highest heating values of the components shown in Table 3.6; the lipids have the next highest heating values. The dominant component in most biomass is cellulose. It has a HHV of 17.51 MJ/kg (7533 Btu/lb).

TABLE 3.5 Typical Compositions and Heating Values of Virgin and Waste Biomass, Peat, and Coal[a]

	Pure cellulose	Pine wood	Kentucky bluegrass[b]	Giant brown kelp[c]	Water hyacinth[d]	Feedlot manure[e]	RDF[f]	Primary biosolids[g]	Reed sedge peat[h]	Bituminous coal[i]
Ultimate analysis (wt %)										
C	44.44	51.8	45.8	27.65	41.1	35.1	41.2	43.75	52.8	69.0
H	6.22	6.3	5.9	3.73	5.29	5.3	5.5	6.24	5.45	5.4
O	49.34	41.3	29.6	28.16	28.84	33.2	38.7	19.35	31.24	14.3
N		0.1	4.8	1.22	1.96	2.5	0.5	3.16	2.54	1.6
S		0	0.4	0.34	0.41	0.4	0.2	0.97	0.23	1.0
Ash		0.5	13.5	38.9	22.4	23.5	13.9	26.53	7.74	8.7
C (maf)	44.44	52.1	52.9	45.3	52.9	45.9	47.9	59.5	57.2	75.6
Proximate analysis (wt %)										
Moisture		5–50	10–70	85–95	85–95	20–70	18.4	90–98	84.0	7.3
Organic matter		99.5	86.5	61.1	77.7	76.5	86.1	73.47	92.26	91.3
Ash		0.5	13.5	38.9	22.4	23.5	13.9	26.53	7.74	8.7
Higher heating value										
MJ/dry kg	17.51	21.24	18.73	10.01	16.00	13.37	12.67[j]	19.86	20.79	28.28
MJ/kg (maf)	17.51	21.35	21.65	16.38	20.59	17.48		27.03	22.53	30.97
MJ/kg carbon	39.40	41.00	40.90	36.20	38.93	38.09		45.39	39.38	40.99

[a]All analyses and HHVs were determined by the Institute of Gas Technology.
[b]Harvested from a residential site in the Midwest.
[c]Macrocystis pyrifera harvested from kelp beds off the California coast.
[d]Eichornia crassipes harvested from a biosolids-fed lagoon in Mississippi.
[e]From a commercial cattle feedlot.
[f]Refuse-derived fuel; i.e., the combustible fraction of municipal solid waste, from a Chicago facility.
[g]From a Chicago Metropolitan Sanitary District facility.
[h]From Minnesota.
[i]From Illinois.
[j]As received with metals.

TABLE 3.6 Typical Carbon Content and Heating Value of Selected
Biomass Components[a]

Component	Carbon (wt %)[b]	Higher heating value (MJ/kg)[b]
Monosaccharides	40	15.6
Disaccharides	42	16.7
Polysaccharides	44	17.5
Crude proteins	53	24.0
Lignins	63	25.1
Lipids	76–77	39.8
Terpenes	88	45.2
Crude carbohydrates	41–44	16.7–17.7
Crude fibers[c]	47–50	18.8–19.8
Crude triglycerides	74–78	36.5–40.0

[a]Adapted from Klass (1994).
[b]Approximate values for dry mixtures.
[c]Contains 15–30% lignins.

Typical lower heating values (LHV, product water in vapor state) of selected biomass species are shown in Table 3.7. Woody and fibrous materials appear to have energy contents between about 19 and 21 MJ/ kg, whereas the water-based algae *Chlorella* has a higher value, undoubtedly because of its higher lipid or protein contents. Oils derived from plant seeds are much higher in energy content and approach the heating value of paraffinic hydrocarbons. High concentrations of inorganic components in a given biomass species can greatly affect its energy content because inorganic materials generally do not contribute to the heat of combustion. This is illustrated by the HHV for giant brown kelp, which leaves an ash residue equivalent to about 40 wt % of the dry weight, as shown in Table 3.3. On a dry basis, the HHV is about 10 MJ/ kg, while on a dry, ash-free basis, the heating value is about 16 MJ/kg.

When the heating values of the waste and virgin biomass samples and even the peat and coal samples listed in Table 3.5 are converted to energy content per mass unit of carbon, it is apparent that they fall within a narrow range. This is usually characteristic of most biomass. The energy value of the total material can be estimated from the carbon analysis and moisture determinations without actual measurement of the heating values in a calorimeter. Manipulation of the data in Table 3.5 leads to a simple equation for calculating the HHV of biomass and also coal and peat with reasonably good accuracy:

HHV in MJ/dry kg = 0.4571(% C on dry basis) − 2.70.

TABLE 3.7 Typical Lower Heating Values of Selected
Biomass and Fossil Materials[a]

Material	Lower heating value (MJ/dry kg)
Trees	
Oak	19.20
Bamboo	19.23
Birch	20.03
Beech	20.07
Oak bark	20.36
Pine	21.03
Fiber	
Bagasse	19.25
Buckwheat hulls	19.63
Coconut shells	20.21
Green algae	
Chlorella	26.98
Seed oils	
Linseed	39.50
Rape	39.77
Cottonseed	39.77
Amorphous carbon	33.80
Paraffinic hydrocarbon	43.30
Crude oil	48.20

[a]Burlew (1953) and Hodgman (1949).

A comparison of the experimental HHVs with the calculated HHVs for the biomass, coal, and peat using the carbon analyses listed in Table 3.5 is shown in Table 3.8. With the exception of the primary biosolids sample, the percentage error of the calculated HHV is relatively small.

C. Enthalpies of Formation

The enthalpies of formation of biomass are quite useful for thermodynamic calculations. The heats of specific reactions that utilize biomass feedstocks can be estimated from the standard enthalpies of formation at 298 K of the combustion products (in kcal/g-mol: CO_2, −94.05; liquid H_2O, −68.37; NO_2, 8.09; SO_2, −70.95), the elemental analyses of the biomass being examined, and its HHV. The enthalpy of formation of a particular biomass sample is equal to the sum of the heats of formation of the products of combustion

TABLE 3.8 Comparison of the Measured and Calculated Higher Heating Values of Biomass, Coal, and Peat[a]

Material	Measured higher heating value (MJ/dry kg)	Calculated higher heating value (MJ/dry kg)	Error (%)
Giant brown kelp	10.01	9.94	−0.70
Cattle feedlot manure	13.37	13.34	−0.19
Water hyacinth	16.00	16.09	+0.54
Pure cellulose	17.51	17.61	+0.59
Kentucky bluegrass	18.73	18.24	−2.64
Primary biosolids	19.86	17.30	−12.90
Reed sedge peat	20.79	21.43	+3.10
Pine wood	21.24	20.98	−1.23
Illinois bituminous coal	28.28	28.84	+1.98

[a]The measured HHVs (bomb calorimeter) and the carbon analyses were determined by the Institute of Gas Technology. The sample of primary biosolids contained an unusually large amount of fatty material. The calculated HHVs are estimated from the formula 0.4571(% C on dry basis) −2.70.

minus the HHV. It is assumed that the ash is inert. For example, a sample of giant brown kelp has the empirical formula $C_{2.61}H_{4.63}N_{0.10}S_{0.01}O_{2.23}$ (dry basis), which is derived from the elemental analysis, and a HHV of 296.1 kcal/g-mol (12.39 MJ/kg) at an assumed molecular weight of 100, including the ash. The stoichiometry for calculating the enthalpy of formation is

$$2.61C + 2.315H_2 + 0.05N_2 + 0.01S + 1.115O_2$$
$$\rightarrow C_{2.61}H_{4.63}N_{0.10}S_{0.01}O_{2.23}Ash_{26.7} \ (\Delta H_f \ -107.5).$$

The enthalpy of formation is −107.5 kcal/g-mol (−4.50 MJ/kg) including the ash for this particular biomass sample. An example of the utilization of this information is illustrated by applying it to the biological gasification process under anaerobic conditions. The stoichiometry of the process is

$$C_{2.61}H_{4.63}O_{2.23} \ (s) + 0.337H_2O(l) \rightarrow 1.326CH_4 \ (g)$$
$$+ 1.283CO_2 \ (g) \ (\Delta H \ -13.85).$$

The enthalpy of the process is estimated to be −13.85 kcal/g-mol (−0.58 MJ/kg) of kelp reacted (Klass and Ghosh, 1977).

It is assumed in these calculations that the inorganic components are carried through the process unchanged, and that the nitrogen and sulfur can be ignored since their concentrations are small. For each kilogram of kelp reacted, the feedstock energy input is 12.39 MJ, and the energy output is 11.81 MJ as methane (0.8903 MJ/g-mol at 298 K). The calculations indicate the process is slightly exothermic. About 95% of the feed energy resides in product methane, and about 5% is lost as heat of reaction.

D. ABUNDANCE OF MAJOR ORGANIC COMPONENTS

Typical organic components in representative, mature biomass species are shown in Table 3.9 along with the corresponding ash contents. With few exceptions, the order of abundance of the major organic components in whole-plant samples of terrestrial biomass is celluloses, hemicelluloses, lignins, and proteins. Aquatic biomass does not appear to follow this trend. The cellulosic components are often much lower in concentration than the hemicelluloses as illustrated by the data for water hyacinth. Other carbohydrates and derivatives are dominant in species such as giant brown kelp to almost complete exclusion of the celluloses. The hemicelluloses and lignins have not been found in this species.

Biomass often undergoes compositional changes, some of which can be subtle or pronounced, during growth and sometimes after harvesting depending on age of the biomass and environmental factors. An example of this phenomenon is the gradual decrease in sugar content and the gradual increase in hydrocarbon content during the maturation of E. lathyris (Ayerbe et al., 1984). Another phenomenon that is quite common during biomass growth is the nonuniform distribution of organic components in various plant parts. For example, the hydrocarbon content in the leaves of E. lathyris is more than twice the amount in the stems (Sachs et al., 1981). All of these factors must be considered in some detail when biomass is utilized for production of certain organic compounds or as a feedstock for conversion to fuels and energy products.

Alpha-cellulose, or cellulose as it is more generally known, is the chief structural element and a major constituent of many biomass species. In trees, cellulose is generally about 40 to 50% of the dry weight. As a general rule, the major organic components on a moisture and ash-free basis in woody biomass are about 50 wt % cellulosics, 25 wt % hemicelluloses, and 25 wt % lignins. However, cellulose is not always the dominant component in the carbohydrate fraction of biomass. As just mentioned and as shown in Table 3.9, it is one of the minor components in giant brown kelp. Mannitol, a hexahydric alcohol that can be formed by reduction of the aldehyde group of D-glucose to a methylol group, and alginic acid, a polymer of mannuronic and glucuronic acids, are the major carbohydrates.

The lipid and protein fractions of plant biomass are normally much less on a percentage basis than the carbohydrate components. The lipids are usually present at the lowest concentration, while the protein fraction is somewhat

TABLE 3.9 Organic Components and Ash in Representative Biomass[a]

Biomass type	Marine	Freshwater	Herbaceous	Woody	Woody	Woody	Waste
Name	Giant brown kelp	Water hyacinth	Bermuda grass	Poplar	Sycamore	Pine	RDF
Component (dry wt %)							
Celluloses	4.8	16.2	31.7	41.3	44.7	40.4	65.6
Hemicelluloses		55.5	40.2	32.9	29.4	24.9	11.2
Lignins		6.1	4.1	25.6	25.5	34.5	3.1
Mannitol	18.7						
Algin	14.2						
Laminarin	0.7						
Fucoidin	0.2						
Crude protein	15.9	12.3	12.3	2.1	1.7	0.7	3.5
Ash	45.8	22.4	5.0	1.0	0.8	0.5	16.7
Total	100.3	112.5	93.3	102.9	102.1	101.0	100.1

[a]All analyses were performed by the Institute of Gas Technology. The crude protein content is estimated by multiplying the nitrogen value by 6.25. RDF is refuse-derived fuel; i.e, the combustible fraction of municipal solid waste.

higher, but still lower than the carbohydrate fraction. Crude protein values can be approximated by multiplying the organic nitrogen analyses by 6.25. This factor is used because the average weight percentage of nitrogen in pure dry protein is about 16%, although the protein content of each biomass species can best be determined by amino acid assay. The calculated crude protein values of the dry biomass species in Table 3.9 range from a low of about 0 wt % for pine wood to a high of about 30 wt % for Kentucky bluegrass. For grasses, the protein content is strongly dependent on the growing procedures used before harvest, particularly the fertilization methods. Some biomass species such as the legumes, however, fix nitrogen from the ambient atmosphere and often contain high protein concentrations.

It is apparent from the data in Table 3.9 that the sulfur content of virgin and waste biomass ranges from very low to about 1 wt % for primary biosolids. This sulfur level is similar to the sulfur content of high-sulfur Illinois bituminous coal. Woody biomass generally contains very little sulfur.

E. CHEMICAL STRUCTURES OF MAJOR COMPONENTS

Knowledge of the chemical structures of the major organic components in biomass is quite valuable in the development of processes for producing derived fuels and chemicals. Information on the chemical structures can often lead to methods of improving existing processes and to development of advanced conversion methods. Somewhat more detailed information on the chemical structures of the major components in biomass is presented here.

Alpha cellulose is a polysaccharide having the generic formula $(C_6H_{10}O_5)_n$ and an average molecular weight range of 300,000 to 500,000. Complete hydrolysis established that the polymer consists of D-glucose units. Partial hydrolysis yields cellobiose (glucose-β-glucoside), cellotriose, and cellotetrose. These results show that the glucose units in cellulose are linked as in cellobiose (Fig. 3.6). Cotton is almost pure α-cellulose, whereas wood cellulose, the raw material for the pulp and paper industry, always occurs in association with hemicelluloses and lignins. Cellulose is insoluble in water, forms the skeletal structure of most terrestrial biomass, and constitutes approximately 50% of the cell wall material. Carefully purified wood cellulose contains a few carboxyl groups which are believed to be esterified in the natural state.

Starches are polysaccharides that have the generic formula $(C_6H_{10}O_5)_n$. They are reserve sources of carbohydrate in some biomass, and are also made up of D-glucose units as shown by the results of hydrolysis experiments. But in contrast to the structure of cellulose, the hexose units are linked as in maltose, or glucose-α-glucoside (Fig. 3.6), as indicated by the results of partial hydrolysis. Another difference between celluloses and starches is that the latter can

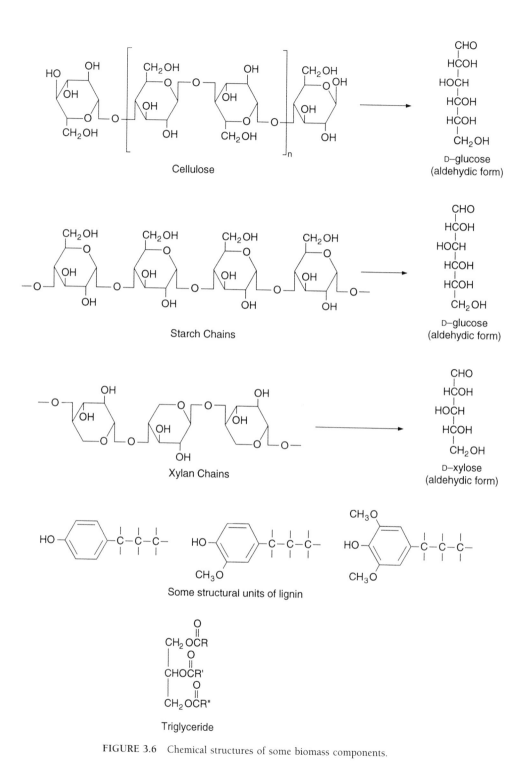

FIGURE 3.6 Chemical structures of some biomass components.

be separated into two fractions by treatment with hot water: a soluble component called amylose (10 to 20%) and insoluble amylopectin (80 to 90%). Amylose and amylopectin have molecular weight ranges of 10,000 to 50,000 and 50,000 to 1,000,000, respectively. Both fractions yield glucose or maltose on hydrolysis, but amylopectin is believed to consist of branched chains. Starches occur in the form of minute granules in seeds, tubers, and other plant parts and are important constituents of corn, beans, potatoes, rice, wheat, and other biomass foodstuffs.

Hemicelluloses are complex polysaccharides that occur in association with cellulose in the cell walls. But unlike cellulose, hemicelluloses are soluble in dilute alkali and consist of branched structures, which vary significantly among different woody and herbaceous biomass species. Many have the generic formula $(C_5H_8O_4)_n$. They are termed pentosans and yield mainly pentoses instead of hexoses on hydrolysis. Some hemicelluloses, however, contain hexose units. Hemicelluloses usually consist of 50 to 200 monomeric units and a few simple sugar residues. The most abundant one, xylan, consists of D-xylose units linked in the 1- and 4-positions (Fig. 3.6). Xylan is closely related to polyglucuronic acid with which it is associated in the natural state, and from which it can be produced by decarboxylation. Other hemicelluloses include the glucomannans, which consist of D-glucose and D-mannose units in the polymeric chains in ratios of about 30 : 70, and galactoglucomannans, which consist of D-galactose, D-glucose, and D-mannose in the polymeric chains in ratios of about 2 : 10 : 30. The pentosans can occur in large amounts (20 to 40%) in corncobs and corn stalks and in biomass straws and brans. The xylans have been found in softwoods and hardwoods up to about 10% and 30% of the dry weight of the species, respectively, whereas mannans are generally present at about 15% of the dry weight in softwoods and only a few percent by weight in hardwoods.

The lignins are highly branched, substituted, mononuclear aromatic polymers in the cell walls of certain biomass, especially woody species, and are often bound to adjacent cellulose fibers to form what has been called a lignocellulosic complex. This complex and the lignins alone are often quite resistant to conversion by microbial systems and many chemical agents. The complex can be broken and the lignin fraction separated, however, by treatment with strong sulfuric acid, in which the lignins are insoluble. The precise structures of the polymers have not been determined because of their diverse nature and complexity. The dominant monomeric units in the polymers are benzene rings bearing methoxyl, hydroxyl, and propyl groups that can be attached to other units (Fig. 3.6). The lignin contents on a dry basis in both softwoods and hardwoods generally range from 20 to 40% by weight, and from 10 to 40% by weight in various herbaceous species such as bagasse, corncobs, peanut shells, rice hulls, and straws.

As previously mentioned, the triglycerides found in biomass are esters of the triol, glycerol, and fatty acids (Fig. 3.6). These water-insoluble, oil-soluble esters are common in many biomass species, especially the oilseed crops, but the concentrations are small compared to those of the polysaccharides and lignins. Many saturated fatty acids have been identified as constituents of the lipids. Surprisingly, almost all the fatty acids that have been found in natural lipids are straight-chain acids containing an even number of carbon atoms. Most lipids in biomass are esters of two or three fatty acids, the most common of which are lauric (C_{12}), myristic (C_{14}), palmitic (C_{16}), oleic (C_{18}), and linoleic (C_{18}) acids. Palmitic acid is of widest occurrence and is the major constituent (35 to 45%) of the fatty acids of palm oil. Lauric acid is the most abundant fatty acid of palm-kernel oil (52%), coconut oil (48%), and babassu nut oil (46%). The monounsaturated oleic acid and polyunsaturated linoleic acid comprise about 90% of sunflower oil fatty acids. Linoleic acid is the dominant fatty acid in corn oil (55%), soybean oil (53%), and safflower oil (75%). Saturated fatty acids of 18 or more carbon atoms are widely distributed, but are usually present in biomass only in trace amounts, except in waxes.

Other classes of organic materials, such as alkaloids, pigments, resins, sterols, terpenes, terpenoids, and waxes, and many simple organic compounds are often present in various biomass species, but are not discussed here because they are usually present in very small amounts. The peptides present in herbaceous biomass are also not discussed here because, although the nitrogen and sulfur contents of the biomass should be assessed for certain microbiological processes, the amino acids that make up the proteins are generally not important factors in conversion processes.

F. Relationship of Biomass Properties and Conversion Processes

Many processes can be used to produce energy or gaseous, liquid, and solid fuels from virgin and waste biomass. In addition, chemicals can be produced from biomass by a wide range of processes. It is evident from the data and information presented in this chapter, however, that the characteristics of potential feedstocks, particularly their moisture and energy contents, can have profound effects on the utility of specific biomass species and waste biomass. Table 3.10 is a summary of the principal feedstock, process, and product types that are considered in developing a synfuel-from-biomass process. There are many interacting parameters and possible feedstock–process–product combinations, but not all are feasible from a practical standpoint. For example, the separation of small amounts of metals present in biomass and the direct

TABLE 3.10 Summary of Feedstock, Conversion Process, and Primary Energy Product Types

Feedstock	Conversion process	Primary energy product

Terrestrial biomass
  Trees
  Plants
  Grasses
	Physical { Densification, Drying, Extraction, Separation, Size reduction }	Energy { Electric, Steam, Thermal }

Aquatic biomass
  Freshwater plants
  Marine plants
  Microalgae
	Thermal { Carbonization, Chemical hydrolysis, Combustion, Cracking, Dehydrogenation, Hydrogenation, Partial oxidation, Pyrolysis, Steam reforming }	Solids { Char, Combustibles, Fabricated solids }

Waste biomass
  Agricultural
  Forestry
  Industrial
  Municipal
	Microbial/Biochemical { Aerobic fermentation, Anaerobic fermentation, Biophotolysis, Composting, Enzymic }	Gases { Hydrogen, Light hydrocarbons, Low-Btu gas, Methane, Synthesis gas }

	Natural	Liquids { Ethanol, Esters, Ethers, Glycerides, Higher alcohols, Higher hydrocarbons, Methanol, Oils, Transesterified acids }
		Chemicals Derivatives

combustion of high-moisture-content algae are technically possible, but energetically unfavorable.

Moisture content of the biomass chosen is especially important in the selection of suitable conversion processes. The giant brown kelp, *Macrocystis pyrifera,* contains as much as 97 wt % intracellular water, so thermal gasification techniques such as pyrolysis and hydrogasification cannot be used directly without first drying the algae. Anaerobic fermentation is preferred because the water does not need to be removed. Wood, on the other hand, can often be processed by several different thermal conversion techniques without drying. Figure 3.7 illustrates the effects of thermal drying on biomass used for production of synthetic natural gas (SNG). A large portion of the feed's equivalent energy content can be expended for drying, so the properties of a potential feedstock must be considered carefully in relation to the conversion process.

Table 3.11 lists the important feedstock characteristics to be examined when developing a conversion process for a specific virgin or waste biomass feedstock. A particular process also may have specific requirements within a given process type. For example, biological gasification and alcoholic fermenta-

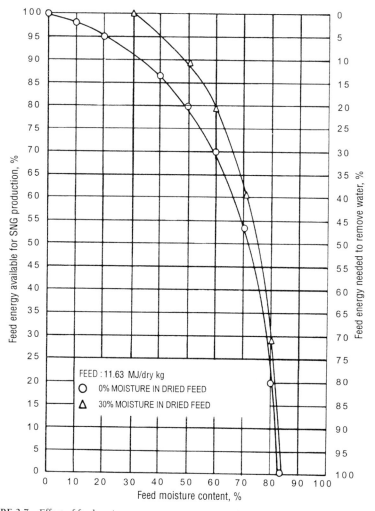

FIGURE 3.7   Effect of feed moisture content on energy available for SNG production. (Example of use: Reduction of an initial moisture content of 70 wt % by thermal drying to 30% requires the equivalent of 37% of the feedstock energy content and leaves 63% of the feedstock energy available for SNG production.)

tion are both microbiological conversion processes, but animal manure, which has a relatively high biodegradability, is not equally applicable as a feedstock for both processes.

In summary, it is not a simple matter to select the proper conversion process for a given biomass feedstock. Both biomass properties and process

TABLE 3.11   Feedstock Characteristics That Affect Suitability of a Conversion Process

Characteristic	Process type			
	Physical	Thermal	Biochemical–Microbial	Chemical
Water content	X	X	X	X
Energy content	X	X		X
Noncombustibles	X	X	X	
Chemical composition		X	X	X
Carbon reactivity		X		X
Bulk component analysis	X	X	X	X
Density	X			
Size/size distribution	X	X	X	X
Biodegradability			X	
Organism content/type			X	
Nutrient content/type			X	

requirements must be examined together and in depth to develop a technically and economically feasible system for producing the desired synfuels and energy products. These subjects will be examined in some detail in subsequent chapters.

REFERENCES

Arnon, D. I., Allen, M. B., and Whatley, F. R. (1954). *Nature* **174**, 394.

Ayerbe, L., Funes, E., Tenorio, J. L., Ventas, P., and Mellado, L. (1984). *Biomass* **5**, 37.

Barber, J. (1995). *Nature* **376**, 388.

Bassham, J. A. (1976). *In* "Clean Fuels from Biomass, Sewage, Urban Refuse, Agricultural Wastes," (F. Ekman, ed.), p. 205. Institute of Gas Technology, Chicago.

Berguson, W., Hansen, E. A., Johnson, W., Morse, C. B., and Zimmerman, D. (1990). *In* "Energy from Biomass and Wastes XIII," (D. L. Klass, ed.), p. 275. Institute of Gas Technology, Chicago.

Boley, C. C., and Landers, W. S. (1969). "Entrainment Drying and Carbonization of Wood Waste," Report of Investigation 7282. U.S. Bureau of Mines, Washington, D.C.

Bowerman, F. R. (1969). "Principles and Practices of Incineration," (R. C. Corey, ed.). John Wiley, Inc., New York.

Bransby, D. I., and Sladden, S. E. (1991). *In* "Energy from Biomass and Wastes XV," (D. L. Klass, ed.), p. 333. Institute of Gas Technology, Chicago.

Burlew, J. S. (1953). "Algae Culture from Laboratory to Pilot Plant," Publication No. 600. Carnegie Institute of Washington, Washington, D.C.

Calvin, M., and Benson, A. A. (1947). *Science* **105**, 648.

Calvin, M. (1983). *Science* **219**, 24.

Chow, P., Rolfe, G. L., Lambert, R. O., Jr., Barrier, J. W., Ehrlinger, H. P., and Lightsey, G. R. (1995). In "Second Biomass Conference of the Americas," p. 244. National Renewable Energy Laboratory, Golden, CO.

Cooper, J. P. (1970). Herb. Abstr. 40, 1.

Enebo, L. (1969). "Engineering of Unconventional Protein Production," Chemical Engineering Progress Symposium Series (H. Bieber, ed.), Vol. 65, No. 93.

Freudenberg, K. (1957). Ind. Eng. Chem. 49, 1384.

Graboski, M., and Bain, R. (1979). In "A Survey of Biomass Gasification," Vol. II, Chapt. 3, p. II-21, SERI/TR-33-239, July. Solar Energy Research Institute, Golden, CO.

Greenbaum, E., Lee, J. W., Tevault, C. V., Blankinship, S. L., and Mets, L. J. (1995). Nature 376, 438.

Hill, R., and Bendall, D. S. (1960). Nature 186, 136.

Hodgman, C. D., ed. (1949). In "Handbook of Chemistry and Physics," 31st Ed., p. 1535. Chemical Rubber Publishing Co., Cleveland, OH.

Jerger, D. E., Conrad, J. R., Fannin, K. F., and Chynoweth, D. P. (1982). In "Energy from Biomass and Wastes VI," (D. L. Klass, ed.), p. 341. Institute of Gas Technology, Chicago.

Klass, D. L., and Ghosh, S. (1977). In "Clean Fuels from Biomass and Wastes," (W. W. Waterman, ed.), p. 323. Institute of Gas Technology, Chicago.

Klass, D. L. (1980). In "Kirk-Othmer Encyclopedia of Chemical Technology," 3rd Ed., Vol. II, p. 334. John Wiley, New York

Klass, D. L. (1984). Science 223, 1021.

Klass, D. L. (1994). In "Kirk-Othmer Encyclopedia of Chemical Technology," 4th. Ed., Vol. 12, p. 16. John Wiley, New York.

Kok, B., and Van Oorschot, J. L. P. (1954). Acta. Bot. Neer. 3, 533.

Lipinsky, E. S. (1978). In "Second Annual Fuels from Biomass Symposium," (W. W. Shuster, ed.), p. 109. Rensselaer Polytechnic Institute, Troy, NY.

Loomis, R. S., and Williams, W. A. (1963). Crop. Sci. 3, 63.

Loomis, R. S., Williams, W. A., and Hall, A. E. (1971). Ann. Rev. Plant Physiol. 22, 431.

McGarry, M. G. (1971). Process Biochem. 6, 50.

Monk, R. L., Miller, F. R., McBee, G. G., Creelman, R. A., and Chynoweth, D. P. (1984). In "Energy from Bioass and Wastes VIII," (D. L. Klass, ed.), p. 99. Institute of Gas Technology, Chicago.

Osmond, C. B. (1978). Ann. Rev. Plant Physiol. 29, 379.

Oswald, W. J. (1969). In "Engineering of Unconventional Protein Production," Chemical Engineering Progress Symposium Series (H. Bieber, ed.), Vol. 65, No. 93.

Paisley, M. A., Litt, R. D., Taylor, D. R., Tewksbury, T. L., Hupp, D. E., and Wood, R. D. (1993). "Phase Completion Report on Operation and Evaluation of an Indirectly Heated Biomass Gasifier." Battelle, Columbus, OH, November 5.

Pober, K. W., and Bauer, H. F. (1977). In "Fuels from Waste," (L. L. Anderson and D. A. Tillman, eds.), p. 73. Academic Press, New York.

Rabinovitch, E. I. (1956). "Photosynthesis," Vols. 1-2. Interscience, New York.

Retovsky, R. (1966). "Theoretical and Methodological Basis of Continuous Culture of Microorganisms." Academic Press, New York.

Rodin, I. E., and Brazilevich, N. I. (1967). "Production and Mineral Cycling in Terrestrial Vegetation." Oliver & Boyd, Edinburgh.

Sachs, R. M., Low, C. B., MacDonald, J. D., Awad, A. R., and Sully, M. J. (1981). California Agriculture 29 July-August.

Sanderson, M. A., Hussey, M. A., Ocumpaugh, W. R., Tischler, C. R., Read, J. C., and Reed, R. L. (1995). In "Second Biomass Conference of the Americas," NREL/CP-200-8098, DE95009230, p. 253. National Renewable Energy Laboratory, Golden, CO, August.

Schneider, T. R. (1973). Energy Convers. 13, 77.

Stanier, R. Y., Ingraham, J. L., Wheelis, M. L., and Painter, P. R. (1986). "The Microbial World," 5th Ed. Prentice-Hall, Englewood Cliffs, NJ.

Tillman, D. A. (1978). "Wood as an Energy Resource." Academic Press, New York.

Wen, C. Y., Railie, R. C., Lin, C. Y., and O'Brien, W. S. (1974). "Advances in Chemistry Series," Vol. 131. American Chemical Society, Washington, D.C.

Westlake, D. F. (1963). *Biol. Rev.* **38**, 385.

Yount, J. L., and Grossman, R. A. (1970). *J. Water Pollut. Contr. Fed.* **42**, 173.

CHAPTER 4

Virgin Biomass Production

I. INTRODUCTION

The manufacture of synfuels or energy products from virgin biomass requires that suitable quantities of biomass chosen for use as energy crops be grown, harvested, and transported to the end user or conversion plant. For continuous, integrated biomass production and conversion, provision must be made to supply sufficient feedstock to sustain conversion plant operations. Since at least 250,000 botanical species, of which only about 300 are cash crops, are known in the world, it would seem that biomass selection for energy could be achieved rather easily. This does not necessarily follow simply from the multiplicity of biomass species that can be considered for energy usage. Compared to the total known botanical species, a relatively small number are suitable for the manufacture of synfuels and energy products. The selection is not easily accomplished in some cases because of the discontinuous nature of the growing season and the compositional changes that sometimes occur on biomass storage. Many parameters must be studied in great detail to choose the proper biomass species or combination of species for operation of the system. They concern such matters as growth area availability; soil type, quality, and topography; propagation and planting procedures; growth cycles; fertilizer, herbicide, pesticide, and other chemical needs; disease resistance of monocultures; insolation, temperature, precipitation and irrigation needs; preharvest management, crop management, and harvesting methods; storage stability of the harvest; solar drying in the field versus in-plant drying in connection with conversion requirements; growth area competition for food, feed, fiber, and other end uses; the possibilities and potential benefits of simultaneous or sequential growth of two or more biomass species for synfuels and foodstuffs; multiple end uses; and transport to the conversion plant gate or end-use site.

As mentioned in earlier chapters, biomass chosen for energy applications, in the ideal case, should be high-yield, low-cash-value species that have short growth cycles and that grow well in the area in which the biomass energy system is located. Fertilization requirements should be low and possibly nil if the species selected fix ambient nitrogen, thereby minimizing the amount of external chemical nutrients that have to be supplied to the growth areas. In areas having low annual rainfall, the species grown should have low consumptive water usage and be able to utilize available precipitation at high efficiencies. For terrestrial energy crops, the requirements should be such that they can grow well on low-grade soils so that the best classes of agricultural or forestry land are not needed. After harvesting, growth should commence again without the need for replanting by vegetative or coppice growth. Surprisingly, several biomass species meet many of these idealized characteristics and appear to be quite suitable for energy applications. This chapter addresses the important factors that affect biomass production for energy applications.

II. CLIMATIC AND ENVIRONMENTAL FACTORS

The biomass species selected as energy crops and the climate must be compatible to sustain operation of the energy or fuel farm under human-controlled conditions. Wild stands of biomass are also amenable to harvesting as energy crops and are still the primary sources of virgin biomass feedstocks because large-scale energy and fuel farms in which dedicated biomass energy crops are grown have not yet been established. The few attempts that have been made to design, build, and operate such farms have not been too successful. The compatibility of biomass and climate is, nevertheless, essential to ensure that these systems can ultimately be operated at a profit on a commercial scale. The three primary climatic factors that have the most influence on the productivity and yields of an indigenous or transplanted biomass species are insolation, precipitation, and temperature. Natural fluctuations of these factors remove them from human control, but the information compiled over the years in meteorological records and from agricultural and forestry practice supplies a valuable data base from which biomass energy systems can be conceptualized and developed. Of these three factors, precipitation has the greatest impact because droughts can wreak havoc on biomass growth. Fluctuations in insolation and temperature during normal growing seasons do not adversely affect biomass growth as much as insufficient water. Ambient carbon dioxide (CO_2) concentration and the availability of macronutrients and micronutrients are also important factors in biomass production.

A. INSOLATION

The intensity of the incident solar radiation at the earth's surface is one of the key factors in photosynthesis, as shown in Chapter 3. Natural biomass growth will not occur without solar energy. Insolation varies with geographic location, time of day, and season of the year, and as is well known, it is high in the tropics and near the equator. The approximate changes of insolation with latitude are illustrated in Table 4.1 (Brinkworth, 1973), and Fig. 4.1 shows how the mean annual insolation varies at the earth's surface with geographic location (Crutchfield, 1974). A more quantitative summary of average total daily insolation values over the continental United States is shown in Table 4.2 (U.S. Dept. of Commerce, 1970). At a given latitude, the incident radiation is not constant and often exhibits large changes over relatively short distances. Although several environmental factors influence biomass productivity, there is usually a relatively good correlation between the annual yields of dry biomass per unit area and the average insolation value (see Table 3.1). All other factors being equal, it is generally true that the higher the insolation, the higher the annual yield of a particular energy crop provided it is adapted to the local environment. C_4 biomass species often exhibit higher productivities in terms

TABLE 4.1 Insolation at Various Latitudes for Clear Atmospheres[a]

Location	Latitude	Maximum		Minimum		Average[b]	
		W/m^2	Btu/ft^2-day	W/m^2	Btu/ft^2-day	W/m^2	Btu/ft^2-day
Equator	0°	315	2400	236	1800	263	2000
Tropics	23.5°	341	2600	171	1300	263	2000
Mid-earth	45°	355	2700	70.9	540	210	1600
Polar circle	65.5°	328	2500	0	0	158	1200

[a]Brinkworth (1973).
[b]Yearly total divided by 365.

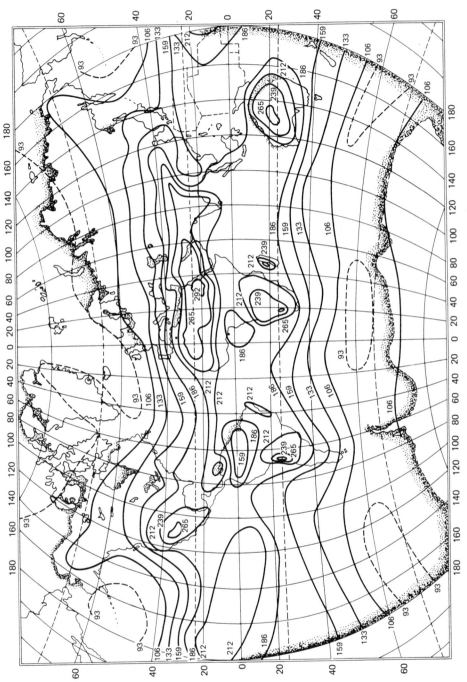

FIGURE 4.1 Mean annual insolation at earth's surface (W/m²).

TABLE 4.2   Average Daily Insolation for Selected U.S. Cities[a]

Location	January (W/m²)	April (W/m²)	July (W/m²)	October (W/m²)	Annual (W/m²)
Arizona—Tucson	146	289	288	208	229
California—Fresno	93	290	338	187	229
Florida—Lakeland	135	260	247	189	210
Indiana—Indianapolis	90	188	242	120	157
Louisiana—Lake Charles	109	215	236	175	191
Minnesota—St. Cloud	76	178	275	104	157
Montana—Glasgow	72	190	299	118	175
Nevada—Ely	108	257	288	176	210
Oklahoma—Oklahoma City	81	212	264	155	183
Texas—San Antonio	113	198	286	182	199
Vermont—Burlington	76	182	208	100	146
Virginia—Sterling	91	173	233	113	159
Washington—Seattle	37	179	276	98	151

[a]U.S. Dept. of Commerce (1970).

of growth rates and annual yields because of their capability to utilize incident solar radiation at higher efficiencies for photosynthesis.

B. PRECIPITATION

Precipitation as rain, or in the form of snow, sleet, or hail, depending on atmospheric temperature and other conditions, is governed by movement of air and is generally abundant wherever air currents are predominately upward. The greatest precipitation should therefore occur near the equator. The average annual precipitation in the continental United States is shown in Fig. 4.2 (Visher, 1954); Table 4.3 (U.S. Dept. of Commerce, 1995) is a summary of the average monthly and annual precipitation at different locations in the United States. The average annual rainfall is about 79 cm.

The moisture needs of aquatic biomass are presumably met in full because growth occurs in liquid water, but the growth of terrestrial biomass is often water-limited. The annual requirements for good growth have been found for many biomass species to be in the range 50 to 76 cm (Roller et al., 1975). Some crops, such as wheat, exhibit good growth with much less water, but they are in the minority. Without irrigation, water is supplied during the growing season by the water in the soil at the beginning of the season and by

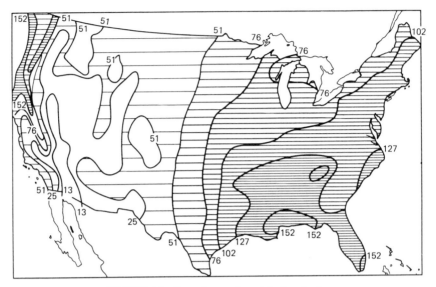

FIGURE 4.2   Normal annual precipitation (cm).

rainfall. Figure 4.3 illustrates the normal precipitation recorded in the continental United States during the growing season, April to September (Visher, 1954). This type of information and the established requirements for the growth of terrestrial biomass can be used to divide the United States into precipitation regions as shown in Fig. 4.4 (Visher, 1954). The regions that are more productive for biomass generally correlate with the precipitation regions, as shown in Figs. 4.5 and 4.6 (Visher, 1954). It should be realized, though, that rainfall alone is not quantitatively related to productivity of terrestrial biomass because of the differences in soil characteristics, water evaporation rates, and infiltration. Also, as suggested in Fig. 4.5, certain areas that have low precipitation can be made productive through irrigation. Some areas of the country that vary widely in precipitation as a function of time, such as many western states, will produce moderate biomass yields and often sufficient yields of cash crops without irrigation to justify commercial production.

   The transpiration of water to the atmosphere through biomass stomata is proportional to the vapor pressure difference between the atmosphere and the saturated vapor pressure inside the leaves. Transpiration is obviously affected by atmospheric temperature and humidity. The internal water is essential for biomass growth. The efficiency of utilizing this water (water-use efficiency, WUE) has been defined as the ratio of biomass accumulation to the water consumed, expressed as transpiration or total water input to the system. Analysis of the transpiration phenomenon and the possibilities for manipulation

TABLE 4.3 Average Monthly and Annual Precipitation for Selected U.S. Cities, 1961 to 1990[a]

Location	January (cm)	April (cm)	July (cm)	October (cm)	Annual (cm)
Alaska—Juneau	11.5	7.0	10.6	19.9	137.9
Arizona—Phoenix	1.7	0.6	2.1	1.7	19.5
California—Los Angeles	6.1	1.8	0.03	0.9	30.5
California—San Francisco	11.0	3.5	0.08	3.1	50.0
Colorado—Denver	1.3	4.3	4.9	2.5	39.1
Florida—Miami	5.1	7.2	14.5	14.3	142.0
Hawaii—Honolulu	9.0	3.9	1.5	5.8	55.9
Indiana—Indianapolis	5.9	9.4	11.4	6.7	101.4
Louisiana—New Orleans	12.8	11.4	15.5	7.7	157.2
Minnesota—Minneapolis	2.4	6.1	9.0	5.6	71.9
Montana—Great Falls	2.3	3.6	3.1	2.0	38.6
Nevada—Reno	2.7	1.0	0.7	1.0	19.1
Oklahoma—Oklahoma City	2.9	7.0	6.6	8.2	84.7
Texas—Dallas-Fort Worth	4.6	8.9	5.6	5.8	85.6
Vermont—Burlington	4.6	7.0	9.3	7.3	87.6
Virginia—Norfolk	9.6	7.8	12.9	8.0	113.4
Washington—Seattle	13.7	5.9	1.9	8.2	94.5

[a]U.S. Dept. of Commerce (1995).

of WUE have led some researchers to conclude that biomass production is inextricably linked to biomass transpiration. Agronomic methods that minimize surface runoff and soil evaporation, and biochemical alterations that reduce transpiration in C_3 plants, have the potential to increase WUE. But for water-limited regions, the fact remains that without additional water, the research results indicate that these areas cannot be expected to become regions of high biomass yields (Sinclair, Tanner, and Bennett, 1984). Irrigation and full exploitation of humid climates are of highest priority in attempting to increase biomass yields in these areas.

C. TEMPERATURE

Most biomass species grow well in the United States at temperatures between 15.6 and 32.3°C (60 and 95°F). Typical examples are corn, kenaf, and napier grass. Tropical grasses and certain warm-season biomass have optimum growth

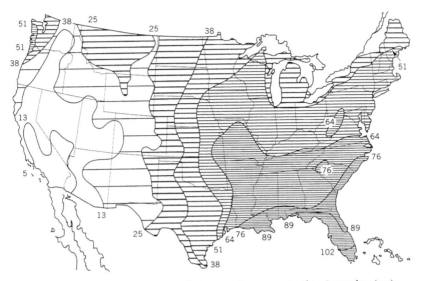

FIGURE 4.3   Normal precipitation during growing season April to September (cm).

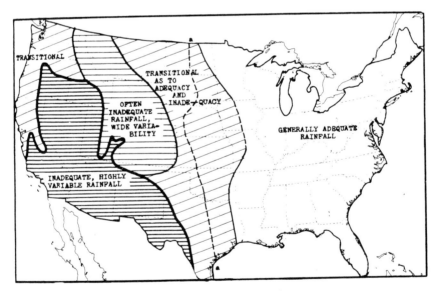

FIGURE 4.4   Precipitation regions.

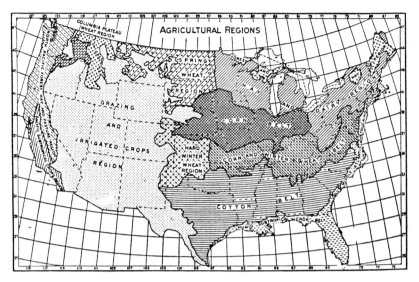

FIGURE 4.5   Agricultural regions.

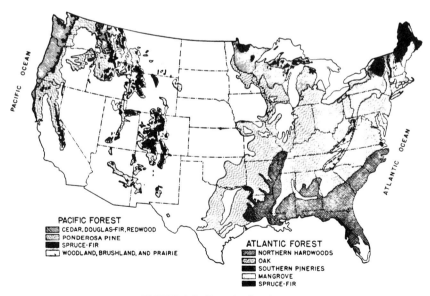

FIGURE 4.6   Forestland regions.

temperatures in the range 35 to 40°C (95 to 104°F), but the minimum growth temperature is still near 15°C (Ludlow and Wilson, 1970). Cool-weather biomass such as wheat may show favorable growth below 15°C, and certain marine biomass such as the giant brown kelp only survive in water at temperatures below 20 to 22°C (North, 1971). The average number of days per year in the continental United States where the temperature is less than 6.1°C (43°F) and essentially no biomass growth occurs are shown in Fig. 4.7 (Visher, 1954). Table 4.4 is a summary of average monthly and annual temperature fluctuations with time and location in the United States (U.S. Dept. of Commerce, 1995). The growing season is clearly longer in the southern portion of the country. In some areas such as Hawaii, the Gulf states, southern California, and the southeastern Atlantic states, the temperature is usually conducive to biomass growth most of the year.

The effect of temperature fluctuations on net CO_2 uptake is illustrated by the curves in Fig. 4.8 (El-Sharkawy and Hesketh, 1964). As the temperature increases, net photosynthesis increases for cotton and sorghum to a maximum value and then rapidly declines. Ideally, the biomass species grown in an area should have a maximum rate of net photosynthesis as close as possible to the average temperature during the growing season in that area.

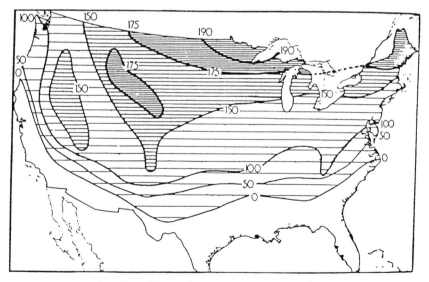

FIGURE 4.7   Annual number of days with temperature less than 6.1°C (43°F) and little or no biomass growth.

TABLE 4.4 Average Monthly and Annual Temperature for Selected U.S. Cities, 1961 to 1990[a]

Location	January (°C)	April (°C)	July (°C)	October (°C)	Annual (°C)
Alaska—Juneau	−4.3	4.3	13.3	5.7	4.8
Arizona—Phoenix	12.0	21.1	34.2	23.6	22.6
California—Los Angeles	13.8	15.6	20.6	19.3	17.2
California—San Francisco	9.3	13.1	17.1	16.1	13.9
Colorado—Denver	−1.3	9.0	23.1	10.8	10.2
Florida—Miami	19.6	24.0	28.1	25.7	24.4
Hawaii—Honolulu	22.7	24.3	26.9	26.4	25.1
Indiana—Indianapolis	−3.6	11.3	24.1	12.6	11.3
Louisiana—New Orleans	10.7	20.3	27.7	20.6	20.1
Minnesota—Minneapolis	−11.2	8.0	23.1	9.3	7.2
Montana—Great Falls	−6.0	6.4	20.1	8.6	7.1
Nevada—Reno	0.5	9.2	22.0	10.4	10.4
Oklahoma—Oklahoma City	2.2	15.8	27.8	16.7	15.6
Texas—Dallas-Fort Worth	6.3	18.6	29.4	13.4	18.6
Vermont—Burlington	−8.7	6.6	21.4	8.8	7.0
Virginia—Norfolk	3.9	13.9	25.7	16.2	15.1
Washington—Seattle	4.5	9.6	18.6	11.6	11.1

[a]U.S. Dept. of Commerce (1995).

D. Ambient Carbon Dioxide Concentration

Many studies show that higher concentrations of CO_2 than normally present in air will promote more carbon fixation and increase biomass yields. In confined environmentally controlled enclosures such as hothouses, carbon dioxide-enriched air is often used to stimulate growth. In large-scale, open-air systems such as those envisaged for biomass energy farms, this is not practical. For aquatic biomass production, CO_2 enrichment of the water phase may be a potentially attractive method of promoting biomass growth if CO_2 concentration is a limiting factor, since biomass growth often occurs by uptake of CO_2 from both the air and liquid phase near the surface.

For some high-growth-rate biomass species, the CO_2 concentration in the air among the leaves of the plant is often considerably less than that in the surrounding atmosphere. Photosynthesis may be limited by the CO_2 concentrations under these conditions when wind velocities are low and insolation is high.

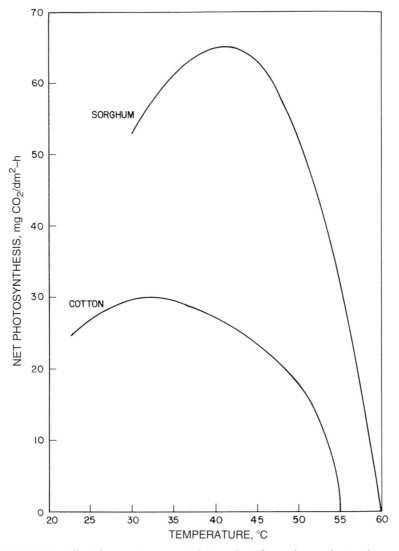

FIGURE 4.8   Effect of temperature on net photosynthesis for sorghum and cotton leaves.

E. NUTRIENTS

All living biomass requires nutrients other than carbon, hydrogen, and oxygen
to synthesize cellular material. Major nutrients are nitrogen, phosphorus, and

potassium; other nutrients required in lesser amounts are sulfur, sodium, magnesium, calcium, iron, manganese, cobalt, copper, zinc, and molybdenum. The last five nutrients, as well as a few others not listed, are sometimes referred to as micronutrients because only trace quantities are needed to stimulate growth. For terrestrial biomass, all of these elements are usually supplied by the soil, so eventually the soil's natural nutrients are depleted if they are not replaced through fertilization. Some biomass species, such as the legumes, are able to meet all or part of their nitrogen requirements through fixation of ambient nitrogen. Marine biomass such as giant brown kelp use the natural nutrients in ocean waters. Freshwater biomass such as water hyacinth is often grown in water rich in nutrients such as municipal wastewaters. The growth of the plant is stimulated and at the same time, the influent wastewater is stabilized because its components are taken up by the plant as nutrients. So-called luxuriant growth of water hyacinth on biosolids, in which more than the needed nutrients are removed from the wastewater, can be used as a substitute wastewater treatment method.

Whole plants typically contain 2 wt % N, 1 wt % K, and 0.5 wt % P, so at a yield of 20 t/ha-year, harvesting of the whole plant without return of any of the plant parts to the soil corresponds to the annual removal of 400 kg N, 200 kg K, and 100 kg P per hectare. This illustrates the importance of fertilization, especially with these macronutrients, to maintain soil fertility. Indeed, biomass growth is often nutrient-limited and yield correlates with increased dose rates. An example is U.S. corn production from 1945 to 1970. Nitrogen fertilizer applications were increased from 8 kg/ha to 125 kg/ha over this period; the corresponding edible corn yields increased from 2132 to 5080 kg/ha (Krummel, 1976). Average nitrogen fertilizer applications for production of wheat, rice, potato, and brussels sprouts are about 73, 134, 148, and 180 kg/ha, respectively, in the United States (Krummel, 1976). Much of the success of the "green revolution" is claimed to be the result of greater fertilizer applications. Estimates of balanced fertilizers needed to produce various land biomass species are shown in Table 4.5 (Roller *et al.*, 1975). Note that alfalfa does not require added nitrogen because of its nitrogen-fixing ability. It is estimated that this legume can fix from about 130 to 600 kg of elemental nitrogen per hectare annually (Evans and Barber, 1977).

Normal weathering processes that occur in nutritious soils release nutrients, but they are often not available at rates that promote maximum biomass yields. Fertilization is usually necessary to maximize yields. Since nitrogenous fertilizers are currently manufactured from fossil fuels, mainly natural gas, and since fertilizer needs are usually the most energy intensive of all the inputs in a biomass production system, a careful analysis of the integrated biomass production-conversion system is necessary to ensure that net energy production is positive. Trade-offs between synfuel outputs, nonsolar energy inputs,

TABLE 4.5   Estimated Fertilizer Requirements of Selected Biomass[a]

Biomass	Required mass per unit weight of whole dry plant, kg/dry t			
	N	P_2O_5	K_2O	CaO
Alfalfa	0	12.3	34.0	20.7
Corn	11.8	5.7	10.0	0
Kenaf	13.9	5.0	10.0	16.1
Napier grass	9.6	9.3	15.8	8.5
Slash pine (5 year)	3.8	0.9	1.6	2.3
Potato	16.8	5.3	28.3	0
Sugar beet	18.0	5.4	31.2	6.1
Sycamore	7.3	2.8	4.7	0
Wheat	12.9	5.3	8.4	0

[a]Roller et al. (1975).

and biomass yields are required to operate a system that produces only energy products.

III. AVAILABILITY OF LAND AND WATER AREAS FOR BIOMASS PRODUCTION

A. LAND AREAS

The availability of land suitable for production of terrestrial biomass can be estimated by several techniques. For the United States, one method relies on a land capabilities classification scheme in which land is divided into eight classes (Table 4.6) (U.S. Dept. of Agriculture, 1966). Classes I to III are suited for cultivation of many kinds of crops; Class IV is suited only for limited production; and Classes V to VIII are useful only for permanent vegetation such as grasses and trees. The U.S. Department of Agriculture surveyed nonfederal land usage for 1987 in terms of these classifications and arrived at the breakdown shown in Table 4.7 (U.S. Dept. of Agriculture, 1989). Out of about 568 million ha, which corresponds to about 60% of the 50-state area, about 43% of the land (246.4 million ha) was in Classes I to III, 13% (75.6 million ha) was in Class IV, and 43% (246.3 million ha) was in Classes V to VIII. The actual usage of this land at the time of the survey is shown in Table 4.8 (U.S. Dept. of Agriculture, 1989). This table shows that of all the land judged suitable for cultivation in Classes I to III, only about 58% of it was

TABLE 4.6   Land Capability Classification by United States Department of Agriculture[a]

Class	Description
I	Few limitations that restrict use.
II	Moderate limitations that reduce the choice of plants or require moderate conservation practice.
III	Severe limitations that reduce the choice of plants or require special conservation practices, or both.
IV	Very severe limitations that reduce the choice of plants or require very careful management, or both.
V	Not likely to erode, but other limitations, impractical to remove, that limit use largely to pasture, range, woodland, or wildlife habitat.
VI	Severe limitations that make soils generally unsuited to cultivation and limit their use largely to pasture or range, woodland, or wildlife habitat.
VII	Severe limitations that make soils unsuited to cultivation and that restrict use largely to pasture or range, woodland, or wildlife habitat.
VIII	Limitations that preclude use for communical plants and restrict use largely to recreation, wildlife habitat, water supply, or to esthetic purposes.

[a]U.S. Dept. of Agriculture (1966).

TABLE 4.7   Land Capability Classification of Nonfederal Rural Land by U.S. Department of Agriculture in 1987[a]

Land class	Area (10^6 ha)	% of total
I	13.47	2.37
II	117.36	20.65
III	115.53	20.33
IV	75.59	13.30
V	13.55	2.38
VI	106.71	18.78
VII	114.86	20.21
VIII	11.22	1.97
Total	568.29[b]	

[a]Adapted from U.S. Dept. of Agriculture (1989). Data are for the 48 contiguous states, Hawaii, and the Caribbean area.
[b]This area is 72.35% of the land surveyed, or 60.55% of the total 50-state area excluding the outlying areas (938.50 million ha). The federal land, water, and developed land areas are 17.43, 2.14, and 3.34% of the total 50-state area, respectively.

TABLE 4.8   Summary of U.S. Nonfederal Rural Land Usage by Use Type, 1987[a]

Land class	Cropland (10⁶ ha)	Pastureland (10⁶ ha)	Rangeland (10⁶ ha)	Forest (10⁶ ha)	Minor uses (10⁶ ha)	Total (10⁶ ha)
I	11.58	0.82	0.17	0.66	0.23	13.47
II	77.31	12.81	6.58	18.00	2.66	117.36
III	54.28	16.00	18.76	24.26	2.23	115.53
IV	18.60	10.30	21.62	23.56	1.51	75.59
V	1.16	1.86	1.99	7.52	1.03	13.55
VI	6.56	6.84	53.34	37.34	2.63	106.71
VII	1.59	3.91	58.40	46.88	4.08	114.86
VIII	0.035	0.067	1.68	1.40	7.81	11.00
Other	0	0	0	0	2.08	2.08
Total	171.12	52.60	162.56	159.62	24.25	570.15
Percent	30.01	9.22	28.51	28.00	4.25	

[a]Adapted from U.S. Dept. of Agriculture (1989). Totals may not be precise summations because of rounding. Cropland is land used for production of crops for harvest alone or in rotation with grasses and legumes. Pastureland is land used for production of adapted, introduced, or native species in a pure stand, grass mixture, or a grass–legume mixture. Rangeland is land on which the vegetation is predominantly grasses, grass-like plants, forbs, or shrubs suitable for grazing or browsing. Forest is land that is at least 10% stocked by trees of any size or formerly having had such tree cover and not currently developed for nonforest use. Other land is land such as farmsteads, strip mines, quarries, and lands that do not fit into the other land use category.

actually used as cropland, the locations of which are shown in Table 4.9 (U.S. Dept. of Agriculture, 1989). Also, the combined areas of pasture, range, and forestlands made up about 66% of the total nonfederal lands. This survey suggests that there is ample opportunity to produce biomass for energy applications on nonfederal land that is not used for foodstuffs production. Large areas of land in Classes V to VIII not suited for cultivation would appear to be available also for biomass energy applications, and sizable areas in Classes I to IV that are not being used for crop production seem to be available. Land now used for crop production could also be considered for simultaneous or sequential growth of biomass for foodstuffs and energy. Portions of the federally owned lands, which are not included in this survey, might also be dedicated to biomass energy applications. Careful design and management of land-based biomass production areas could very well result in improvement or upgrading of lands to higher land capability classifications.

B. Water Areas

The production of marine biomass in the ocean, even on the largest scale envisaged for energy applications, would require only a very small fraction of

TABLE 4.9 Summary of U.S. Cropland Capability Classification by Region, 1987[a]

Region	No. of states	Classes I–III (10^6 ha)	Class IV (10^6 ha)	Classes V–VIII (10^6 ha)	Total (10^6 ha)
Northeast	11	5.567	0.875	0.375	6.817
Appalachian	5	7.252	1.138	0.712	9.102
Southeast	4	5.915	0.786	0.372	7.073
Delta states	3	7.897	0.561	0.342	8.800
Corn Belt	5	34.868	2.210	0.846	37.924
Southern Plains	2	15.314	1.532	0.760	17.606
Northern Plains	4	32.688	3.746	2.256	38.690
Lake states	3	15.255	1.865	0.745	17.865
Mountain states	8	11.674	3.981	2.247	17.902
Pacific	3	6.578	1.860	0.593	9.031
Hawaii, Caribbean	1	0.166	0.051	0.096	0.313
Total	49	143.17	18.60	9.34	171.12
Percent of total		83.7	10.9	5.5	

[a]Adapted from U.S. Dept. of Agriculture (1989). Alaska excluded.

the available ocean areas. For example, it is estimated that, depending on biomass yield, a square area about 320 to 540 km on each edge off the coast of California may be sufficient to produce enough giant brown kelp for conversion to methane to supply all U.S. natural gas needs (Bryce, 1978). This is a large area, but it is very small when compared with the total area of the Pacific Ocean. Also, the benefits to other marine life from large kelp plantations have been well documented. Any conflicts that might arise would be concerned primarily with ocean traffic. With the proper plantation design for marine biomass and precautionary measures to warn approaching ships, it is expected that marine biomass growth could be sustained over long periods.

Freshwater biomass could in theory be grown on the 20 million ha of fresh water in the United States. But there are several difficulties that mitigate against large-scale freshwater biomass energy systems. About 80% of the fresh water in the United States is located in the northern states, whereas several of the freshwater biomass species considered for energy applications require a warm climate such as that found in Gulf states. The freshwater areas suitable for biomass production in the southern states, however, are much smaller than those in the North, and the density of usage is higher in southern inland waters. Overall, these characteristics make small-scale aquatic biomass production systems more feasible for energy applications. In the future, it may be advisable to examine the possibility of constructing large man-made lakes for this purpose, but this does not seem practical at this time except possibly where an aquatic biomass species is used for wastewater treatment.

IV. SELECTION OF VIRGIN BIOMASS SPECIES FOR ENERGY APPLICATIONS

A. TERRESTRIAL BIOMASS

Much effort to evaluate terrestrial biomass for energy applications has been expended (for the United States, see Hohenstein and Wright, 1994; Ferrell, Wright, and Tuskan, 1995). In general, this work has been aimed at selecting high-yield biomass species, characterizing their physical and chemical properties, defining their growth requirements, and rating their energy use potential. Several species have been proposed specifically for energy usage, whereas others have been recommended for multiple uses, one of which is as an energy resource. The latter case is exemplified by sugarcane; bagasse, the fibrous material remaining after sugar extraction, is used in several sugar factories as a boiler fuel. It is probable that most land-based biomass plantations operated for energy production or synfuel manufacture will also yield products for nonenergy markets. Large-scale biomass energy plantations that produce single energy products will probably be the exception rather than the rule. Land-based biomass for energy production can be divided into forest biomass, grasses, and cultivated plants.

Forest Biomass

About one-third of the world's land area is forestland. Broad-leaved evergreen trees are a dominant species in tropical rain forests near the equator (Spurr, 1979). In the northern hemisphere, stands of coniferous softwood trees such as spruce, fir, and larch dominate in the boreal forests at the higher latitudes, while both the broad-leaved deciduous hardwoods such as oak, beach, and maple and the conifers such as pine and fir are found in the middle latitudes. Silviculture, or the growth of trees, is practiced by five basic methods: exploitative, conventional extensive, conventional intensive, naturalistic, and short-rotation (Spurr, 1979). The exploitative method is simply the harvesting of trees without regard to regeneration. The conventional extensive method is the harvesting of mature trees so that natural regeneration is encouraged. Conventional intensive silviculture is the growing and harvesting of commercial tree species in essentially pure stands such as Douglas fir and pine on tree farms. The naturalistic method has been defined as the growth of selected mixed tree species, including hardwoods, in which the species are selected to match the ecology of the site. The last method, short-rotation silviculture (short-rotation woody crop or SRWC, short-rotation intensive culture or SRIC), has been suggested as the most suitable method for energy applications. In this technique, trees that grow quickly are harvested every few years, in contrast to once every 20 or more years. Fast-growing trees such as cottonwood, red alder, and aspen are intensively cultivated and mechanically harvested every 3 to 6 years when they are 3 to 6 m high and only a few centimeters in diameter. The young trees are converted into chips for further processing or direct fuel use and the small remaining stems or stumps form new sprouts by vegetative growth (coppicing) and are intensively cultivated again. SRWC production affords dry yields of several tons of biomass per hectare annually, often without large energy inputs for fertilization, irrigation, cultivation, and harvesting, so that the energy balance is positive.

It should be noted that although the prime purpose is to produce wood fiber for the manufacture of paper products, the pulp and paper companies have operated large tree plantations that yield energy as a by-product for decades. Heat, steam, and electricity are produced from wood wastes and also black liquor, which is generated in the paper manufacturing process (Chapter 5). Almost two-thirds of all renewable fuels consumed by the U.S. industrial sector is accounted for by the industry's use of black liquor (U.S. Dept. of Energy, 1995). The pulp and paper industry produces well over half of its own energy needs and clearly has a great interest in sustained-yield forestry. In the United States, several pulp and paper companies are developing SRWC technology to provide improved methods for supplying fiber to pulp mills and by-product energy (Stokes and Hartsough, 1994). In 1994, approximately 20,000 ha of SRWC systems were operated in the United States by the pulp and paper industry; 40,000 to 80,000 ha were projected to be operated by the year 2000.

Historically, trees are important resources and still serve as major energy resources in many developing countries. No fewer than 1.5 billion people in developing countries derive at least 90% of their energy requirements from wood and charcoal, and at least another billion people meet at least 50% of their energy needs this way (National Academy of Sciences, 1980). Hundreds of species in the seven genera *Acacia, Casuarina, Eucalyptus, Pinus, Prosopis, and Trema* are used as fuelwood in developing countries (Little, 1980). Several studies of temperate forests indicate productivities from about 9 to 28 t/ha-year, while the corresponding yields of tropical forests are higher, ranging from about 20 to 50 t/ha-year (Nichiporovich, 1967). These yields are obtained using conventional forestry methods over long periods of time, 20 to 50 years or more. Productivity is initially low in a new forest, slowly increases for about the first 20 years, and then begins to decline. Coniferous forests will grow even in the winter months if the temperatures are not too low; they do not exhibit the yield fluctuations characteristic of deciduous forests.

One of the tree species that has been studied in great detail as a renewable energy resource is the eucalyptus (Mariani, 1978), evergreen hardwood trees that belong to the myrtle family, Myrtaceae, and the genus *Eucalyptus*. There are approximately 450 to over 700 identifiable species in the genus. The eucalyptus is a rapidly growing tree native to Australia and New Guinea, and is widely grown in the United States, especially in Southern California and Hawaii for a variety of construction purposes. High-density plantings (17,790 trees/ha) in Southern California of *E. grandis* harvested twice annually have been reported to yield in excess of 22 dry t/ha-year (Sachs, Gilpin, and Mock, 1980). It appears to be a prime candidate for energy use because it reaches a size suitable for harvesting in about 7 years. Several species have the ability to coppice after harvesting, and as many as four harvests can be obtained from a single stump before replanting is necessary. In several South American countries, eucalyptus trees are converted to charcoal and used as fuel. Eucalyptus wood has also been used to power integrated sawmill, wood distillation, and charcoal-iron plants in western Australia. Several large areas of marginal land in the United States may be suitable for establishing eucalyptus energy farms. These areas are in the western and central regions of California and the southeastern United States.

Various species and hybrids of the genus *Populus* are some of the more promising candidates for SRWC growth and harvesting as an energy resource (Sajdak *et al.*, 1981). The group has long been cultivated in Europe and more recently in the eastern United States and Canada. *Populus* hybrids are easily developed and the resulting progeny are propagated vegetatively using stem cuttings. Consequently, there are hundreds of numbered or named clones established throughout the eastern United States. Summaries of record SWRC small-plot yields for *Populus* hybrids have shown production levels of 15-20 dry t/ha-year (Hansen, 1988) and yields of 30-40 dry t/ha-year have been projected as attainable goals through genetic engineering (Ranney, Wright, and Layton, 1987). SRWC growth of hybrid poplar clone D-01 has been reported to afford yields of biomass that range as high as 112-202 green t/ha-year (56-101 dry t/ha-year at 50 wt % moisture) (Dula, 1984). These results were reported with very high-density plantings that have been termed wood-grass in which the crop is grown like grass and is harvested several times each growing season. However, there is some dispute regarding the benefits of woodgrass growth vs SRWC growth (Wright *et al.*, 1989). Several investigators have not been able to reproduce these results (*cf.* DeBell and Clendenen, 1991), although the high-density planting technique seems to have some potential benefits.

It was concluded from early studies that deciduous trees are preferred over conifers for the production of woody biomass for conversion to biofuels (InterTechnology Corp., 1975). Conifers are used as fuel in many parts of the world, including the United States, but the long-term research effort to develop woody species as dedicated energy crops emphasizes mostly deciduous species (*cf.* Ferrell *et al.*, 1993; Wright, 1994; Ferrell, Wright, and Tuskan, 1995). Several deciduous species can be started readily from clones, resprout copiously and vigorously from their stumps at least five or six times without loss of vigor, and exhibit rapid initial growth. They can also be grown on sites with slopes as steep as 25%, where precipitation is 50 cm or more per year. It has been estimated that yields between about 18 and 22 dry t/ha-year are possible on a sustained basis almost anywhere in the Eastern and Central time zones in the United States from deciduous trees grown in dense plantings. Table 4.10 lists deciduous trees that were judged in early work to have desirable growth characteristics for plantation culture and that have been shown to grow satisfactorily at high planting densities for short and repeated harvest cycles.

Grasses

Grasses are very abundant forms of biomass (U.S. Dept. of Agriculture, 1948). About 400 genera and 6000 species are distributed all over the world and grow in all land habitats capable of supporting higher forms of plant life. Grasslands cover over one-half the continental United States, and about two-thirds of this land is privately owned. Grass, as a family (Gramineae), includes the great fruit crops, wheat, rice, corn, sugarcane, sorghum, millet, barley, and oats. Grass also includes the many species of sod crops that provide forage or pasturage for all types of farm animals. In the concept of grassland agriculture, grass also includes grass-related species such as the legumes family—the clovers, alfalfas, and many others. Grasses are grown as farm crops, for decorative purposes, for preserving the balance of productive capacity of lands by crop rotation, for controlling erosion on sloping lands, for the protection of water sheds, and for the stabilization of arid areas. Many advances in grassland agriculture have been made since the 1940s through breeding and the use of improved species of grass, alone or in seeding mixtures; cultural practices, including amending the soil to promote herbage growth best suited for specific purposes; and the adoption of better harvesting and storage techniques. Until the mid-1980s, very little of this effort had been directed to energy applications. A few examples of energy applications of grasses can be found such as the combustion of bagasse for steam and electric power, but many other opportunities exist that have not been developed.

Perennial grasses have been suggested as candidate feedstocks for conversion to synfuels. Most perennial grasses can be grown vegetatively, and they reestab-

TABLE 4.10 Representative Deciduous Trees for Plantation Culture and Locations Where They Have Been Shown to Grow Well in North America in Managed Plantings[a]

Locality state/province	Hybrid poplar	Aspen and hybrids	Black cottonwood	Red alder	Sycamore	Pin cherry	Plains cottonwood	Eastern cottonwood	Silver maple	European black alder	Green ash	Sweetgum	Eucalyptus
Alabama					X			X		X	X	X	X
Florida													X
Georgia					X						X	X	X
Illinois								X		X			
Indiana								X					
Kansas						X		X	X				
Louisiana								X					
Minnesota		X											
Mississippi					X			X					
Nebraska								X					
New Hampshire		X				X							
North Dakota							X						
Ohio										X			
Pennsylvania	X												
Texas								X					
Washington	X		X	X									
Wisconsin	X	X	X										
British Columbia				X									
Manitoba		X											
Nova Scotia	X												
Ontario	X												
Saskatchewan	X												

[a]InterTechnology Corp. (1975).

lish themselves rapidly after harvesting. Also, more than one harvest can usually be obtained per year. The warm-season grasses are preferred over the cool-season grasses because their growth increases rather than declines as the temperature rises to its maximum in the summer months. In certain areas, rainfall is adequate to permit harvesting every 3 to 4 weeks from late February into November, and yields between about 18 and 24 t/ha-year of dry grasses may be obtainable in managed grasslands. Some tropical and semitropical grasses are very productive and can yield as high as 50 to 60 t/ha-year on good sites (Westlake, 1963). The tropical fodder grass *Digitaria decumbens* has been grown at yields of organic matter as high as 85 t/ha-year (Westlake, 1963). Table 4.11 lists some promising grasses that have been proposed as energy resources in the United States (Cushman and Turhollow, 1991).

An example of a tropical grass that has been grown commercially as a combination foodstuff and fuel crop for many years is sugarcane (*Saccharum*

TABLE 4.11 Average Annual Yields of Most Productive Herbaceous Species in Field Trials in U.S. Southeast and Midwest/Lake States[a]

Biomass type and species	Southeast (dry t/ha-year)[b]	Midwest/Lake states (dry t/ha-year)[b]
Annuals		
Warm-season		
Sorghums[c]	0.2–19.0	1.9–29.1
Cool-season		
Winter rye[d]	0.0–7.2	2.4–6.1
Perennials[e]		
Warm-season		
Switchgrass[d]	2.9–14.0	2.5–13.4
Weeping lovegrass[d]	5.4–13.7	
Napiergrass-energy	20.4–28.3	
canes[c]		
Cool-season		
Reed canary grass[d]		2.7–10.8
Legumes		
Alfalfa		1.6–17.4
Flatpea	2.1–12.9	3.9–10.2
Sericea lespedeza	1.8–11.1	

[a]Cushman and Turhollow (1991).
[b]Averaged by site; data are for range of sites.
[c]Thick-stem grass.
[d]Thin-stem grass.
[e]Productivity rates after 1- to 2-year extablishment period.

spp.), but rising production costs, alternative sweeteners, and the nebulous mixture of changing social, political, and agricultural policy issues have not been kind to insular sugar planters (Alexander, 1993). A great deal of information has been accumulated about sugarcane, and it might well be used as a model tropical herbaceous crop for other biomass energy systems. It grows rapidly and produces high yields; the fibrous bagasse is used as boiler fuel for the generation of electric power; and sugar-derived ethanol is used as a motor fuel in gasoline blends (gasohol). Sugarcane plantations and the associated sugar processing and ethanol plants are in reality biomass fuel farms. About one-half of the organic material in sugarcane is sugar and the other half is fiber. Total cane biomass yields have been reported to range as high as 80 to 85 dry t/ha-year. Normal cultivation provides yields of about 50 to 59 dry t/ha-yr (Westlake, 1963). Studies on sugarcane managed specifically as an energy crop have been underway in Puerto Rico for several years. "First-generation energy cane" consisting of conventional varieties managed for optimal growth with irrigation averaged 186 green t/ha-year of whole cane including detached trash, whereas second generation yields exceeded 269 green t/ha-year (Alexander, 1983). At an average of 40 wt % dry matter, these yields range from 74 to 108 dry t/ha-year. Presuming the energy content of energy cane is about 18.5 GJ/dry t, the energy yields correspond to 1369 to 1998 GJ/ha-year, or the equivalent of 232 to 338 BOE/ha-year, a very high yield.

In moderate climates, switchgrass (*Panicum virgatum*) has been recommended as one of the model biomass energy crops for North America because of its high yield potential, adaptation to marginal sites, and tolerance to water and nutrient limitations (Sanderson *et al.*, 1995). It is a warm-season grass native to much of North America and is a major species in tall prairie grasses. Average yields are reported to range from 5.5 to 11.3 dry t/ha-year in the midwestern and eastern United States (Wright, 1994). In the southwestern United States, evaluation of eight switchgrass cultivars showed that for six locations in Texas, single harvests of fertilized plots of the Alamo cultivar afforded the highest average yields, 10.7 to 15.7 dry t/ha-year (Sanderson *et al.*, 1995).

Other productive grasses that have been given serious consideration as raw materials for the production of energy and synfuels include the perennials Reed canary grass, tall fescue, crested wheatgrass, weeping lovegrass, and Bermuda grass, the annual sorghum and its hybrids, and others. It is apparent that there are many grasses and related biomass species that can be considered for energy applications. They have many of the desirable characteristics needed for terrestrial biomass energy systems.

Other Cultivated Crops

Many other terrestrial biomass species have been proposed as renewable energy resources for their high-energy components that can be used as fuels, for the components capable of conversion to biofuels and chemicals, or for the contained energy (*cf.* Buchannan and Otey, 1978; Cherney *et al.*, 1989; DeLong *et al.*, 1995; Gavett, Van Dyne, and Blase, 1993; Klass, 1974; McLaughlin, Kingsolver, and Hoffmann, 1983; Nemethy, Otvos, and Calvin, 1981; O'Hair, 1982; Schneider, 1973; Shultz and Bragg, 1995; Stauffer, Chubey, and Dorell, 1981; Taylor, 1993). Among them are kenaf (*Hibiscus cannabinus*), an annual plant reproducing by seed only; sunflower (*Helianthus annuus* L.), an annual oil seed crop grown in several parts of North America; *Eurphorbia lathyris,* a sesquiterpene-containing plant species that grows in the semiarid climates of the Southwest and California; Buffalo gourd (*Curcurbita foetidissima),* a perennial root crop native to arid and semiarid regions of the southwestern United States; other root crops such as Jerusalem artichoke (*Helianthus tuberosus*), fodder beet (*Beta vulgaris*), and cassava (*Manihot esculenta*); alfalfa (*Medicago sativa*), a perennial legume that grows well on good sites in many parts of North America; soybean (*Glycine max*) and rapeseed (*Brassica campestris*), oilseed crops that produce high-quality oil and protein; and many other biomass species that are potentially suitable as renewable energy resources or multipurpose crops including energy and biofuels. Kenaf, for example, is highly fibrous and exhibits rapid growth, high yields, and high cellulose content. It is a pulp crop and is several times more productive than pulpwood trees. Maximum economic growth usually occurs in less than 6 months, and consequently two croppings may be possible in certain regions of the United States. Without irrigation, heights of 4 to 5 meters are average in Florida and Louisiana, but 6-m plants have been observed under near-optimum growth conditions. Yields as high as 45 t/ha-year have been observed on experimental test plots in Florida, and it has been suggested that similar yields could be achieved in the Southwest with irrigation. Another example is the sunflower, which is a good candidate for biomass energy applications too because of its rapid growth, wide adaptability, drought tolerance, short growing season, massive vegetative production, and adaptability to root harvesting. Dry yields have been projected to be as high as 34 t/ha per growing season. Rapeseed is another example; its seeds normally yield 38 to 44 wt % high quality protein and over 40 wt % oil, which affords high-quality biodiesel fuel at the rate of 750 to 900 L/ha-year on extraction and transesterification. Still another example is alfalfa, a well-known and widely-planted herbaceous crop that offers environmental and soil conservation advantages when grown as a 4-year segment in a 7-year rotation with corn and soybeans. With alfalfa yields of about 9 dry t/ha-year and the alfalfa leaf fraction sold as a high-value animal feed, the remaining alfalfa stem fraction can be used as feedstock for power production.

B. Aquatic Biomass

The average net annual productivities of dry organic matter on good growth sites for terrestrial and aquatic biomass are shown in Table 4.12. With the exception of phytoplankton, which generally has lower net productivities,

TABLE 4.12   Average Net Annual Biomass Yields on Fertile Sites[a]

Average Net Yield (dry t/ha-year)	Climate	Ecosystem type	Remarks
1	Arid	Desert	Much more if hot and irrigated
2		Ocean phytoplankton	
2	Temperate	Lake phytoplankton	Little human influence
3		Coastal phytoplankton	Probably higher in some polluted estuaries
6	Temperate	Polluted lake phytoplankton	In agricultural and sewage runoffs
6	Temperate	Freshwater submerged macrophytes	
12	Temperate	Deciduous forests	
17	Tropical	Freshwater submerged macrophytes	
20	Temperate	Terrestrial herbs	Possibly more if grazed
22	Temperate	Agriculture—annuals	
28	Temperate	Coniferous forests	
29	Temperate	Marine submerged macrophytes	
30	Temperate	Agriculture—perennials	
30		Salt marsh	
30	Tropical	Agriculture—annuals	Including perennials in continental climates
35	Tropical	Marine submerged macrophytes	
38	Temperate	Reedswamp	
40	Subtropical	Cultivated algae	More if CO_2 supplied
50	Tropical	Rainforest	
75	Tropical	Agriculture—perennials, reedswamp	

[a]Westlake (1963).

aquatic biomass seems to exhibit higher net organic yields than most terrestrial biomass. Aquatic biomass species that are considered to be the most suitable for energy applications include the unicellular and multicellular algae, freshwater plants, and marine species.

Algae

Microalgae have long been under development as renewable energy resources and other useful products (Benemann and Weissman, 1993). Almost 20,000 species are known. Unicellular algae such as the species *Chlorella* and *Scenedesmus* have been produced by continuous processes in outdoor light at high photosynthesis efficiencies. *Chlorella* has been reported to be produced at a rate as high as 1.0 dry t/ha-day. This corresponds to an annual rate of 401dry t/ha-year presuming growth can be sustained (Retovsky, 1966). These figures are probably in error, but there is no theoretical reason why yields cannot achieve very high values because the process of producing algae can be almost totally controlled. Also, production is not composed only of surface growth. Algae are produced as slurries in lakes, ponds, and custom-designed raceways so that the depth of the biomass-producing area as well as plant yield per unit volume of water are important parameters. The nutrients for algae production can be supplied by municipal biosolids and other wastewaters. It should be pointed out that most unicellular algae are grown in fresh water, which tends to limit their energy applications to small-scale algae farms. The high water content of unicellular algae also tends to limit the conversion processes to biological methods. But this can be an advantage in some cases where the particular microalgae exudes triglycerides without cell destruction so that the product oil is continuously formed and can be easily recovered from the water surface.

Macroscopic multicellular algae, or seaweeds, have also been considered as renewable energy resources for many years. Some of the candidates are the giant brown kelp *Macrocystis pyrifera* (Bryce, 1978; North, 1971; North, Gerard, and Kubawabara, 1981), the red benthic alga *Gracilaria tikvahiae* (LaPointe and Hanisak, 1985; Ryther and DeBusk, 1982), and the floating, brown pelagic algae *Sargassum natans* and *S. fluitans* (LaPointe and Hanisak, 1985). Giant brown kelp has been studied in great detail and is harvested commercially off the California coast. Because of its high potassium content, giant brown kelp was used as a commercial source of potash during World War I and is used today as a commercial source of organic gums, thickening agents, and alginic acid derivatives. Off the East Coast, *Laminaria* seaweed is harvested for the manufacture of alginic acid derivatives. In tropical seas not cooled by upwelled water, species of the *Sargassum* variety of algae may be suitable as renewable energy sources. Several species of *Sargassum* grow naturally around reefs surrounding the Hawaiian Islands. Unfortunately, only a small amount of research has been done on *Sargassum* and little detailed information is available about this alga. A considerable amount of data on yields and growth requirements are available, however, on the *Macrocystis* and *Laminaria* varieties. Again, the very high water content of macroscopic algae suggests that biological conversion processes rather than thermochemical conversion processes should be used for synfuel manufacture. The manufacture of co-products from macroscopic algae, such as polysaccharide derivatives, along with biofuel might make it feasible to use thermochemical processing techniques on intermediate process streams.

Water Plants

The productivity of some salt marshes is similar to that of seaweeds. *Spartina alterniflora* has been grown at net annual yields of about 33 dry t/ha-year, including underground material, on optimum sites (Westlake, 1963). Other emergent communities in brackish water, including mangrove swamps, appear to have annual organic productivities of up to 35 t/ha-year (Westlake, 1963), but insufficient information is available to judge their value in biomass energy systems. Freshwater swamps are believed to be highly productive and offer opportunities for energy production. Both the reed *Arundo donax,* and bulrush *Scirpus lacustris* appear to produce 57 to 59 t/ha-year yields (Westlake, 1963); if these can be sustained, they should be suitable candidates for biomass energy usage. Cattail (*Typha* spp.) is a wetland biomass that has been proposed as an energy resource (Pratt and Andrews, 1980). It grows naturally in monocultures, is highly productive, has few insect pests, and can be grown on marginal lands. Managed stands are reported to yield 25 to 30 dry t/ha-year of cattail in the northern climates of the United States (Minnesota).

A strong aquatic biomass candidate for energy applications is the water hyacinth (*Eichhornia crassipes*) (Klass, 1974). This biomass species is highly productive, as might be expected because it grows in warm climates and has submerged roots and aerial leaves like reedswamp plants. It has been estimated that water hyacinth could be produced at rates up to about 150 t/ha-year if the plants were grown in a good climate, the young plants always predominated, and the water surface was always completely covered (Westlake, 1963). Some evidence has been obtained to support this growth rate (McGarry, 1971; Yount and Grossman, 1970). If such yields can be maintained on a steady-state basis, water hyacinth could possibly turn out to be a prime aquatic biomass candidate as a nonfossil carbon source for synfuels manufacture as well as other potential applications such as the manufacture of paper. Water hyacinth currently has no competitive uses and is considered to be an undesirable species on inland waterways. Many attempts have been made to rid navigable streams in Florida of water hyacinth without success; the plant is a very hardy, disease-resistant species (Del Fosse, 1977).

V. THE ECONOMICS OF VIRGIN BIOMASS PRODUCTION

A. EFFECTS OF FOSSIL FUEL PRICES

The practical value of biomass energy ultimately depends on the costs of salable energy and biofuels to the end users. Consequently, many economic analyses have been performed on biomass production, conversion, and integrated biofuels systems. Conflicts usually abound when attempts are made to compare the results developed by two or more groups for the same biomass feedstock or biofuel because the methodologies are not the same. The assumptions made by each group are sometimes so different that valid comparisons cannot be made even when the same economic ground rules are employed. Comparative analyses, especially for hypothetical processes conducted by an individual or group of individuals working together, should be more indicative of the economic performance and ranking of biomass energy systems. However, several generalizations can be made that are quite important. The first is that fossil fuel prices are well documented and can be considered to be the primary competition for biomass energy. Table 4.13 summarizes U.S. tabulations of average, consumption-weighted, delivered fossil fuel prices by end-use sector in the mid-1990s (U.S. Energy Information Administration, December 1995). It is evident that the delivered price of a given fossil fuel is not the same to each end-use sector. The residential sector normally pays more for fuels than the other sectors, and the large end users pay less.

In the context of virgin biomass energy costs, dry woody and fibrous biomass species have an energy content on a dry basis of approximately 18.5 MJ/kg (7959 Btu/lb) or 18.5 GJ/t (16 MBtu/ton). For comparison purposes, if such types of biomass were available at delivered costs of $1.00/GJ ($1.054/MBtu), or $18.50/dry t ($16.78/dry ton), biomass on a strict energy content basis without conversion would cost less than most of the delivered fossil fuels listed in Table 4.13. The U.S. Department of Energy has set cost goals of delivered virgin biomass energy crops at $1.90-2.13/GJ ($2.00-2.25/MBtu), which corresponds to $35.15 to 39.41/dry t of virgin biomass (Fraser, 1993).

In the mid-1990s, few virgin biomass species were grown and harvested in the United States specifically for energy or conversion to biofuels, with the possible exceptions of feedstocks for fuel ethanol and a few tree plantations. This is not difficult to understand from an economic standpoint, especially if conversion costs are included. The nominal price of natural gas in the United

TABLE 4.13   U.S. Delivered Fossil Fuel Prices to End Users by Sector, 1993[a]

Fossil fuel	Residential ($/GJ)	Commercial ($/GJ)	Industrial ($/GJ)	Transportation ($/GJ)	Utility ($/GJ)
Coal	2.85	1.69	1.57		1.32
Natural gas	5.68	4.77	2.99		2.43
Petroleum	7.46	4.80	4.49	7.61	2.32
LPG	9.74	8.32	4.50	8.00	
Kerosine	7.18	5.07	5.06		
Distillate fuel	6.49	4.83	4.53	7.64	
Motor gasoline		8.90	8.60	8.60	
Aviation gasoline				7.82	
Jet fuel				4.07	
Residual fuel		2.61	2.29	1.88	
Heavy oil					2.25
Light oil					4.25
Petroleum coke					0.34

[a]Adapted from U.S. Energy Information Administration (December 1995). All figures are consumption-weighted averages for all states in nominal dollars and include taxes. Heavy oil includes Grade Nos. 4, 5, and 6. Light oil includes Grade Nos. 1 and 2, kerosine, and jet fuel.

States in 1994 at the wellhead (not end-use cost) was estimated to be $1.74/GJ ($1.83/MBtu) (U.S. Energy Information Administration, July 1995). For virgin biomass to compete as a feedstock for methane production on an equivalent basis, it would have to be grown, harvested, and gasified to produce methane at the same or lower cost. Assuming a gasification cost of zero and biomass conversion to substitute natural gas at 100% thermal efficiencies, both assumptions of which are totally unrealistic but which will help illustrate the best-case economics, the maximum market price of the biomass feedstock cost at the conversion plant gate including profit is then $32.19/dry t (at 18.5 GJ/dry t × $1.74/GJ). At an optimistic yield of 22.4 dry t/ha-year (10 dry ton/ac-year), the biomass producer who supplies the gasification plant with feedstock would then realize not more than $721/ha-year ($292/ac-year), a marginal amount to permit a net return on an energy crop without other incentives. Similar calculations for the production and conversion of virgin biomass to liquid petroleum substitutes at zero conversion cost and 100% thermal efficiencies at the average U.S. nominal wellhead price of crude oil of $13.19/bbl ($2.234/GJ) in 1994 (U.S. Energy Information Administration, July 1995) correspond to a maximum market price for virgin biomass feedstock of $41.33/dry t ($37.49/dry ton). The average price of hay, for example, received by farmers across the United States in 1994 was \$95.59/t (\$86.70/ton) (U.S. Dept. of Commerce, 1996). This indicates that the production of hay, and probably most grasses, as energy crops for conversion to liquid biofuels in direct competition with petroleum liquids was not economically feasible at that time. These simplistic calculations emphasize the effect of fossil fuel prices on dedicated biomass energy crops. Inclusion of gasification or liquefaction costs and conversion efficiency factors by the processor would result in still lower market prices that the processor would be willing to pay for biomass feedstocks. Negative feedstock costs (wastes), substantial by-product credits, captive uses, other markets and uses, environmental credits, and/or tax incentives would be needed to justify dedicated energy crop production on strict economic grounds.

B. Biomass Production Costs

An example of the detailed production costs in the mid-1990s of two commercial herbaceous crops grown without irrigation in the Corn Belt of the U.S. Midwest, the perennial alfalfa and the annual corn, is shown in Table 4.14 (University of Illinois Urbana-Champaign FaRM Lab, 1995). The economics are shown for the maintenance and harvesting of established alfalfa. The cost of planting in the first year is therefore excluded. For corn, no-till, no-rotation planting is used. This technique affords the lowest production cost, although attention must eventually be given to counteract any adverse effects on soil chemistry. Alfalfa has been proposed as a dedicated energy crop and corn is a commercial feedstock for fuel ethanol production. The analysis showed that the annual loss in nominal dollars was about \$115/ha for alfalfa and \$101/ha for corn. It is evident that at the production costs, reported yields, and market prices at that time, production of either crop could have led to a significant loss for the farmer. It is also evident that the major variable cost factors are chemicals and harvesting labor, and the major fixed cost is land rent.

It is immediately apparent from this assessment that situations can exist that would make alfalfa and corn production profitable. If the land is rented at much lower cost than indicated in Table 4.14 or is owned by the farmer with no outstanding debt, or the crops are grown on one or more family farms where resident labor is available, the economics can be quite different and favorable. Many scenarios can be envisaged that will improve the net return. The point is that what may appear to be uneconomic at first is subject to change when the details are analyzed and appropriate actions can be taken to improve profitability. The difficulty of accurately predicting market prices is another factor that complicates matters further. Indeed, it was only a few months after this analysis that the market price of corn began to increase at a rapid rate and to reach an all-time high of over \$5.00/bu, which effectively

TABLE 4.14 Production Costs of Alfalfa Hay and Corn In Northern and Central Illinois[a]

Biomass	Alfalfa hay			Corn		
Production method	Maintain and harvest			No till and no rotation		
Time of costs	As of 11/11/95			As of 11/11/95		
Yield/growing season	9 t/ha-year			358 bu/ha-year[b]		
Market price	$77/t			$2.35/bu[b]		
Variable costs	Unit/ha	Price/Unit	Cost/ha	Unit/ha	Price/Unit	Cost/ha
Fertilizer						
Anhydrous NH$_3$	0 kg	$ 0.44	$ 0.00	91 kg	$ 0.44	$ 83.84
P$_2$O$_5$	54 kg	$ 0.53	$ 28.61	67 kg	$ 0.53	$ 35.51
K$_2$O	224 kg	$ 0.29	$ 64.96	44 kg	$ 0.29	$ 12.76
Lime	1.12 t	$ 14.33	$ 16.05	1.12 t	$ 14.33	$ 16.05
Total fertilizer			$ 109.62			$ 148.16
Herbicides	Multiple	Multiple	$ 59.30	Multiple	Multiple	$ 84.02
Insecticides	Multiple	Multiple	$ 12.36	Multiple	Multiple	$ 34.59
Total pesticides			$ 71.66			$ 118.61
Seed	In-place	In-place	$ 0.00	69.2 k	$ 0.85	$ 58.82
Crop insurance			$ 0.00			$ 12.36
Mach. fuel, repairs			$ 0.00			$ 12.36
Labor			$ 0.00	0.72 h	$ 10.00	$ 7.20
Preharvest interest	3 mo.	9%	$ 4.94	7 mo.	9%	$ 19.77

Harvest costs						
Mach. fuel, repairs	15.39 h	$ 10.00			$ 76.60	$ 14.45
Labor			3.14 h	$ 10.00	$ 153.90	$ 31.40
Trucking			358 bu	0.02	$ 0.00	$ 7.16
Drying			3.3 L/bu	0.042/L	$ 0.00	$ 49.62
Storage					$ 0.00	$ 0.00
Total variable costs					$ 416.72	$ 479.91
Fixed costs						
Mach: cap., taxes, ins.		$ 81.91		$ 153.27	$ 81.91	$ 153.27
Land rent		$ 309.00		$ 309.00	$ 309.00	$ 309.00
Total fixed costs					$ 390.91	$ 462.27
Total costs					$ 807.63	$ 942.18
Total revenue					$ 693.00	$ 841.30
Net return (loss)						
Per ha-year over variable					$ 276.28	$ 361.39
Per unit over variable					$ 30.70/t	$ 1.01/bu
Per ha-year over total					($ 114.63)	($ 100.88)
Per unit over total					($ 12.74/t)	($ 0.28/bu)

[a] Adapted from University of Illinois Urbana-Champaign FaRM Lab (1995).
[b] One bushel (0.03524 m^3) of corn is approximately 25.4 kg (56 lb).

doubled the farmer's revenue. Careful consideration of all cost factors is obviously necessary, but there is no approach to the elimination of all risk when growing a dedicated energy crop, or any other crop for that matter.

An economic analysis of the delivered costs of virgin biomass energy in 1990 dollars has been performed for candidate virgin herbaceous and woody biomass for different regions of the United States (Fraser, 1993). The analysis was done for each decade from 1990 to 2030 for Class I and II lands, but only the results for biomass grown on Class II lands for the years 1990 and 2030 are shown in Table 4.15. The total production costs for biomass were projected with discounted cash flow models, one for the herbaceous crops switchgrass, napier grass, and sorghum, and one for the short-rotation production of sycamore and hybrid poplar trees. The delivered costs are shown in Table 4.15 in 1990 $/dry t and 1990 $/MJ and are tabulated by region and biomass species.

TABLE 4.15   Estimated U.S. Delivered Costs for Candidate Biomass Energy Crops in 1990 and 2030[a]

Region and species	1990			2030		
	Yield (dry t/ha-year)	Cost ($/dry t)	Cost ($/GJ)	Yield (dry t/ha-year)	Cost ($/dry t)	Cost ($/GJ)
Great Lakes						
Switchgrass	7.6	104.07	5.26	15.5	61.32	3.60
Energy sorghum	15.5	62.56	3.17	30.9	36.79	2.16
Hybrid poplar	10.1	113.79	5.76	15.9	72.82	4.29
Southeast						
Switchgrass	7.6	105.89	5.36	17.3	52.91	3.11
Napier grass	13.9	63.72	3.22	30.9	33.31	1.96
Sycamore	8.1	88.61	4.49	14.3	53.19	3.13
Great Plains						
Switchgrass	5.4	74.32	3.77	10.3	44.05	2.59
Energy sorghum	6.3	91.73	4.65	13.7	48.07	2.83
Northeast						
Hybrid poplar	8.1	105.26	5.33	11.9	71.69	4.26
Pacific Northwest						
Hybrid poplar	15.5	66.69	3.56	23.8	44.73	2.63

[a]Adapted from Fraser (1993). Discounted cash-flow models account for the use of capital, income taxes, time value of money, and operating expenses. Real after-tax return is assumed to be 12.0%. Short-rotation model used for sycamore and poplar. Herbaceous model used for other species. The costs are in 1990 dollars. The yields in 1990 are on Class II lands. The average total field yields are for the entire region on prime to good soil, less harvesting and storage losses. The yields in 2030 are assumed to be attained through research and genetic improvements. Short-rotation woody crops (hybrid poplar and sycamore) are grown on 6-year rotations on six independent plots. Net income is negative for first 5 years for each SRWC plot.

The yield figures for 1990 were obtained by the analysts from the literature and the projected yields for 2030 were assumed to be achievable from continued research. The annual, dry biomass yields per unit area have a great influence on the final estimated costs, as would be expected. This analysis indicates that the lowest-cost energy crop of those chosen can be different for different regions of the country. A few of the biomass-region combinations appear to come close to providing delivered biomass energy near the U.S. Department of Energy cost goal. But realizing that there are many differences in the methodologies and assumptions used to compile the 1990 costs for delivered fossil fuels in Table 4.13 and delivered virgin biomass energy in Table 4.15, it appears that many of the biomass energy costs are competitive with those of fossil fuels in several end-use sectors, even without incorporating the yield improvements that are expected to evolve from continued research on biomass energy crops.

However, it is essential to recognize several other factors in addition to the basic cost of virgin biomass and its conversion when considering whether the economics are competitive with the costs of other energy resources and fuels. Some potential biomass energy feedstocks have negative values; that is, waste biomass of several types such as municipal biosolids, municipal solid wastes, and certain industrial and commercial wastes that must be disposed of at additional cost by environmentally acceptable methods. These biomass feedstocks will be discussed in the next chapter, but suffice it to say at this point that many generators of waste biomass will pay a service company for removing and disposing of the wastes, and many of the generators will undertake the task on their own. These kinds of feedstocks often provide an additional economic benefit and revenue stream that can support commercial use of biomass energy.

Another factor is the potential economic benefit that may be realized from the utilization of both waste and virgin biomass as energy resources due to current and future environmental regulations. If carbon taxes are ever imposed on the use of fossil fuels in the United States as they have been in a few other countries to help reduce undesirable automobile and power plant emissions to the atmosphere, additional economic incentives will be available to stimulate development of new biomass energy systems. Certain tax credits and subsidies are already available for commercial use of specific types of biomass energy systems (Klass, 1995).

C. LOCAL AND REGIONAL ECONOMIC EFFECTS OF BIOMASS ENERGY

In addition to the costs of biomass energy, another and probably more important factor should be considered when assessing market prices for biomass energy and biofuels. It is the accumulated, tangible socioeconomic benefits of the commercial utilization of a local or regional energy resource for the local or regional economy. A detailed assessment of these benefits is perhaps best illustrated by the results of a projection done for the state of Wisconsin on the impacts of a 75% increase in Wisconsin's biomass energy use by the year 2010 (Clemmer and Wichert, 1994). The study was referred to briefly in Chapter 2; more detail is presented here. Using indigenous biomass feedstocks, the projection consists of the impacts of 775 MW of new generating capacity and 379 million liters per year of new fuel ethanol production. This amount of biomass energy could supply electricity to 500,000 Wisconsin homes and 10 vol % ethanol–90 vol % gasoline blends (gasohol) to 45% of Wisconsin's automobiles. Investment under this projection generates about three times more jobs, earnings, and output (sales) in Wisconsin than the same level of imported fossil fuel usage and investment. This incremental increase in biomass energy alone is equivalent to 63,234 more job-years of net employment, $1.2 billion in higher wages, and $4.6 billion in additional output. Over the operating life of the technologies analyzed, about $2 billion in avoided payments for imported coal, natural gas, and petroleum fuels could remain in Wisconsin to pay for state-supplied renewable resources, technologies, and labor. Collecting and distributing the wood, corn, and waste feedstocks correspond to 47% of the total net new employment for this industry and create permanent forestry, agriculture, and transportation jobs in Wisconsin's rural communities. Operating and maintaining biomass energy technologies produce 27% of the net employment growth, and installing and manufacturing these technologies generate 13% of the new jobs on a temporary basis. Net savings in consumer income, with environmental regulations, account for the remaining 13% of new jobs. Five of the 11 biomass technologies analyzed for power production are less expensive to operate than a new baseload coal plant, without considering incentives for environmental costs. When federal incentives and potential environmental regulation costs are included, 9 of the 11 biomass technologies cost less than a new coal-fired plant. Ethanol produced by established technologies competes with gasoline at the federal incentive levels in place in the 1990s. Investing in biomass energy instead of fossil fuels in Wisconsin could save the state's residents about $700 million in avoided environmental regulations to control CO_2 and SO_2 emissions from fossil fuels and $250 million in personal income. So it is evident there is more to development of a biomass energy industry that can compete with conventional fossil fuels than the basic costs of biomass energy.

VI. RESEARCH ON VIRGIN BIOMASS PRODUCTION

A large variety of virgin biomass feedstock developments for the production of energy, biofuels, and chemicals is in the research stage in Canada, the

United States, and many other countries. Research is progressing to develop and select special species and clones of trees and herbaceous crops and to develop advanced growth and management procedures for dedicated energy crops. This work is being done in the laboratory and in the field and is aimed at reducing the cost of biomass and increasing the efficiency of production. Research on short-rotation tree growth methods and the screening of woody and herbaceous biomass continues, generally on small-scale test plots. The North American effort has focused on hybrid poplar, willow, switchgrass, and a few other species. The emphasis in South America is on species such as *Eucalyptus* that grow well in semitropical and tropical climates. Larger scale field trials in which dedicated biomass production is integrated with conversion are beginning to evolve in the United States from the research done with small systems. But most of the continuing research in the United States on the selection of suitable biomass is limited to laboratory studies and small-scale test plots. Many of the research programs on feedstock development were started in the 1970s and early 1980s. Based on the research data accumulated in this work, some of the herbaceous and woody biomass species that appear to be good models for energy feedstock production are shown by region in Fig. 4. 9 for the continental United States and Hawaii (Wright, 1994).

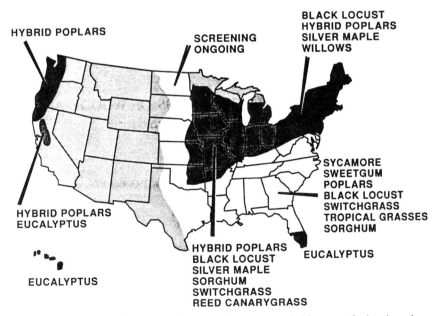

FIGURE 4.9   Woody and herbaceous biomass species recommended for energy feedstock production. From Wright (1994).

A. Herbaceous Biomass

Considerable research has been conducted to screen and select herbaceous plants as potential biomass candidates that are mainly unexplored in the Continental United States. Other research has concentrated on cash crops such as sugarcane and sweet sorghum, and still other research has emphasized tropical grasses. In the late 1970s, a comprehensive screening study of plants grown in the United States generated a list of 280 promising candidates from which up to 20 species were recommended for field experiments in each region of the country (Saterson and Luppold, 1979). The four highest-yielding species recommended for further tests in each region are listed in Table 4.16. Since many of the plants in the original list of 280 species had not been grown for commercial use, the production costs were estimated as shown in Table 4.17 for the various classes of herbaceous species. The results were used in conjunction with yield and other data to develop the recommendations in Table 4.16.

A large number of research projects directed to small-scale field tests of potential herbaceous energy crops have since been carried out. The productivity ranges for some of the promising species for the U.S. Midwest and Southeast are shown in Table 4.11. The results of this research helped to establish a strategy that herbaceous biomass energy crops should be primarily grasses and legumes produced by use of management systems similar to those used for conventional forage crops. It was concluded that the ideal selection of herbaceous energy crops for these areas would consist of at least one annual species, one warm-season perennial species, one cool-season perennial species, and one legume. Production rates, cost estimates, and environmental considerations indicate that perennial species are preferred to annual species on many sites, but annuals may be more important in crop rotations.

In greenhouse, small-plot, and field-scale research tests conducted to screen tropical grasses as energy crops, three categories emerged, based on the time required to maximize dry-matter yields: short-rotation species (2 to 3 months), intermediate-rotation species (4 to 6 months), and long-rotation species (12 to 18 months) (Alexander, 1991). A sorghum-sudan grass hybrid (Sordan 70A), the forage grass napier grass, and sugarcane were outstanding candidates in these categories. Minimum-tillage grasses that produced moderate yields with little attention were wild *Saccharum* clones and Johnson grass in a fourth category. The maximum yield observed was 61.6 dry t/ha-year for sugarcane propagated at narrow row centers over 12 months. The estimated maximum yield is of the order of 112 dry t/ha-year using new generations of sugarcane and the propagation of ratoon (regrowth) plants for several years after a given crop is planted.

Overall, the research that has been completed in the United States on the development of herbaceous biomass energy crops shows that a wide range of suitable species exist from which good candidates can be chosen for each area.

TABLE 4.16 Reported Maximum Productivities in United States for Recommended Herbaceous Plants[a]

Region[b]	Species	Yield (dry t/ha-year)
Southeastern prairie delta and coast	Kenaf	29.1
	Napier grass	28.5
	Bermuda grass	26.9
	Forage sorghum	26.9
General farm and North Atlantic	Kenaf	18.6
	Sorghum hybrid	18.4
	Bermuda grass	15.9
	Smooth bromegrass	13.9
Central	Forage sorghum	25.6
	Hybrid sorghum	19.1
	Reed canary grass	17.0
	Tall fescue	15.7
Lake states and Northeast	Jerusalem artichoke	32.1
	Sunflower	20.0
	Reed canary grass	13.7
	Common milkweed	12.3
Central and southwestern plains and plateaus	Kenaf	33.0
	Colorado River hemp	25.1
	Switchgrass	22.4
	Sunn hemp	21.3
Northern and western Great Plains	Jerusalem artichoke	32.1
	Sunchoke	28.5
	Sunflower	19.7
	Milkvetch	16.1
Western range	Alfalfa	17.9
	Blue panic grass	17.9
	Cane bluestem	10.8
	Buffalo gourd	10.1
Northwest/Rocky Mountain	Milkvetch	12.1
	Kochia	11.0
	Russian thistle	10.1
	Alfalfa	8.1
California subtropical	Sudan grass	35.9
	Sudan–sorghum hybrid	31.6
	Forage sorghum	28.9
	Alfalfa	19.1

[a]Satterson and Luppold (1979).
[b]As defined by U.S. Dept. of Agriculture (1972); excludes Alaska and Hawaii.

TABLE 4.17   Production Costs for Annual Herbaceous Biomass[a]

Plant group	Model crop used	Whole plant yield (dry t/ha-year)	Cost ($/t)
Tall grasses	Corn	17.3	19.1
Short grasses	Wheat	9.9	17.2
Tall broadleaves	Sunflower	15.0	12.7
Short broadleaves	Sugar beet	13.9	77.1
Legumes	Alfalfa[b]	13.7	20.9
Tubers	Potatoes	9.2	136

[a]Saterson and Luppold (1979).
[b]Perennial.

B. SHORT-ROTATION WOODY CROPS

Research to develop trees as energy crops in the United States via short-rotation intensive culture made significant progress in the 1980s and 1990s. Projections indicate that yields of organic matter can be substantially increased by coppicing techniques and genetic improvements. Advanced designs of whole-tree harvesters, logging residue collection and chipping units, and automated planters for rapid planting have been developed to the point where prototype units have been evaluated in the field and some are being manufactured for commercial use. It is expected that several additional devices will be offered for commercial use. The on going research is also leading to significant changes in forestry harvesting techniques. Clear-cutting is being phased out and partial harvesting or thinning operations are being phased in. New thinning technologies have been proposed for testing in the forests of the Northwest after successful tests in California. The California research data show that the thinning of overgrown stands reduces tree mortality, provides healthier stands, and may offer biomass fuels at a cost that make it possible to operate wood-fueled power plants on a stand-alone basis at a profit in competition with market prices for electric power.

Some of the tree species that have been targeted for continuing research are red alder, black cottonwood, Douglas fir, and ponderosa pine in the Northwest; *Eucalyptus,* mesquite, Chinese tallow, and the leucaena in the West and Southwest; sycamore, Eastern cottonwood, black locust, catalpa, sugar maple, poplar, and conifers in the Midwest; sycamore, sweetgum, European black alder, and loblolly pine in the Southwest; and sycamore, poplar, willow, and sugar maple in the East. Generally, tree growth in research plots is studied in terms of soil type and the requirements for site preparation, planting density, irrigation, fertilization, weed control, disease control, and nutrients. Harvesting methods are equally important, especially in the case of coppice growth for SRWC hardwoods. Although three species native to the region are usually included in the experimental designs, nonnative and hybrid species have often been tested in research plots as well. Advanced biochemical methods and techniques such as tissue culture propagation, genetic transformation, and somaclonal variation are being used in this research to clonally propagate individual geno-types and to regenerate genetically modified species.

After an intensive research effort over about 10 years, SRWC yields in the United States, based on accumulated data are projected to be 9, 9, 11, 17, and 17 dry t/ha-year in the Northeast, South/Southeast, Midwest/Lake, Northwest, and Subtropics, respectively (Wright, 1992). The corresponding research goals are 15, 18, 20, 30, and 30 dry t/ha-year. Hybrid poplar, which grows in many parts of the United States, and *Eucalyptus,* which is limited to Hawaii, Florida, southern Texas, and part of California, have shown the greatest potential thus far for attaining exceptionally fast growth rates. Both have achieved yields in the range of 20 to 43 t/ha-year in experimental trials with selected clones. Continuing research indicates that other promising species are black locust, sycamore, sweetgum, and silver maple.

Research on hybridizing techniques seem to be leading to super trees that have short growth cycles and that yield larger quantities of biomass. Fast-growing clones are being developed for energy farms in which the trees are ready for harvest in as little as 10 years and yield up to 30 m^3/ha-year. Genetic and environmental manipulation has also led to valuable techniques for the fast growth of saplings in artificial light and with controlled atmospheres, humidity, and nutrition. The growth of infant trees in a few months is equiva-lent to what can be obtained in several years by conventional techniques.

Chemical injections into pine trees have been found to have stimulatory effects on the natural production of resins and terpenes and may result in high yields of these valuable chemicals. Combined oleoresin-timber production in mixed stands of pine and timber trees is under development, and it appears that when short-rotation forestry is used, the yields of energy products and timber can be substantially higher than the yields from separate operations.

One of the largest research projects on SRWC in the Western World, LEBEN or the Large European Bioenergy Project, was reported to be scheduled for initiation in the Abruzzo region of Italy in the mid-1980s and to be established near the end of that decade (Grassi, 1987; Klass, 1987). This project integrates SRWC production, the production of herbaceous energy crops and residues, and biomass conversion to biofuels and energy. About 400,000 t/year of biomass, consisting of 260,000 t/year of woody biomass from 700 ha and 120,000 t/year of agricultural residues from 700 ha of vineyards and olive and fruit orchards, will be used. Later, 110,000 t/year of energy crops from 1050 ha will be utilized. The energy products include liquid fuels (biomass-derived oil), charcoal, 200 million kWh/year of electric power, and waste heat for injection into the regional agroforestry and industrial sectors. This project is still in the start-up stages in the mid-1990s.

One of the largest demonstration programs in the United States was started in 1993 in Minnesota where hybrid poplar is grown under short-rotation conditions on a few sites that total 2000 ha. As the results of this program are reported, a much more rigorous analysis of the potential of SWRC for energy will be possible. The ultimate approach to perfecting this technology, however, is to integrate large-scale biomass production with conversion. Little research of this type has been done. The assumptions and projections that have been made to evaluate the technology are based primarily on small-scale laboratory results, what others have reported as research results, or predictions about individual steps that make up the overall system. But this situation is starting to change as government-industry support of integrated biomass production and conversion research make it possible to examine the sustainability of these systems in detail. In the United States, several research projects in which virgin biomass production is integrated with conversion have been selected for field demonstration in plots that are expected to be a minimum of 405 ha (1000 ac) in size (Klass, 1996). This research will provide first-hand experience in operating integrated systems on a sustained basis in which a dedicated biomass feedstock is supplied to a conversion plant. The first group of biomass energy technologies to be scaled up consist of alfalfa production integrated with a gasifier-combined-cycle power plant in Minnesota, switchgrass production integrated with a power plant in Iowa in which biomass and coal are co-fired, hybrid willow production integrated with a power plant in New York in which biomass and coal are co-fired, and an innovative whole tree production system integrated with a power plant in Minnesota (Spaeth and Pierce, 1996). As these projects are implemented, others are expected to be added to the program.

C. Aquatic Biomass

Aquatic biomass, particularly micro- and macroalgae, is more efficient at converting incident solar radiation to chemical energy than most other biomass species. For this reason and the fact that most aquatic plants do not have commercial markets, research was performed in the United States in the late 1970s up to the 1990s to evaluate several species as energy crops. The overall goals of the research were generally directed either to biomass production, often with simultaneous waste treatment, for subsequent conversion to fuels by fermentation, or to species that contain valuable products. The aquatics that have been studied and their main applications are microalgae for liquid fuels, the macrophyte water hyacinth for wastewater treatment and conversion to methane, and marine macroalgae for specialty chemicals and conversion to methane.

Research in the United States on microalgae has focused on the growth of these organisms under conditions that promote lipid formation. This eliminates the high cost of cell harvest because the lipids can sometimes be separated by simple flotation or extraction. The research program supported by the U.S. Department of Energy on microalgae in the 1980s was one of the largest of its kind. It consisted of several projects and emphasized the isolation and characterization of the organisms and the development of microalgae that afford high oil yields. The research included projects on siting studies; collection, screening, and characterization of microalgae; the growth of certain species in laboratory and small-scale production systems; exploration of innovative approaches to microalgae production; and innovative methods for increasing oil formation. Some microalgae, such as *Botryococcus braunii*, have been reported to produce lipid yields that are 40-50% of the dry cell weight under nitrogen-limited conditions (*cf.* Klass, 1985). However, in other research, *B. braunii* has been reported to yield 20-52% of the dry cell weight as liquid hydrocarbons (*cf.* Klass, 1987).

Much of the research on aquatic biomass species as energy resources in the United States has since been terminated or reduced in scope, but it continues in other countries with emphasis on the production of high value products rather than energy.

REFERENCES

Alexander, A. G. (1983). *In* "Energy from Biomass and Wastes VII," (D. L. Klass, ed.), p. 185. Institute of Gas Technology, Chicago.

Alexander, A. G. (1991). *In* "Energy from Biomass and Wastes XIV," (D. L. Klass, ed.), p. 367. Institute of Gas Technology, Chicago.

Alexander, A. G. (1993). *In* "Energy from Biomass and Wastes XVI," (D. L. Klass, ed.), p. 269. Institute of Gas Technology, Chicago.

Benemann, J. R., and Weissman, J. C. (1993). *In* "First Biomass Conference of the Americas," NREL/CP-200-5768, DE93010050, p. 1427. National Renewable Energy Laboratory, Golden, CO, August–September.

Brinkworth, B. J. (1973). "Solar Energy for Man." John Wiley, New York.

Bryce, A. (1978). *In* "Energy from Biomass and Wastes," (W. W. Waterman and D. L. Klass, eds.), p. 353. Institute of Gas Technology, Chicago.

Buchannan, R. A., and Otey, F. O. (1978). Paper presented at 19th Annual Meeting, Society for Economic Botany, St. Louis, MO, June 11–14.

Cherney, J. H., Lowenberg-DeBoer, J., Johnson, K. D., and Volenec, J. J. (1989). *In* "Energy from Biomass and Wastes XII," (D. L. Klass, ed.), p. 289. Institute of Gas Technology, Chicago.

Clemmer, S., and Weichert, D. (1994). "The Economic Impacts of Renewable Energy Use in Wisconsin," Department of Administration, Division of Energy and Intergovernmental Affairs. Wisconsin Energy Bureau, Madison, WI.

Crutchfield, H. J. (1974). *In* "General Climatology," 3rd Ed., p. 22. Prentice-Hall, Englewood Cliffs, NJ.

Cushman, J. H., and Turhollow, A. F. (1991). *In* "Energy from Biomass and Wastes XIV," (D. L. Klass, ed.), p. 465. Institute of Gas Technology, Chicago.

DeBell, D. S., and Clendenen, G. W. (1991). *In* "Energy from Biomass and Wastes XV," (D. L. Klass, ed.), p. 411. Institute of Gas Technology, Chicago.

DeLong, M. M., Swanberg, D. R., Oelke, E. A., Hanson, C., Onischak, M., Schmid, M. R., and Wiant, B. C. (1995). *In* "Second Biomass Conference of the Americas," NREL/CP-200-8098, DE95009230, p. 1582. National Renewable Energy Laboratory, Golden, CO, August.

Del Fosse, E. S. (1977). *In* "Clean Fuels from Biomass and Wastes," (F. Ekman, ed.), p.73. Institute of Gas Technology, Chicago.

Dula, J. C. (1984). *In* "Energy from Biomass and Wastes VIII," (D. L. Klass and H. H. Elliott, eds.), p. 193. Institute of Gas Technology, Chicago.

El-Sharkawy, M., and Hesketh, J.D. (1964). *Crop Sci.* **4**, 514.

Evans, H. J., and Barber, L. E. (1977). *Science* **197**, 332.

Ferrell, J., Tardif, M-L., Couto, L., Garca, L. R., Betters, D., and Ashworth, J. (1993). *In* "First Biomass Conference of the Americas," NREL/CP-200-5768, DE93010050, p. 162. National Renewable Energy Laboratory, Golden, CO, August.

Ferrell, J. E., Wright, L. L., and Tuskan, G. A. (1995). *In* "Second Biomass Conference of the Americas," NREL/CP-200-8098, DE95009230, p. 197. National Renewable Energy Laboratory, Golden, CO, August.

Fraser, M. D. (1993). *In* "Energy from Biomass and Wastes XVI," (D.L. Klass, ed.), p. 295. Institute of Gas Technology, Chicago.

Gavett, E. E., Van Dyne, D., and Blase, M. (1993). *In* "Energy from Biomass and Wastes XVI," (D. L. Klass, ed.), p. 709. Institute of Gas Technology, Chicago.

Grassi, G. (1987). *In* "Energy from Biomass and Wastes X," (D.L. Klass, ed.), p. 1545. Institute of Gas Technology, Chicago.

Hansen, E. A. (1988). *In* "Economic Evaluations of Short-Rotation Biomass Energy Systems," p. 197. International Energy Agency, Duluth, MN.

Hohenstein, W. G., and Wright, L. L. (1994). *Biomass & Bioenergy* **6** (3), 161.

InterTechnology Corp. (1975). "Solar SNG, Final Report," AGA Project IU-114-1. American Gas Association, Washington, D.C., October.

Klass, D. L. (1974). *Chemtech* **4**, 161.

Klass, D. L. (1985). *In* "Energy from Biomass and Wastes IX," (D. L. Klass, ed.), p. 34. Institute of Gas Technology, Chicago.

Klass, D. L. (1987). *In* "Energy from Biomass and Wastes X," (D. L. Klass, ed.), p. 13. Institute of Gas Technology, Chicago.

Klass, D. L. (1995). *Energy Policy* **23** (12), 1035.

Klass, D. L. (1996). "Statement Submitted to the House Committee on Appropriations, Subcommittee on Energy and Water Development," Washington, D.C., February 29, 5 pp.

Krummel, J. (1976). *In* "Clean Fuels from Biomass, Sewage, Urban Refuse, Agricultural Wastes," (F. Ekman, ed.), p. 359. Institute of Gas Technology, Chicago.

LaPointe, B. E., and Hanisak, M. D. (1985). *In* "Energy from Biomass and Wastes IX," (D. L. Klass, ed.), p. 111. Institute of Gas Technology, Chicago.

Little, E.L., Jr. (1980). "Common Fuelwood Crops." Communi-Tech Associates, Morgantown, WV.

Ludlow, M. M., and Wilson, G. L. (1970). *J. Australian Inst. of Ag. Science* **43**, March.

Mariani, E. O. (1978). *In* "Energy from Biomass and Wastes," (D. L. Klass and W. W. Waterman, eds.), p. 29. Institute of Gas Technology, Chicago.

McGarry, M. G. (1971). *Process Biochem.* **6**, 50.

McLaughlin, S. P., Kingsolver, B. E., and Hoffmann, J. J. (1983). *Economic Botany* **37** (2), 150.

National Academy of Sciences (1980). "Firewood Crops, Shrub and Tree Species for Energy Production," Washington, D.C.

Nemethy, E. K., Otvos, J. W., and Calvin, M. (1981). In "Fuels from Biomass and Wastes," (D. L. Klass and G. H. Emert, eds.), p. 405. Ann Arbor Science Publishers, Ann Arbor, MI.

Nichiporovich, A. A. (1967). "Photosynthesis of Productive Systems." Israel Program for Scientific Translations, Jerusalem.

North, W. J. (1971). "The Biology of Giant Kelp Beds (Macrocystis) in California." J. Cramer, Lehrte, Germany.

North, W. J., Gerard, V. A., and Kuwabara, J. S. (1981). In "Biomass as a Nonfossil Fuel Source," ACS Symposium Series 144 (D. L. Klass, ed.), p. 77. American Chemical Society, Washington, D.C.

O'Hair, S. K. (1982). In "Energy from Biomass and Wastes VI," (D. L. Klass, ed.), p. 135. Institute of Gas Technology, Chicago.

Pratt, D. C., and Andrews, N. J. (1980). In "Energy from Biomass and Wastes IV," (D. L. Klass and J. W. Weatherly III, eds.), p. 43. Institute of Gas Technology, Chicago.

Ranney, J.W., Wright, L.L., and Layton, P.A. (1987). J. Forestry 85 (9), 17.

Retovsky, R. (1966). "Theoretical and Methodological Bases of Continuous Culture of Microorganisms," (I. Malch and Z. Fenel, eds.). Academic Press, Inc., New York.

Roller, W. L., Keener, H. M., Kline, R. D., Mederski, H. J., and Curry, R. B. (1975). "Grown Organic Matter as a Fuel Raw Material Resource," NASA Report CR-2608, Contract NGL-36-007-001, Ohio Agricultural Research and Development Center. National Aeronautics and Space Administration, Washington, D.C., October.

Ryther, J. H., and DeBusk, T. A. (1982). In "Energy from Biomass and Wastes VI," (D. L. Klass ed.), p. 221. Institute of Gas Technology, Chicago.

Sachs, R. M., Gilpin, D. W., and Mock, T. (1980). In "Energy from Biomass and Wastes IV," (D. L. Klass and W. W. Waterman, eds.), p. 107. Institute of Gas Technology, Chicago.

Sajdak, R. L., Lai, Y. Z., Mroz, G. D., and Jurgensen, M. F. (1981). In "Biomass as a Nonfossil Fuel Source," ACS Symposium Series 144 (D.L. Klass, ed.), p. 21. American Chemical Society, Washington, D.C.

Sanderson, M. A., Hussey, M. A., Ocumpaugh, W. R., Tischler, C. R., Read, J. C., and Reed, R. L. (1995). In "Second Biomass Conference of the Americas," NREL/CP-200-8098, DE95009230, p. 253. National Renewable Energy Laboratory, Golden, CO, August.

Saterson, K. A., and Luppold, M. W. (1979). In "3rd Annual Biomass Energy Systems Conference Proceedings," SERI/TP-33-285, p. 245. Solar Energy Research Institute, Golden, CO, June 5–7.

Schneider, T. R. (1973). Energy Convers. 13, 77.

Shultz, E. B., and Bragg, W. G. (1995). In "Second Biomass Conference of the Americas," NREL/CP-200-8098, DE95009230, p. 313. National Renewable Energy Laboratory, Golden, CO, August.

Sinclair, T. R., Tanner, C. B., and Bennett, J. M. (1984). BioScience 34(1), 36.

Spaeth, J. J., and Pierce, L. K. (1996). In "Bioenergy '96, Proceedings of the Seventh National Bioenergy Conference," Vol. I, p. 52. Tennessee Valley Authority, The Southeastern Regional Biomass Energy Program, Muscle Shoals, AL.

Spurr, S. H. (1979). Sci. Am. 240, 76.

Stauffer, M. D., Chubey, B. B., and Dorrell, D. G. (1981). In "Fuels from Biomass and Wastes," (D. L. Klass and G. H. Emert, eds.), p. 79. Ann Arbor Science Publishers, Ann Arbor, MI.

Stokes, B., and Hartsough, B. (1994). In "Bioenergy '94, Proceedings of the Sixth National Bioenergy Conference," Vol. I, p. 309. The Western Regional Biomass Energy Program, Golden, CO.

Taylor, C. S. (1993). In "First Biomass Conference of the Americas," NREL/CP-200-5768, DE93010050, p. 1472. National Renewable Energy Laboratory, Golden, CO, August–September.

University of Illinois Urbana-Champaign FaRM Lab (1995-1996). In "Continuous Corn—No Till, Northern/Central Illinois, Grain 3," and "Hay: Maintain and Harvest, Northern/Central Illinois,

Cover 2," Revised November 11, 1995. Illinois Cooperative Extension Service, Urbana-Champaign, IL.

U.S. Dept. of Agriculture (1948). "Grass The Yearbook of Agriculture 1948." U.S. Government Printing Office, Washington, D.C.

U.S. Dept. of Agriculture (1966). "Land Capability Classification, Agricultural Handbook 210." Soil Conservation Service, Washington, D.C.

U.S. Dept. of Agriculture (1989). "Summary Report 1987 National Resources Inventory," No. 790. Soil Conservation Service, Washington, D.C.

U.S. Dept. of Commerce (1970). "Climatological Data, National Summary," Vol. 21, Nos. 1–12. U.S. Government Printing Office, Washington, D.C.

U.S. Dept. of Commerce (1995). "Statistical Abstract of the United States," U.S. Government Printing Office, Washington, D.C.

U.S. Dept. of Commerce (1996). "Statistical Abstract of the United States." U.S. Government Printing Office, Washington, D.C.

U.S. Dept. of Energy (1995). In "Technology Partnerships," DOE/GO-10095-170, DE95004086, April, p. 58. Office of Energy Efficiency and Renewable Energy, Washington, D.C.

U.S. Energy Information Administration (July 1995). "Annual Energy Review 1994," DOE/EIA-0384(94). U.S. Dept. of Energy, Washington, D.C.

U.S. Energy Information Administration (December 1995). "State Energy Price and Expenditure Report 1993," DOE/EIA 0376(93). U.S. Dept. of Energy, Washington, D.C.

Visher, S. S. (1954). "Climatic Atlas of the United States." Harvard University Press, Cambridge, MA.

Westlake, D. F. (1963). Biol. Rev. 38, 3385.

Wright, L. L., DeBell, D. S., Strauss, C. H., and Geyer, W. A. (1989). In "Energy from Biomass and Wastes XII," (D.L. Klass, ed.), p. 261. Institute of Gas Technology, Chicago.

Wright, L. L. (1992). In "Biomass Energy Production in the United States: Situation and Outlook," (L. L. Wright and W. G. Hohenstein, eds.), Chapt. 2. Oak Ridge National Laboratory, Oak Ridge, TN.

Wright, L. L. (1994). Biomass and Bioenergy 6 (3), 191.

Yount, J. L., and Grossman, R. A. (1970). J. Water Pollut. Control Fed. 42, 173.

CHAPTER 5

Waste Biomass Resource Abundance, Energy Potential, and Availability

I. INTRODUCTION

Up to the mid-1990s, only a few commercial virgin biomass energy systems in which dedicated biomass is grown for use as an energy resource were in operation in industrialized countries. The technology is available or under development and is slowly being incorporated into regional, national, and world energy markets. More rapid deployment awaits the inevitable effects on renewable energy usage of fossil fuel depletion and environmental issues. Most of the contribution of biomass to primary energy demand in the 1990s comes from waste biomass. Waste biomass is energy-containing materials that are discarded or disposed of and that are mainly derived from or have their origin in virgin biomass. They are lower in cost than virgin biomass and often have negative costs. Some are quite abundant, and some can be disposed of in a manner that provides economic benefits to reduce disposal costs. Waste biomass is generated by anthropological activities and some natural events. It includes municipal solid waste (urban refuse); municipal biosolids (sewage); wood wastes and related residues produced in the forests and logging and forestry operations; agricultural wastes such as animal manures and crop residues produced in farming, ranching, and related operations; and the wastes produced by certain industries such as the pulp and paper industry and those involved with processing foodstuffs. In this chapter, the production of these wastes and their energy potential and availabilities are addressed. The United States is used as the model country because U.S. data are available on most waste biomass to illustrate its role in the development of biomass energy on a national basis. But the conclusions regarding waste biomass as an energy resource in the United States are generally similar for other industrialized countries.

II. MUNICIPAL WASTES

There are basically two types of municipal waste that offer opportunities for combined waste disposal and energy recovery—municipal solid waste (MSW, urban refuse, garbage) and biosolids (sewage, sludge). Each has its own distinctive set of characteristics as a biomass energy resource.

A. MUNICIPAL SOLID WASTE

Abundance

As the populations of urban areas grow, the production of MSW increases, sometimes in a disproportionate way. To illustrate, the generation of MSW in the United States increased from about 80 million tonnes in 1960 to 180 million tonnes in 1990 and shows no sign of reaching a plateau. During this same period, the corresponding per-capita generation of MSW in 10-year increments was 1.23 kg/person-day in 1960, 1.49 in 1970, 1.65 in 1980, and 1.97 in 1990. The associated difficulties of MSW disposal have become serious problems that do not bode well for future generations of city dwellers and areas that have high population densities. Governments often mandate the use of more environmentally acceptable methods of MSW disposal while limiting and sometimes phasing out some of the more traditional disposal methods. The collection and disposal costs increase and proper disposal becomes more difficult to achieve with the passage of time. The average "tipping fees" of MSW in the United States, for example, increased from about $11 per tonne in 1982 to about $32 per tonne in 1992. In some highly populated areas, the tipping fee is over $90 per tonne. At the same time, the loss of natural resources in the MSW occurs if no effort is made to recover them. The opportunities for combined waste disposal and energy recovery are evident.

Table 5.1 is a detailed summary of relevant data on MSW generation, disposal, and recovery in the United States from 1960 to 1993. Several conclusions can be reached from examination of this data. In the 1960s and 1970s, combined disposal-energy recovery systems did not exist to any significant extent even though 20 to 30% of the MSW generated was disposed of by burning. No effort was made to recover the heat evolved on combustion of the MSW. Since then, energy recovery systems have been incorporated into some of the disposal processes so that by the mid-1990s, about 15% of the MSW generated and disposed of by combustion includes energy recovery operations. Throughout this period, the bulk of MSW continued to be disposed of by landfilling. This process will be discussed in some detail in later chapters, but suffice it to say at this point that a medium-energy fuel gas containing about 50 mol % methane is emitted by MSW landfills. The recovery of this gas over long periods of time from many landfills is a well-established commercial technology.

It is apparent from the compositional data on the raw MSW in Table 5.1 that the combustible materials make up the bulk of the MSW on a weight percentage basis, about 85 wt % in the 1990s. The amount of the individual MSW components recovered since 1960, presumably for sale of marketable components, has increased to about one-fifth of the total amount generated. The largest components by weight in the recovered material include paper and paperboard and noncombustible metals and glass. Much of this material is recycled.

Availability

MSW is collected for disposal by urban communities in all industrialized countries, so there is no question regarding its physical availability as a waste biomass feedstock in centralized locations in these countries. The question is how best to utilize this material if it is regarded as an "urban ore" rather than an urban waste. The data in Table 5.1 show that in the mid-1990s, a large portion of the MSW generated in the United States was available as feedstock for additional energy recovery processing. As indicated above, landfilled MSW can provide energy as fuel gas for heat, steam, and electric power production over long time periods. Surface-processing of MSW can also provide energy for the same end uses when MSW is used as a fuel or a feedstock.

Energy Potential

At a higher heating value of 12.7 MJ/dry kg of MSW (Table 3.3), the energy potentially available from the MSW generated in the United States in the 1990s is in the range of 2.5 EJ/year. Presuming the total combustibles in the recovered

TABLE 5.1 Municipal Solid Waste Generation, Recovery, and Disposal in United States, 1960–1993[a]

Parameter and units	1960	1970	1980	1990	1993
MSW generated, recovered, and disposed of:					
Total generated, 10^6 t	79.6	110.6	137.4	179.6	187.7
Total generated, kg/person-day	1.2	1.5	1.7	2.0	2.0
Recovered, 10^6 t	5.4	7.8	13.2	29.8	40.8
Recovered, % of total generated	6.7	7.1	9.6	16.6	21.7
Combustion with energy recovery, % of total generated	(NA)	0.4	1.7	15.0	15.1
Combustion without energy recovery, % of total generated	30.8	20.3	7.3	1.1	0.8
Disposal by landfilling or other method, % of total generated	62.3	72.3	81.4	67.3	62.3
Distribution of components generated, % of generation:					
Paper and paperboard	34.1	36.3	36.1	36.7	37.6
Plastics	0.5	2.5	5.2	8.5	9.3
Yard wastes	22.8	19.0	18.2	17.7	15.9
Wood wastes	3.4	3.3	4.4	6.2	6.6
Food wastes	13.9	10.5	8.7	6.7	6.7
Rubber and leather	2.3	2.6	2.8	3.0	3.0
Textiles	1.9	1.6	1.7	3.3	3.0
Ferrous metals	11.3	10.3	7.6	6.2	6.2
Aluminum	0.5	0.6	1.2	1.4	1.4
Other nonferrous metals	0.2	0.5	0.7	0.6	0.6
Glass	7.7	10.4	9.9	6.7	6.6
Miscellaneous	1.6	2.2	3.4	3.1	3.1
Components recovered, % of generation:					
Paper and paperboard	18.1	16.7	21.8	27.9	34.0
Plastics				2.2	3.5
Yard wastes				12.0	19.8
Other wastes	1.5	2.4	1.9	4.5	6.4
Ferrous metals	1.0	0.8	3.4	13.7	26.1
Aluminum			16.7	35.3	35.4
Other nonferrous metals		42.9	45.5	66.4	62.9
Glass	1.5	1.6	5.3	20.0	22.0
Distribution of components recovered, % of recovered:					
Paper and paperboard	91.5	86.0	82.1	61.7	58.9
Plastics				1.2	1.6
Yard wastes				12.8	14.4
Other wastes	5.1	7.0	4.1	6.1	6.7
Ferrous metals	1.7	1.2	2.8	5.2	7.6
Aluminum			2.1	3.0	2.4
Other nonferrous metals		3.5	3.4	2.1	1.8
Glass	1.7	2.3	5.5	7.9	6.7

[a]Adapted from Franklin Associates (1994). Sums of individual figures may not equal totals because of rounding. The data in this table are for postconsumer residential and commercial MSW, which makes up the major portion of typical collections. Excludes mining, agricultural, and industrial processing, demolition and construction wastes, municipal biosolids, and junked autos and equipment wastes. Based on material-flows estimating procedure and wet weight as generated. Other wastes are predominantly foodstuffs, leather, rubber, textiles, and wood.

fractions of MSW are utilizable and that the energy recovery systems in operation continue to be used, the data for the United States indicate that about 60 to 65% of the MSW generated could have supplied up to an additional 1.6 EJ/year in the mid-1990s. New energy recovery plants supplied with MSW feedstock could also provide an additional benefit by increasing the life of landfills. Only the unrecyclable inorganic materials in the ash would be landfilled if thermal processing of the MSW is employed. Some of the ash itself could be used as material of construction, such as in roadbeds and other applications.

B. BIOSOLIDS

Abundance

Municipal wastewater treatment plants in industrialized countries receive wastewaters from residential sources, industry, groundwater infiltration, and stormwater runoff. The pollutants associated with these sources include a wide range of suspended and dissolved compounds and oxygen-demanding materials, many of which are toxic. Pathogenic components are present, including certain bacteria, viruses, organic compounds, inorganic nutrients, and heavy metals. The purpose of most wastewater treatment processes is to remove or reduce these components, other pollutants, and biological oxygen demand before discharge to receiving waters. In the 1970s and 1980s, about 70 to 75% of the U.S. population was served by wastewater treatment facilities (U.S. Environmental Protection Agency, 1985; 1990). In 1992, more than 20,000 treatment and collection facilities served 180.6 million people or 71% of the population (U.S. Environmental Protection Agency, 1993). Of the 20,000 facilities, 15,613 provided treatment; the design capacity was 149 billion L/day. The need for new wastewater treatment capacity is expected to increase with growth of the sewered population. The need to treat 172 billion L/day is projected for 2012.

Primary biosolids (settleable and suspended solids) are present at a level of a few percent in the influent wastewater and are produced at a rate of about 0.091 dry kg/person-day (0.20 dry lb/person-day). Per million population, this corresponds to the production of 33,200 t/year of primary biosolids. After conventional primary and secondary wastewater treatment, the digested, dewatered biosolids are reduced to the equivalent of about 0.063 dry kg/person-day (0.14 dry lb/person-day), or 23,000 dry t/year per million population. For the United States in 1995, primary and treated biosolids production were about 8.6 and 5.9 million dry tonnes.

About 40 to 45% of the treated biosolids are disposed of in municipal landfills, 30% is applied to land or distributed or marketed as fertilizer, 20%

is incinerated, and the remainder is disposed of in dedicated landfills or by a few other methods.

Availability

Again, there is no question of the physical availability of biosolids. They are collected in municipal wastewater systems and are therefore available in centralized locations. But in this case, treatment is essential for health reasons and protection of the public. Unless processes exist that can be used to treat and stabilize the waste and at the same time recover energy, it does not make much sense to use untreated biosolids as a waste biomass feedstock. In fact, such processes exist and will be discussed in some detail in later chapters. The other option to consider is the utilization of treated biosolids as a waste biomass.

Energy Potential

The energy potential of municipal biosolids is small. At an average higher heating value of 19 MJ/dry kg (Table 3.3), the energy content of all the primary and treated biosolids produced in 1995 in the United States can be estimated to be 0.163 and 0.113 EJ/year, both of which are much less than the energy potential of MSW. This relationship is a permanent one because of the nature of MSW and biosolids generation. Nevertheless, several combined treatment-energy recovery processes alluded to earlier for extracting energy from biosolids are in commercial use in many wastewater treatment facilities. The recovered energy is generally utilized on site as a captive fuel for the facility.

III. AGRICULTURAL SOLID WASTES

The U.S. Department of Agriculture's 1938 Yearbook of Agriculture contains these statements: "One billion tons of manure, the annual product of livestock on American farms, is capable of producing $3,000,000,000 worth of increase in crops. The potential value of this agricultural resource is three times that of the nation's wheat crop and equivalent to $440 for each of the country's 6,800,000 farm operators." Since then, animal wastes have been transformed from a definite asset to a liability. By 1965, the disposal of animal excreta had become a serious problem (American Chemical Society, 1969). At any given time, an estimated 11 million cattle were on feedlots, the capacities of which ranged from 1000 to 50,000 head.

The problem has become much more severe today. Application of animal wastes to land is one of the most economical choices for disposal as well as providing fertilizing benefits. However, the utilization of livestock and poultry manures as waste biomass resources for energy applications could help mitigate pollution and at the same time open new markets. This possibility is examined here. Agricultural crop residues are included in the assessment.

A. LIVESTOCK AND POULTRY MANURES

Abundance

Intuitively, high populations of specific animals would be expected to offer the greatest opportunity to serve as sources of waste biomass because waste generation is maximized. Because of the relationship of waste productivity and animal size, this is not always the case as will be shown here. Domestic farm animals and those confined to feedlots are appropriate choices. In addition, commercial poultry production systems, some of which have bird populations over 200,000, would be expected to provide large accumulations of manures in one location. The animals that produce large, localized quantities of excreta are cattle, hogs and pigs, sheep and lambs, and poultry. U.S. populations of these animals in the mid-1990s, the estimated total, annual manure production for each species, and the human population equivalents in terms of solid waste generation are shown in Table 5.2. Several observations can be derived from

TABLE 5.2 Livestock and Poultry Manures Generated in the United States and Their Human Population Equivalent[a]

| Livestock/Poultry | Population (10⁶) | Manure production | | Human population equivalent | |
		(dry kg/head-day)	(10⁶ dry t/year)	Factor	(10⁶)
Cattle	103.3	4.64	174.9	16.4	1694
Hogs and pigs	59.6	0.564	12.3	1.90	113
Sheep and lambs	8.9	0.756	2.5	2.45	22
Chickens	377.5	0.0252	3.5	0.14	53
Commercial broilers	7018	0.0403	103.2	0.14	983
Turkeys	289	0.101	10.7	0.14	40

[a]U.S. Dept. of Agriculture (1995) for population data. Populations of cattle, hogs and pigs, and sheep and lambs are for 1995; remaining populations are for 1994. With the exception of the commercial broiler population, other populations are assumed to be steady-state values because the variations are relatively small for each of the preceding 10 years. Commercial broiler production was approximately 20% higher in 1995 than in 1990, and 57% higher than in 1985. Daily manure production factors on a dry basis include ash and were calculated from Stanford Research Institute (1976). The factors for converting animal populations to human population equivalents in terms of waste generation are from Wadleigh (1968).

these data assuming that they represent reasonably steady-state conditions. With the exception of the commercial broiler population, the animal populations are assumed to be steady-state values because the variations are relatively small for each of the preceding 10 years. Commercial broiler production was approximately 20% higher in 1995 than in 1990, and 57% higher than in 1985. But it is assumed that the population is relatively constant for purposes of this assessment because the population increases each year. This should tend to eliminate some of the fluctuations in manure production throughout the year because of animal growth and marketing cycles.

Commercial broilers had the highest population, about 7 billion, and they produced the second largest amount of manure; cattle, which had a population of about 100 million, produced the largest amount. No differentiation was made in this assessment between dairy and beef cattle. The daily production rate per head of dry manure solids used was the arithmetic average of dairy cattle and beef cattle, since the dairy-to-beef cattle ratio of the reported manure production rates is about 1.45 (Stanford Research Institute, 1976). Some assessments indicate that the ratio of productivities is 3.1 (Jaycor, 1990). The effects of daily manure production per animal are evident as shown in Table 5.2. When published waste production factors are used for conversion of the animal populations to human equivalents in terms of solid waste generation as shown in the table, the total human population equivalent of these animals is estimated at almost 3 billion people. This is approximately 12 times the U.S. population in the mid-1990s, a ratio considerably less than reported in 1968, when it was estimated to be 20 times that of the human population (American Chemical Society, 1969). But the latter ratio included liquid wastes as well. The ratio of total animal excreta to total municipal biosolids generation calculated here, both of which exclude liquid wastes, is about 36 on a mass basis (307.1 dry t/8.6 million dry t).

Energy Potential

The energy potential of each category of animal excreta was estimated by adjusting the measured heating values of air-dried samples including ash (Stanford Research Institute, 1976) to dry waste. The adjusted heating values were used to calculate the energy potentials shown in Table 5.3. Cattle and commercial broilers are the two largest energy producers, and the total energy potential of all animal excreta is about 4.6 EJ/year, an amount 28 times that of the energy content estimated here for primary biosolids production in the United States.

Availability

The availability of animal excreta for use as waste biomass is difficult to estimate without a detailed inventory of how much is collected, dropped in the field

TABLE 5.3   Energy Potential of Livestock and Poultry Manures Generated in United States[a]

Livestock/Poultry	Production (10^6 dry t/year)	Heating value (MJ/dry kg)	Energy potential (EJ/year)
Cattle	174.9	15.73	2.751
Hogs and pigs	12.3	16.99	0.209
Sheep and lambs	2.5	17.82	0.0446
Chickens	3.5	13.53	0.0474
Commercial broilers	103.2	13.53	1.396
Turkeys	10.7	13.49	0.144
Total			4.592

[a]Production data are from Table 5.2. Heating values are from Stanford Research Institute (1976).

by the animal, collected and returned to the soil as fertilizer, or used for other purposes. Seasonality is not usually a factor with animal excreta production because as long as the populations do not undergo major changes, production is continuous and fluctuations in supply are generally small. On dairy cattle farms in the Midwest, many farmers use as much of the collected manures as possible as fertilizer to minimize chemical purchases. A few large cattle feedlot operators in the Southwest collected the manure in the 1970s and early 1980s for use as biomass feedstock to produce fuel gas by biological gasification. Hog and poultry manures were also used as biomass feedstocks, but for smaller systems. Several dairy cattle, hog, and poultry farm operations continue to produce fuel gas in the United States from collected manures (Lusk, 1994). Much larger usage of manure gasification technology occurs in such countries as India and China, where blends of biosolids and animal wastes are employed as feedstocks in both small- and large-scale conversion systems. This technology is discussed in later chapters.

One assessment carried out on a county-by-county basis for all 3069 counties of the contiguous United States in the 1970s found that about 70% of animal manures is returned to the soil, 14% is sold, and the remainder is wasted (Stanford Research Institute, 1976). In a similar analysis of the four states Arizona, Colorado, New Mexico, and Utah, it was found that 30 to 40% of animal manures was used by the feeder or farmer for land applications or dried for sale or reuse as animal feed; the remainder was either sold or bartered with corn silage farmers who return silage to the feedlot operator (Burford and Varani, 1976). The majority of cattle manure production was found on only a few cattle feedlots. In Colorado, eight locations containing at least 50,000 head of cattle feeding capacity each were within a 24-km radius of each other and represented 77% of all the state's feeding cattle. Five locations were found in Arizona which contained 78% of the state's cattle feeding capacity, and three locations were found in New Mexico which had 65% of the state's cattle feeding capacity. In an analysis of 13 states in the Southeast, the major producers of animal excreta were Missouri (17.3%), Virginia (13.2%), and North Carolina (11.5%), whereas dairy cattle accounted for 65% of the total animal excreta produced in confined areas (Jaycor, 1990). The cattle feedlot industry is obviously small in the Southeast.

The data presented in Table 5.3 indicate that the most promising locations to consider for the acquisition of animal wastes for conversion to energy and fuels are large cattle feedlots and commercial broiler farms. Availability should not be an insurmountable problem in those cases where waste biomass feedstock applications can compete with alternative uses. However, the confinement of the animals and collectability of the wastes are important factors. The assessment of the 13-state area in the Southeast (Jaycor, 1990) indicated the percentages of the animals confined and the collectible wastes for dairy cows, beef cattle, pigs and hogs, and chickens vary over a wide range:

Dairy cows	50% (confined) and 85% (collectible)
Beef cattle	10% and 65%
Pigs-hogs	85% and 100%
Chickens	100% and 80%

The corresponding percentages of the collectible excreta in a confined area are then 42.5% for dairy cows, 6.5% for beef cattle, 85% for pigs and hogs, and 80% for chickens. The factor of 6.5% for beef cattle clearly does not hold for the Southwest.

In any case, the information and data discussed here show that although the amounts of certain animal wastes are substantial and represent a large energy potential, careful assessment of availabilities is necessary to develop strategies for commercial development of animal excreta as waste biomass feedstocks.

B. Agricultural Crop Residues

Abundance

Agricultural crop residues are those left in the field or accumulated during sorting and cleaning of produce. Because of the discontinuity in growing seasons, the many crops that are grown, the differences between specific crops, variations in crop yields in different areas, the difficulty of acquiring reliable data, and long-term time effects, an inventory of the annual production of agricultural crop residues and their disposition might seem to be an impossible task. Fairly reliable data can be obtained, however, for small and large regions of a country. Somewhat detailed commentary is justified for one of these studies because it is a good example of how the task was addressed in a reasonably scientific manner for all 3069 counties of the continental United States (Stanford Research Institute, 1976). Although the assessment was done some time ago, the methods used still appear to be valid.

Hay and forage crops were excluded, since little residue accumulates from cultivation of these crops. Food processing wastes were excluded except for bagasse and sugarbeet pulp. With the exception of hay and forage crops, yields on a dry basis of the harvested crops and the areas harvested were tabulated for essentially all other cash crops, about 60, over a 3-year period, 1971 to 1973, and averaged for each county by year and by quarter. At the same time, data were collected regarding what was done with the residue: returned to soil, sold, used as feed or fuel, and wasted. To estimate the quantity of residue generated, a residue factor for each cash crop was developed that, when multiplied by the county yield total for that crop, gave the total mass of residue generated. The residue factors as used in the Stanford assessment were the ratio of field weight of residue per mass unit of crop yield and differed somewhat from those used by others in subsequent work (Table 5.4). It is important to

TABLE 5.4 Comparison of Agricultural Grain
Residue Factors

Crop	Residue factor	
	SRI[a]	Heid[b]
Barley	2.50	1.5
Corn (<95 bu/ac)	1.10	1.0
Corn (>95 bu/ac)	1.10	1.5
Cotton	2.45	1.5[c]
Oats	3.01	1.4
Rice	1.43	1.5
Rye	2.50	1.5
Sorghum	1.57	1.5
Soybeans	2.14	1.5
Wheat, spring	2.53	1.3
Wheat, winter	2.53	1.7

[a]From Stanford Research Institute (1976). These factors are ratios of the fresh weight of residue to the grain weight at field moisture.
[b]From Heid (1984). These factors are the ratios of the dry weight of the residue to the grain weight at field moisture.
[c]Excludes off-farm ginning wastes.

note these differences. Experimental measurements were made to determine the residue factors in the Stanford assessment by collecting field, packing shed, and mill residues immediately before or following crop harvest, determining fresh and oven-dry residue weights, converting the production figures from whatever the standard units were, such as bushels, to mass units, and calculating the residue factors. For example, for field corn, the residue factor was 1.10 mass units of fresh residue per mass unit of corn yield. The weights of the residues were converted to dry weights later in the calculations. The residue factors as defined in the study were applied nationwide, except for two cases, the assumption being that geographical variation in residue generation in a given crop was accounted for by the geographical variation in the yield of that crop. The two exceptions were cotton gin trash and mint, wherein specific regional variations in residue production were considered, which led to development of separate regional residue factors. It is evident that this type of assessment is not an everyday event. Massive amounts of data are compiled, the easy manipulation of which requires computational methods.

A summary of the results of this assessment of residue generation from agricultural crops for the contiguous United States is presented here to provide an idea of the availability of agricultural residues for the contiguous United States in the mid-1970s. The data indicated a total of 292 million dry tonnes of residue were generated annually; about 252 million dry tonnes were judged to be collectible during normal operations. About 74% was returned to the soil, 19% was used as animal feed, 4% was sold, 3% was used as fuel, and a small amount was wasted. While there is controversy over what fraction of the residue returned to the soil could be utilized as waste biomass feedstock without adverse environmental impacts, it represented the largest portion of the collectible residue. Of the total available crop residues, 48% was estimated to be from small grains and grasses.

The top five states for agricultural residue production in millions of dry tonnes per year were Iowa, 25; Illinois, 24; Oregon, 23; California, 22; and Kansas, 21; the sum of these is 39% of total residue generation. Interestingly, only six counties in the continental United States averaged more than 350.2 dry t/km^2-year (1,000 dry ton/mi.^{2}-year). In units of dry t/km^2-year, these were Lewis County, Idaho, 537; Delaware County, Indiana, 378; Lane County, Oregon, 363; Polk County, Oregon, 357; and Cook County, Georgia, 353. Also, only 55 counties averaged 2.24 dry t/ha-year (1.0 dry ton/ac-year) or more in available agricultural residue.

Energy Potential

To estimate the energy potential of U.S. crop residues, several parameters are necessary. Annual crop production, and the residue, availability, dry weight, and ash factors are needed. The collectible dry residue for a given crop is the product of the annual production, residue factor, avilability factor, and dry weight factor. The energy content of the collectible dry residue is then the product of the amount of residue, its percent organic content, and the heating value of the organic material. This methodology is used here to estimate the energy potential of U.S. crop residues. The residue, dry weight, and ash factors determined in the Stanford assessment are used. The availability factors for agricultural crop residues that can be removed from the field have been esti- mated for 13 states in the Southeast (Jaycor, 1990; Purdue University, 1979). These factors exclude residues left in the field because of collection difficulties and necessary soil protection. Averages of the availabilities for small grains, corn, sorghum, rice, and sugarcane are used here for the 48 contiguous United States because the variations are generally small to none. The availability factors are 0.6 for small grains (9 states had factors of 0.6, 3 were 0.7, and Missouri was 0.38); 0.6 for corn (5 states were 0.6, 5 were 0.7, Kentucky was 0.4, Missouri was 0.42, and Tennessee was 0.5); 0.64 for sorghum (all 11 states rated were estimated to have a factor of either 0.6 or 0.7); 1.0 for rice (each of the 4 states rated had factors of 1.0); and 1.0 for sugarcane (Florida was the only state rated). These factors reflect the assumption that residue can only be considered available for removal if the predicted losses from the Universal Soil Loss Equation in the Corn Belt or the Wind Erosion Equation in the Great Plains are less than or equal to the maximum amount of soil loss that is considered safe for continued long-term maximum productivity of the soil. The other availability factors used here are from the Stanford assessment. The assessment developed here from these factors is presented in Table 5.5 using 1993 and 1994 crop production data (U.S. Dept. of Agriculture, 1995). Only crops that were produced at an annual rate of more than one million tonnes are included in the assessment.

The total collectible crop residues are estimated to be about 257 million dry t/year, and the energy potential is estimated to be about 4.2 EJ/year. Collectible crop residues with an energy potential of at least 0.1 EJ/year in decreasing order were corn, wheat, soybeans, barley, rice, and sorghum. The collectible corn and wheat residues together were estimated to be equivalent to 61% of the total collectible residues generated and the energy potential. The states that produce the largest quantities of these crops are obvious choices for collectible agricultural waste biomass systems. The Corn Belt in the Midwest and a few neighboring states appear to have the potential of providing the largest quantities of collectible agricultural waste biomass.

Availability

Availability as such is a difficult subject to evaluate for agricultural crop residues. The answer to the simple question of whether the crop residues are

TABLE 5.5 Agricultural Crop Production, Collectible Dry Residues, and Their Energy Potential for the Contiguous United States[a]

Crop	Site	Production (10^6 t/year)	Residue factor	Availability factor	Percent dry wt.	Collectible (10^6 dry t/year)	Organic (%)	Energy content (EJ/year)
Barley	Field	8.68	2.50	0.6	91	11.85	85	0.187
Beans, dry	Field	1.32	1.58	0.8	40	0.67	90	0.0112
Cabbage	Field	1.33	0.91	0.8	10	0.10	78	0.00145
Corn	Field	254	1.10	0.6	53	88.85	90	1.487
Corn, sweet	Field	4.34	1.41	0.6	18	0.66	88	0.0108
Cotton	Field	4.29	2.45	0.6	50	3.15	81	0.0475
Grapes	Field	5.45	0.16	0.98	25	0.21	80	0.00312
Lettuce	Field	3.08	0.96	0.8	6	0.14	68	0.00177
Oats	Field	3.00	3.01	0.6	90	4.88	85	0.0772
Onions	Field	2.86	0.30	0.95	18	0.15	95	0.0265
Potatoes	Field	20.8	1.14	0.9	11	2.35	90	0.0393
Rice, rough	Field	8.98	1.43	1.0	80	10.27	78	0.149
Sorghum	Field	16.6	1.57	0.64	40	6.67	84	0.104
Soybeans	Field	69.6	2.14	0.8	40	47.67	85	0.754
Sugarbeets	Mill	23.8	0.07	0.9	93	1.39	80	0.0207
Sugarcane	Field	28.2	0.52	1.0	20	2.93	84	0.0458
Tomatoes	Field	1.61	0.98	0.98	13	0.20	78	0.0145
Tomatoes	Proc.	10.5	0.20	0.98	23	0.47	75	0.0656
Watermelon	Field	17.2	1.51	0.9	25	5.84	77	0.0836
Wheat	Field	63.0	2.53	0.6	72	68.86	86	1.101
Total:						257.31		4.231

[a]Production data are from U.S. Dept. of Agriculture (1995) and are for 1993 or 1994. Only crops with yields greater than 1.0 million tonnes per year are listed. The moisture, dry weight, and ash factors are from Stanford Research Institute (1976). The availabilities are also from Stanford Research Institute except for small grains, corn, sorghum, rice, and sugarcane, which are averages from Jaycor (1990). The energy content of each collectible residue is estimated by assuming the higher heating value of the ash-free residue is 18.6 GJ/dry t (16 million Btu/dry ton).

available in sufficient quantities to justify consideration of collectible agricultural residues as waste biomass is illustrated by the data presented in Table 5.5. The answer, of course, is affirmative. But alternative uses of crop residues can clearly have an adverse impact on the realistic availabilities of waste biomass. Also, waste biomass in the field must be collected, often from dispersed locations, and transported to conversion or end-use facilities. Many of the residues have high moisture contents, so if the residue being collected cannot be solar-dried, the collection and transportation costs include the contained water. The normal end uses of agricultural residues, namely, to provide soil conditioning, erosion control, and nutrient values, are strong competition for fuel and feedstock end uses. A few state governments require that some of the erodible lands be covered during winter months to reduce erosion. In contrast, several available crop residues have been utilized for their energy content for many years. Sugarcane bagasse is perhaps the best example. Bagasse is used as fuel for raising steam and electric power production in many parts of the world and is often considered to be a valuable co-product by the sugarcane industry. For example, sugarcane bagasse provided as much as 10% of the electric power consumed in Hawaii until sugarcane production ceased on the islands of Oahu and Hawaii and was reduced on Maui and Kauai (Phillips *et al.*, 1995). In contrast, only about 10% of the fuel used for biomass power production in the mid-1990s in the United States was agricultural residues; the remaining 90% was wood and wood residues (*cf.* U.S. Dept. of Energy, 1996). The majority of the facilities fueled with agricultural residues are located in California.

The many possibilities for expanding agricultural waste biomass usage in the United States as fuel and feedstock are apparent from the data in Table 5.5. Similar opportunities are available in other countries that have substantial agricultural operations in place.

IV. FORESTRY RESIDUES

Forestry residues consist of slash left on the forest floor following logging operations; stems, stumps, tops, foliage, and damaged trees that are not merchantable; and wood and bark residues accumulated at primary wood manufacturing plants during production of lumber. Underground tree roots can also be included in the list of forestry residues. The difficulty of accurately assessing the amounts of forestry residues that are and can be realistically collected and utilized as waste biomass for an entire country has been encountered by almost all who have embarked on the task. In the United States, many federal, state, and regional forestry offices and many of the companies in the logging and lumber manufacturing businesses do not keep and maintain detailed records of residue production and its disposition. Surveys are done only periodically and they vary widely from state to state. The survey of an entire country for a given time period is therefore subject to considerable error. Nevertheless, such assessments provide valuable information and an overall indication of the energy potential of forestry residues as waste biomass.

Abundance

The national inventory of forestry residues generated in the United States in the mid-1970s is one of the first comprehensive assessments to be done on a county-by-county basis for a large country (Stanford Research Institute, 1976). Data were compiled on residues from logging operations and on wood and bark residues produced at primary wood processing mills. The regional offices of the U.S. Forest Service supplied most of the data on the residues generated by the mills. Most of the data on logging residues was obtained by applying residue factors to industrial roundwood production figures for each county. The factors used for different regions of the country were obtained from published reports of actual logging residue studies conducted in the field. For cases in which specific county data were not provided, it was necessary to apportion multicounty regional data or total state data among the counties having primary wood processing mills. Detailed data on bark residues were sometimes limited. The results of this survey indicated that a total of 105 million dry t/year of forestry residues are generated in the contiguous United States. The percentage distribution of forestry residues by region was 33.2% in the Pacific area, 6.4% in the Mountain area, 13.0% in the West South Central area, 14.2% in the East South Central area, 19.8% in the South Atlantic area, 2.2% in the West North Central area, 5.2% in the East North Central area, and 6.1% in the New England and Mid-Atlantic area. The Southeast and the Northwest are the areas that produce most of the U.S. timber and hence most of the forestry residues. Of the total forestry residues, 35.4 million dry t/year or 33.7% consisted of logging slash, and 69.6 million dry t/year or 66.3% consisted of bark and wood mill residues.

In another inventory of various sources of wood wastes in the United States (McKeever, 1995), it was found that for 1991, 26.0 million dry tonnes of bark and 74.5 million dry tonnes of wood residues were generated at primary lumber processing mills. Only 5% of the bark and 6% of the wood residue were wasted and not used. The projection for 1993 based on these findings was that 5.7 million dry tonnes of bark and wood residues were available for recovery and use as an energy resource. The total of 100.5 million dry t/year generated at the mills is about 44% higher than the mill residues found in the Stanford assessment.

Energy Potential

Assuming the results of the Stanford assessment of logging slash residues in the continental United States would provide approximately the same results today, the energy potential of 35.4 million dry t/year of slash is about 0.66 EJ/year unadjusted for availability. Similarly, 100.5 million dry t/year of mill residue is about 1.86 EJ/year.

Availability

Of the forestry residues collected during normal operations in the United States in the mid-1970s according to the Stanford assessment, most of it was wood and bark residues that accumulated at primary lumber mills. It was reported that of the total forestry residues generated, about 33% was sold, 16% was used as fuel, and 51% was wasted. Most of the wasted residue was generated by logging operations. The fuel usage was represented to be an energy contribution of 0.32 to 0.42 EJ/year to primary energy demand. The lumber and wood products industry has consistently used about 0.35 to 0.50 EJ/year of forestry wastes as fuel for the past few decades (Klass, 1990) and is expected to continue to use them at about the same rate up to at least 2040 (Skog, 1993). If adjustments are made to the total energy potential of U.S. forest residues for the amounts not used but available as a waste biomass resource, most of the logging slash is available (0.66 EJ/year) and 5.7 million dry t/year of available, unused mill residues correspond to about 0.11 EJ/year, or a total from available forestry residues of 0.77 EJ/year. Since the mass of logging residue varies from about 25 to 45% of the wood cut (Howlett and Gamache, 1977), it is highly probable that an energy availability estimate of only 0.66 EJ/year from logging residues is too low. A detailed assessment of the recoverable energy potential of forestry residues in only one state, Georgia, is estimated to be 0.20 EJ/year for logging slash and 0.08 EJ/year for wood manufacturing residues, or a total of 0.28 EJ/year (Riall and Bouffier, 1990). These data suggest that a more realistic estimate of the available energy value from logging slash should be at least 1.0 EJ/year.

V. INDUSTRIAL WASTES

Industry uses more than one-third of all energy consumed in the United States. Sizable amounts of waste are generated, but only the pulp and paper industry generates large quantities of waste biomass. The food processing industry also generates waste biomass. It is not discussed here because of the difficulty of compiling the amounts and disposition of the residues produced by a disaggregated industry composed of many small, medium, and large companies involved in diverse activities. However, the utilization of solid residues and some aqueous wastewaters as a captive energy resource by many of the food processing companies is well known.

A. THE PULP AND PAPER INDUSTRY

Abundance

The U.S. pulp and paper industry consumes about 2.6 EJ (2.5 quad) of energy per year, or about 2.9% of total annual U.S. energy consumption (U.S. Dept. of Energy, 1995). Energy consumption per tonne of paper produced is about 36 GJ or 6 BOE, a disproportionately large amount, especially when considered in terms of the energy content of conventional paper products, about 17.6 GJ/dry t. In the mid-1990s, U.S. paper production capacity was about 30% of the world's total capacity and accounted for 40% of all electric power cogenerated by U.S. manufacturing. Yet the paper industry still spent $5.5 billion on energy in 1991, or about 4.3% of the value of its shipments (U.S. Dept. of Energy, 1995). It is not unexpected, therefore, that the paper industry has made a great effort to become as energy self-sufficient as possible.

Black liquor is a major waste biomass resource and a by-product of the paper industry. In the pulping step of the paper manufacturing process, cellulosic fibers are separated from the debarked, chipped wood. The dominant process used for pulping in the United States is the kraft process, which involves cooking the chips at elevated temperature and pressure with a solution of sodium hydroxide and sodium sulfide. Most of the lignins are dissolved in this process, and the resulting pulp is washed to remove the chemicals before further processing of the pulp into paper. The mixture of dissolved wood components and used pulping chemicals in the extract is called black or spent liquor. It is currently burned in recovery boilers to recover the pulping chemicals and to generate steam. The wood residues serve as boiler fuel and the spent chemicals are the bottoms that are processed for reuse in the pulping step.

In 1993, the U.S. paper industry manufactured about 77.1 million tonnes of paper and paperboard products. The average usable energy yield of black liquor for the industry then corresponds to about 14 GJ/t of paper manufactured (2.6 EJ/[77.1 × 10^6 t]), or about 2.3 BOE/t. This clearly illustrates the need to recover energy from the black liquor, if only to minimize production costs. Black liquor can almost be considered to be a co-product rather than a waste by-product. If boiler efficiencies were included in this calculation, the need to recover and use the energy in the paper manufacturing process would be even more apparent. However, these figures also suggest that there is considerable room for improving the efficiency of the chemical pulping process. The fact remains that approximately one-half of the annual primary energy consumed as biomass in the United States in the 1990s is attributed to the paper industry. Black liquor is available in large quantities and is the mainstay of this consumption pattern.

Energy Potential

The energy potential of black liquor, 1.05 EJ/year, is used commercially and is not potential in the sense that it has not been realized yet. If use of the kraft process increases, production of black liquor and its use as a boiler fuel will also increase.

Availability

The pulp and paper industry is the third largest energy-consuming industry in the United States. It relies heavily on its captive sources of waste biomass—black liquor, bark, and wood residues—for more than one half of its energy needs. In the mid-1990s, black liquor supplied about 40% (1.05 EJ/year), and wood and bark supplied about 16% (0.42 EJ/year) of these needs. Fossil fuels and purchased electricity made up the remainder, 36 and 8%, respectively. Black liquor is a product of the kraft process, so as long as this process is used, black liquor will continue to be available. There appears to be no technology on the horizon for the manufacture of paper that will change this relationship. But research is in progress to improve the efficiency of the pulping process, successful development of which might increase the effective energy potential of black liquor because less energy will then be required per unit of paper manufactured.

VI. FUTURE ROLE OF WASTE BIOMASS AS AN ENERGY RESOURCE

Waste biomass contributes a substantial amount of energy to primary demand in the United States (see Table 2.9). The energy consumption pattern is similar in many other industrialized countries. The energy potentials and availabilities discussed here for the United States are summarized in Table 5.6. It is evident that additional contributions to primary energy demand can be realized. At least 4 EJ/year of incremental energy usage appears to be feasible based on the energy potential and available energy estimates. This is equivalent to displacing about 1.9 million BOE/day in oil consumption. At a market price

TABLE 5.6　Energy Potentials and Availabilities from Waste Biomass in United States

Waste biomass	Energy potential[a] (EJ/year)	Energy availability (EJ/year)	Remarks on availability
Municipal			
Biosolids	0.16	0.16	Reduced on treatment
MSW	2.5	1.6	Excludes landfills
Agricultural			
Cattle	2.75	0.5	Estimate
Other livestock	0.25		
Commercial broilers	1.40	0.7	Estimate
Other poultry	0.19		
Corn	1.49	0.3	Estimate
Soybeans	0.75	0.1	Estimate
Wheat	1.10	0.2	Estimate
Barley	0.19	0.05	Estimate
Rice, rough	0.15	0.05	Estimate
Sorghum	0.10	0.02	Estimate
Other crop residues	0.45	0.15	Estimate
Forestry			
Logging slash	>0.66	>0.2	Minimum
Bark and wood	1.86	0.11	0.42 EJ/year already used
Industrial			
Black liquor	1.05		1.05 EJ/year already used
Total	15.05	4.14	

[a]The energy potentials of agricultural crops have been adjusted to take adverse soil impacts into account (Table 5.5).

of $20/bbl, this corresponds to about $13.61 billion/year in avoided cost for imported oil.

The new, incremental energy contributions that can be obtained from waste biomass will depend on future government policies, on the rates of fossil fuel depletion, and on extrinsic and intrinsic economic factors, as well as the availability of specific residues in areas where they can be collected and utilized. Environmental regulations will affect how the producers dispose of waste biomass and whether energy applications can be justified. Extrinsic economic factors include the costs of competitive energy resources, the costs of existing disposal methods, the costs of any mandated disposal methods, and in some cases, the markets for recyclables. The intrinsic economic factors include the costs of collection and transport of the waste biomass to end-use site or market, and conversion costs to energy or fuel. All of these factors should be examined in some detail to evaluate the development of incremental energy contributions from waste biomass.

REFERENCES

American Chemical Society (1969). *In* "Cleaning Our Environment, The Chemical Basis for Action," p. 97. Washington, D.C.

Burford, J. L., Jr., and Varani, F. T. (1976). "Energy Potential Through Bio-Conversion of Agricultural Wastes, Final Report to Four Corners Regional Commission," Technical Assistance Grant FCRC No. 651-366-075. Bio-Gas of Colorado, Inc., October.

Franklin Associates, Ltd. (1994). "Characterization of Municipal Solid Waste in the United States," Report to U.S. Environmental Protection Agency. Washington, D.C.

Heid, W. J., Jr. (1984). "Turning Great Plains Crop Residues and Other Products into Energy," Agricultural Economic Report No. 523. U.S. Dept. of Agriculture, Washington, D.C., November.

Howlett, K., and Gamache, A. (1977). "Silvicultural Biomass Farms, Vol. 2, Biomass Potential of Short Rotation Farms," Technical Report 7347. Mitre Corp., McLean, VA.

Jaycor (1990). "Regional Assessment of Nonforestry-Related Biomass Resources, Summary Volume," Report No. 684-0035a/90. Southeastern Regional Biomass Energy Program, Muscle Shoals, AL, March 19.

Klass, D. L. (1990). *In* "Energy from Biomass and Wastes XIII," (D. L. Klass, ed.), p. 11. Institute of Gas Technology, Chicago.

Lusk, P. (1994). "Methane Recovery from Animal Manures: A Current Opportunities Casebook," NREL/TP-421-7577, DE95004003. National Renewable Energy Laboratory, Golden, CO, December.

McKeever, D. B. (1995). *In* "Second Biomass Conference of the Americas, Energy, Environment, Agriculture, and Industry," p. 77, NREL/CP-200-8098, DE95009230. National Renewable Energy Laboratory, Golden CO, August 21–24.

Phillips, V. D., Tvedten, A. F., Liu, W., and Merriam, R. A. (1995). *In* "Second Biomass Conference of the Americas, Energy, Environment, Agriculture, and Industry," p. 55, NREL/CP-200-8098, DE95009230. National Renewable Energy Laboratory, Golden CO, August 21–24.

Purdue University (1979). *In* "The Potential of Producing Energy from Agriculture, Final Report," p. 11. West Lafayette, IN, May.

Riall, W., and Bouffier, C. G. (1990). *In* "Energy Potential from Biomass Sources in Georgia: Residues in Forestry, Agriculture, and Municipal Solid Wastes," p. 3. Georgia Forestry Commission, May.

Skog, K. E. (1993). *In* "First Biomass Conference of the Americas, Energy, Environment, Agriculture, and Industry," Vol. I, p. 29, NREL/CP-200-5768, DE93010050. National Renewable Energy Laboratory, Golden, CO, August 30–September 2.

Stanford Research Institute (1976). "An Evaluation of the Use of Agricultural Residues as an Energy Feedstock," Vol. I, National Science Foundation Grant No. AER74-18615 A03, NSF/RANN/SE/GI/18615/FR/76/3. Washington, D.C., July.

U.S. Dept. of Agriculture (1995). "Agricultural Statistics 1995." National Agricultural Statistics Service, Washington, D.C.

U.S. Dept. of Energy (1995). "Technology Partnerships—Enhancing the Competitiveness, Efficiency, and Environmental Quality of American Industry," DOE/GO-10095-170, DE95004086. Office of Energy Efficiency and Renewable Energy, Washington, D.C., April.

U.S. Dept. of Energy (1996). "DOE Biomass Power Program, Strategic Plan, 1996-2015," DOE/GO-10096-345, DE97000081. Washington, D.C., December, 13 pp.

U.S. Environmental Protection Agency (1985). "1984 Need Survey Report to Congress, Assessment of Needed Publicly Owned Wastewater Treatment Facilities in the United States," APA 430/9-84-011. Washington, D.C., February 10.

U.S. Environmental Protection Agency (1990). "National Water Quality Inventory, 1988 Report to Congress," APA 440-4-90-003. Washington, D.C., April.

U.S. Environmental Protection Agency (1993). "1992 Need Survey Report to Congress," EPA 832-R-93-002. Washington, D.C., September.

Wadleigh, C. H. (1968). "Wastes in Relation to Agriculture and Industry," Miscellaneous Publication No. 1065. U.S. Department of Agriculture, Washington, D.C.

CHAPTER 6

Physical Conversion
Processes

I. INTRODUCTION

As discussed in previous chapters, there are numerous aquatic and terrestrial virgin biomass species and many types of waste biomass that are potential fuels or feedstocks. With the exception of microalgae and some high-moisture-content biomass, essentially all are solid materials. They contain organic compounds, minerals, and moisture. Some of the compositional differences are large. The aquatics, municipal biosolids, and animal manures are high in moisture content; the terrestrial species contain relatively small amounts of moisture. The ash contents of woody biomass species are small; some aquatics and agricultural crops contain large amounts of ash. On a moisture- and ash-free basis, the heating value of most biomass is in the same range, but on a dry basis, these materials can exhibit wide variations.

Because of these broad differences, many of the possible feedstock-process-energy product combinations are not feasible. For example, untreated municipal biosolids contain very large amounts of moisture and are normally unsuitable for thermochemical conversion. Such feedstocks do not support self-sustained combustion under conventional conditions unless the moisture is reduced by a considerable amount, a high-cost process in wastewater treatment plants. Biosolids are more suited for microbial conversion in aqueous systems where a liquid water medium is essential. In contrast, woody biomass is often suitable for direct use as a solid fuel or as a feedstock for thermochemical conversion. Predrying to remove some of the moisture, if needed, is readily accomplished at low cost.

Chapters 6 to 12 address specific groups of processes and methods employed for converting biomass to energy and fuels. In this chapter, the physical processes employed to prepare biomass for use as fuel or as a feedstock for a conversion process are discussed. The processes examined are dewatering and drying, size reduction, densification, and separation. The physical process, a few specific examples of the process, and its relationship to the thermochemical or microbial process that may be used for subsequent conversion are described.

II. DEWATERING AND DRYING

A. FUNDAMENTALS

Dewatering refers to the removal of all or part of the contained moisture from biomass as a liquid. Drying is a similar process, except that the moisture is removed as vapor. In some cases where a waste or virgin biomass feedstock is thermally processed directly for energy recovery, it may be necessary to partially dry the raw feed before conversion. Otherwise, more energy might be consumed by the conversion process than would be produced in the form of energy or fuel. Open-air solar drying is usually the lowest cost drying method, if it can be used. Raw materials that are not sufficiently stable to be dried by solar methods can be dried more rapidly using industrial dryers such as spray dryers, drum dryers, and convection ovens if cost permits. For large-scale drying applications, forced-air furnaces and drying systems designed to use hot stack gases are sometimes just as efficient.

Since a large portion of a feedstock's equivalent energy content can be expended for drying, there is a balance between the cost of moisture removal, the incremental improvement in efficiency on conversion, and the advantages of handling drier feedstock. The key biomass property that should obviously be examined, in addition to conversion process requirements, is the moisture content of the fresh biomass, the methods available for its partial or total removal, and the effects, if any, on the properties of the remaining biomass. The moisture content of biomass is as variable as the multitude of biomass species available as potential feedstocks.

Aquatic biomass is one category of feedstock that can be classified as high-water content biomass. Freshwater, marine, and microalgal biomass, such as water hyacinth, giant brown kelp, and *Chlorella,* respectively, contain large amounts of intracellular moisture. The water content is usually over 95 wt % of the fresh biomass (see Appendix A, Part A.8, for the definitions of wet and dry weight percentages). Several types of municipal, industrial, and farm animal wastes are also produced in association with water and are potential feedstocks. One example is untreated municipal biosolids. Because of the nature of the collection systems (i.e., dilution with water to facilitate localized disposal and transport in municipal lines to wastewater treatment plants), raw municipal biosolids contain over 95 wt % water. They are stabilized at the wastewater treatment plant by subjecting them to various primary and secondary treatments which reduce volatile solids, BOD, and COD, and are dewatered for disposal. Examples of high-water-content industrial wastes are pulp and paper mill sludges and alcoholic beverage industry sludges. They are often subjected to biological stabilization and dewatering processes similar to those used for municipal biosolids. The major farm animal wastes that are potential feedstocks for conversion processes are cattle, hog, and poultry manures. The moisture contents of most of these wastes range up to 80 wt %.

Terrestrial biomass, as freshly harvested green biomass, generally contains 40 to 60 wt % moisture. These potential feedstocks include most herbaceous species, softwoods, and hardwoods. Agricultural crop residues that have been exposed to open-air solar drying contain less moisture, often in the 15-wt % range or less. Straws are good examples. Other examples consist of by-products such as corn cobs, rice hulls, and nut shells from the processing of agricultural and orchard crops. Many of these residuals have even lower moisture levels. As-received municipal solid waste (MSW) usually contains 10 to 30 wt % moisture depending upon the season of the year and geographic location where it is collected. All of these terrestrial biomass species and residuals are suitable feedstocks for one or more different thermochemical conversion processes. Microbial conversion is also feasible in many cases after suitable pretreatment. Unlike many high-water-content biomass species, most terrestrial virgin biomass and residuals are sufficiently stable to undergo solar or thermal drying.

Woody biomass is the largest source of standing biomass and is the preferred feedstock for large-scale, integrated, biomass production-conversion systems. So it is worthwhile to consider how moisture accumulates in these materials internally during growth and from humid air. Water is contained in all tissues of a tree, both dead and alive (Mirov, 1949). Young leaves and roots contain up to 90 wt % moisture; tree trunks contain as much as 50 wt %. In the transpiration process, water is absorbed by the roots from the soil, pushed into the sapwood, and then pulled up to the leaves above ground where it is given off to the atmosphere by evaporation through the stomata, the same openings that admit CO_2 (Chapter 4). Various osmotic and diffusional forces drive the water from the soil to the leaves through semipermeable membranes and capillary passages. About one-half of the solar energy falling on the leaves supplies the energy to facilitate transpiration, which is necessary for photosynthesis to occur. The stoichiometric photosynthesis of 100 kg of cellulose requires about 55.6 kg of water, but during the process, the tree transpires about 100,000 kg of water to the atmosphere.

Wood also absorbs moisture from humid air and is the equivalent of an elastic gel that exhibits limited swelling as water vapor is taken up from the air. Two different mechanisms are operative: *adsorption* and *absorption* (Nikitin *et al.*, 1962). In adsorption, moisture is transferred from air to the wood surfaces and results from the attraction between polar water molecules and the negatively charged surfaces of the wood. The negative charges involve functional groups on the surface that can carry full or partial negative charges or organic molecules that can exist as dipoles with the negative ends clustered on the surface. The amount of moisture adsorbed on wood surfaces is relatively small; it ranges up to about 5 to 6 wt % of the wood at 20°C and 100% relative humidity. In absorption, water molecules are drawn into the permeable pores of the wood by spongelike processes due to diffusional and osmotic forces followed by capillary condensation. A large number of fine capillaries in the wood fibers facilitate this process. The amount of moisture absorbed within the woody structures depends upon the pore diameters and distribution of the capillaries. In spruce wood pulps, for example, the amount of water vapor absorbed at 20°C and 100% relative humidity is about 25 wt %. The maximum total amount of water taken up from air at ambient conditions by adsorption and absorption is about 30 wt % of the wood, but can reach 200 wt % or more if the wood is soaked in liquid water (Nikitin *et al.*, 1962). Exposure to precipitation is the third mechanism that raises the moisture content of green wood to an average of about 50 wt %. It is evident from this explanation of the various mechanisms of water uptake by wood that transpiration of soil moisture is essential for tree growth to occur, that the temperature and humidity of the surrounding air affects the moisture content of the wood if sufficient time is allowed to reach equilibrium moisture levels, and that wood is hygroscopic. The same principles are applicable to most terrestrial biomass because of the vascular structure typical of many species.

B. BIOMASS MOISTURE CONTENT AND CONVERSION REQUIREMENTS

It is normally not necessary to reduce the water content of high-moisture-content or wet biomass feedstocks for microbial conversion processes. This contrasts with thermal conversion processes such as combustion. Dry biomass burns at higher temperatures and thermal efficiencies than wet biomass. For example, the flame temperatures of green wood containing 50 wt % moisture and dry wood in conventional combustors that supply boiler heat are about 980°C and 1260 to 1370°C, respectively (cf. FBT, Inc., 1994). Flame temperature is directly related to the amount of heat necessary to evaporate the moisture contained in the wood—the lower the moisture content, the lower the amount of energy needed to remove the water and the higher the boiler efficiency. Although flame temperature is not the actual bed temperature in advanced-design, fluid-bed combustors, the effects on temperature and efficiency are the same. The maximum amounts of acceptable moisture in wood fuels for conventional furnace systems are illustrated in Table 6.1, and the typical moisture contents and heating values of several biomass fuels for combustion in commercial fluid-bed, grate, and suspension firing units are presented in Table 6.2. With the exception of suspension firing units, for which the moisture content of the fuel is usually in the 20-wt % range, the maximum moisture content range is 55 to 65 wt %. Indeed, combustion of biomass containing 65 wt % moisture in conventional grate-type systems can result in lowering of the adiabatic flame temperature to the point where self-sustained combustion does not occur.

Many of the large-scale biomass combustion systems for producing heat, hot water, or steam accept biomass fuels containing relatively large amounts of moisture and are operated without much apparent concern for the effects of moisture content of the fuel on the combustion process itself. One of the

TABLE 6.1 Comparison of Conventional Wood-Burning Furnace Characteristics[a]

Furnace type	Maximum fuel moisture (wt %)	Suitable wood fuel	Maximum steam output rating	Ash entrainment rating	Fuel bed type
Gravity fed Dutch oven	65	Unprepared, limited max. size	Medium	Low	Stationary
Spreader stoker pinhole grate	65	Hogged fuel	High	Low	Stationary
Spreader stoker travelling grate	55	Hogged fuel	High	High	Moving
Gravity-fed inclined grate	65	Partially hogged fuel	Medium	Medium	Moving
Suspension firing	20	Chips, shavings, and sander dust	Low	High	Suspension
Cyclone furnace	65	Hogged fuel	Low	Medium	Stationary

[a]Ismail and Quick (1991).

TABLE 6.2   Typical Moisture Contents and Heating Values of Waste Biomass for
Combustion in Fluid-Bed, Grate, and Suspension Firing Units

Fuel	Moisture (wt %)	Higher heating value (MJ/kg)
Orchard prunings[a]	30.0	13.64
Secondary wood[a]	20.0	15.45
Almond shells[a]	8.7	17.78
Cotton stalks[a]	30.0	12.76
Bark[b]	48–40	10.5–12.1
General wood wastes[b]	48–25	10.5–15.1
Sawdust, shavings, and sanderdust[b]	48–12 .	10.5–17.6
Bagasse[b]	55–20	8.4–15.1
Peat[b]	60–50	9.2–11.7
Coffee grounds[b]	55–40	11.3–15.1
Nut hulls[b]	25–18	17.2–18.8
Rice hulls[b]	18–15	12.1–15.1
Corn cobs[b]	16–12	18.4–19.2

[a]Murphy (1991). For fluid-bed units.
[b]Routly (1991). For grate or suspension firing.

largest biomass-fueled power plants equipped with traveling grates operates
very well with wood chips containing an average of 50 wt % moisture, although
a few initial handling and storage problems caused by high-moisture fuel
supplies had to be solved (Tewksbury, 1987). Another power plant equipped
with traveling grates operates very well on fuel containing about the same
amount of moisture and consisting of a mixture of about 80% hogged mill
wastes and 20% wood chips (Ganotis, 1988). Some air drying of stringy bark
fuels is needed in the spring to eliminate fuel handling problems. Still another
power plant equipped with a fluidized, bubbling bed combustor operates well
with a mixture of about 40% whole tree chips, 20% sawmill residues, and
40% agricultural residue from almond orchards. The fluid-bed combustors are
designed to operate with fuels having a variable moisture content up to about
50 wt % (Normoyle and Gershengoren, 1989).

Thus, biomass fuels containing up to about 50 to 55 wt % moisture do not
require pre-drying for acceptable performance in combustion systems that are
designed for such fuels. However, a moisture content of 15 wt % is generally
recognized as optimum for efficient thermochemical gasification of biomass
(Miles, 1984), although several thermochemical gasification processes satisfactorily convert feedstocks containing up to about 30 to 35 wt % moisture.
Also, in certain types of thermochemical gasification processes such as steam

gasification, water is a reactant and the contained water in the feedstock can be beneficial (Chapter 8).

To illustrate more quantitatively the effect of moisture content on the performance of a thermochemical process, consider the direct combustion of sugarcane bagasse in a conventional boiler to raise steam and the effects on boiler efficiency of bagasse moisture content relative to the other sources of efficiency losses. The results of a complex series of calculations to examine boiler losses and efficiencies are shown in Table 6.3 (Institute for Energy Studies, 1977). A typical amount of excess air used in bagasse-fired boilers was chosen as 30%, the moisture content of the bagasse ranged from 0 to 60 wt % in 10-wt % increments, and the stack gas temperatures ranged from 177°C (450 K) to 260°C (533 K) in approximately 28°C increments. Boiler efficiency is 100 minus the sum of the boiler losses in percentage units. The boiler losses, other than those caused by the moisture content of the bagasse, include those due to dry gases of combustion, which refers to nonuseful heat losses; those due to moisture in air; those due to the moisture formed on bagasse combustion; and other losses. Plots of bagasse moisture content against boiler efficiency and the losses due to bagasse moisture content alone at stack gas temperatures of 450 K are shown in Fig. 6.1. The analysis shows that when the moisture content of the bagasse is more than about 35 wt %, it has a greater impact on boiler efficiency losses than the other moisture sources. Note that the incremental improvement in boiler efficiency for drying the bagasse from about 35 to 0.0 wt % moisture increases boiler efficiencies by only about 7%. It was concluded by the analysts who performed the study that sugar mills can increase boiler efficiency about 5% by drying the bagasse from its typical level of 48 wt % moisture to about 35 wt % moisture. Bagasse moisture reductions by solar drying and drying with stack gases were suggested as low-cost approaches to increasing boiler efficiencies. It was also concluded that the higher temperature generated using dryer bagasse can increase heat transfer efficiencies in addition to reducing stack gas losses. This efficiency increase results from both higher temperature differentials and lower furnace gas velocities. Historically, the drying of bagasse to improve boiler efficiencies was proposed in the early 1900s. The same principles apply generally to improving the efficiencies of biomass combustion processes.

Predrying of biomass has sometimes been justified in the past only for large-scale operations, or where low-cost energy is available as waste heat. It is important to realize, however, that the absence of any capability to predry feedstock for thermochemical conversion has sometimes caused severe operating problems, particularly for gasification processes. In one of the early fluid-bed gasification plants fueled with wood chips and sawdust to produce low-energy gas as an on-site boiler fuel, it was very difficult to control combustion. The industrial gas burners installed in the plant did not function satisfactorily

TABLE 6.3 Boiler Efficiency and Losses Due to Moisture on Bagasse Combustion[a]

Bagasse moisture (wt %)	Stack gas temperature (°C [°F])	Combustion dry gases (%)	Moisture in air (%)	Moisture in bagasse (%)	Moisture of combustion (%)	Other losses (%)	Boiler efficiency (%)
				Sources and amount of boiler losses			
0	177 [350]	5.86	0.15	0.00	8.07	5.00	80.92
	204 [400]	6.95	0.18	0.00	8.23	5.00	79.64
	232 [450]	8.04	0.21	0.00	8.39	5.00	78.36
	260 [500]	9.12	0.24	0.00	8.55	5.00	77.09
10	177 [350]	5.86	0.15	1.64	8.19	5.00	79.16
	204 [400]	6.94	0.18	1.67	8.35	5.00	77.86
	232 [450]	8.03	0.21	1.70	8.51	5.00	76.55
	260 [500]	9.11	0.24	1.73	8.67	5.00	75.25
20	177 [350]	5.88	0.15	3.51	8.30	5.00	77.16
	204 [400]	6.96	0.18	3.58	8.46	5.00	75.82
	232 [450]	8.05	0.21	3.65	8.62	5.00	74.47
	260 [500]	9.14	0.24	3.72	8.78	5.00	73.12
30	177 [350]	5.86	0.15	6.55	8.42	5.00	74.02
	204 [400]	6.94	0.18	6.68	8.59	5.00	72.61
	232 [450]	8.03	0.21	6.81	8.76	5.00	71.19
	260 [500]	9.11	0.24	6.94	8.92	5.00	69.79

	5.86	0.15	9.83	8.52	5.00	70.64
	6.95	0.18	10.02	8.68	5.00	69.17
	8.04	0.21	10.21	8.85	5.00	67.69
	9.12	0.24	10.41	9.02	5.00	66.21
	5.87	0.15	13.46	8.70	5.00	66.82
	6.96	0.18	13.72	8.88	5.00	65.26
	8.05	0.21	13.98	9.05	5.00	63.71
	9.13	0.24	14.25	9.22	5.00	62.16
	5.84	0.15	23.40	8.99	5.00	56.62
	6.92	0.18	23.86	9.16	5.00	54.88
	8.00	0.21	24.32	9.34	5.00	53.13
	9.08	0.24	24.78	9.52	5.00	51.38

40   177 [350], 204 [400], 232 [450], 260 [500]
50   177 [350], 204 [400], 232 [450], 260 [500]
60   177 [350], 204 [400], 232 [450], 260 [500]

[a]Adapted from Institute for Energy Studies (1977). Various assumptions are used for this analysis. The stack gases exhibit perfect gas behavior with constant specific heat equal to 0.558 kJ/kg. Superheated steam has a specific heat of 1.093 kJ/kg and is constant with temperature. The fuel inlet air temperature is 26.7°C (80°F) and the relative humidity of the air is 60% at 101.3 kPa. The combustion efficiency is 98%. The unburned bagasse fraction is 0.03. The ash sediment fraction is 0.015 in bone-dry bagasse. The psychrometry of the stack gases is not significantly different than that of air because the partial pressure of CO_2 in the stack gases is small. The calculated value is 15.86 kPa (2.3 psia) and is approximately constant for constant excess air, independent of the bagasse moisture content. "Other losses" include losses due to unburned combustibles (about 3%), losses due to radiation (about 0.5%), and unaccounted-for losses (about 1.5%).

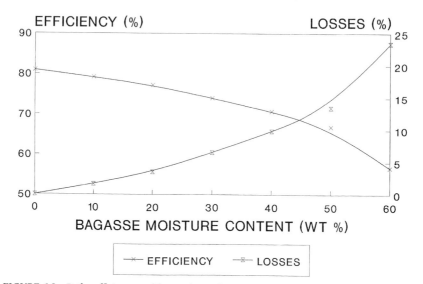

FIGURE 6.1   Boiler efficiency and losses due to bagasse moisture content. (Stack gas temperature = 450 K.)

with the product gas (Bircher, 1982). These problems were attributed to large variations in the quality of the gas caused by accepting wood feedstock at any moisture content up to 50 wt %, which in turn resulted in large swings in gas heating values from about 3 to 8 MJ/m^3 (about 80 to 200 Btu/ft^3). Drying of the feedstock has been found to be extremely important in wood gasification because it is only through the availability of a uniform feedstock that consistent gas quality can be assured (Miller, 1987).

C. DEWATERING METHODS

Dewatering methods are available for most high-water-content virgin and waste biomass. This suggests that the moisture content of such feedstocks can be readily adjusted before conversion. This is not the case, however, because it is often difficult to reduce moisture content to the level desired at reasonable cost.

The equipment used for dewatering includes filters and screening devices of various types, centrifuges, hydrocyclones, extrusion and expression presses, water extractors, and thickening, clarifying, and flotation hardware. The processing methods encompass a broad range of water-removal techniques. They can also incorporate the use of chemical flocculants and surfactants and high- and low-temperature treatments. The drawbacks to the dewatering of high-water-content biomass by most of these methods are numerous. Direct physical separation of the occluded moisture in aquatic species by dewatering is normally not feasible unless the biomass is subjected to physical processes that disrupt the cell walls. Solar drying in open air is a low-cost option for moisture reduction as already pointed out, but most high-water-content biomass species begin to decompose, some quite rapidly and often with a relatively large loss in carbon and energy content, when dried under these conditions. In contrast, municipal biosolids are often dewatered to 5 to 20 wt % solids content, and some of the advanced dewatering methods are capable of increasing the solids content to as high as 50 wt % or more. The drying methods used commercially in wastewater treatment plants facilitate final disposal, but they are costly and afford products that are still far from the preferred moisture content range of feedstocks for thermochemical conversion.

Strict physical processing of high-water-content biomass for partial removal of moisture can sometimes be accomplished by combined use of shearing or cutting devices and mechanical pressing. Some of the dewatered products produced by these techniques can sustain their own combustion, can be combined with low-moisture feedstock for thermochemical conversion, or can be fabricated into briquettes or pellets for use as fuels. Overall consideration of the difficulties of dewatering high-water-content biomass suggests that microbial conversion processes should be used so the feedstock does not have to be dewatered or dried and can be used as such.

Although it is relatively costly, one drying method deserves special mention because it is used commercially for several high-water-content waste biomass streams such as brewery and fermentation industry wastes, food and dairy industry wastes, and primary and secondary municipal biosolids. The technique is based on the equivalent of multiple-effect evaporation and vapor recompression so that most of the water exits the process as liquid, except in the last effect, to avoid losing the latent heat of vaporization. Several advanced processes have been developed to separate water from solids at lower energy inputs than conventional, single-effect drying systems. With the advent of large centrifugal compressors in the 1960s, it became possible to mechanically recompress the water vapor from an evaporative stage to drive that same stage rather than another as in multiple-effect evaporation. Using standard technology, the dewatering of high-water-content biomass can utilize mechanical vapor recompression to raise the solids concentration to near 30 wt %, followed by multiple-effect evaporation to raise the solids concentration to near 50 wt %, followed by a rotary dryer if desired for further moisture reduction (cf. Crumm and Crumm, 1984). The efficiency of the segment of evaporation that mechanical vapor recompression accomplishes is very high. At an energy equivalent of

10.5 MJ/kWh (10,400 Btu/kWh), mechanical vapor recompression can vaporize 1 kg of water for less than 0.46 MJ (1.0 lb for less than 200 Btu). The Carver-Greenfield process is based on combining mechanical vapor recompression with multiple-effect evaporation to dry high-water-content biomass and other solid suspensions. Many full-scale units have been placed in operation since the first facility was installed in 1961. One unit was used at the Hyperion wastewater treatment plant in Los Angeles from 1987 to early 1995 to dry 40 t/day of biosolids wetcake to 99+% total solids content (Haug, Moore, and Harrison, 1995). The process has since been replaced by rotary steam dryers because it was not possible to reach the design capacity of the unit.

D. Drying Methods

The mechanisms of water uptake by trees suggest several methods of drying terrestrial biomass. The most obvious method is to expose biomass to circulating, low-humidity air that is heated. Open-air solar drying meets these requirements and has been used for hundreds of years to season or cure woods and grasses. The final moisture content of the air-dried biomass is usually in the 35-wt % range or less. The advantage of this partial drying method is that it is low in cost. The disadvantages are several. The process is slow and depends on the local climate. Some labor is required to arrange the freshly harvested biomass in suitable piles or windrows to facilitate exposure to sunlight and air circulation. Periodic turning of the windrows may be necessary to allow drying of plant parts in direct contact with the soil and to prevent fungal infection of wet biomass. Natural precipitation may require excessive drying times. Forage crops have traditionally been partially dried in open air to this moisture level so they can be removed from the field and stored without significant deterioration and loss of nutrient value. Solar drying also facilitates densification of hay by baling.

In a field study in Florida of the tall grasses elephantgrass (*Pennisetum purpureum*) and energycane (*Saccharum spontaneum* L.), which are good candidates as biomass energy crops, the air drying in windrows of mature crops of 2- to 4-cm stem diameters required about 7 to 10 days without rainfall to reach moisture levels of 15 to 20 wt % (Mislevy and Fluck, 1993). The seasoning of freshly harvested mature trees by air drying requires longer time periods to reduce the moisture level to about 25 to 35 wt % because of the larger diameter trunks and pieces. Decay fungi that may be present progress rarely, if ever, at moisture contents below 25 wt %. Green wood chips can be air dried in less time because of their smaller size. In a study of the use of hybrid willow harvested at 3-year rotations as fuel for a direct wood-fired, gas turbine power plant, it was projected that air-dried willow bundles would reach 30 to 35 wt % moisture at the same cost as green wood chips at 50 wt % moisture content (Ismail and Quick, 1991). The cost is the same because what is saved in not chipping the wood is spent on bundling and storage for 6 months to air-dry the bundles.

Kiln drying under controlled conditions is commonly employed to improve the stability and physical characteristics of lumber products used as materials of construction or for manufacturing furniture, whereas open-air drying is traditionally employed for the curing or seasoning of tree parts and roundwoods to be used as fuel. Kiln drying promotes the removal of moisture by circulating heated air by natural draft or with fans or blowers through the wood, which is carefully piled in the kiln to promote the drying process. Heat is transferred from hot air heated by steam coils supplied by a boiler, or from hot stack gases heated by the burning of waste biomass or other fuels through manifolds. In the batch-drying of large volumes of wood, the temperature of the air can be gradually increased; the final temperatures and humidities are usually near 90°C and 15%. Kiln drying is rapid compared to the rate of open-air solar drying, but it is too slow for some continuous, thermochemical conversion processes unless the dryers and storage facilities are sized to handle the demand for predried feedstock. The continuous drying of wood chips, wood chunks, and hog fuel with industrial dryers or in drying ducts installed prior to the conversion unit is the approach that is often used when predrying is judged to be sufficiently beneficial. Continuous, direct-heat drying, in which hot air or stack gas contacts the biomass as it is fed to the conversion reactor, and indirect-heat drying, in which heat is transferred by convection and radiation from conducting surfaces to the biomass, can be utilized. Many commercial drying ovens and dryers such as rotary drum dryers, which have been effectively used for many years for drying wood and other biomass, are available. The use of superheated steam for drying rather than burning some of the feedstock as a heat source may allow further improvements in efficiency (cf. Wiltsee, McGowin, and Hughes, 1993). The direct-heat systems are generally lower in cost than the indirect-heat systems if commercial drying units are used. Thermochemical conversion reactors can also be designed so that incoming fresh feed is dried to the desired level by heat transfer from the hot reaction products. The simple addition of enclosed drying tunnels for passage of hot air or stack gases over and through incoming fresh feed can sometimes suffice to reduce moisture to the desired level and preheat the feed without the need to install industrial driers.

Note, however, that stack gases from biomass-fired boilers contain about 15 wt % moisture, and that at temperatures below 250°C, only a small amount of additional moisture can be absorbed before the gas becomes fully saturated. This is evident from the following equation (Routly, 1991):

$$WG = (2940 \ M)/T_i - T_o$$

where WG = drying gas weight, kg/h
$\quad M$ = water evaporated, kg/h
$\quad T_i$ = temperature of drying gas entering, °C
$\quad T_o$ = temperature of drying gas leaving, °C

This equation indicates that large fans and motors are required for circulation of the drying gases when low-temperature gas is used as the drying medium. To obtain sufficient heat for drying purposes, some of the stack gas may have to be extracted upstream of the boiler heat recovery equipment, which can have an adverse effect on steam generation. Stack gas drying should therefore be evaluated for each application to determine whether it is technically and economically feasible. For most thermochemical conversion systems that process green biomass, a balance is usually struck among the optimum moisture range needed for conversion, the feedstock demand rate, the drying requirements, the size of the feedstock storage facility, feedstock stability on storage, and the cost of supplying predried feedstock.

The transpirational drying in open air of whole trees felled in the forest has been evaluated, but has not been widely adopted (McMinn, 1986). However, the drying of whole trees has been incorporated as part of the whole-tree-burning concept for power production (Chapters 7 and 14; also see Ostlie and Drennan, 1989). Whole trees including branches are dried in large buildings equipped with heat exchangers supplied with warm water at temperatures up to 50°C. Additional higher temperature waste heat is available from the power plant for peaking. Fans along the base of the drying buildings draw outside air over the heat exchangers and circulate it through piles of whole trees. The resulting warm, moist air is drawn out of the buildings through vents. For optimal drying conditions, the relative humidity levels are kept below 35%. After approximately 30 days of storage in the drying buildings, the moisture content of the whole trees is reduced to 25 wt % or less. Experimental testing of whole tree drying provided several interesting and perhaps unexpected results. The two tree species tested, aspen and eastern cottonwood, dried significantly faster with the leaves intact than without the leaves. It was also found that logs do not appear to dry more quickly than whole trees, and that the branches of the trees tested were drier than the corresponding trunks.

Whenever it is necessary to remove moisture from virgin or waste biomass feedstocks, air drying, mechanical dewatering, and drying with waste heat or stack gases should be evaluated first. The lower costs of these methods compared to the costs of thermal drying in which external fuel or a portion of the feedstock supplies heat may justify their use.

III. SIZE REDUCTION

A. FUNDAMENTALS

Reduction in physical size is often required before biomass is used as a fuel or feedstock. Size-reduction techniques are employed to prepare biomass for direct fuel use, fabrication into fuel pellets, cubes, and briquettes, or conversion. Smaller particles and pieces of biomass reduce its storage volume, facilitate handling of the material in the solid state and transport of the material as a slurry or pneumatically, and sometimes permit ready separation of components such as bark and whitewood. The size of the pieces or particles can be critical when drying is used because the exposed surface area, which is a function of physical size, can determine drying time and the methods and conditions needed to remove moisture. There are a few exceptions where size reduction is not needed, such as in whole-tree burning.

The physical dimensions of the feedstock are also related to the conversion method that is used. Particle size should satisfy the requirements of supplying feedstock to the conversion reactor and of the conversion process itself. For combustion systems, the combustion chamber and heat exchanger designs, the operating conditions, and the methods of delivering solid fuel and removing the ash determine the optimum size characteristics of the fuel. For thermal gasification and liquefaction processes, particle size and size distribution can influence the rate of conversion, the operating conditions of the process, and product yields and distributions. Biological processes are also affected by the physical size of the feedstock. In general, the smaller the substrate particles, the higher the reaction rate because more surface area is exposed to the enzymes and microorganisms that promote the process.

If the particle size of the biomass fuel or feedstock is not predetermined by its history, as is the case for sawdusts, nutshells, and a few other waste biomass materials, size reduction is usually carried out with one or more units that make up the "front end" of the total processing system. Many different kinds of machines are employed. Generally, the size of the feed is reduced by grinding, cutting, or impact mechanisms. Not all of the designs are suitable for biomass energy applications because the equipment is customized for certain uses or the cost of size reduction is excessive. Agricultural crops and woody biomass are also usually processed by different types of machines. A brief review of the basic types of machines that are or have been used for biomass follows to illustrate the variety of size-reduction equipment and their biomass applications.

B. Hardware and Some Biomass Applications

Dry shredders are commercially used for reducing the size of biomass. The two most common types of machines are vertical and horizontal shaft hammermills. Metal hammers on rotating shafts or drums reduce particle size by impacting the feed material until the particles are small enough to drop through grate openings. Hammermills are commonly used in MSW-processing systems to reduce the size of the components before separation of RDF (i.e., refuse-derived fuel or the combustible fraction of MSW), and other materials. Hammermills are also used as agricultural choppers and tree chippers. Rotating cutters equipped with knife blades that reduce particle size by a cutting or shearing action are used for the same applications, although they usually have smaller capacities than hammermills.

Early biomass grinders were used to produce wood pulps from roundwood for the manufacture of paper (Riegel, 1933). Logs are positioned with the sides against a rotating grindstone so that damage to the fibers is minimized. Water is passed over the stone to wash the wood meal into a storage tank. In the original designs, the logs were held against the grindstone by springs. Later versions were hydraulic magazine grinders that automatically replaced the ground logs with fresh logs. In the 1950s, attrition mills were introduced to produce mechanical pulps from wood chips. Rotating, opposing discs, either one stationary or both moving in opposite directions, are used. The chips are fed to the mill near the central, rotating shaft and move outward to the periphery of the discs through a series of successively smaller channels that progressively reduce the feed to pulp-size particles.

Hydropulpers are wet shredders in which a high-speed cutting blade pulverizes a water suspension of the feed over a perforated plate. The pulped material passes through the plate and the nonpulping materials are ejected. The action is similar to that of a kitchen waste disposal unit. Hydropulpers can also be used for the simultaneous size reduction and separation of the combustible fraction of MSW from the inorganic materials. But since the product is a water slurry of small particles, which can have the consistency of a heavy cream, the hydropulper is quite suitable for the preparation of RDF for microbial conversion after passage through a liquid cyclone to remove gritty, mostly inert material. Fiber recovery operations where long fibers are removed for resale can be performed before microbial processing. Experimental studies have shown that hydropulpers can also supply good feedstocks for microbial processing from other biomass. Maintenance costs for wet shredders are lower than those for dry shredders.

Agricultural choppers that are operated as stationary cutters and as moving choppers in the field separately from the harvesters or that are part of forage harvesters that chop the crop during the harvesting process are commonly employed for preparing hay and other forage crops for ensiling. When a chopper is used in the field, more stems pass through the chopper lengthwise and a shorter average cut is obtained than when the same setting is used on a stationary chopper. Forage chopping in the field usually supplies material 25 mm or less in length. Silage systems offer several advantages for the size reduction of herbaceous biomass energy crops. The major disadvantage for high-yielding biomass species is the inability of the forage chopper to effectively harvest severely lodged (fallen) crops or plants (Coble and Egg, 1989). Sugar-cane harvesters, which are designed for harvesting high-yield, thick-stemmed, lodged biomass, have been used to harvest such crops. A separate size-reduction step is needed prior to storage or conversion.

In contrast to mechanical pulps, chemical wood pulps are often made from bark-free wood chips. One form of chipper has four knives fastened at 90° from each other on a rotating disc in such a way that only the edge of the blade projects beyond the disc. Each knife cuts a thin strip about 10 mm thick from the logs, which are fed along their long axis to the chipper at about a 38° angle to the disc. The mechanical action is similar to that of a sausage slicer. The denser hardwoods produce thinner chips than softwoods, but both types of chips are suitable feedstocks for most fixed- and fluid-bed gasifiers and most other thermal conversion processes.

Chipping has been the traditional mechanical method of size reduction to prepare wood fuels for direct combustion. It is an energy-intensive operation, but it does improve bulk density, handling, and transportation costs. Disc chipping and hogging are two preferred means of preparing wood fuels (Suadi-cani and Heding, 1992). Hammer hogs with free-swing hammers break the feed into small pieces, whereas knife hogs cut the feed with blades. The least desirable option seems to be chipping in the field at the time of harvest, which requires that a power chipper accompany the harvester through the field. Whole-tree chips are also reported to lose approximately 10% of their oven dry weight after storage for 6 months (*cf.* Curtin and Barnett, 1986). A variety of machines are available for producing wood chips in the field. One of the notable developments in North America is a swath harvester called "Jaws" that produces fuel chips while clearing 2.4-m wide paths through young, crowded stands of pine (Ranney *et al.*, 1985). Another is the large mobile wood chipper, "Chiparvestor," that can handle trees up to 0.76 m in diameter, and the medium-size unit that chips trees 0.56 to 0.69 m in diameter (Biomass Energy Research Association, 1990). The latter unit can produce 544 t of chips in 8 h, while another model for small-diameter trees can produce 91 t of chips in 8 h. Commercial wood chippers, both mobile and stationary, and chip harvesting methods are far advanced in Finland, where the usage of wood chip fuels has increased greatly (Seppanen, 1988).

Among the other options that can be considered for producing wood are chunking, billeting, and crushing. For smaller trees, chunking and billeting are similar processes that cut stems and limbs into 6- to 20-cm long pieces. Machines have been designed that chop small-diameter stems into smaller chunks or slightly longer billets. Chunkwood is approximately fist-sized. The production of chunkwood requires less energy than chip production, but it is not certain that the cost is competitive (Suadicani and Heding, 1992). Crushing is carried out by passing the stems between two or more metal rolls of varying size, rotational speed, and surface. Tests have shown that crushing rates of approximately 15 linear m/min can be achieved on stems up to 21 cm in diameter using only 11.2 kW (15 HP) of power (cf. Ranney et al., 1987). The crushing and bunching of wood may offer significant advantages over chipping. This technique is flexible and is able to process large stems and stem lengths to yield bolts of crushed wood that exhibit relatively rapid drying. For reactor feeding purposes, however, further size reduction would be necessary. The feedstock characteristics required for the combustion or conversion process used determine which of these methods of size reduction may be applicable.

C. STEAM EXPLOSION

The treatment of wood chips with steam at elevated pressures and temperatures for short time periods followed by rapid decompression changes the physical state of the woody structure by defibration. Although some chemical changes occur with the hemicelluloses and lignins in this process, the particle sizes are reduced and surface areas and pore volumes are increased, so some discussion of the technique is warranted. The process was originally developed in 1925 and has been extensively used in the manufacture of hardboard (Spalt, 1977). The commercial process involves pressurization with saturated steam at pressures up to about 7 MPa. The process has also been proposed for the pretreatment of lignocellulosic feedstocks in the production of fermentation ethanol (Chapter 11) because of the large increase in accessibility of the cellulosic fraction to enzymatic hydrolysis (cf. Schultz, Biermann, and McGinnis, 1983; Mes-Hartree, Hogan, and Saddler, 1987; Foody and Foody, 1991).

In a series of steam-explosion experiments with different wood chips, sugarcane bagasse, ground corn stover, and ground rice hulls at selected temperatures between 190 and 250°C and treatment times of 1 or 2 min, the results were as follows (Schultz and McGinnis, 1984): All material was defibrated; the hemicelluloses were at least partially degraded, while the remaining hemicelluloses were extractable with hot water; the lignins were depolymerized by cleavage of the ether linkages, while at the higher temperatures, they became moderately condensed; and the enzymatic hydrolysis rates of the steam-exploded material increased dramatically for all materials except corn stover. Corn stover hydrolyzed at high rates without treatment. This study and others on steam explosion suggest that the technique can be used for several different biomass applications ranging from modifying the fibrous structure and particle sizes alone at the lower temperatures to a combination of physical and chemical changes at the higher temperatures.

IV. DENSIFICATION

A. FUNDAMENTALS

Baling has long been used to densify hays, straws, and other agricultural crops such as cotton to simplify removal from the field and to reduce storage space and transportation costs. Baled straw has a density of 70 to 90 kg/m³ at 10 to 15 wt % moisture content, whereas the bulk density of piled straw is about 5 to 15% of this density range. When straws are compressed to form pellets, briquettes, or cubes in specially designed dies and presses, the density can be increased to 350 to 1200 kg/m³. In contrast, dried wood has a density of 600–700 kg/m³ and a bulk density of about 350 to 450 kg/m³, whereas the bulk densities and densities of wood briquettes are 700 to 800 kg/m³ and up to 1400 kg/m³, respectively.

One of the original uses of biomass pellets in the United States was as fodder. Alfalfa, other grasses, and some straws were pelletized and sold as livestock feed. Biomass densification appears to have the greatest use for upgrading agricultural and forestry residues that might otherwise be lost or that require disposal at additional cost. The potential advantages for energy and feedstock applications of densified waste biomass are evident. High-density, fabricated biomass shapes simplify the logistics of handling and storage, improve biomass stability, facilitate the feeding of solid biomass fuels to furnaces and feedstocks to reactors, and offer higher energy density, cleaner burning solid fuels that in some cases can approach the heating value of coals. However, the basic problem often encountered in the use of densified biomass fuels and feedstocks is production cost. Some of the economic factors are discussed in Part D.

The heating value depends on the moisture and ash contents of the densified material and is usually in the range of 15 to 17 MJ/kg. The use of asphaltic binders or pelletizing conditions that result in some carbonization can yield densified products that have higher heating values. Pellets, briquettes, and logs have been manufactured by densification methods from biomass for many years. "Prestologs" made from waste wood and sawdust were marketed before 1940 in North America, and the market for pellet fuels made from wood sawdust, shavings, and chips for residential pellet-burning stoves has grown significantly since the 1980s (Pickering, 1995; Folk and Govett, 1992). Numerous commercial processes for production of densified fuels in the form of logs, briquettes, and pellets from a wide range of biomass provide domestic fuels for space heating; industry uses the pellets and briquettes as boiler fuels (Edwards, 1991). In Europe, briquettes made from waste biomass are commercially available and are used for both residential and industrial applications. In Spain, households consume 80% of the total production of briquettes for use in furnaces, fireplaces, and barbecues. Bakeries use them in furnaces, and small industries such as ceramic plants use them as boiler fuel (Ortiz, Miguez, and Granada, 1996).

B. COMPACTION METHODS AND HARDWARE

Numerous devices and methods of fabricating solid fuel pellets and briquettes from a variety of biomass, especially RDF, wood, and wood and agricultural residues, have been developed and patented. The pellets and briquettes are manufactured by extrusion and other techniques. A binding agent such as a thermoplastic resin may be incorporated during fabrication. A ring-die extrusion or die and roller mill is the most widely used machine type in wood pelleting, although punch and die technology has been developed (Folk and Govett, 1992). Other types of pelleting machines include disk pelletizers, drum and rotary cylinder pelletizers, tablet presses, compacting and briquetting rolls, piston-type briquetters, cubers, and screw extruders.

An exemplary method for production of pellets was developed in 1977 (Gunnerman, 1977). A raw material of random particle size such as sawdust or other wood residue, from which rocks, tramp metal, and other foreign materials are removed, is conveyed to a hammermill where particle size is adjusted to a uniform maximum dimension that is about 85% or less of the minimum thickness of the pellets desired. The milled product is then dried in a rotary drum dryer to a moisture content of about 14 to 22 wt % and fed through a ring-shaped die capable of generating pressures between 55 and 275 MPa to afford the desired shape and diameter. The pellet mill die and roller assembly must be capable of producing sufficient compression within the die to raise the temperature of the material to about 160-177°C. The products from the mill have a low, uniform moisture content, a maximum cross-sectional dimension of 13 mm, a density of 400 kg/m^3, and a heating value of 19.8 to 20.9 MJ/kg. It is not necessary to add a binder to the particles, providing the pressure during pelleting produces the necessary temperature increase. During extrusion, the lignins in the biomass migrate to the pellet surface and form a skin on cooling that protects the pellet from shattering and from any rapid change in moisture content before use. This same basic procedure has been used over the years in several different hardware designs.

Briquettes are formed by similar procedures except the products are usually larger in diameter and length than pellets. Briquetting is described to consist of subjecting wood residues containing 8 to 15 wt % moisture at a maximum particle size of 0.5 to 1.0 cm to a pressure of about 200 MPa, which increases the temperature about 100-150°C (Ortiz, Migues, and Granada, 1996). The major machine types used to manufacture briquettes are impact, extrusion, hydraulic, pneumatic, and double-roll presses, and die presses that can also be used for pellet production. Briquette production rates are 200 to 1500 kg/h for impact presses, but some models can produce 2000 to 6000 kg/h; 500 to 2500 kg/h for extrusion presses; and up to 5000 kg/h for hydraulic and pneumatic presses. The pellet machines suitable for pellet or briquette production contain annular or flat dies. The production rates are as high as 25,000 kg/h.

A few examples of typical biomass densifiers, feedstocks, and densified products are shown in Table 6.4. The first six examples in this table are commercial or commercially available systems, the last of which, Biotruck 2000, is unique (Sutor, 1995). It is a moving vehicle of special design that

TABLE 6.4   Typical Biomass Densification Hardware, Feedstocks, and Products[a]

| Machine | Feedstock | | Densified bulk product | | |
	Type	Moisture (wt %)	Size (cm)	Density (kg/m^3)	Moisture (wt %)
Impact press[a]	Wood residues	15–17	Briquettes	990–1200	8–15
Extrusion press[a]	Wood residues	10–20	Briquettes	1300–1400	8–15
Hydraulic press[a]	Wood residues		Briquettes	590–800	8–15
Briquetting machine[a]	Wood residues		Briquettes	>990	8–15
Pelleting machine[a]	Wood residues	8–15	0.5–2.5 dia.	800	8–15
Biotruck 2000[b]	Hay and straws		6 × 1.4 × 4	800–1200	22
Extruder[c]	Hogged bark, some wood	56.5	5.7 dia.	1070	34.8
Extruder[c]	Western hemlock sawdust	64.2	5.7 dia.	1100	36.5
Flat die press[d]	Fine straws	10–20	0.6–2.0 dia.	450–650	10–15

[a]Ortiz and Gonzalez (1993), Ortiz, Miguez, and Granada (1996).
[b]Sutor (1995).
[c]Edwards (1991).
[d]Wilen et al. (1987).

continuously performs all of the operations in the field from harvesting agricultural virgin biomass to pellet production. The operating sequence consists of the integration into one machine of continuous crop harvesting, size reduction to about 0.6-mm pieces, heating the pieces to temperatures between 80 and 120°C using the waste heat of the engine, and compressing the heated pieces in a toothed-wheel pelleting press. No binder is used. The production rate of pelletized cereal crops is about 8000 kg/h and the bulk density is 500 to 700 kg/m³. In addition to cereal crops, the agricultural biomass suitable for harvesting and conversion to pellets by this system include grasses such as Chinese silvergrass, switchgrass, and hays and straws. Pellets for both feed and fuel applications are produced with Biotruck 2000.

Another unique example of densification listed in Table 6.4 is the production of high-density, moisture-resistant briquettes from wet wood residues without predrying or the use of binders (Edwards, 1991). The briquettes do not disintegrate when wet and retain a maximum of about 40 wt % moisture after immersion in water. They are made from wood and bark alone or from mixtures in a pilot extruder at operating ram pressures typically ranging from 30 to 50 MPa at a maximum surface temperature of about 210°C. Moisture-resistant briquettes were made in tests from Western hemlock sawdust, a 50 : 50 mixture of Western hemlock and red cedar sawdusts, and Western hemlock bark hog fuel. The feed contains up to about 65 wt % moisture and must be sized so that the maximum size is less than 80% of the barrel diameter. The key to using wet biomass appears to be the simultaneous removal of excess moisture in the initial portion of the extruder while the feedstock is heated under pressure as it moves through the barrel, and the reduction of the temperature to less than 100°C before the briquettes leave the barrel to avoid the risk of explosive flash evaporation. Briquettes made in this manner contained about 35 wt % moisture. Over a 24-h period of immersion in water, they exhibited 0% swelling and only small increases in density and moisture content. The upper limit of the moisture content after immersion was consistently near 40 wt %.

Although not listed in Table 6.4, the combustible fraction of municipal solid waste, RDF, is commercially available as pellets that are similar to those produced from agricultural and woody residues (cf. Davis and Koep, 1990). The pellets have heating values of about 16.3 to 18.6 MJ/kg, moisture contents of 8 to 10 wt %, less than 10 wt % ash, and densities of 600 to 700 kg/m³.

Development of other densification methods for certain agricultural residues is expected to lead to improvements in soil growth characteristics as well as advanced residue recovery systems for energy applications. For example, cotton is a major crop in the state of Arizona. State law requires that cotton plant residue must be buried to prevent it from serving as an overwintering site for insect pests such as the pink bollworm. Research is underway to develop two systems for collecting and densifying this residue to facilitate removal from the field (Coates, 1995). The stalks are first pulled with an implement developed for the purpose. They are then baled using equipment that produces large round bales, or chopped with a forage harvester and converted into modules. The bales are either 1.2 m in diameter x 1.2 m long, or 1.8 m in diameter x 1.5 m long, depending on the baler used. The modules measure 2.1 m x 2.2 m in cross section and are up to 9.6 m long. The densities of the round bales are 93 to 168 kg/m³. The modules have densities of 168 to 252 kg/m³. The energy required to harvest and densify the residues is 9.2 kWh/t for the bales, and 8.6 kWh/t for the modules, and the heating values of the densified residues are about the same as those of wood. The module system produced a denser package than the baling system, and also made loading easier using truck-mounted module movers.

C. Standards for Biomass Pellet Fuels

U.S. standards for biomass pellet fuels have been developed and recommended by the Pellet Fuels Institute in the United States; they are shown in Table 6.5. The older standards includedrecommendations for moisture content and heating value, but these do not. Instead, it is recommended that the heating value be certified by the pellet manufacturer, so whatever the pellet material

TABLE 6.5   Recommended U.S. Residential Pellet Fuel Standards[a]

Parameter	Premium grade	Standard grade
Material	Disclose; i.e., wood, paper, ag residue, etc.	Same
Maximum moisture	Not specified	Same
Minimum heating value	Not specified, should be certified by manufacturer	Same
Inorganic ash	Less than 1%	Less than 3%
Sodium	Disclosed	Same
Maximum fines	0.5 wt % through a 2.8 mm (1/8 in. screen)	Same
Minimum bulk density	641 kg/m³ (40 lb/ft³)	Same
Length	None longer than 3.38 cm (1.5 in.)	Same
Diameter	Not specified	Same

[a]Pellet Fuels Institute (1995). The maximum moisture content, heating value, and diameter recommended in pellet standards published in the 1984 and 1988 were 8 to 10 wt %, 18.6 to 19.1 MJ/kg (8000 to 8200 Btu/lb), and 6 to 8.9 mm (0.235 to 0.350 in.), respectively. These parameters are not specified in the standards published in 1995, which are the current recommendations (7-97).

and its moisture content, the consumer should be able to estimate the energy cost. The national standards in Table 6.5 make it possible for the manufacturers of pellet stoves, most newer versions of which auger-feed the pellets from the top, to produce units designed to accept the standardized pellets.

D. ECONOMIC FACTORS

The wholesale cost in the United States of wood waste pellets is in the range of $85 to $140/t (mid-1997). This cost range effectively precludes their use as feedstocks for most conversion processes, and it limits residential fuel applications. The production cost exclusive of biomass cost is estimated to be about 30 to 60% of the wholesale cost and depends on production rates and the amount of processing needed. For example, in Spain, the increase in electric energy consumption required to mill wood wastes to 5- to 8-mm sizes is almost totally compensated for by the decrease in electric energy consumption during densification (Ortiz and Gonzalez, 1993). Exclusive of wood cost, the cost of manufacturing densified wood residues in small units operated by one person is about $22/t at a production rate of 1250 t/year. Smaller particles in the 2-mm size range can increase production rates by 50% or more, but the energy cost is excessive. Industrial manufacturing costs in Spain of densified wood wastes exclusive of wood cost are about $32 to $48/t at production rates of 1.0 t/h (Ortiz, Miguez, and Granada, 1996). In Finland, the cost of producing straw fuel pellets on farms in small, portable pelletizers is estimated to be about $54 to $84/t (Wilen *et al.*, 1987). Note that the hardware cost can be a major factor in the cost of producing densified biomass. Biotruck 2000, described earlier, for producing pellets or briquettes from agricultural wastes in Europe has a production rate of about 8 t/h in the field and costs about $400,000 (700,000 DM) (Sutor, 1995).

V. SEPARATION

A. FUNDAMENTALS

It is sometimes desirable to physically separate potential biomass feedstocks into two or more components for different applications. The subject is quite broad in scope because of the wide range of biomass types processed and the variety of separation methods that are used. Even the harvesting of virgin biomass involves physical separation technologies. Examples are the separation of agricultural biomass into foodstuffs and residues that may serve as fuel or as a raw material for synfuel manufacture, the separation of forest biomass into the darker bark-containing fraction and the pulpable components, the separation of marine biomass to isolate various chemicals, the separation of urban refuse into RDF and metals, glass, and plastics for recycling, and the separation of oils from oilseeds. Common operations such as screening, air classification, magnetic separation, extraction, mechanical expression under pressure, distillation, filtration, and crystallization are often used as well as industry-specific methods characteristic of farming, forest products, and specialized industries. Since the biomass types are so numerous and the physical separation methods are usually customized, some details of a few specific examples are described here to illustrate the scope of the subject, and how separation is performed. A few potential applications of physical separation methods are also described.

B. MUNICIPAL SOLID WASTE

MSW is a complex mixture of inorganic and organic materials (Table 5.1). Efficient separation and economic recovery of RDF and the components that can be recycled is the ultimate challenge to engineers who specialize in designing resource recovery equipment for the large-scale processing of solid wastes generated by urban communities. Unfortunately, the number of MSW plants designed to recover recyclables and RDF make up only about 20 to 25% of the total MSW-to-energy facilities in the United States. This is probably caused by the success of mass burn technologies (Chapter 7) and the fluctuating markets for recyclables. Nevertheless, the processing schemes and hardware employed to separate MSW are innovative and justify some elaboration. Literally hundreds of hardware designs and machines have been developed to separate and recover most of the components in MSW. Some resource recovery facilities have even installed equipment for recovering coinage, which is just a small fraction of the total mass of MSW.

One of the first comprehensive resource recovery plants in the world was built in Dade County, Florida (Todd, 1984; Berenyi and Gould, 1988). A brief description of this facility when it was in full-scale operation to recover recyclables and RDF is informative. The plant was designed to process 2720 t/day (3000 ton/day) of MSW, but it frequently processed over 3630 t/day (4000 ton/day), and could process 4540 t/day (5000 ton/day) if only household garbage were received. It was designed to accept, in addition to household garbage, a wide variety of solid wastes including trash, garden clippings, trees, tires, plastics, pathological wastes, white goods (i.e., stoves, refrigerators, air conditioners, etc.), and industrial, commercial, and demolition wastes. RDF and shredded tires, approximately 1000/day, were burned for on-site power generation in a 77-MW power plant, and glass, aluminum, ferrous metals, and other materials including the ash and flyash were recovered and sold. The plant achieved a 97% volumetric reduction compared to as-received MSW. Only 6 wt % of the total incoming MSW remained as unsalable residue; this was disposed of in a landfill. The plant also conformed to all effluent, leachate, emissions, noise, and odor requirements. Impressive results such as these depended on the availability and reliability of efficient separation methods.

A simplified description of the first comprehensive materials recovery facility of its type in the United States illustrates how one plant was designed to accomplish some of these separations (Waste Management, Inc., 1977). The plant, called Recovery 1, was built in New Orleans, Louisiana, to process 590 t/day of MSW. The waste was delivered and unloaded at one of two receiving pit conveyors, and transported by conveyors to the first separation unit, a 13.7-m long by 3-m diameter rotating trommel that contained circular holes 12 cm in diameter. Plastic and paper bags tumbling in the trommel were broken open by lifters. The smaller, heavier objects such as heavy metal and glass bottles that fell through the holes were transported directly to a magnetic ferrous recovery station and an air classifier. The larger and lighter materials such as paper, textiles, and aluminum containers that passed through the trommel were conveyed to a 746-kW primary shredder. This shredded material was then conveyed to the ferrous recovery station and the air classifier. In the air classifier, a high-speed air current blows the light materials out of the top of the classifier. This fraction, RDF, consists of shredded paper, plastic, wood, yard wastes, and food wastes. The heavy fraction is essentially glass, aluminum, other nonferrous metals, and some organic material. It was routed to the recovery building for further processing. A secondary, 746-kW shredder system handled oversized, bulky wastes without passage through the trommel. The output was also conveyed to the air classifier, where RDF was obtained as the overhead, and the heavy fraction was conveyed to the recovery building. Each shredder system was sized to process 590 t of MSW in about 12 h to ensure operating reliability.

Three modules were located in the recovery building. The first module consisted of a vibrating screen to separate the shredded material by particle size, a drum magnet to separate residual ferrous material, an eddy current separator to remove the nonmagnetic aluminum and other nonferrous metals, and a small hammermill to further shred the aluminum fraction to increase its bulk density. The output from the first module consisted of the ferrous fraction, the aluminum fraction, and a fraction that contained primarily glass and some nonferrous metals. The glass fraction containing some residual nonferrous metal was conveyed to the second recovery module, which consisted of a crusher, another vibrating screen, a rod mill, and a two-deck, fine-mesh vibrating screen. The glass fraction was crushed and screened in the second module. The smaller fraction was treated with a pulsed water stream that separated the light fraction, which was discarded. The heavier glass fractions were pumped as slurries to the bottom deck of the fine-mesh second screen to separate the larger particles for crushing in the rod mill. Recycling of the milled material back to the top deck of the fine-mesh screen yielded a glass cullet fraction for further treatment in the third module, and a nonferrous metal fraction which was removed from the second screen. The third module contained a hydrocyclone, a froth flotation tank, and a glass dryer. The glass cullet fraction from the second module was mixed with clean water in a prefloat tank to remove any remaining organic particles, separated from the slurry through centrifugal separation and froth flotation, dried, and conveyed to the loadout building for shipment. RDF was recovered from the air classifier, and the ferrous, aluminum, and glass fractions were recovered from the "bottoms" of the classifier.

This is a simplified description of how MSW is separated into recyclables and fuel. There are many refinements of these operations.

C. VIRGIN BIOMASS

The production of virgin biomass for food and feed has progressed from very labor-intensive, low-efficiency agricultural practices over the 1800s and 1900s to what some consider to be a modern miracle. The invention of numerous agricultural machines in the late 1700s and 1800s that can seed the earth and reap the harvests with minimal labor and energy inputs made it possible to continuously produce biomass in quantity to help meet the massive demand for foodstuffs and other farm products caused by the growing population. In the United States today, only a few percent of the population living on farms is sufficient to produce enough food to meet all the nation's demands for foodstuffs as well as supply surplus amounts for export. Farm equipment is available so that almost all row and grain crops can be continuously planted and harvested and separated into foodstuffs, feed, and residual materials. Eli Whitney's cotton gin and Cyrus McCormick's reaper are just two of the devices that helped mechanize agriculture and change the course of history by providing non-labor-intensive methods of physically separating the desired products, cotton and grain for these particular inventions, from biomass. As candidate energy crops evolve, such as several of the thick-stemmed grasses that are difficult to harvest at high growth densities, new agricultural equipment designs and adaptations of existing machinery are expected to solve these problems also.

Simultaneously with the advancement of agriculture, although not via the same pathway, new hardware and improved methodologies were developed for the planting, managing, and harvesting of trees that made large-scale commercial forestry operations more economic and less dependent on labor. Better methods of land clearing, thinning, and growth management, and improved hardware for harvesting, such as feller-bunchers, which were first used in the early 1970s, resulted in a modern forest products industry that supplies commercial and industrial needs for wood and wood products. As the use of trees for energy and feedstocks expands, it is expected that much of the existing commercial hardware and improvements will be applied to meet these needs.

Equipment and methods for the harvesting of the smaller short-rotation woody crops at low cost are also expected to be developed. Much of the ongoing work to design improved equipment for SRWC is directed to feller-bunchers that perform severing, bunching, and off-loading functions. The results from systems analysis studies indicate that prototype feller-buncher harvesters can be balanced with two or three small grapple skidders to move bunched SRWC to a landing for chipping or just loading in the case of whole trees (Perlack *et al.*, 1996). However, because of the high skidding costs, a whole-tree, direct-load system for use with a track-type feller-buncher is preferred.

A few of the nonmanual separation methods used for woody biomass processing that have use in energy applications are briefly described here. Delimbing and debarking of trees is an old technology. For the smaller trees where fiber in the form of white wood chips is the desired product, the trees can be debarked and delimbed by the use of chain flails, which remove the outer bark layer, leaving the white wood behind. Hammermilling then yields a homogeneous product (Hudson and Mitchell, 1992). In most thermochemical energy applications, however, separation of the bark and wood is not necessary. But where it is necessary to remove the bark, some efforts have been made to recover the residues for fuel from flail machines by using them together with tub grinders (Stokes, 1992). A tub grinder operating simultaneously with a chain flail was successfully used to comminute the residues (Baughman, Stokes, and Watson, 1990). The green weight of the fuel residues was about one-fourth to one-third of the total clean chip-plus-fuel weight.

In a few installations that burn hogged wood, disc and shaker screens have been employed to separate preselected, oversize pieces for subsequent size reduction and return to the fuel stream. Finely divided wood fuels such as sawdust and sanderdust are also sometimes screened to remove the larger pieces.

By-product hulls from the production of rice, cotton, peanut, soybean, and similar crops that have outer shells covering small seed or fruit are sometimes used directly as fuels or feedstocks. After the shells are fractured, most of the hulls can be separated with vibrating screens or rotating trommels having appropriately sized openings. The by-product hulls that have high ash contents and bulk densities present a few difficulties on direct combustion or gasification, but specially designed systems are available to eliminate these problems (*cf.* King and Chastain, 1985; Bailey, 1990; Bailey and Bailey 1996).

D. Extraction

Solvent extraction of biomass, its derived ash, or biomass parts such as the seeds has been or is currently used commercially to isolate and separate certain chemicals or groups of related compounds that are present. Inorganic salts are found in some biomass species at concentrations that may justify extraction and purification (Chapter 3). Aqueous extraction of the ash from giant brown kelp and the spent pulp of sugar beet and fractional crystallization of the extract, for example, were commercial processes for the manufacture of potassium compounds in the early 1900s. Examples of some of the organic compounds that are extracted with solvents are triglycerides, terpenes, and lignins. Water and water in mixtures with polar solvents have been used for extraction of several of the low-molecular-weight, water-soluble sugars. Some detail on the extraction of lignins illustrates how solvent extraction processes might be developed.

Aqueous organic solvents are effective for the selective extraction of lignins in biomass. Lignins can also be extracted from biomass by use of dilute aqueous alkali under mild conditions (*cf.* Lawther, Sun, and Banks, 1996), but aqueous alcohols alone such as 50% ethanol solubilize lignins in wood, leaving relatively pure undecomposed cellulose (Aronovsky and Gortner, 1936; Nikitin *et al.*, 1962). Deciduous trees are delignified by aqueous ethanol extraction to a greater extent than conifers. Lignin is also readily extracted by mixtures of butanol or amyl and isoamyl alcohols with water. Separation of the lignins from the extracts yields tarlike substances that become brittle on cooling. Since one of the prime objectives of producing chemical pulps from wood is delignification without changing the cellulosic fibers, the data accumulated on the solvent extraction of wood suggests that high-quality paper pulps could be manufactured by solvent extraction of hardwoods and softwoods as well as other biomass species. The lignins in the extracts might provide the starting point for the production of new lignin derivatives and polymers.

Solvent extraction of biomass under relatively mild conditions to remove lignins by a strictly physical process without the addition of other chemicals would seem to offer several advantages over chemical pulping methods. Solvent recoveries approaching 100% should permit solvent recycling with minimal losses. A continuous process for the pulping of wood with aqueous *n*-butanol, which was found to be the most effective solvent, has been proposed for the pulping of wood and the separation of the lignins (Hansen and April, 1981).

This type of process, which would be expected to be environmentally benign, does not seem to have been commercialized to any extent by the pulp industry (*cf.* U.S. Dept. of Energy, 1995).

REFERENCES

Aronovsky, S. I., and Gortner, R. A. (1936). *Ind. Eng. Chem.* 28 (11), 1270.

Bailey, R. W. (1990). *In* "Energy from Biomass and Wastes XIII," (D. L. Klass, ed.), p. 655. Institute of Gas Technology, Chicago.

Bailey, R. W., and Bailey, R., Jr., (1996). *In* "Bioenergy '96" Proceedings of the Seventh National Bioenergy Conference," Vol. I, p. 284. The Southeastern Regional Biomass Energy Program, Muscle Shoals, AL.

Baughman, R. K., Stokes, B. J., and Watson, W. F. (1990). *In* "Harvesting Small Trees and Forest Residues Workshop," (B.J. Stokes, ed.), p. 21, Proceedings for International Energy Agency, Task VI, Activity 3 Workshop, May 28, 1990. U.S. Department of Agriculture, Forest Service, Southern Forest Experiment Station, Auburn, AL.

Berenyi, E., and Gould, R. (1988). "1988–89 Resource Recovery Yearbook." Governmental Advisory Associates, Inc., New York.

Biomass Energy Research Association (1990). "A Directory of U.S. Renewable Energy Technology Vendors." Produced for the U.S. Agency for International Development, Washington, D.C.

Bircher, K. G. (1982). *In* "Energy from Biomass and Wastes VI," (D. L. Klass, ed.), p. 707. Institute of Gas Technology, Chicago.

Coates, W. (1995). *In* "Second Biomass Conference of the Americas: Energy, Environment, Agriculture, and Industry," p. 130, NREL/CP-200-8098, DE95009230. National Renewable Energy Laboratory, Golden, CO.

Coble, C. G., and Egg, R. (1989). *In* "Energy from Biomass and Wastes XII," (D. L. Klass, ed.), p. 361. Institute of Gas Technology, Chicago.

Crumm, C. J., II, and Crumm, K. A. (1984). *In* "Energy from Biomass and Wastes VIII," (D. L. Klass and H. H. Elliott, eds.), p. 273. Institute of Gas Technology, Chicago.

Curtin, D. T., and Barnett, P. E. (1986). "Development of Forest Harvesting Technology: Application in Short Rotation Intensive Culture (SRIC) Woody Biomass," Technical Note B58, TVA/ ONRED/LER-86/7. Tennessee Valley Authority, Muscle Shoals, AL, January.

Davis, R. A., and Koep, M. (1990). *In* "Energy from Biomass and Wastes XIII," (D. L. Klass, ed.), p. 583. Institute of Gas Technology, Chicago.

Edwards, W. (1991). *In* "Energy from Biomass and Wastes XV," (D. L. Klass, ed.), p. 503. Institute of Gas Technology, Chicago.

FBT, Inc. (1994). "Fluidized Bed Combustion and Gasification: A Guide for Biomass Waste Generators," Report for the Southeastern Regional Biomass Energy Program. Tennessee Valley Authority, Muscle Shoals, AL, July.

Folk, R. L., and Govett, R. L. (1992). "A Handbook for Small-Scale Densified Biomass Fuel (Pellets) Manufacturing for Local Markets." U.S. Department of Energy, Bonneville Power Administration, Portland, OR, July.

Foody, B. E., and Foody, K. J. (1991). *In* "Energy from Biomass and Wastes XIV," (D. L. Klass, ed.), p. 1225. Institute of Gas Technology, Chicago.

Ganotis, C. G. (1988). *In* "Energy from Biomass and Wastes XI," (D. L. Klass, ed.), p. 247. Institute of Gas Technology, Chicago.

Gunnerman, R. (1977). U.S. Patent 4,015,951, U.S. Patent Office, Washington, D.C.

Hansen, S. M., and April, G. C. (1981). In "Fuels from Biomass and Wastes," (D. L. Klass and G. H. Emert, eds.), p. 327. Ann Arbor Science Publishers, Ann Arbor, MI.

Haug, R. T., Moore, G. L., and Harrison, D. S. (1995). In "Second Biomass Conference of the Americas: Energy, Environment, Agriculture, and Industry," p. 734, NREL/CP-200-8098, DE95009230. National Renewable Energy Laboratory, Golden, CO.

Hudson, J. B., and Mitchell, C. P. (1992). Biomass and Bioenergy 2(1–6), 121.

Institute for Energy Studies (1977). In "Biomass Energy for Hawaii," Vol. II, (C. Beck, ed.), Appendix E, p. 4-107. Stanford University, Stanford, CA.

Ismail, A., and Quick, R. (1991). In "Energy from Biomass and Wastes XV," (D. L. Klass, ed.), p. 1063. Institute of Gas Technology, Chicago.

King, D. R., and Chastain, C. E. (1985). In "Energy from Biomass and Wastes IX," (D. L. Klass, ed.), p. 437. Institute of Gas Technology, Chicago.

Lawther, J. M., Sun, R.-C. Sun, and Banks, W. B. (1996). Ind. Crops and Products 5 (2), 97.

McMinn, J. W. (1986). Forest Products J. 36 (3), 25; "Transpirational Drying of Piedmont Hard-woods" Research Paper No. 63, Georgia Forestry Commission, Athens, GA, April.

Mes-Hartree, M., Hogan, C. M., and Saddler, J. N. (1987). In "Energy from Biomass and Wastes X," (D. L. Klass, ed.), p. 879. Institute of Gas Technology, Chicago.

Miles, T. R. (1984). In "Thermochemical Processing of Biomass," (A. V. Bridgwater, ed.), p. 69. Butterworths & Co., Ltd., London.

Miller, B. (1987). In "Energy from Biomass and Wastes X," (D. L. Klass, ed.), p. 723. Elsevier Applied Science Publishers, London, and Institute of Gas Technology, Chicago.

Mirov, N. T. (1949). In "Trees The Yearbook of Agriculture," (A. Stefferud, ed.), p. 1. U.S. Department of Agriculture, Washington, D.C.

Mislevy, P., and Fluck, R. C. (1993). In "Energy from Biomass and Wastes XVI," (D. L. Klass, ed.), p. 479. Institute of Gas Technology, Chicago.

Murphy, M. L. (1991). In "Energy from Biomass and Wastes XV," (D.L. Klass, ed.), p. 1167. Institute of Gas Technology, Chicago.

Nikitin, N. I. et al. (1962). "The Chemistry of Cellulose and Wood" (Translated in 1966 from Russian by J. Schmorak, Israel Program for Scientific Translations, Jerusalem, Israel). Academy of Sciences of the USSR, Institute of High Molecular Compounds, Moscow-Leningrad.

Normoyle, G. B., and Gershengoren, E. (1989). In "Energy from Biomass and Wastes XII," (D. L. Klass, ed.), p. 673. Institute of Gas Technology, Chicago.

Ortiz, L., and Gonzalez, E. (1993). In "First Biomass Conference of the Americas: Energy, Environment, Agriculture, and Industry," Vol. I, p. 649, NREL/CP-200-5768, DE93010050. National Renewable Energy Laboratory, Golden, CO.

Ortiz, L., Miguez, J. L., and Granada, E. (1996). In "Bioenergy '96," Proceedings of the Seventh National Bioenergy Conference, Vol. II, p. 747. The Southeastern Regional Biomass Energy Program, Muscle Shoals, AL.

Ostlie, L. D., and Drennen, T. E. (1989). In "Energy from Biomass and Wastes XII," (D. L. Klass, ed.), p. 621. Institute of Gas Technology, Chicago.

Pellet Fuels Institute (1995). "Residential Pellet Fuel Standards." Arlington, VA, April.

Perlack, R. D., Walsh, M. E., Wright, L. L., and Ostlie, L. D. (1996). Bioresource Technology 55 (3), 223.

Pickering, W. H. (1995). In "Second Biomass Conference of the Americas: Energy, Environment, Agriculture, and Industry," p. 1190, NREL/CP-200-8098, DE95009230. National Renewable Energy Laboratory, Golden, CO.

Ranney, J. W., Wright, L. L., Trimble, J. L., Perlack, R. D., Dawson, D. H., Wenzel, C. R., and Curtin, D. T. (1985). "Short Rotation Woody Crops Program: Annual Progress Report for 1984," Publication No. 2541, ORNL-6160. Oak Ridge National Laboratory, Oak Ridge, TN, August.

Ranney, J. W., Wright, L. L., Layton, P. A., McNabb, W. A., Wenzel, C. R., and Curtin, D. T. (1987). "Short Rotation Woody Crops Program: Annual Progress Report for 1986," Publication No. 2839, ORNL-6348. Oak Ridge National Laboratory, Oak Ridge, TN, November.

Riegel, E. R. (1933). "Industrial Chemistry," The Chemical Catalog Company, Inc., New York.

Routly, W. L. (1991). *In* "Energy from Biomass and Wastes XV," (D. L. Klass, ed.), p. 1141. Institute of Gas Technology, Chicago.

Schultz, T. P., and McGinnis, G. D. (1984). "Final Report, Evaluation of a Steam-Explosion Pretreatment for Alcohol Production from Biomass," USDA-SEA Grant No. 59-2281-1-2-098-0, September 15, 1981 to October 1, 1984. Mississppi State University, Mississippi State, MS.

Schultz, T. P., Biermann, C. J., and McGinnis, G. D. (1983). *I&EC Product Research & Development* **22**, 344.

Seppanen, V. (1988). *In* "Energy from Biomass and Wastes XI," (D. L. Klass, ed.), p. 103. Elsevier Applied Science Publishers, London, and Institute of Gas Technology, Chicago.

Spalt, H. A. (1977). *ACS Symposium Series 43* p. 193. American Chemical Society, Washington, D.C.

Stokes, B. J. (1992). *Biomass and Bioenergy* **2** (1–6), 131.

Suadicani, K., and Heding, N. (1992). *Biomass and Bioenergy* **2** (1–6), 149.

Sutor, P. (1995). *In* "Second Biomass Conference of the Americas: Energy, Environment, Agriculture, and Industry," p. 1228, NREL/CP-200-8098, DE95009230. National Renewable Energy Laboratory, Golden, CO.

Tewksbury, C. (1987). *In* "Energy from Biomass and Wastes X," (D. L. Klass, ed.), p. 555. Elsevier Applied Science Publishers, London, and Institute of Gas Technology, Chicago.

Todd, J. H. (1984). *In* "Energy from Biomass and Wastes VIII," (D. L. Klass and H. H. Elliott, eds.), p. 353. Institute of Gas Technology, Chicago.

U.S. Dept. of Energy (1995). "Technology Partnerships," DOE/GO-10095-170, DE95004086. Office of Industrial Technologies, Washington, D.C., April.

Waste Management, Inc. (1977). "Recovery 1 Operational Outline," May; *The American City & County,* "Solid Waste Recovers Land for Industry Use," April.

Wilen, C., Ståhlberg, P., Sipilä, K., and Ahokas, J. (1987). *In* "Energy from Biomass and Wastes X," (D. L. Klass, ed.), p. 469. Elsevier Applied Science Publishers, London, and Institute of Gas Technology, Chicago.

Wiltsee, G. A., Jr., McGowin, C. R., and Hughes, E. E. (1993). *In* "First Biomass Conference of the Americas: Energy, Environment, Agriculture, and Industry," Vol. I, p. 347, NREL/CP-200-5768, DE93010050. National Renewable Energy Laboratory, Golden, CO.

CHAPTER 7

Thermal Conversion: Combustion

I. INTRODUCTION

The simple act of burning biomass to obtain heat, and often light, is one of the oldest biomass conversion processes known to mankind. The basic stoichoimetric equation for the combustion of wood, represented by the empirical formula of cellulose, $(C_6H_5O_5)_n$, is represented by

$$(C_6H_{10}O_5)_n + 6nO_2 \rightarrow 6nCO_2 + 5nH_2O.$$

Carbon dioxide (CO_2) and water are the final products along with energy. If most biomass did not sustain its own combustion to make heat readily available in preindustrial times when and where it was needed, our historical development would not have reached its present state, and would probably have taken a different course. Up to the early 1900s, much of industrialized society utilized biomass combustion and a few related thermal processes for a wide range of applications including heating, cooking, chemical and charcoal production, and the generation of steam and mechanical and electric power.

The science of combustion has advanced a great deal since then and improved our understanding of the chemical mechanisms involved. Improved combustion processes are available for conversion of virgin biomass and complex waste biomass feedstocks to heat, steam, and electric power in advanced combustion systems and in co-combustion systems supplied with both biomass and fossil fuels. Small-scale catalytic woodstoves have been developed that operate at higher overall thermal efficiencies with low emissions. Medium- to large-scale incinerators have been designed with heat recovery capability for efficient combustion and disposal of municipal solid wastes (MSW) with minimal emissions. And modern boiler systems are available for wood, municipal solid waste, refuse-derived fuel (RDF or the combustible fraction of MSW), and other biomass fuels for municipal and utility use. The technology has progressed far beyond the basic idea of just using biomass as a solid, combustible fuel.

It is noteworthy that of all the processes that can be used to convert biomass to energy or fuels, combustion is still the dominant technology. More than than 95% of all biomass energy utilized today is obtained by direct combustion.

In this chapter, the basic chemistry of direct biomass combustion, developments that have made it possible to improve operating efficiencies and environmental performance, and state-of-the-art systems that have been or are expected to be commercialized are examined. Improvements needed to overcome some of the operating problems and advancements that are expected from ongoing research are also discussed.

II. FUNDAMENTALS

A. DEFINITION

Complete combustion (incineration, direct firing, burning) of biomass consists of the rapid chemical reaction (oxidation) of biomass and oxygen, the release of energy, and the simultaneous formation of the ultimate oxidation products of organic matter—CO_2 and water. Chemical energy is released, usually as radiant energy and thermal energy, the amount of which is a function of the enthalpy of combustion of the biomass. In the idealized case, stoichiometric amounts of biomass and oxygen are present and react so that perfect combustion occurs; that is, each reactant is totally consumed and only CO_2 and water are formed. Under normal conditions, such combustion does not occur with most carbon-containing solid fuels, including biomass.

When biomass is combusted under normal conditions, a flame is produced as visible radiation, provided oxidation occurs at a sufficient rate. By use of thermodynamic data, the theoretical temperature at which the products of combustion form under adiabatic, reversible conditions can be calculated. The theoretical flame temperature for the combustion of wood of various moisture contents with excess air is shown in Fig. 7.1 (Tewksbury, 1991). Green wood generally contains about 50% moisture by weight in the field, and excess air is used to promote complete combustion. It is apparent from Fig. 7.1 that both high fuel moisture levels and excess air significantly reduce the theoretical flame temperature. It is also apparent that to achieve maximum flame temperature, dry fuel and small amounts of excess air are required. In actual practice, however, combustion is not adiabatic and the reactions that occur are irreversible, so the actual flame temperature is less than the theoretical value.

B. PROCESS STEPS

An extremely large amount of basic research has been carried out on the combustion of solid fuels, including fossil, biomass, and inorganic solid fuels, to ascertain the mechanisms and kinetics of the process. Each category of fuel combusts under different conditions which are determined by a variety of intensive chemical and physical properties of the solids and external ambient factors. An empirical view of biomass combustion involves the evaporation of the high-energy volatiles such as the terpenes, which burn in the gas phase with flaming combustion (cf. Shafizadeh and DeGroot, 1976). The lignocellulosics in the solid biomass, under the influence of high temperature or a sufficiently

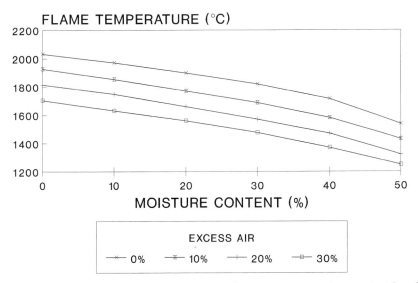

FIGURE 7.1 Theoretical flame temperature vs wood moisture content and excess air. Adapted from Tewksbury (1991).

strong energy source, decompose to form pyrolysis products, which also burn in the gas phase with flaming combustion. The residual char burns at a lower rate by surface oxidation or glowing combustion. The cellulosics are converted mainly to combustible and noncombustible volatiles, including water and CO_2, while the lignins contribute mainly to the char fraction.

The temperature within the flame is a function of reaction time, combustion intensity, flame velocity, and energy transferred to the surroundings. The flame temperatures measured on combustion of acetylene, gasoline, hydrogen, and natural gas in air under controlled conditions are 2319, 2310, 2045, and 1880°C, respectively (Reed, 1983). The combustion of biomass does not reach these temperature levels because of the lower energy density of biomass and the mechanism of biomass combustion. Simplistically, the mechanism involved in the combustion of solid biomass can be viewed as a stepwise process, although all steps occur simultaneously in the combustion chamber. First, the increasing temperature dries the incoming, fresh biomass. The physically contained moisture in the biomass is vaporized. At about 150 to 200°C, thermal decomposition and devolatilization of the solid biomass then begins on the biomass surface and volatile organic compounds are evolved as a gas, which burns in the combustion chamber. (Note that the discussion of biomass pyrolysis in Chapter 8 addresses some of the important research on the mechanisms and kinetics of biomass devolatilization which also apply to biomass combustion.) The remaining fuel components in the carbonaceous residue are combusted by diffusion of oxygen to the solid surface at temperatures of about 400 to 800°C and greater. This temperature range is attained by absorption of radiant energy from the hot combustion products and the combustion chamber surfaces. Temperatures as high as 1500°C have been recorded when the incoming fresh fuel is dry and the combustion process is carefully controlled. The use of preheated air permits similar temperatures in some systems even with biomass that contains some moisture.

C. EMISSIONS

During combustion, the chemically bound carbon and hydrogen in the various organic components of biomass are oxidized. Incomplete combustion can result in excessive emissions of particulate matter and partially oxidized derivatives, some of which are toxic. Chemically bound nitrogen and sulfur that may be present in the biomass are oxidized to nitrogen and sulfur oxides—mostly sulfur dioxide, SO_2, but some sulfur trioxide, SO_3; and mostly nitric oxide, NO, but some nitrogen dioxide, NO_2. Air is the usual source of the oxidant, oxygen, for biomass combustion. Small amounts of nitrogen in the air are also converted to nitrogen oxides at combustion temperatures according to the reactions

$$N_2 + O_2 \rightarrow 2NO$$
$$2NO + O_2 \rightarrow 2NO_2.$$

The equilibrium concentrations of NO formed from equimolar amounts of nitrogen and oxygen at various temperatures are shown in Table 7.1. It is evident that the higher the combustion temperature, the higher the NO concentration. The concentrations of chemically bound nitrogen and sulfur are zero to very low in most woody biomass species, but some biomass can contain relatively large amounts of these elements (Chapter 3). Elements such as chlorine, which can be present at relatively high concentrations in biomass such as MSW and RDF, but which are present in very small concentrations in woody biomass, are converted to chlorine compounds such as hydrogen chloride. Most of the chlorine derivatives are considered to be pollutants. Carbon monoxide, acid gases, unburned hydrocarbons, partially oxidized organic compounds, polycyclic aromatic derivatives, trace metal oxides, nitrogen and sulfur oxides, chlorine derivatives, particulate carbon, and flyash are found in the flue gases of poorly controlled systems. The amounts of ash formed on oxidation of the metallic elements in biomass can be minor or major combustion products, depending on the composition of the biomass fuel. Biomass combustion systems should be designed to approach complete combustion under controlled conditions as closely as possible to extract the maximum amount of thermal energy, minimize undesirable emissions, and meet environmental regulations.

The stoichiometric amount of oxygen is the minimum amount needed for complete combustion of the fuel. A limited amount of excess oxygen is often

TABLE 7.1 Thermodynamic Equilibrium Concentrations of Nitric Oxide from Equimolar Amounts of Nitrogen and Oxygen[a]

Temperature (°C)	Nitric oxide (ppm)
727	43
827	89
927	251
1027	500
1127	1000
1227	1630
1527	5460

[a]Tewksbury (1991).

used with solid fuels to promote complete combustion. Since ambient air contains about 79 mol % nitrogen and is the usual source of oxygen under normal conditions, nitrogen is a major constituent of the flue gas. The temperature attained in the combustion chamber depends on the rate of heat release, its dissipation and transfer, and the quantity of combustion gases. So the increase in combustion temperature is substantially less with air as the oxygen source compared to pure oxygen because of dilution of the combustion gases with nitrogen. The air-to-biomass mass ratio is therefore an important parameter because it affects the rate of combustion and the final temperature of the combustion gases. Oxygen-enriched air and the use of small fuel particles or powders have been employed to maximize combustion temperatures.

D. STOICHIOMETRIC MODEL

A theoretical model of the combustion of biomass is illustrated by the complete oxidation of giant brown kelp. Note that kelp, for which complete analytical data were available, is used here simply to illustrate the utility of the model, which is applicable to all biomass species. Based on the empirical formula derived from the elemental analysis of dry kelp at an assumed molecular weight of 100, the combustion stoichiometry is

$$C_{2.61}H_{4.63}N_{0.10}S_{0.01}O_{2.23} \, ash_{26.7} + 2.7625O_2 \rightarrow 2.61CO_2 + 0.10NO_2 + 0.01SO_2$$
$$+ 2.315H_2O + 26.7ash.$$

The experimentally measured ash content is assumed to be present in the original biomass and to be carried through the process unchanged. This is not strictly true since oxygen is chemically taken up as metal oxides are formed during standard ash determinations. The ash content is calculated as the difference between the weight of the residue after ashing the sample and the original sample weight, so it does not correspond to the actual ash-forming, metallic elements in the original, dry sample. But for purposes of illustrating the stoichiometry of complete combustion, this equation is adequate. The heat evolved by combustion of this particular sample of kelp is 12.39 MJ/kg (296.1 kcal/g-mol) with product water in the liquid state (Chapter 3). Since on the average, air is 20.95 mol % oxygen, the stoichiometric air requirement for complete combustion is 13.19 mol of air per mole of kelp, or an air-to-kelp mass ratio of 3.805. The ultimate concentration of CO_2 in the dry flue gas is 19.85 mol %.

Except for submerged combustion processes that are used for treatment of aqueous dissolved and suspended biosolids and a few other special combustion processes, the combustion of virgin and waste biomass involves solid fuels. Stoichiometric combustion data for four types of biomass, two coals, and one coke are compared in Table 7.2. Each of the biomass fuels is assumed to contain 15.0 wt % moisture. The stoichiometric air requirements are considerably less for biomass than for coals and cokes. The reason for this is that the C-to-H mass ratios of biomass are much less than those of fossil fuels. Also, most of the carbon in biomass is, effectively, already partially oxidized. Less oxygen is needed for complete oxidation. For the data in Table 7.2, it is assumed that organic nitrogen and sulfur in each solid fuel are oxidized to NO_2 and SO_2 and that nitrogen in air is inert. The calculated amounts of NO_2 and SO_2 formed on complete combustion are more than might be expected for a biomass fuel. The relatively high concentrations of organic nitrogen and sulfur in each biomass sample, except the pine wood sample, could potentially cause air pollution problems that require NO_x and SO_x removal from the combustion products before the flue gases are exhausted to the atmosphere. This will be discussed later. It is sufficient to state here that agricultural and forestry residues, wood chips, bagasse generated in sugarcane plantations, MSW, and RDF have been used as fuels for combustion systems for many years.

E. PHYSICAL PARAMETERS

The moisture content of green biomass can be quite high and can adversely affect the combustion process. If the moisture content is excessive, the combustion process may not be self-sustaining and supplemental fuel must be used, which could defeat the objective of producing energy by biomass combustion for captive use or market. High moisture can also cause incomplete combustion, low overall thermal efficiencies, excessive emissions, and the formation of products such as tars that interfere with operation of the system. Predrying of the fuel or blending it with dry fuel to reduce the equivalent moisture content before combustion may be necessary in these cases. Woody biomass fuels containing 10 to 20 wt % moisture are generally preferred for conventional biomass combustion systems. This moisture content range permits a close approach to complete combustion without incurring the costs of further biomass drying and allows temperatures in the combustion chamber to reach 750 to 1000°C. As already mentioned, lower moisture contents in biomass fuels can facilitate attainmant of even higher combustion temperatures.

Another factor in biomass combustion is fuel particle size and particle size distribution. The furnace design often determines the optimum ranges of these parameters. But in general, the smaller the fuel particles, the more rapid and complete the combustion process. The larger particles require longer residence times in the combustion chamber at a given temperature. In commercial systems, the capital and operating costs of fuel particle size reduction and predry-

TABLE 7.2 Estimated Stoichiometric Combustion Data for Some Solid Fuels[a]

Parameter	Pine wood	Kentucky bluegrass	Feedlot manure	RDF	Bituminous coal	Anthracite coal	Coke
Moisture, wt %	15.0	15.0	15.0	15.0	3.1	5.2	0.8
Higher heating value, MJ/kg	18.05	15.92	11.36	12.51	32.61	29.47	29.50
C/H wt ratio	8.2	7.8	6.6	7.5	16.0	33.6	106
Air/fuel wt ratio	5.37	5.51	3.97	4.25	10.81	9.92	10.09
Product CO_2, wt/wt fuel	1.90	1.68	1.29	1.51	2.94	2.96	3.12
Product H_2O, wt/wt fuel	0.56	0.53	0.47	0.49	0.49	0.22	0.07
N_2 from air, wt/wt fuel	4.85	4.97	3.58	3.83	8.26	7.58	7.73
CO_2 in dry flue gas, mol %	19.9	17.4	18.4	20.0	18.5	19.9	20.4
NO_2 in dry flue gas, mol %	0.032	1.55	1.13	0.21	0.0	0.0	0.266
SO_2 in dry flue gas, mol %	0.0	0.054	0.075	0.035	0.086	0.101	0.089

[a]The data for pine wood, Kentucky bluegrass, feedlot manure, and RDF were calculated from empirical formulas derived from the data in Table 3.5. Each biomass fuel was assumed to contain 15.0 wt % moisture. The data for the coals and coke are from Reed (1983), except for the data on NO_2 and SO_2 in the dry flue gas, which were calculated from the empirical formulas derived from the elemental analyses in Reed (1983). The overall assumptions are that combustion is complete, the ash and nitrogen in air are inert, and all organic nitrogen and sulfur are oxidized to NO_2 or SO_2.

ing are weighed against their beneficial effects on combustion and furnace design and costs.

III. EQUIPMENT AND APPLICATIONS

A. HARDWARE

The purposes of solid fuel-burning equipment are to proportion and mix the fuel and air, to initiate and maintain ignition, to volatilize the fuel, to position the flames in areas of useful heat release, and to supply fuel and air at the proper rates and pressures to facilitate each of these functions (Reed, 1983). The specific equipment appropriate for most biomass combustion and energy recovery systems depends on the types, amounts, and characteristics of the biomass fuel; the ultimate energy form desired (heat, steam, electric or cogenerated power); the relationship of the system to other systems in the plant (independent, integrated); whether recycling or co-combustion is practiced; the disposal methods needed for residues; and environmental factors. The design of efficient, large-scale biomass combustion systems requires detailed analysis of many parameters and hardware components. Among them are the numerical values and variability of moisture, volatile matter content, ash content, composition, and energy content of the biomass fuel; biomass handling, drying, and grinding equipment; the furnace design and associated heat transfer requirements and materials of construction; combustion and emissions controls; the amounts, composition, fusion temperature, agglomerating characteristics, and disposal of ash; and flue gas compositions and treatment that may be needed to meet emissions limitations.

In conventional biomass combustion equipment, combustion of the solid fuel takes place on horizontal or inclined steel grates or in shallow suspension above the grate. The grate is a stationary, vibrating, reciprocating, or traveling platform, and the fuel is supplied in the batch, semicontinuous, or continuous mode. Many furnace designs have been used such as pile-burning systems (Dutch ovens), fixed- and moving-bed furnaces, multiple hearth furnaces, stationary and rotating horizontal and inclined kilns, overfeed, underfeed, and spreader stokers, and pulverized fuel burners. The principal difference between conventional solid fuel-burning equipment and liquid-fuel- or gas-burning equipment is that furnaces for solid fuels must allow for additional fuel residence time for the slower burning chars to combust after all gases and volatile liquids have been driven off. One of the principal methods of expediting this process is by burning smaller fuel particles. Advanced combustion designs such as fluid-bed and cyclonic combustors further improve biomass combustion and are discussed in Part E.

The differences in the furnaces suitable for biomass combustion reside mainly in the design of the combustion chambers, the operating temperatures, and the heat transfer mechanisms. Refractory-lined furnaces operating at about 1000°C were standard until the introduction a few years ago of water-wall incinerators. Ash buildup can occur rapidly in refractory-lined furnaces, and excess air must be introduced to limit the wall temperature. The water-wall incinerator has combustion chamber walls containing banks of tubes through which water is circulated, thereby reducing the amount of cooling air needed. Heat is transferred directly to the tubes to produce steam. There are numerous configurations, but the basic concept has not changed for many years, apart from operating conditions and materials improvements to improve heat transfer and thermal efficiencies. Considerable advancements have been made, however, in ancillary hardware designs to control the combustion process and reduce emissions, to remove ash, and to remove flyash and emissions from the stack gas. Improvements have also been made in the methods used to recover sensible heat from the stack gases and heat from the condensate and boiler blowdown. Other overall efficiency improvements have resulted from advances in predrying hardware for moisture reduction in the incoming biomass fuel.

B. Residential and Small Commercial Systems

Residential biomass fuels are usually chunks and pieces of wood and logs that are burned in small woodstoves and fireplaces, in contrast to medium- and large-scale municipal, industrial, commercial, and power-generating facilities that burn a large variety of virgin and waste biomass fuels such as MSW, RDF, sawdust, bagasse, rice hulls, wood chips, and industry-specific wastes. The residential wood fuels range from about 15 to 50% moisture content depending on the extent of air drying. The moisture content of seasoned firewood is typically about 20%. Most of the older woodstoves consist of conventional downdraft, updraft, or crossdraft fireboxes with fixed steel grates. Airflow is adjusted manually and the flue connections are sized for maximum loadings. Electric air blowers are sometimes used in the larger wood-burning appliances. The ash collects in a compartment below the grate. A common characteristic of the older woodstoves is that they permit long fuel residence times in the firebox to maximize fuel usage. Over the years, hundreds of woodstove models of different design have been marketed, many of which are claimed to have superior performance. Some of the modifications did indeed effect small increases in thermal efficiency, but many were strictly aesthetic changes.

With few exceptions, it was not until the 1980s when major advances were made in woodstoves in response to government mandates to reduce pollution.

Catalytic woodstoves with secondary combustion chambers are good examples of the application of modern technology to improve the operating efficiencies. These appliances are at the high end of the efficiency scale as compared to noncatalytic woodstoves; fireplaces are at the low end. A comparison of the operating efficiencies of woodstoves is shown in Table 7.3 (Long and Weaver, 1985). The woodstove efficiency in this comparison corresponds to the usable heat over the energy content of the fuel input unadjusted for moisture content. The efficiencies range from a low of 13% for conventional fireplaces to 75% for airtight, catalytic woodstoves with a secondary combustion chamber.

As discussed in Chapter 6, pellet-burning stoves for residential use, or pellet stoves as they are generally called, and pellet fuels made from wood, wood wastes, straws, RDF, waste paper, and other waste biomass have been commercially available for several years. Residential pellet standards have been proposed by the Pellet Fuels Institute in the United States (Table 6.5), where the annual market has averaged about 35,000 pellet stoves over a 10-year period (Pickering, 1995, 1996). Pellet stoves are marketed in both free-standing and fireplace-insert models. These stoves are equipped with hoppers which hold about 20 kg or more of pellets that are auger-fed, usually from the top, into the combustion chamber. The advanced design units employ forced air flow past the pellets, and passage of the hot combustion products through a heat

TABLE 7.3 Comparison of Woodstove Efficiencies[a]

Woodstove type	Efficiency (%)	Remarks
Airtight catalytic	75	An airtight stove with a catalytic afterburner and a smoke chamber such as a double-drum stove or a design that provides for improved burning in a secondary combustion chamber.
Average airtight	63	Any of the openable stoves with tight fitting doors that would not fall into the airtight catalytic category.
Parlor, box, potbelly	50	Non-airtight stoves with flue connections smaller than 20.3 cm (8 in.).
Franklin (doors closed)	38	Non-airtight stove with a 20.3-cm flue connection.
Fireplace	25	Equipped with heatalator or similar device to improve heating effect.
Fireplace	13	Without heatalator.

[a]Long and Weaver (1985). The efficiency does not include a correction for moisture content. When adjustments are made for moisture, the efficiency of the average airtight woodstove or fireplace for wood fuel with 20% moisture, for example, would be 50% rather than 63%, or 10% rather than 13%.

exchanger to heat circulating room air and then through a simple flue to the outside. The advanced stoves are relatively complex and require several motors, fans, and electronic modules to control fuel and combustion air.

Small furnaces and boilers for wood fuels with and without backup oil or natural gas have been designed for burning logs, wood chips, sawdust, and pelletized wood wastes for central space heating in northern climates (cf. Brandon, 1981; Sahrman, 1983). The loading systems are gravity-fed hoppers or screw-fed hoppers (stokers) for chips, pellets, and sawdust. Logs and split logs are loaded manually. When thermal energy storage is employed, provisions are generally made for hot water storage and circulation through radiators when heat is required. One example of the controls for a conventional furnace and boiler is those used for one type of manually loaded round wood system (Brandon, 1981). Regulation of the heat output is achieved by varying the amount of combustion air available. Demand for heat from the building thermo-stat activates a control motor that opens a primary air vent to the furnace; when no heat is required, the vent closes and the fire dies down. If the wood fire is not controlled by the primary air vents, the heat is dumped to the building by switching fans or water circulators. This ensures safe operation by preventing overheating of the furnace or the boiler. Automated feeding systems for particulate fuels are often controlled so that both combustion air and fuel feeding are adjusted with the demand for heat. Controls for advanced space-heating units usually take into account the fact that wood combustion cannot be controlled by instantaneous on-off devices, as an oil or gas burner can. Combustion can be sustained in a low-level, standby mode, or a fast-start device such as an auxiliary oil burner can be employed to renew combustion when heat is needed. There are many variations in designs and controls for these systems. In comparative tests of 10 commercially available units for residential use, the overall efficiencies in terms of wood fuel input over useful output during the heating season ranged from about 40 to 50% for 9 of the 10 units evaluated (Brandon, 1981). This is quite high for small systems.

C. Solid Waste Incineration

Landfilling of MSW is the preferred disposal method. But the shortage of suitable sites and the regulations and controls now applicable to the construc-tion and operation of new landfills, the operation of existing landfills, and the closure methods and subsequent monitoring requirements have led to renewed interest in incineration. MSW disposal by open-air burning and incineration in small- and large-scale facilities without energy recovery has been practiced for many years. Some small- and many large-scale MSW incineration systems now incorporate energy recovery systems for steam and thermal energy. Some produce electric power as discussed below. MSW incineration plants with energy recovery span a large throughput range—about 50 to 4,000 t/day of MSW.

In the United States, three main technologies are used for waste-to-energy facilities: mass burn for MSW, modular mass burn, and RDF (Berenyi, 1995). Many of the modern plants in operation are based on European combustion hardware and utilize a waste-heat boiler or a waterwall system to produce steam. In mass burn technology, which is used in the majority of facilities, MSW is combusted as received or with minimum processing to reduce the size of the pieces and clumps present in the mixture and to separate some of the material. At most locations, large appliances, car batteries, and hazardous materials are removed at the tipping floor. Most mass burn plants use waterwall incineration technologies; some use refractory-lined furnaces, rotary combustors, and a few other configurations. Modular mass burn facilities often use one or more small-scale combustion units to process smaller quantities of MSW than the waterwall systems. Steam is commonly generated from the hot flue gases in many modular plants using a two-chamber furnace design. Final combustion occurs in the second chamber. For plants based on RDF technologies, MSW is first shredded and then separated into the combustible fraction or RDF, and selected recyclables, such as the ferrous fraction, aluminum, and glass. RDF is usually burned in semisuspension or suspension-fired furnaces or cofired with other fuels. It is also cofired in minor amounts with coal to produce steam in some of the larger power plants or with dewatered municipal biosolids in some plants. Separation of the RDF and recyclables is accomplished with various combinations of magnets, eddy-current devices, air classifiers, trommel screens, rotary drums, flotation devices, and pulping devices (see Chapter 6). Some processes involve the production of powdered RDF or pelletized material for use as fuels.

Numerous industrial solid wastes are disposed of in incinerators that have energy recovery capability. Most of these systems are smaller than MSW incinerators. The compositions of specific industrial wastes are more uniform than those of MSW, but the range of waste categories is so broad that special hardware and furnaces must sometimes be used. Rotary kilns, multihearth furnaces, and fluidized-bed incinerators have been employed for industrial waste incineration systems.

D. ELECTRIC POWER PRODUCTION

Before discussion of advanced, biomass combustion systems, it is in order to consider electric power generation with biomass fuels because several advanced technologies are being used or are planned for this application. A typical utility boiler consists of a furnace, where heat is transferred to enclosed water-cooled tubes, and a convection section, where more heat is transferred to the water tubes. Steam superheating can occur, and various economizers and recuperators may be installed. The steam is produced at rates of about 100 to 4500 t/h and converted to electric power in high-speed steam turbine generators, which range in capacity from about 20 to 1300 MW. In the United States, a very high percentage of electric power generation by utilities is by turbine-generator systems in which steam is expanded in variations of the Rankine cycle (Miller and Allen, 1985). This cycle, originally developed with steam engines, closely approximates the Carnot cycle when used with low-pressure steam. As pressures increase to obtain higher saturated steam temperatures, the Rankine cycle does not improve as much as the Carnot cycle because the low-temperature heat being added to bring the condensate back to boiler saturation temperature becomes a major portion of the total heat content of the saturated steam. But with regenerative feedwater heating and reheating the steam after it has been partially expanded through the turbine, Rankine efficiencies can approach Carnot efficiencies. Overall thermal efficiencies for power production usually range from about 28 to 34%. Some plants have been reported to operate at up to 40% overall efficiencies. The thermal energy in terms of fuel consumption needed to generate 1000 kWh of electricity is assumed in most U.S. tabulations in non-SI units to be about 1.8 bbl of crude oil, 0.47 ton of coal, 0.6 ton of dry biomass, or 10,000 ft^3 of natural gas. This is equivalent to thermal energy consumption of about 11 MJ/kWh (10,400 Btu/kWh). Fossil-fueled steam-electric plants typically use about 10.5 to 12.7 MJ (10,000 to 12,000 Btu) of fuel input per kilowatt-hour generated.

At full load, one of the largest single-boiler, stoker grate, wood-fueled, electric utility plants—a central station power plant in Burlington, Vermont that generates 50 MW of net production—consumes dry equivalent wood at a rate of 925 t/day (Tewksbury, 1987). Net electrical production was reported to be 280,137,900 kWh for a total green wood fuel consumption of 394,612.9 tonne (435,060.7 ton) at 50 wt % moisture over a 1-year period. At an average energy density of 18.6 GJ/t (16.0 million Btu/ton) for dry wood, this corresponds to thermal energy consumption of 13.10 MJ/kWh (12,424 Btu/kWh) generated and a thermal efficiency of 27.5%. The opportunities for fuel savings in conventional electric power generation facilities are obvious. Note that at 100% resistance heating efficiencies, 1.0 kWh is equivalent to 3.60 MJ (3412 Btu) of thermal energy independent of the generating process.

Modern fossil-fired plants typically have capacities from 300 to 900 MW; 600 MW is the approximate average for U.S. utilities (Miller and Allen, 1985). Some plants have been built with capacities of 1300 MW. Steam conditions

have effectively been standardized at 16,500 kPa and initial temperatures of
538°C with reheat to 538°C. Some plants utilize supercritical pressures of
approximately 24,100 kPa, mostly with steam temperatures at 538°C/538°C.
Some plants also utilize double-reheat and steam temperatures up to 565°C.
A few advanced plants were designed to operate with steam pressures up to
34,500 kPa and steam temperatures up to 650°C. The net heat rate and the
labor cost and investment per kilowatt-hour decrease with increasing plant
size, so larger plants are desirable.

Biomass-fired boilers are typically limited to steam production rates up to
227 to 273 t/h (250 to 300 ton/h) according to some analysts because of
fuel availability, fuel cost considerations, and materials handling difficulties
associated with low-density fuels (Tillman, 1985). This restriction in turn
limits the maximum economical pressure to about 10,300 kPa compared to
coal-fired units, which range from 16,500 to 24,100 kPa, increases the steam
rate requirement, and limits the number of feedwater heaters to 1 to 4, compared to the 8 feedwater heaters commonly associated with fossil-fired units.
The characteristics of biomass power plants shown in Table 7.4 illustrate how
these limitations can affect the technology. Biomass-fired cogeneration power
plants usually have capacities in the range 5 to 25 MW, whereas condensing
power plants have capacities up to 60 MW. Cogeneration is the simultaneous
conversion of thermal energy into electrical energy and some other form of

TABLE 7.4   Biomass Power Plant Characteristics[a]

	Generation mode	
Characteristic	Cogeneration	Condensing power
Size, MW		
Minimum	1	10
Maximum	35	50
Typical throttle steam pressure, atm		
Minimum	30	40
Maximum	100	100
Typical steam rate, kg/kWh		
Minimum	7.7	3.6
Maximum	13.6	5.4
Typical heat rate chargeable to power, MJ/kWh		
Minimum	4.9	13.2
Maximum	6.3	21.1

[a]Tillman (1985). The generation mode is for a back-pressure turbine. The minimum heat rate is
based on large systems and biomass fuel containing 15 wt % moisture. The maximum heat rate
is based on small systems and biomass fuel containing 50 wt % moisture.

energy. For example, steam produced in a boiler drives a steam turbine to generate electric power, and the waste heat is recovered and used for heat or process steam production. The overall thermal efficiency is higher because of the recovery of additional, useful energy. From a practical standpoint, the availability of fuel at a sustainable, competitive price is probably the most important factor that determines plant size.

The 50-MW plant in Burlington, Vermont, was limited in capacity by the wood fuel available within the area circumscribed by a radius of 80 km (50 mi.) from the plant. This is considered by most energy specialists to be the maximum distance that wood fuel can be obtained and economically transported to the plant by truck or rail. For captive sources of biomass fuels, the capacity can be larger. One example is the 60-MW, wood waste-fueled power plant located in Williams Lake, British Columbia (Baker, 1995). This plant is located in the center of a major lumber industry region that has five large sawmills located within 5 km of each other. The mills produce more than 540,000 green tonnes of bark, sawdust, and other wood waste products per year.

Efficiency improvements in the conversion of thermal energy to electric power are a direct route to increasing power plant capacity. Several techniques have been developed that offer large improvements in efficiency. Among them is the combined cycle configuration. In one configuration of a combined cycle plant, a combustion turbine drives a generator and the hot exhaust is fed to a heat recovery steam generator. The steam from this unit drives a steam turbine generator and the exhaust is used to provide process steam or is condensed and returned to the heat recovery steam generator. There are many variations of this design. Integrated gasification-combined cycle (IGCC) configurations for coal-fired systems are an example. IGCC systems are also applicable to biomass feedstocks and will be discussed in Chapter 9. The systems can be designed to operate at an overall energy conversion efficiency considerably larger than the sum of the efficiencies of separate systems that convert the same total quantity of fuel to electric power. Some projections indicate that overall thermal efficiencies as high as 70% might be possible.

Another approach to increasing power plant efficiencies is to use a nonthermal conversion method for power production, such as fuel cells. Fuel cells rely on electrochemical conversion of the chemical energy in the fuel to electric power. In the cogeneration mode, these systems have been reported to be operable at overall efficiencies as high as 85% (Schora, 1991). Large-scale power plants based on fuel cells have not been developed yet and are not expected to be available for generating central station power until well into the twenty-first century.

The U.S. Department of Energy has developed a strategic plan that delineates how electric power generation from biomass can be significantly increased in the U.S.A. (U.S. Dept. of Energy, 1996). The U.S. biomass power industry in the mid-1990s represented an investment base of $15 billion and supported about 66,000 jobs. DOE's projections indicated the potential for biomass power to grow to an industry of 30,000 MW employing 150,000 persons in mainly rural areas and producing 150 to 200 billion kWh by the year 2020. This would require 127 million tonnes of dry biomass fuel annually according to DOE's estimates, which is equivalent on a gross energy content basis to about 2.36 EJ/year and the annual gross generation of about 223 billion kWh at 85% availability and 33% overall thermal efficiency. If the required fuels were all dedicated biomass energy crops, 80,940 km² (8.1 million ha) of growth area would be required at a conservative yield of 15.7 dry t/ha-year. Various strategies have been proposed to achieve the 30,000-MW target, including the cofiring of biomass fuels and coals as a bridging strategy. For example, the cofiring of wood wastes in coal-fired utility boilers has the potential to reduce fuel costs, support local economic development, and address environmental concerns (cf. Tillman et al., 1995).

The most realistic approach to attainment of 30,000 MW of on-line biomass power in the United States within the next few decades is to develop large-scale, integrated biomass production-conversion systems that operate at high overall thermal and net energy production efficiencies (Chapter 14). This is perhaps the only practical approach, although efficiency improvements in power generation via advanced combined-cycle schemes and high-efficiency, nonthermal generation with fuel cells will help reduce the amounts of dedicated energy crops needed. Energy crop yields and costs are most certainly primary factors in achieving the 30,000-MW target.

A contrasting viewpoint is that large-scale, biomass-fueled power generation systems are unlikely to be economically competitive with natural gas or coal-fired generation, but that they can fill important niche markets, especially via distributed generation (Whittier, Haase, and Badger, 1996). Distributed generation is defined as any modular technology that is sited throughout a utility's service area—interconnected to the distribution or subtransmission system—to lower the cost of service. They typically have capacities less than 50 MW. Distributed generation is claimed to provide multiple benefits to utilities and end users, including lower capital costs and reduced financial risk compared to those of the larger generation systems; deferral of upgrades to substations; provision of power in increments that match projected demand patterns; and various forms of grid support. Other advantages are that the logistics of sustaining operations are simplified and most of the biomass conversion technologies qualify as distributed generation candidates.

Assessments of commercial biomass power technologies indicate that opportunities exist, particularly for niche market applications, when the business conditions are right. Federal legislation can have a large impact on these opportunities. For example, the U.S. Public Utilities Regulatory Policies Act of 1978 (PURPA, PL 95-617) created a utility market for independent, nonutility power producers by requiring public utilities to purchase power from them at the so-called avoided cost, or the utility's cost of purchasing or generating the power itself. Many small power producers and cogenerators took advantage of this arrangement by generating power for on-site use and selling the surplus to the local utility. One technology that fared quite well under PURPA, when Standard Offer contracts in California allowed independent power producers to lock in payments that started at $0.08 to $0.09/kWh, is bubbling, fluid-bed combustion. It was a key technology that allowed plants to achieve favorable economics in a changing regulatory and fuel price environment. The flexibility of 3 such plants (net capacities of 10, 10, and 25 MW), for example, permitted them to accept a very wide range of biomass fuels and to meet California's strict emissions requirements using ammonia injection and limestone additive for NO_x and SO_x control over a 7-year operating history (Ferris, 1996). When scheduled shutdowns and reduced loads were required by the utility that purchased the power, the advanced designs of these plants made it possible for them to be operated as peaking units after the utility offered payments up to $0.06/kWh under curtailment contracts.

E. Advanced Combustion Systems

Other advanced combustion systems for solid biomass fuels also offer considerable advantages over conventional designs and are in commercial use or under development. A few of them are described here.

Combustion of waste biomass is often employed not for energy recovery, but for waste disposal purposes. One of the most difficult of biomass solids to combust is municipal biosolids (sewage). Its high moisture content of 95% or more and its chemical and physical properties require special dewatering techniques and furnace designs when combustion is used as the primary disposal method. Supplemental fuels are usually required, but it is possible to use dewatered biosolids for self-sustained combustion. In one plant, thickened biosolids at a concentration of about 4% solids is dewatered to about 38% solids, and then combusted in a six-hearth incinerator (U.S. Environmental Protection Agency, 1985). The dewatered material contains about 70% volatile solids, but only has a net heating value of 1.7 MJ/kg. Yet stable autogenous combustion is obtained by automatically controlling the injection of primary air into the bottom stage of the furnace to take advantage of the draft effect that changes according to the load to the furnace and the biosolids properties. The temperature in the hottest hearth is held between 700 and 900°C. The control measures used prevent unstable combustion, high air-fuel ratios, and discharge of unburned biosolids; they also minimize clinker and slag formation. Autogeneous combustion was attained with a small amount of heavy oil at a rate of 8 L/t of dewatered biosolids. Oil consumption is commonly 170 L/t of biosolids.

The disposal of waste automobile tires is a major problem. In the United States, it is estimated that more than 200 million tires per year are disposed of in some form or recycled for retreading or reuse. About 75% are disposed of in landfills. Combustion of whole tires and tire chips is already being practiced to provide supplemental fuel for the combustion of high-moisture wood residue fuels. But emissions of metal oxides, volatile organic compounds, and sulfur oxides from the tires have precluded the use of high ratios of tire fuel in conventional combustors. The ability to handle the high steel wire concentrations, which can be as much as 10% of the total weight of the tires, has limited waste tire usage as fuel. A circulating fluidized-bed combustion system has been designed to combust tires with nearly 100% conversion of the carbon, good emissions characteristics, and the capability of separating the wire (Murphy, 1988). Carbon monoxide levels of 25 ppm in the flue gases have been readily maintained with excess air. Sulfur oxide capture with limestone in the fluidized bed and ash recycle can be as high as 80%. The sand is withdrawn from the bottom of the unit and after the temperature is reduced to about 315°C, the material is passed over a rotating drum magnet for wire removal. The dewired sand is screened to remove any oversized particles before return to the combustor.

Fluid-bed combustion has been given a great deal of attention in recent times because of its advantages, particularly in large-scale systems (*cf.* Murphy, 1991). Typically, combustion takes place in a cylindrical vessel in which air is dispersed through an orifice plate at the bottom of the unit. The air then passes through a bed of an inert refractory, pieces and particles of fuel, and ash and residual inorganic particles remaining from combustion, thereby causing the effective volume of the bed to increase and the bed to become "fluidized." Small particles burn rapidly above the fluidized bed while larger particles filter into the bed where they are dried and gasified. Most of the residual char is burned in the fluidized bed while volatiles burn both in and above the bed. The fuel is fed to this rapidly mixed bed, where flameless combustion occurs at about 650°C. This temperature can be substantially below flame temperature. Because of the lower heat input requirements, many high-moisture-content fuels can be combusted without supplemental fuel. Materials such as limestone are often added to the bed to minimize pollutants in the flue gases. The constant motion of the fluidized bed ensures good mixing and intimate contact of the air and fuel, improves combustion, reduces emissions, and makes it possible to combust a wide range of fuels having different shapes, sizes, moisture contents, and heating values. Excellent heat transfer rates to boiler tubes or materials immersed in the bed can be obtained. Bubbling and circulating fluidized-bed designs are the principal hardware configurations. The combination of fluidized-bed technology and cyclonic combustion has led to the development of innovative two-stage systems for disposal of waste biomass with heat recovery (Rehmat and Khinkis, 1991). The first stage is a sloped-grid, agglomerating fluidized-bed reactor that can operate under either substoichiometric or excess air conditions. When municipal biosolids are burned, the noncombustibles are agglomerated to form a vitrified, glassy matrix that is removed from the bottom of the fluidized bed. The inert agglomerate can be safely used in construction applications and is reported to meet leachability standards in landfills. The amount of supplemental fuel required to maintain temperatures of about 815 to 1100°C in the bed depends on the heating value of the fuel. The second stage is a cyclonic combustor where flue gas from the fluidized bed is further combusted. The cyclonic combustor provides sufficient residence time at operating conditions to oxidize all carbon monoxide and organic compounds to CO_2 and water. The combined system is reported to have a destruction and removal efficiency for organic materials greater than 99.99%. The system is used mainly for waste disposal, but can be operated in the autogenous mode with dry waste biomass feedstocks.

Direct-fired gas turbines are another innovative development in biomass combustion (McCarroll and Partanen, 1995). The compressor section of the gas turbine provides pressurized combustion air to burn biomass in an external, pressurized combustor capable of operating at pressures required by the gas turbines. Hot combustion gases are ducted through a cyclonic separator into the hot section of the gas turbine to drive a generator. Hot exhaust gas from the turbine at about 480°C can be either used directly as a source of thermal energy or fed into a heat recovery steam generator to produce process steam. Full utilization of both types of energy in the cogeneration mode is expected to allow system efficiencies in excess of 70%. This type of direct-fired turbine is believed suitable for small and medium-sized industrial and commercial applications up to 5 MW in capacity. Low-ash, debarked wood particles less than 0.3 cm long and containing less than 15% moisture are the preferred fuel, but other processed biomass can also be used. A similar 3-MW, direct-fired, gas turbine system used dried sawdust fuel containing 12 to 25% moisture as it entered the combustor (Hamrick, 1987). This system has been modified and upgraded, and a 5-MW commercial plant was built in Tennessee to demonstrate the technology (Rizzie, Picker, and Freve, 1996). The power will be sold to the Tenessee Valley Authority. The plant is fueled with fresh sawdust from local sawmills, and will later be used with other biomass fuels. Fine-tuning of this plant is expected to produce a net output of up to 6.6 MW in the open cycle mode at a heat rate of 14.2 MJ/kWh.

Pulsed combustion is another advanced technology under development for biomass (Buchkowski and Kitchen, 1995). A pulse combustor consists of a combustion chamber in the form of a short pipe with an air and fuel admitting valve at one end and a length of reduced-diameter pipe at the opposite end. The valve, which allows flow in only one direction, admits air from a blower to the combustion chamber, where it mixes with the fuel to form an explosive mixture. Ignition is provided by a spark plug and a rapid increase in pressure follows. The gases are driven out through the small-diameter tail pipe. A vacuum follows the explosion and a new charge of fuel and air are drawn into the combustion chamber. The cycle is repeated many times per second. Although fuel gases are suitable fuels, pulverized hog fuel and sawdust with less than 15% moisture may be suitable alone as fuels after the system is operational. A wood fuel feed auger was employed for the initial studies. Pulsed combustion was achieved momentarily, which indicates that a practical design is possible. Pulsed combustion is reported to offer high heat transfer rates, efficient combustion, low nitrogen oxide emissions, and a source of kinetic energy for providing the motive force for a drying system.

An innovative approach to large-scale biomass combustion for power generation is the whole-tree-burning concept in which whole trees, including branches, are supplied directly to the combustion chamber using conveyors and rams (Ostlie and Drennen, 1989). The whole trees are stored in large piles in drying buildings for 30 days before combustion. Condenser waste heat supplies dry, heated air to these buildings. The combustion chamber is a two-stage combustion unit. In the first stage, a water-cooled grate supports the pile of trees. Burning releases gases which combust above the pile at temperatures reported to be as high as 1480°C. Temperatures within the pile are reported to be 100°C. The second stage of combustion occurs below the bed as char falls through openings in the grate. Ash collects at the bottom of the second stage for removal through an ash discharge. Underfire air at approximately 340°C enters the secondary combustion chamber and is used for control. Raising or lowering the flow rate and the temperature of the air raises and lowers the combustion rate of the trees and the release of volatiles. Introduction of secondary air above the pile assures complete burning of the volatiles, while the boiler sections installed above the primary combustion chamber ensure maximum steam production.

F. Some Operating Problems

The fouling of heat exchanger surfaces can be a major problem with solid biomass fuels, especially straws and herbaceous residues. Fouling occurs because of formation in the conversion zone of low-fusion point alkali metal salt eutectics such as the alkali metal silicates. If the temperature is above the fusion point of the salts, particulates form in the combustion gases that can stick to heat exchanger surfaces when the gases leave the zone. The problem can be severe in biomass combustion systems, but is usually not severe in biomass gasifiers (Chapter 9). Furnace-boiler systems for solid biomass fuels are often designed to keep the temperature in the combustors below about 900°C to reduce slagging and formation of molten agglomerates. Careful design of the internals is necessary to avoid contact of the hot gases that may contain low-fusion point particulates with higher temperature surfaces.

Another method of eliminating this problem with solid biomass fuels when the combustors and gasifiers are operated above slagging temperatures is to remove the ash from the bottom of the units as molten slag. This technique is well established with coal fuels, which often have higher ash contents than biomass, and seems to be quite effective. It is important to note that some biomass, although high in mineral matter, may be low in alkali metals. Fouling by sticky particulates should therefore be far less with this type of biomass. An obvious approach to the reduction of alkali metal fouling is to remove the alkali metals from the fuel before conversion. Extraction of the water-soluble salts has been evaluated, but unless it is effective and low in cost, it adds unnecessary complexity and expense to the process.

A slagging index developed by the coal industry has been used to rate solid fuels for fouling. This index corresponds to the mass of alkali metal as oxides (K_2O + Na_2O) per energy unit in the fuel and is useful for rating biomass feedstocks too. The calculation is made by

$$0.1[(\% \text{ ash})(\% \text{ alkali in ash})](\text{MJ/dry kg})^{-1} = \text{kg alkali/GJ}$$

An index range of 0 to 0.17 kg/GJ (0 to 0.4 lb/MBtu) is a low slagging risk; 0.17 to 0.34 kg/GJ (0.4 to 0.8 lb/MBtu) indicates the material will probably slag; and an index greater than 0.34 kg/GJ indicates virtual certainty of slagging (Miles et al., 1993). When applied to hybrid poplar, pine, and switchgrass by use of the data in Tables 3.3 and 3.4, the corresponding indexes are 0.11, 0.009, and 0.85 kg alkali/GJ, respectively.

Another fouling mechanism that can occur is corrosion of boiler tubing and erosion of refractories due to formation of acids and their buildup in the combustion units from conversion of sulfur and chlorine present in the fuel. Fortunately, the amounts of these elements in most biomass are nil to small. The addition of small amounts of limestone to the media in fluidized-bed units or the blending of limestone with the fuel in the case of moving-bed systems are effective methods of eliminating this problem. Other sorbents such as dolomite, kaolin, and custom blends of aluminum and magnesium compounds are also effective (Coe, 1993).

IV. ENVIRONMENTAL ISSUES

A. SMALL-SCALE WOOD BURNING

Wood-burning units for cooking, water heating, and space heating are used worldwide, particularly in developing countries. Eighty percent of the energy required for cooking and water heating in India in the 1980s, for example, was derived from burning firewood (Jayaraman, Bhatt, and Rao, 1988). At that time, about 130 million tonnes of firewood were consumed annually in about 140 million ovens and woodstoves. The equipment used on an everyday basis in homes and on farms had average thermal efficiencies as low as 8 to 12%. The escape of smoke and unburned volatiles and charcoal buildup was common. Radiation losses from the openings were large, while thermal conduction through the thick walls and floor consumed nearly 25 to 30% of the heat input. Unfortunately, the equipment has not changed much and continues to be used today; the environmental problems still exist.

The situation is very different in many industrialized countries. Environmental improvements started to be made in the 1980s. Residential firewood usage in the United States has been concentrated in the Northeast and North Central States. In 1983, about 9 to 11% of U.S. space heating input was from firewood (Lipfert and Dungan, 1983). Other more localized data showed that in 1983, wood fuel accounted for 33.8% of residential heating in Vermont (Clendinen, 1983) and was a major heat source for 18% of New York state's households (Lassaie *et al.*, 1983). The early 1980s was the period when oil prices peaked, and U.S. sales of woodstoves were in the range of one million units per year. In the mid-1990s, total annual usage of residential wood fuel for the 10 million wood-burning stoves in the United States (Pickering, 1996) was about 30.5 million dry tonnes (Energy Information Administration, 1995). About 4% of the total number of woodstoves were pellet stoves, and the annual wood pellet fuel sales were about 550,000 dry tonnes (Pickering, 1995).

The growing markets that began in the early 1980s for woodstoves led to concerns with air pollution and resulted in the enactment of legislation and ordinances by several governmental bodies to control emissions (*cf.* Klass, 1984). Considerable attention was given in the United States to the reduction of emissions and improving the combustion efficiency. A detailed comparison of the performance of a conventional closed airtight stove, a conventional open-door Franklin stove, and a catalytic stove is presented in Table 7.5 (Shelton, 1984). It is apparent that the catalytic woodstove, which was one of the new-design stoves when tested, had the lowest emissions and the highest operating efficiencies. Of particular interest is the heat transfer efficiency as defined in this table; that is, the useful heat output expressed as a fraction of the heat generated in combustion.

TABLE 7.5 Comparison of Emissions from Catalytic and Noncatalytic Woodstoves[a]

Parameter	Closed stove		Franklin stove		Advanced stove	
	Conventional airtight		Conventional open door		Catalytic	
Operating conditions						
Load size, kg	6.92	7.03	3.55	5.64	10.1	10.0
Wood moisture, wt %	22.0	22.0	19.0	21.2	21.0	21.0
Burn rate, kg/h	2.1	5.8	2.0	4.9	0.9	3.2
Excess air ratio	0.33	0.12	8.7	3.4	1.07	0.27
Performance						
Flue temperature, °C	145	385	133	237	45	190
Overall energy efficiency, %	45.6	44.0	37.4	39.7	76.8	65.4
Combustion efficiency, %	63.1	71.1	88.0	90.7	90.4	88.6
Heat transfer efficiency, %	72.2	61.9	42.5	43.7	85.0	73.9
Particulates, kg/TJ	7.1	3.3	<1.5	<1.5	<1.5	<1.5
Carbon monoxide, kg/TJ	22.7	32.7	8.0	6.1	1.3	6.7

[a]Adapted from Shelton (1984). The advanced stove is an American Eagle brand unit. The wood fuel is northern red oak logs cut into four different lengths and split to specified dimensions. The loads were usually between 50 and 75% of maximum capacity, and usually consisted of 4 to 7 pieces. The overall energy efficiency is useful heat output/wood energy input. The combustion efficiency is heat generated in combustion (useful heat output + sensible heat stack loss + latent heat of water vapor stack loss)/wood energy input. The heat transfer efficiency is the useful heat output/heat generated in combustion.

The effort to improve woodstove characteristics in the 1980s in the United States resulted from some of the state initiatives taken to mandate improved stove performance. New woodstoves sold in Oregon after 1986 had to reduce particulates and carbon monoxide in the flue gas with designs and devices such as dual-combustion chambers and catalytic converters. The Oregon Environmental Quality Commission standards set an emission rate for particulates of 15 g/h for non-catalytic wood-stoves, and 6 g/h for woodstoves equipped with catalytic combustors. The standards for particulates became mandatory on July 1, 1986 and were further reduced to 9 and 4 g/h on July 1, 1988. Local ordinances were also enacted. The city of Aspen, Colorado, passed an ordinance that allowed only one traditional fireplace per new structure, including apartment buildings. Another Colorado community required an on-line heat sensor in each fireplace chimney to warn the community computer that the fire was not out when excessive haze made it necessary for the community to order all fires doused by means of a red signal light next to each fireplace. This Orwellian approach to a community problem is probably the ultimate in pollution control.

The U.S. Environmental Protection Agency promulgated a New Source Performance Standard for woodstoves in 1988 (Barnett and Morgan, 1991). Woodstoves that had passed Oregon's 1988 regulations were "grandfathered" into the U.S. standards. The final 1990 standards for particulates were 4.1 g/h for catalytic stoves and 7.5 g/h for noncatalytic stoves. Fireplaces, old uncertified woodstoves, EPA-certified woodstoves, and wood pellet stoves, all fueled with wood, are estimated to produce an average of 47, 42, 6.0, and 1.2 g/h of particulate emissions, respectively (Hearth Products Association, 1997). Interestingly, the wood pellet stoves have the lowest emissions. The overall impact of the regulations imposed in various parts of the United States and the national standards was that the woodstove manufacturing industry redesigned their products and incorporated new control systems to make them less polluting and more efficient. Emissions from modern woodstoves are now up to 80 to 90% lower than those of the older woodstoves, and the overall thermal efficiencies are notably higher than those of the older stoves.

B. BIOMASS BURNING REGULATIONS

Uncontrolled burning of biomass in the open air, as encountered in slash-and-burn agriculture, sugarcane trash burning, forest fires, and the clearing of forest land for development, emits large numbers of pollutants to the atmosphere. Some 345 chemicals have been identified as being released to the atmosphere from such fires (Khalil and Rasmussen, 1995; Graedel, Hawkins, and Claxton, 1986). Many of these compounds are greenhouse gases and ozone-depleting compounds that are believed to affect the earth's climate. Some national, state, and local governments have enacted legislation or regulations to attempt to limit large-scale burning of biomass in open air, but most of what occurs is beyond human control. Some local governments have prohibited or limited even small-scale biomass burning such as leaf burning.

In contrast to uncontrolled biomass burning, regulations exist or are being developed in industrialized countries that apply to controlled burning of bio-mass and other fuels. Many different types of emissions- and ash-removal systems have been installed and operated in large-scale commercial biomass combustion plants to meet national and regional regulations and emissions limits. Worldwide, regulations that govern the use of virgin and waste biomass combustion processes range from nonexistent to very detailed and complex. The United States is in the latter category. The state programs complicate matters further because they are often at odds with federal programs and are sometimes more severe than the national requirements. Some regulatory systems in the United States are designed to meet California's South Coast regulations, which are more constrictive than those of the U.S. Environmental Protection Agency and are believed to be among the most stringent in the world (cf. Moore and Cooper, 1990). Some of the U.S. federal requirements are reviewed here to illustrate how national mandates can affect commercial biomass combustion technologies. Basically, federal regulations that affect biomass combustion are divided into two groups—one concerned with emis-sions and one concerned with residue or ash disposal.

The development of U.S. federal pollution policies began in 1970 when the U.S. Environmental Protection Agency was established. The details of pollution legislation have been undergoing continual revision since then. Included in this legislation is the Clean Air Act of 1970, the New Source Performance Standards of 1971, the Resource Conservation and Recovery Act of 1976, the Pollution Prevention Act of 1990, the Clean Air Act Amendments of 1990, and the Revised New Source Performance Standards of 1991. The Clean Air Act Amendments of 1990 have been called the most complicated and far-reaching environmental legislation ever enacted. Titles II, III, IV, and VI of this Act deal with mobile sources, hazardous air pollutants, acid deposition control, and stratospheric ozone protection, respectively. Title V of this legisla-tion established a program for issuing operating permits to all major sources of air pollution in the United States. A "major source" was defined in Part D of Title I of the Act as having "the potential to emit." As originally proposed, the standards applied only to "furnaces and boilers" with a heat input of 250 million Btu/h used in the process of burning fossil fuels for the primary purpose of producing steam by heat transfer.

Subsequent revisions expanded the source category to cover some steam generators firing nonfossil fuels and those used in commercial and institutional applications, and quantified emission limits for sulfur and nitrogen oxides and particulates (Dykes, 1989). An air pollution program was formulated for the entire country. The Hazardous and Solid Waste Amendments of 1984 amended the Resource Conservation and Recovery Act of 1976 and set as national policy that "wherever feasible, the generation of hazardous waste is to be reduced or eliminated as expeditiously as possible" (U.S. General Accounting Office, 1994). The Pollution Prevention Act of 1990 broadened the scope of the policy by stating that pollution "should be prevented or reduced at the source whenever feasible" for all environmental media—air, land, and water. A comprehensive pollution prevention strategy has thus become a national policy of the United States. This strategy affects a multitude of sources of emissions and residues, including virgin and waste biomass combustion systems.

A potential barrier to the adoption of solid biomass fuels is the problem posed by the disposal and utilization of the ash (cf. McGinnis et al., 1995). Users must have a means to dispose of the ash that is cost effective and does not degrade the environment or become a regulatory burden for the ash producer. Soil amendment with the ash would avoid the problems created by landfill disposal, but there is uncertainty regarding the variability of ash composition and the fate of the heavy metals and toxic organic compounds present in the ash. At the federal level, wood ash disposal may be regulated, but certain exemptions may apply, depending on the nature of the facility, the fuel, and the ash (Rughani et al., 1995). Determination of whether wood ash is to be treated as a hazardous waste is based on the corrosiveness of the ash and whether it exceeds certain threshold concentrations for toxic contaminants as measured by a Toxic Characteristic Leaching Procedure (TCLP). The regulatory concentration limits in the leachate by the TCLP are shown in Table 7.6. The standard for corrosiveness of aqueous ash leachate is a pH of 2 or less or 12.5 or more. Ash exceeding these thresholds must be handled and disposed of in a designated hazardous waste facility (cf. McGinnis et al., 1995). If the ash is not so designated, soil amendment is a preferred disposal option among large producers, whereas landfilling is the preferred option for small producers. Most wood-fueled facilities that burn "clean wood," or wood that has not been treated with chemicals, and "treated wood," or wood that has been treated with chemicals, have not experienced problems with ash being designated as hazardous materials. The federal regulations for ash disposal from municipal waste combustion facilities are more complicated (cf. Malloy and McAdams, 1994; Clearwater and Hill, 1989). Because of the composition of municipal wastes and the large variations, strict application of the threshold values in Table 7.6 to the ashes from MSW and RDF combustion facilities would appear to make it more difficult for them to be classified as nonhazardous materials.

Thermal Conversion: Combustion

TABLE 7.6   U.S. Regulatory Threshold Concentrations in Leachate by Toxic Characteristic Leaching Procedure Analysis[a]

Analyte	Threshold (mg/L)	Analyte	Threshold (mg/L)
Arsenic	5.0	1,1-Dichloroethylene	0.7
Barium	100.0	2,4-Dinitrotoluene	0.13
Cadmium	1.0	Endrin	0.02
Chromium	5.0	Heptachlor	0.008
Lead	5.0	Hexachlorobenzene	0.13
Mercury	0.2	Hexachlorobutadiene	0.5
Selenium	1.0	Hexachloroethane	3.0
Silver	5.0	Lindane	0.4
Benzene	0.5	Methoxychlor	10.0
Carbon tetrachloride	0.5	Methyl ethyl ketone	200.0
Chlorobenzene	100.0	Nitrobenzene	2.0
Chlordane	0.03	Pentachlorophenol	100.0
Chloroform	6.0	Pyridine	5.0
o-Cresol	200.0	Tetrachloroethylene	0.7
m-Cresol	200.0	Toxaphene	0.5
p-Cresol	200.0	Trichloroethylene	0.5
Cresol (total)	200.0	2,4,5-Trichlorophenol	400.0
2,4-D	10.0	2,4,6-Trichlorophenol	2.0
1,4-Dichlorobenzene	7.5	2,4,5-TP (Silvex)	1.0
1,2-Dichloroethane	0.5	Vinyl chloride	0.2

[a]Adapted from McGinnis et al. (1995).

Energy facilities that fire virgin and waste biomass, or that cofire these fuels with fossil fuels, may be subject to some of the emission standards under subpart D of the Code of Federal Regulations, Part 60 (Dykes, 1989). In 1984, the U.S. Environmental Protection Agency expanded the standards, which initially pertained only to fossil-fuel-fired systems, to include all steam generators firing nonfossil fuels in industrial-commercial-institutional steam generating units. Biomass-fired or cofired boilers became subject to selected emission limits under Subpart D. Fuels designated under the standards included fossil fuels, wood, municipal-type solid waste, and chemical by-product fuels.

Because of their complexity, the U.S. environmental laws and revisions of various performance standards for industrial-commercial-institutional steam generating units must be studied in detail to ascertain how they apply to existing and new systems. With regard to biomass combustion systems specifically, in addition to municipal waste combustors and some biomass-fueled systems which are already subject to pollution control regulations (cf. U.S. Environmental Protection Agency, 1986, 1987, 1995), it appears that both virgin and waste biomass-fueled plants will eventually be covered in depth (cf. U.S. Environmental Protection Agency, 1996). Pollutants to be considered for regulation include CO, SO_2, NO_x, HCl, lead, cadmium, mercury, particulate matter, and dioxins and furans.

C. EMISSION CONTROLS

Perfect combustion of biomass and most other fuels would help minimize the emissions of particulate matter, and gaseous and toxic compounds. In practice, this can be difficult to attain, even with natural gas, the dominant component of which is methane, the simplest of all organic fuels (Chisholm and Klass, 1966). The carbon monoxide content of the flue gas is a good indicator of the completeness of combustion. The less carbon monoxide, the more complete the combustion process and the lower the emissions of organic compounds. As already indicated, a wide range of variables and independent parameters affect how closely perfect combustion can be approached for solid biomass fuels: particle size range and distribution, moisture content, composition, and heating value; air-fuel ratios, mixing, and reaction time; temperature and sometimes pressure range in the combustion chamber; and furnace design and heat-transfer methods. A compromise is often made between the operating parameters that promote complete combustion and those that minimize the emissions of inorganic derivatives. Under the best of conditions, a solid biomass fuel should consist of small, uniform particles, be low in moisture and ash contents, and have zero to very low chlorine, nitrogen, and sulfur contents. With the proper processing, virgin biomass such as wood can approximate these characteristics, whereas municipal solid waste is representative of biomass fuels that might be the furthest removed from them.

Emissions from biomass-fueled boilers can be controlled by a variety of methods. The control systems needed depend mainly on the composition of the feedstock. First, good combustion control is essential to maximize combustion and to minimize emissions of unburned hydrocarbons and carbon monoxide. Efficient removal of particulate matter in the flue gases can be achieved by various combinations of cyclonic separation, electrostatic precipitation, agglomeration, and filtration. Removal of acid gas emissions can be achieved by flue gas scrubbing and treatment with lime. There are several approaches to the control of NO_x emissions (Clearwater and Hill, 1991). Combustion control techniques include use of staged combustion, low excess air, and flue gas recirculation. Staged combustion involves reduction of the maximum attainable flame temperature and the control of residence time. In the primary stage of combustion, the maximum flame temperature and thermal NO_x formation are reduced by the transfer of heat, which is not returned, or by combustion with substoichiometric amounts of air. The formation of NO from chemically bound nitrogen is also largely avoided in the primary stage under these conditions if the residence time is sufficient to permit nitrogen to form. Combustion is then completed in the second stage with excess secondary air at short residence times to minimize NO formation. Potential add-on controls include selective noncatalytic and catalytic reduction and natural gas reburning techniques. Selective noncatalytic reduction, such as ammonia injection, is one of the preferred methods because it has been effectively employed in several MSW combustion plants and has been shown to afford NO_x emissions in the range of 120 to 200 ppmv. As discussed at the end of this section, staged combustion can be carried out without actually using hardware designed for staged combustion. The technique is quite effective for minimizing NO_x in the Burlington, Vermont power plant.

Toxic polychlorinated dibenzo-p-dioxins ("dioxins" or PCDDs) and polychlorinated dibenzofurans ("furans" or PCDFs) can form on combustion of chlorine-containing biomass and be emitted in the flue gases and possibly in adsorbed form on flyash or in particulate matter. The isomer 2,3,7,8-tetrachlorodibenzo-p-dioxin, a strong carcinogen and a contaminant found in the defoliant Agent Orange used in the Vietnam War, is claimed to be the most lethally toxic of man-made chemicals when administered to guinea pigs (Esposito, Tiernan, and Dryden, 1980). The compound is destroyed when the temperature is 1000°C for as little as a millisecond or at lower temperatures for longer periods (Barnes, 1983). At stack gas temperatures below 200°C., the PCDDs and PCDFs are predominantly found on the particulate matter. These data suggest that control of combustion and downstream temperatures coupled with particulate matter reduction measures can reduce or eliminate PCDD and PCDF emissions.

MSW-fueled boiler systems present greater air pollution problems than most other biomass-fueled plants. Some of the advanced emission control systems used with a 350-t/day plant that meets California's stringent South Coast Air Quality Requirements are described here to illustrate how effective the controls are. This particular plant is believed to have the lowest emissions of any refuse-to-energy plant in the world (Moore and Cooper, 1990). It is designed to generate 52,200 kg/h of steam at 435°C and 4483 kPa to supply a turbine generator developing 11.4 MW of electrical power. Combustion of MSW results in the formation of acid gases derived from chemically bound chlorine, fluorine, and sulfur in the refuse. These gases and particulate carryover from the boiler must be removed before the flue gases are exhausted to the atmosphere. The flue gases enter the bottom of a dry scrubber through a cyclonic section designed to remove flyash particles larger than 150 μm. From the cyclone, the gases flow upward through a spray section where atomized lime slurry is introduced. The lime reacts with the acid gases to produce nonacidic salts. Water in the slurry is completely evaporated by the flue gas, which lowers the temperature of the gas leaving the unit. The resulting dry reaction product falls out into a scrubber hopper. The remaining particulate matter in the flue gas is conditioned by a material (Tesisorb) that promotes particle agglomeration for subsequent removal by glass fabric filters. The removal efficiencies of the acid gases and particulates are in the high 90s on a percentage basis. Nitrogen oxides are reduced by ammonia injection above the combustion zone, and careful control of combustion conditions minimizes carbon monoxide, dioxin, and furan emissions. The carbon monoxide and nitrogen oxide emissions ranged from 18 to 25 ppmv and 48 to 69 ppmv, respectively. The dioxin emissions meet California's requirements, assuming the lower analytical detection limit is the actual emission.

It cannot be emphasized enough that the combustion process in biomass-fueled power plants should always be controlled with the objective of maximizing boiler efficiency and minimizing stack gas emissions. These goals might be considered to be contradictory, since high-efficiency combustion generally means higher flame temperatures, which can result in higher NO_x emissions. However, in a power steam generator firing whole tree chips, it is quite possible to achieve rated boiler efficiency and low NO_x formation at the same time. Operation of the 50-MW plant in Burlington, Vermont under the proper conditions with 100% green, whole tree chips containing 40 wt % moisture afforded NO_x emissions as low as 0.062 kg/GJ while still achieving a boiler efficiency in the range 68 to 73%, which is in the high end of the design range, without the use of postcombustion treatment or flue gas recirculation (Tewksbury, 1991). At full load, this plant is designed to burn 90.7 t/h of green wood fuel; the nominal steam capacity is 217,687 kg/h at 8828 kPa and 510°C. This performance was achieved at full load by careful control of the fuel distribution on the grates and the air-fuel ratio, and by balancing the overfire air and underfire air. Substoichiometric firing of the wood on the grates kept flame temperature and NO_x formation low, but generated a high level of CO. A second level of combustion higher in the fire box occurs when additional air is added to complete the combustion process at a temperature where little or no further NO_x is created. This operating mode simulated a two-stage combustion system. When the plant must be operated at minimum load, about 33% of normal load, a similar operating mode provided the best results, although the NO_x emissions were slightly higher than before.

REFERENCES

Baker, B. Z. (1995). *In* "Second Biomass Conference of the Americas: Energy, Environment, Agriculture, and Industry," p. 1345, NREL/CP-200-8098, DE95009230. National Renewable Energy Laboratory, Golden, CO.

Barnes, D. G. (1983). *In* "Energy from Biomass and Wastes VII," (D. L. Klass and H. H. Elliott, eds.), p. 291. Institute of Gas Technology, Chicago.

Barnett, S. G., and Morgan, S. J. (1991). *In* "Energy from Biomass and Wastes XIV," (D. L. Klass, ed.), p. 191. Institute of Gas Technology, Chicago.

Berenyi, E. B. (1995). *In* "Second Biomass Conference of the Americas: Energy, Environment, Agriculture, and Industry," p. 88, NREL/CP-200-8098, DE95009230. National Renewable Energy Laboratory, Golden, CO.

Brandon, R. J. (1981). *In* "Energy from Biomass and Wastes V," (D. L. Klass and J. W. Weatherly, III, , eds.), p. 175. Institute of Gas Technology, Chicago.

Buchkowski, A. G., and Kitchen, J. A. (1995). *In* "Second Biomass Conference of the Americas: Energy, Environment, Agriculture, and Industry," p. 410, NREL/CP-200-8098, DE95009230. National Renewable Energy Laboratory, Golden, CO.

Chisholm, J. A., Jr., and Klass, D. L. (1966). "Trace Organic Compounds in Natural Gas Combustion," *American Chemical Society Division of Fuel Chemistry Preprints* **10** (3), 105.

Clearwater, S. W., and Hill, M. A. (1991). *In* "Energy from Biomass and Wastes XIV," (D. L. Klass, ed.), p. 317. Institute of Gas Technology, Chicago.

Clendinen, D. (1983). *New York Times* **132** (45,612), 11, March 9.

Coe, D. R. (1993). *In* "First Biomass Conference of the Americas: Energy, Environment, Agriculture, and Industry," Vol I, p. 406, NREL/CP-200-5768, DE93010050. National Renewable Energy Laboratory, Golden, CO.

Dykes, R. M. (1989). *In* "Energy from Biomass and Wastes XII," (D. L. Klass, ed.), p. 379. Institute of Gas Technology, Chicago.

Energy Information Administration (1995). "Renewable Energy Annual 1995," DOE/EIA-0603(95). U.S. Dept. of Energy, Washington, D.C., December.

Esposito, M. P., Tiernan, T. O., and Dryden, F. E. (1980). "Dioxins," EPA-600/2-80-197. U.S. Environmental Protection Agency, Washington, D.C.

Ferris, J. M. (1996). *In* "Bioenergy '96," Vol. I, p. 409. Southeastern Regional Biomass Energy Program, Tennessee Valley Authority, Muscle Shoals, AL.

Graedel, T. E., Hawkins, D. T., and Claxton, L. D. (1986). *In* "Chemical Compounds in the Atmosphere," p. 512. Academic Press, New York.

Hamrick, J. T. (1987). *In* "Energy from Biomass and Wastes X," (D. L. Klass, ed.), p. 517. Institute of Gas Technology, Chicago.

Hearth Products Association (1997). "Clearing the Air, How to Build a Clean Fire," Arlington, VA.

Jayaraman, S., Bhatt, M. S., and Rao, J. R. (1988). *In* "Energy from Biomass and Wastes XI," (D. L. Klass, ed.), p. 305. Institute of Gas Technology, Chicago.

Khalil, M. A. K., and Rasmussen, R. A. (1995). *In* "Second Biomass Conference of the Americas: Energy, Environment, Agriculture, and Industry," p. 1397, NREL/CP-200-8098, DE95009230. National Renewable Energy Laboratory, Golden, CO.

Klass, D. L. (1984). *In* "Energy from Biomass and Wastes VIII," (D. L. Klass, ed.), p. 1. Institute of Gas Technology, Chicago.

Lassaie, J. P., Provencher, R. W., Goff, G. R., and Brown, T. L. (1983). "Fire Safety Appraisal of Residential Wood and Coal Stoves in New York State, Final Report," NYSERDA- 83-3. New York State Energy Research and Development Authority, Albany, NY, April.

Lipfert, F. W., and Dungan, J. L. (1983). *Science* **219** (4591), 1425.

Long, H. C., and Weaver, T. F. (1985). *In* "Handbook of Energy Systems Engineering Production and Utilization," (L. C. Wilbur, ed.), p. 1347. John Wiley, New York.

Malloy, M. G., and McAdams, C. L. (1994). "The U.S. & International Municipal Waste Combustion Industry," *Waste Age* 25 (11), 89.

McCarroll, R. L., and Partanen, W. E. (1995). *In* "Second Biomass Conference of the Americas: Energy, Environment, Agriculture, and Industry," p. 400, NREL/CP-200-8098, DE95009230. National Renewable Energy Laboratory, Golden, CO.

McGinnis, G., Rughani, J., St. John, W., Diebel, J., Shetron, S., and Jurgensen, M. (1995). "Wood Ash in the Great Lakes Region: Production, Characteristics and Regulation." Institute of Wood Research, Michigan Technological University, Houghton, MI, September.

Miles, T. R., Miles, T. R., Jr., Baxter, L. L., Jenkins, B. M., and Oden, L. L. (1993). *In* "First Biomass Conference of the Americas: Energy, Environment, Agriculture, and Industry," Vol I, p. 406, NREL/CP-200-5768, DE93010050. National Renewable Energy Laboratory, Golden, CO.

Miller, E. H., and Allen, R. P. (1985). *In* "Handbook of Energy Systems Engineering Production and Utilization," (L. C. Wilbur, ed.), p. 1413. John Wiley, New York.

Moore, S. R., and Cooper, M. J. (1990). *In* "Energy from Biomass and Wastes XIII," (D. L. Klass, ed.), p. 557. Institute of Gas Technology, Chicago.

Murphy, M. L. (1988). *In* "Energy from Biomass and Wastes XI," (D. L. Klass, ed.), p. 371. Institute of Gas Technology, Chicago.

Murphy, M. L. (1991). *In* "Energy from Biomass and Wastes XV," (D. L. Klass, ed.), p. 1167. Institute of Gas Technology, Chicago.

Ostlie, L. D., and Drennen, T. E. (1989). *In* "Energy from Biomass and Wastes XII," (D. L. Klass, ed.), p. 621. Institute of Gas Technology, Chicago.

Pickering, W. H. (1995). *In* "Second Biomass Conference of the Americas: Energy, Environment, Agriculture, and Industry" p. 1190, NREL/CP-200-8098, DE95009230. National Renewable Energy Laboratory, Golden, CO; *idem* (1996). *In* "Bioenergy '96," Vol. II, p. 838, Southeastern Regional Biomass Energy Program, Tennessee Valley Authority, Muscle Shoals, AL.

Reed, R. J. (1983). "North American Combustion Handbook," Ed. 2. North American Mfg. Co., Cleveland, OH.

Rehmat, A., and Khinkis, M. (1991). *In* "Energy from Biomass and Wastes XV," (D. L. Klass, ed.), p. 1111. Institute of Gas Technology, Chicago.

Rizzie, J. W., Picker, F. M., and Freve, W. W., Jr. (1996). *In* "Bioenergy '96," Vol. I, p. 393. Southeastern Regional Biomass Energy Program, Tennessee Valley Authority, Muscle Shoals, AL.

Rughani, J., Diebel, J., McGinnis, G., and St. John, W. (1995). *In* "Second Biomass Conference of the Americas: Energy, Environment, Agriculture, and Industry," p. 168, NREL/CP-200-8098, DE95009230. National Renewable Energy Laboratory, Golden, CO.

Sahrman, K. (1983). *In* "Energy from Biomass and Wastes VII," (D. L. Klass and H. H. Elliott, eds.), p. 265. Institute of Gas Technology, Chicago.

Schora, F. C., Jr. (1991). *In* "Asian Natural Gas—For a Brighter '90s," (D. L. Klass and T. Ohashi, eds.), p. 231. Institute of Gas Technology, Chicago.

Shafizadeh, F., and DeGroot, W. F. (1976). *In* "Thermal Uses and Properties of Carbohydrates and Lignins," (F. Shafizadeh, K. V. Sarkanen, and D. A. Tillman, eds.), p. 1. Academic Press, New York.

Shelton, J. M. (1984). "Advanced Wood Heating Technologies," NYSERDA Report 84-15. New York State Energy Research and Development Authority, Albany, NY, August.

Tewksbury, C. (1987). *In* "Energy from Biomass and Wastes X," (D. L. Klass, ed.), p. 555. Elsevier Applied Science Publishers, London, and Institute of Gas Technology, Chicago.

Tewksbury, C. (1991). *In* "Energy from Biomass and Wastes XV," (D. L. Klass, ed.), p. 95. Institute of Gas Technology, Chicago.

Tillman, D. A. (1985). *In* "Handbook of Energy Systems Engineering Production and Utilization," (L. C. Wilbur, ed.), p. 1252. John Wiley, New York.

Tillman, D., Stahl, R., Bradshaw, D., Chance, R., Jett, J. R., Reardon, L., Rollins, M., and Hughes, E. (1995). *In* "Second Biomass Conference of the Americas: Energy, Environment, Agriculture, and Industry," p. 382, NREL/CP-200-8098, DE95009230. National Renewable Energy Laboratory, Golden, CO.

U.S. Dept. of Energy (1996). "DOE Biomass Power Program, Strategic Plan 1996-2015," Draft September 10, 1996. Washington, D.C.

U.S. Environmental Protection Agency (1985). *In* "Municipal Wastewater Sludge Combustion Technology," p. II-26, EPA/625/4-85/015. Cincinnati, OH, September.

U.S. Environmental Protection Agency (1986). "Fact Sheet, New Source Performance Standards for Industrial-Commercial-Institutional Steam Generating Units, (PM/NO$_x$ Promulgation)." Washington, D.C., November.

U.S. Environmental Protection Agency (1987). "Fact Sheet, Promulgation of New Source Performance Standards for Industrial-Commercial-Institutional Steam Generating Units (Subpart Db)." Washington, D.C., November.

U.S. Environmental Protection Agency (1995). "Fact Sheet, Existing Municipal Waste Combustors—Subpart Cb Emissions Guidelines (1995)." Washington, D.C., November; "Fact Sheet, New Municipal Waste Combustors—Subpart Eb Standards of Performance (1995)." Washington, D.C., November.

U. S. Environmental Protection Agency (1996). "Industrial Combustion Coordinated Rulemaking, Proposed Organizational Structure and Process," Office of Air Quality Planning and Standards, Research Triangle Park, NC, June.

U.S. General Accounting Office (1994). "Pollution Prevention: EPA Should Reexamine the Objectives and Sustainability of State Programs," GAO/PEMD-94-8. Washington, D.C., January.

Whittier, J., Haase, S., and Badger, P. C. (1996). *In* "Bioenergy '96," Vol. II, p. 598. Southeastern Regional Biomass Energy Program, Tennessee Valley Authority, Muscle Shoals, AL.

CHAPTER 8

Thermal Conversion: Pyrolysis and Liquefaction

I. INTRODUCTION

As discussed in Chapter 7, the final products of biomass combustion are CO_2, water, and energy. This is the case, of course, for the combustion of all organic matter. It is not known how much time passed after the utility of biomass combustion was discovered by man until biomass pyrolysis was discovered. But when it was, it is probable that a new era in biomass usage evolved, and quite rapidly. Biomass pyrolysis can be described as the direct thermal decomposition of the organic components in biomass in the absence of oxygen to yield an array of useful products—liquid and solid derivatives and fuel gases. Eventually, pyrolysis processes were utilized for the commercial production of a wide range of fuels, solvents, chemicals, and other products from biomass feedstocks. Improvements continue to be made today to perfect the technology. (It is important to note at the outset that any organic material can be pyrolyzed. Indeed, the pyrolysis of coal has been in commercial use for many years, and still is in several areas of the world for the production of fuel gases, cokes, tars, and chemicals.)

Knowledge of the effects of various independent parameters such as biomass feedstock type and composition, reaction temperature and pressure, residence time, and catalysts on reaction rates, product selectivities, and product yields has led to development of advanced biomass pyrolysis processes. The accumulation of considerable experimental data on these parameters has resulted in advanced pyrolysis methods for the direct thermal conversion of biomass to liquid fuels and various chemicals in higher yields than those obtained by the traditional long-residence-time pyrolysis methods. Thermal conversion processes have also been developed for producing high yields of charcoals from biomass.

In this chapter, the basic chemistry of the direct pyrolysis of biomass and the state-of-the-art systems that have been or are expected to be commercialized are discussed. Pyrolysis in the presence of hydrogen (hydropyrolysis) and methane (methanolysis) are also addressed. Energy recovery in the form of liquid fuels and chars is emphasized. Another group of processes for direct thermal conversion of biomass employs a liquid medium for conversion of biomass to liquid fuels. For convenience, these processes are grouped together in this chapter as miscellaneous thermal liquefaction methods. Some discussion of biomass gasification is included in this chapter when needed to clarify pyrolysis chemistry. High-temperature pyrolysis, which yields gaseous fuels and feedstocks such as low- and medium-energy fuel gases, hydrogen, and synthesis gases, is discussed in more detail in Chapter 9.

II. FUNDAMENTALS

A. CONVENTIONAL AND FAST PYROLYSIS

Conventional pyrolysis (carbonization, destructive distillation, dry distillation, retorting) consists of the slow, irreversible, thermal degradation of the organic components in biomass, most of which are lignocellulosic polymers, in the absence of oxygen. Slow pyrolysis has traditionally been used for the production of charcoal. Detailed studies of biomass pyrolysis beginning in the 1970s have led to methods of controlling the selectivities and yields of the gaseous, liquid, and solid products by controlling the pyrolysis temperature and heating rate (cf. Stevens, 1994). Short-residence-time pyrolysis (flash, rapid, ultra pyrolysis) of biomass at moderate temperatures can afford up to 70 wt % yields of liquid products (cf. Bridgwater and Bridge, 1991). Pyrolysis conditions can be used that provide high yields of gas or liquid products and char yields of less than 5%. One configuration of an advanced biomass pyrolysis system, for example, involves an ablative, vortex reactor for pyrolysis at biomass residence times of fractions of a second coupled to a downstream vapor cracker (Diebold and

Scahill, 1988). The products can be varied to yield up to about 56% liquids (dry) or 90% gases; the char yields are about 10-15% in each case. Advanced pyrolysis processes are discussed in more detail in Section III.

B. Mechanisms

Many dehydration, cracking, isomerization, dehydrogenation, aromatization, coking, and condensation reactions and rearrangements occur during pyrolysis. The products are water, carbon oxides, other gases, charcoal, organic compounds (which have lower average molecular weights than their immediate precursors), tars, and polymers. When cellulose is slowly heated at about 250 to 270°C, a large quantity of gas is produced consisting chiefly of carbon dioxide and carbon monoxide. Table 8.1 illustrates how the carbon oxides, hydrocarbons, and hydrogen in the product gas vary with increasing temperature as the slow, dry distillation of wood progresses (Nikitin *et al.*, 1962). Initially, small amounts of hydrogen and hydrocarbon gases and larger amounts of carbon oxides are emitted. The hydrocarbons in the product gas then increase with further temperature increases until hydrogen is the main product. The carbon oxides and most other products owe their formation to secondary and further reactions.

Pyrolysis of cellulose yields the best-known of the 1,6-anhydrohexoses, β-glucosan or levoglucosan (1,6-anhydro-β-D-glucopyranose), in reasonably good yields (Shafizadeh, 1982) (Fig. 8.1). A novel technique based on flash devolatilization of biomass and direct molecular-beam, mass-spectrometric analysis has shown that levoglucosan is a primary product of the pyrolysis of pure cellulose (Evans and Milne, 1987a, 1987b, 1988). However, the yield of levoglucosan on pyrolysis of most biomass is low even though the cellulose

TABLE 8.1 Composition of Gases from the Slow, Dry Distillation of Wood[a]

Process	Temperature (°C)	H_2 (mol %)	CO (mol %)	CO_2 (mol %)	HCs (mol %)
Elimination of water	155–200	0	30.5	68.0	2.0
Evolution of carbon oxides	200–280	0.2	30.5	66.5	3.3
Start of hydrocarbon evolution	280–380	5.5	20.5	35.5	36.6
Evolution of hydrocarbons	380–500	7.5	12.3	31.5	48.7
Dissociation	500–700	48.7	24.5	12.2	20.4
Evolution of hydrogen	700–900	80.7	9.6	0.4	8.7

[a]Nikitin *et al.* (1962). "HCs" are hydrocarbons.

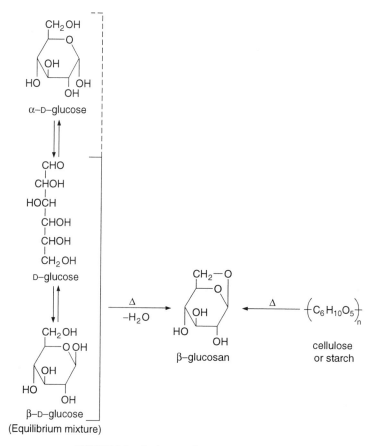

FIGURE 8.1   β-Glucosan formation on pyrolysis.

content is about 50 wt %. Also, when pure cellulose is treated with only a small amount of alkali, levoglucosan formation is inhibited and a different product slate composed of furan derivatives is produced.

Levoglucosan is also obtained directly on pyrolysis of glucose and starch. The compound has the same empirical formula as the monomeric building block of cellulosic polymers, $(C_6H_{10}O_5)$. Some investigators suggest that these observations support a mechanism wherein the initial pyrolysis reaction yields glucose as an intermediate. This is equivalent to the sequential hydrolysis of cellulose by addition of water to form glucose, and elimination of water by dehydration of glucose to form the anhydride. It seems more probable that if levoglucosan is the initial intermediate, a thermally induced, depolymerization-internal displacement reaction occurs to form the pyranose directly by a concerted mechanism.

In early work on the mechanisms and kinetics of biomass pyrolysis, measurement of the weight change as a function of time over a 1000-h period during the pyrolysis of pure cellulose at temperatures up to 260°C in a vacuum led to a multistep mechanism consistent with the experimental data (Broido, 1976). A two-path mechanism was proposed in which one involved depolymerization and led to completely volatile products, and the other involved a sequence of steps leading to char formation. Most investigators now generally recognize at least two pathways for cellulose pyrolysis (Fig. 8.2). One involves dehydration and charring reactions via anhydrocellulose intermediates to form chars, tars, carbon oxides, and water, and one involves depolymerization and volatilization via levoglucosan intermediate to form chars and combustible volatiles (*cf.* Zaror and Pyle, 1982; Antal and Varhegyi, 1995). The first pathway would be expected to occur at lower temperatures where dehydration reactions are dominant. The second pathway results in the formation of oligomeric species as well as their degradation products, which immediately enter the vapor phase (Antal *et al.*, 1996). If permitted to quickly escape the reactor, the vapors form condensed oils and tars. If held in contact with the solid biomass undergoing devolatilization within the reactor, the vapors degrade further to form chars, various gases, and water. The two competitive pathways help to explain the effects of pyrolysis conditions on product yields and distributions. Note that although these pathways may be dominant, there are undoubtedly many other pathways that are operative with actual biomass species. Thermal treatment converts hemicelluloses to furanoses and furans, the lignins to mononuclear and condensed aromatic and phenolic compounds, and the proteins to a wide range of nitrogen-containing aliphatic and heterocyclic compounds.

C. Kinetics

Considerable experimental work has been done with cellulose to clarify the kinetics of biomass pyrolysis. Most kinetic studies on cellulose pyrolysis have

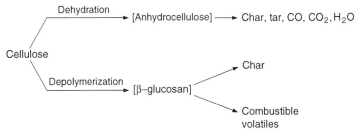

FIGURE 8.2　Pathways for cellulose pyrolysis.

built on the multistep model proposed in the early work with cellulose and described the evolution of volatiles by a single, pseudo-first-order reaction of the type (cf. Broido, 1976; Bradbury, Sakai, and Shafizadeh, 1979; Zaror and Pyle, 1982; Lai, and Krieger-Brockett, 1993; Antal and Varhegyi, 1995)

$$dm/dt = A \exp(-E/RT)(1 - m),$$

where A is the preexponential factor (time^{-1}), E is the apparent activation energy (J/mol), R is the ideal gas constant (J/mol-K), T is the absolute temperature (K), and m is the fraction of volatiles produced at time, t. This expression should be quite useful for designing advanced pyrolysis systems, but unfortunately, the reported kinetic factors vary widely and several results are in conflict. However, it was found that experimental thermogravimetric data obtained on prolonged, low-temperature, isothermal treatment of pure cellulose fit a slightly modified "Broido-Shafizadeh" model (Antal and Varhegyi, 1995). Under conditions of commercial interest, small samples of pure cellulose are reported to undergo thermal decomposition by a single, first-order, high-activation-energy, rate-determining step. The activation energy is estimated to be 238 kJ/g-mol. As our knowledge of biomass pyrolysis expands, it is expected that detailed kinetic parameters for more biomass components will be elucidated and applied to fine-tune practical biomass pyrolysis systems and designs.

Most of the experimental evidence accumulated over many years of study to understand biomass pyrolysis indicates that there are four basic kinds of processes. All of them involve the formation of large amounts of water vapor. One type is dominant at low heating rates and relatively low pyrolysis temperatures below about 250 to 300°C; chars, tars, and dehydration products are the primary products and some volatiles and gases are formed. One type is dominant at conventional pyrolysis conditions of intermediate heating rates and temperatures in the 300 to 600°C range; chars, tars, volatiles, and gases are formed in reasonable yields. The third type is dominant at fast heating rates and temperatures in the range 450 to 600°C; volatile liquids are the primary products, but some chars, tars, and gases are formed. The fourth type of pyrolysis occurs at temperatures above 600°C where gasification reactions begin to dominate. Devolatilization of biomass is important in each of these pyrolysis types.

D. THERMODYNAMICS

The pyrolysis of biomass feedstocks may be endothermic or exothermic, depending on the temperature of the reactants. For most biomass containing highly oxygenated hemicellulosics and cellulosics as the major components, pyrolysis is endothermic at temperatures below about 400 to 450°C and exothermic at higher temperatures. Once the necessary temperature has been reached in a properly designed system, little or no external heat is needed to sustain the process. The principal exothermic reactions that occur in the gas and solid phases during biomass pyrolysis are shown in Table 8.2. These reactions include the reduction of the carbon oxides to methane and methanol, the water gas shift reaction, and carbonization of celluloses. Substantial quantities of hydrogen are required for reduction of carbon oxides to methane and methanol, but hydrogen is not required for the water gas shift, which produces hydrogen, and the char formation reactions listed in Table 8.2. The pyrolysis temperature should be high enough to generate the requisite hydrogen for reduction of the carbon oxides. The water formed on biomass pyrolysis and the vaporization of the physically contained moisture in the fresh feed can participate in the water gas shift reaction. Interestingly, the exothermicity of cellulose carbonization is quite high per monomeric unit ($C_6H_{10}O_5$). Since the char yields on conventional biomass pyrolysis range up to about 35 wt %, and the fixed carbon contents of the chars are high, char formation would be expected to be a dominant driving force for biomass pyrolysis at the lower temperatures at which autogenous pyrolysis begins, but which generate less hydrogen. At these temperatures, pyrolysis is generally reaction-rate controlled, and at higher temperatures, the process becomes mass-transfer controlled. Energy to initiate biomass pyrolysis can be supplied by an external heat source or a portion of the pyrolysis products such as char and low-energy gas.

TABLE 8.2 Exothermic Reactions on Pyrolysis of Cellulose

Process	Reaction	Enthalpy, kJ/g-mol carbon converted at[a]	
		300 K	1000 K
Methanation	$CO + 3H_2 \rightarrow CH_4 + H_2O$	−205	−226
	$CO_2 + 4H_2 \rightarrow CH_4 + 2H_2O$	−167	−192
Methanol formation	$CO + 2H_2 \rightarrow CH_3OH$	−92	−105
	$CO_2 + 3H_2 \rightarrow CH_3OH + H_2O$	−50	−71
Char formation	$0.17C_6H_{10}O_5 \rightarrow C + 0.85H_2O$	−81	−80
Water gas shift	$CO + H_2O \rightarrow CO_2 + H_2$	−42	−33

[a]The standard enthalpy of formation of cellulose was calculated from its heat of combustion.

E. Products and Yields

General

Whatever the actual mechanisms of biomass pyrolysis, many reactions take place and many products are formed. The older biomass pyrolysis processes were carried out in the batch mode over long periods of time for char production. As the technology developed, other processes were designed to operate in the batch or continuous modes over shorter residence times at moderately higher temperatures. Depending on the pyrolysis temperature, the char fraction contains inorganic materials ashed to varying degrees, any unconverted organic solids, and carbonaceous residues produced on thermal decomposition of the organic components. The liquid fraction is a complex mixture of water and organic chemicals having lower average molecular weights than the feedstock components. For highly cellulosic biomass feedstocks, the liquid fraction usually contains acids, alcohols, aldehydes, ketones, esters, heterocyclic derivatives, and phenolic compounds. The tars contain native resins, intermediate carbohydrates, phenols, aromatics, aldehydes, their condensation products, and other derivatives. The pyrolysis gas is a low- to medium-energy gas having a heating value of about 3.9 to 15.7 MJ/m^3 (n) (100 to 400 Btu/SCF). It contains carbon dioxide, carbon monoxide, methane, hydrogen, ethane, ethylene, minor amounts of higher gaseous organics, and water vapor.

It is apparent that if one wishes to obtain pure chemicals by biomass pyrolysis, further processing to separate the reaction mixture is necessary. As will be shown later, this did not hinder commercial use of biomass pyrolysis for the manufacture of specific chemicals. The slow, destructive distillation of biomass was commercial technology for the production of several commodity chemicals long before fossil fuels became the preferred feedstocks. Hardwood pyrolysis once served as an important commercial source of methanol, acetic acid, ketones, and other chemicals.

Conventional Slow Pyrolysis

Chars, gases, light and heavy liquids, and water are formed in varying amounts on pyrolysis of biomass. The yields depend particularly on the feed composition, dimensions of the feed particles, heating rate, temperature, and reaction time. When hardwoods are heated in the absence of air, they decompose and are converted into charcoal and a volatile fraction that partly condenses on cooling to a liquor called pyroligneous acid, which separates into a dark heavy oil as the lower layer in about 10 wt % yields, and an upper aqueous layer. Dry distillation of softwoods such as pine yield similar products in about the same amounts as well as lighter pine oils and terpene liquids such as turpentines. The supernatant layer contains methanol, acetic acid, traces of acetone, allyl alcohol, and other water-soluble compounds. Methanol is formed from the lignin components that bear methoxyl groups. The heavy oil contains tars, higher viscosity pitches, and some char. The wood tars and pitches are complex mixtures in which hundreds of organic compounds have been identified, primarily acidic and heterocyclic compounds.

The data in Table 8.3 show how the gas, pyrolytic oil, and char yields vary with pyrolysis temperature and different biomass feedstocks (Epstein, Kostrin, and Alpert, 1978). Extensive depolymerization of the celluloses starts at about 300°C and usable charcoal formation (carbon content about 75 wt %) starts at about 350°C (Zaror and Pyle, 1982). Higher temperatures and longer residence times promote gas production, while higher char yields are obtained at lower temperatures and slow heating rates. The product slate is similar for each feedstock at a given temperature, although the yields of gas, pyrolytic oil, and char can be quite different. The cellulosics and hemicellulosics are the main sources of volatiles in biomass feedstocks, but yield only about 8 to 15% of their weight as charcoal under conventional pyrolysis conditions (i.e., slow heating rate, atmospheric pressure, and maximum temperatures of 400 to 450°C). The lignins yield nearly 50% of their weight as charcoal under these conditions (Zaror and Pyle, 1982). A more detailed distribution of specific products on long-term pyrolysis of three woody biomass feedstocks, birch, pine, and spruce wood, to a final temperature of 400°C is shown in Table 8.4 (Nikitin et al., 1962; Bagrova and Kozlov, 1958). Both the individual product distributions and the yields of carbon, pyroligneous distillate, and gases are similar for each wood species. It is evident that the product mixture is complex and that selectivities for specific chemicals are low. The order of decreasing yield on a weight basis by product group from highest to lowest is pyroligneous distillate, charcoal, and gaseous products. This might be expected because of the relatively low pyrolysis temperatures and the 8-h period over which these experiments were performed. However, if water is excluded from the yield calculations, the order of decreasing yield is charcoal, pyroligneous distillate, and gaseous products.

The pyrolysis of the combustible fraction of MSW at higher temperatures is illustrated by the data in Table 8.5. These data show how temperature affects product yields and gas and char compositions on pyrolysis at temperatures up to 900°C (Hoffman and Fitz, 1968). Gas yield increases as the temperature is increased from 500 to 900°C. Although the heating value of the product gas remains about the same, significant increases in gas yield on a weight percent and energy yield basis and in hydrogen occur with increasing temperature. Interestingly, as the temperature increases, the char yields and volatile matter content of the chars decrease as expected, but the energy value of the chars is relatively constant.

TABLE 8.3 Product Yields from Different Biomass Feedstocks as a Function of Pyrolysis Temperature[a]

Feedstock	Low-energy gas (wt % at °C)			Pyrolytic oil (wt % at °C)			Charcoal (wt % at °C)		
	500°C	700°C	900°C	500°C	700°C	900°C	500°C	700°C	900°C
Biosolids	10	26		10	2		12	11	
Corncobs	17	65	52	22	7	3	26	14	17
Manure	20	30	42	18	7	2	28	14	11
MSW	23	36	50	11	6	3		24	13
Paper	16	45	70	47	8	3	10	6	4
Wood chips	23	35	53	19	6	2	27	20	22

[a]Epstein, Kostrin, and Alpert (1978). The feedstock was pyrolyzed in a 0.5-m ID fluid-bed reactor containing sand and an inert gas generated from compressed air-natural gas combustion with a slight excess of air (about 0.2 to 0.6%). The fluidizing velocities were 0.3 to 1 m/s. The products were low-energy gas (5.89–11.78 MJ/m^3 (n)), pyrolytic oil (23.3–27.9 MJ/kg), and charcoal. Feed rates were 50–200 kg/h. The moisture contents of the feedstocks were not specified. The balance of the yield for each feedstock is water.

TABLE 8.4   Product Yields from Thermal Decomposition of Birch, Pine, and Spruce Woods Heated over an 8-Hour Period to Final Temperature of 400°C[a]

Products	Birch (wt %)	Pine (wt %)	Spruce (wt %)
Gases			
H_2	0.03	0.03	0.03
CO	4.12	4.10	4.07
CO_2	11.19	11.17	10.95
CH_4	1.51	1.49	1.59
C_2H_4	0.21	0.14	0.15
Subtotal:	17.06	16.93	16.79
Charcoal	33.66	36.40	37.43
Pyroligneous oil			
Water	21.42	22.61	23.44
Settled tar	3.75	10.81	10.19
Soluble tar	10.42	5.90	5.13
Volatile acids	7.66	3.70	3.95
Alcohols	1.83	0.89	0.88
Aldehydes	0.50	0.19	0.22
Esters	1.63	1.22	1.30
Ketones	1.13	0.26	0.29
Subtotal:	48.34	45.58	45.40
Losses	0.94	1.09	0.38

[a]Nikitin et al. (1962); Bagrova and Kozlov (1958). Volatile acids are calculated as acetic acid. Alcohols are calculated as methanol. Aldehydes are calculated as formaldehyde. Esters are calculated as methyl acetate. Ketones are calculated as acetone.

Since biomass pyrolysis product mixtures are very complex and selectivities are low for specific products, considerable effort has been devoted to improving selectivities. Selectivities can sometimes be increased by addition of coreactants or catalysts, or by changing the pyrolysis conditions (cf. Nikitin et al., 1962). For example, the pyrolysis of maplewood impregnated with phosphoric acid increased the yield of methanol to 2.2 wt % of the wood as compared to 1.3 wt % obtained on dry distillation of the untreated wood. Addition of sodium carbonate to oak and maple increased the yield of methanol by 100 and 60%, respectively, compared to pyrolysis yields without sodium carbonate. Other weakly alkaline reagents exhibited a similar effect. Pyrolysis of wood in a stream of benzene, xylene, or kerosine increased the yields of acetic acid, aldehydes, and phenols and reduced the yield of tars. Optimization of pyrolysis conditions will be shown later to have large effects on product distributions and yields.

TABLE 8.5   Effects of Temperature on Product Yields and Gas and Char Compositions from Pyrolysis of the Combustible Fraction in MSW[a]

| | Pyrolysis temperature | | | |
Parameter	500°C	650°C	800°C	900°C
Product yields and recovery				
Gases, wt %	12.3	18.6	23.7	24.4
m³ (n)/kg	0.114	0.166	0.216	0.202
MJ/kg	1.39	2.63	3.33	3.05
Liquids, wt %	61.1	59.2	59.7	58.7
Charcoal, wt %	24.7	21.8	17.2	17.7
Recovery, wt %	98.1	99.6	100.6	100.8
Gas composition and HHV				
H_2, mol %	5.56	16.6	28.6	32.5
CO, mol %	33.5	30.5	34.1	35.3
CO_2, mol %	44.8	31.8	20.6	18.3
CH_4, mol %	12.4	15.9	13.7	10.5
C_2H_6, mol %	3.03	3.06	0.77	1.07
C_2H_4, mol %	0.45	2.18	2.24	2.43
HHV, MJ/m³ (n)	12.3	15.8	15.4	15.1
Char composition and HHV				
Fixed carbon, wt %	70.5	70.7	79.1	77.2
Volatile matter, wt %	21.8	15.1	8.13	8.30
Ash, wt %	7.71	14.3	12.8	14.5
HHV, MJ/kg	28.1	28.6	26.7	26.5

[a]Hoffman and Fitz (1968). "HHV" is higher heating value.

Fast Pyrolysis

Processes categorized as fast pyrolysis systems are continuously operated at temperatures generally in the range of 400 to 650°C and residence times of a few seconds to a fraction of a second. Manipulation of these parameters permits the bulk product yields to be changed from those of conventional pyrolysis systems within a wide range, but the products are still chars, liquids, and gases plus water. Fast pyrolysis is characterized by high heating rates and rapid quenching of the liquid products to terminate additional conversion of the products downstream of the pyrolysis reactor. The selectivity for specific chemicals is usually low, as in the case of conventional pyrolysis. Very rapid heating of biomass results in the fragmentation of the polymeric components in biomass to afford 60 to 70 wt % primary vapor products composed of oxygenated monomers and polymer fragments (Diebold *et al.*, 1987). Rapid, efficient quenching of the product streams and short residence times tend to "freeze" the product compositions so that they correspond more closely with the chemicals formed initially on biomass pyrolysis. More details are presented in Section III on fast pyrolysis.

F. Fixed Carbon

The carbonaceous residues from biomass pyrolysis are in the charcoal fraction. These residues are called "fixed carbon" by most energy specialists. The generally accepted definition of fixed carbon was originally promulgated by coal chemists. It is the amount of combustible material remaining in a sample of coal, coke, or bituminous material after removal of moisture, volatile matter, and ash, and is expressed as a percentage of the original material. American Society for Testing and Materials (ASTM) procedures have been developed for determination of each of these parameters: moisture (ASTM D 3173; 104-110°C for 1 h), volatile matter (ASTM D 3175; 950°C for 7 min), and ash (ASTM D 3174; gradual heating to redness and finishing ignition at 750°C). Fixed carbon is the difference between 100 and the sum of these determinations (ASTM D 3172) and is essentially the elemental carbon in the original coal sample plus the carbonaceous residue formed on heating the coal sample at 950°C for 7 min. As the temperature rises above 300°C, coals emit volatile matter that consists of gases, oils, and tars. The residues contain elemental carbon, some of the higher molecular weight polynuclear aromatic hydrocarbons formed in the process, and a few other high-molecular-weight compounds that are also formed in the process. Peat, which is derived from biomass, is categorized by some specialists as "young coal" and does contain some elemental carbon. In contrast, all carbon in biomass is fixed carbon in the same sense that organic nitrogen is fixed nitrogen, but elemental carbon is not present in biomass. The photosynthetic fixation of CO_2 results in the formation of fixed carbon. But since there is no elemental carbon as such in biomass, its fixed carbon content can also be considered to be zero. In other words, the terminology "fixed carbon" in biomass is a misnomer.

If biomass is subjected to the ASTM D 3172 procedure for determination of fixed carbon, chemical transformation of a portion of the organic carbon in biomass into carbonaceous material occurs as described here. All of the fixed carbon determined by the ASTM procedure is therefore generated by the analytical method. Furthermore, the amount of fixed carbon generated depends on the heating rate used to reach biomass pyrolysis temperatures and the time the sample is subjected to these temperatures. Nevertheless, such analyses are valuable for the development of thermal conversion processes for biomass feedstocks. But application of the ASTM procedures to biomass might more properly be called a method for determination of pyrolytic carbon or coking yields. In the petroleum industry, the Conradson carbon (ASTM D 189, differential heating with a gas burner for total of 30 min to final temperature of cherry-red crucible) and the Ramsbottom carbon (ASTM D 524, 549°C for 20 min) procedures are used to determine the coking tendency on pyrolysis of petroleum products. Use of these procedures with biomass would be expected to give somewhat different results for fixed carbon than ASTM D 3172.

Table 8.6 is a tabulation of the fixed carbon, volatile matter, and ash analyses of selected biomass species, biomass derivatives, and coals as determined by the ASTM D 3172 procedure. The data for the wood species, wood barks, and herbaceous biomass species show that significant quantities of pyrolytic carbon are produced by this method. The pyrolytic chars listed, which already contain substantial amounts of elemental carbon because of the nature of the pyrolysis process, contain more fixed carbon than the coal samples listed in this table. In contrast, MSW and the papers in MSW, which are high in celluloses, contain considerably less fixed carbon suggesting that the lignins in biomass contribute more to production of fixed carbon than the other components. This is expected because of the nature of the chemical structures of the lignins, and the fact that papers are low in lignins. The times and temperatures used for the ASTM D 3172 procedure coupled with the data in Table 8.6 suggest that for maximum yields of noncarbonaceous products to be obtained on biomass pyrolysis, short reaction times should be used at relatively low pyrolysis temperatures. These conditions would be expected to yield smaller amounts of charcoal, tars, and gases, and larger amounts of liquid products. As will be discussed later, optimum conditions have been developed for separately maximizing charcoal and liquid product yields in biomass pyrolysis.

Because of the amounts of sample, labor, and time required to perform the ASTM D 3172 procedure, it is recommended that thermogravimetry (TG) and differential thermogravimetry (DTA) be used for moisture and proximate analysis of biomass and the rapid estimation of their thermal conversion characteristics. Application of these techniques shows that the proximate analyses of standard coal samples agree closely or match the values obtained with the ASTM procedure (cf. Kumar and Pratt, 1996). Thermogravimetric procedures are used to determine the thermal stabilities and properties of inorganic and organic materials and can be carried out with small samples in laboratory equipment. The results are reasonably accurate and reproducible. TG and DTG employ sensitive thermobalances and automated data processors to measure weight loss and the rate of weight loss of the samples as a function of temperature, respectively. A TG curve (thermogram) records the weight of the sample with time at a preset temperature or a programmed heating rate in an inert or reactive gaseous atmosphere. Some laboratory instrumentation has also been designed to operate at elevated pressures (cf. Johnson, 1979). Differentiated TG data with time or temperature provides the rate of weight loss. A TG curve is used for moisture and proximate analyses, and a DTG curve can be used to

TABLE 8.6 Proximate Analysis of Selected Biomass and Fuels[a]

Material	Fixed carbon (dry wt %)	Volatile matter (dry wt %)	Ash (dry wt %)
Oven-dry woods			
Western hemlock	15.0	84.8	0.2
Douglas fir	13.7	86.2	0.1
White fir	15.1	84.4	0.5
Ponderosa pine	12.8	87.0	0.2
Redwood	16.1	83.5	0.4
Cedar	21.0	77.0	2.0
Eucalyptus globulus	17.3	81.6	1.1
Eucalyptus grandis	16.9	82.6	0.5
Casuarina	19.7	78.9	1.4
Poplar	16.4	82.3	1.3
Oven-dry barks			
Western hemlock	24.0	74.3	1.7
Douglas fir	27.2	70.6	2.2
White fir	24.0	73.4	2.6
Ponderosa pine	25.9	73.4	0.7
Redwood	27.9	71.3	0.8
Cedar	13.1	86.7	0.2
Herbaceous biomass			
Alfalfa seed straw	20.1	72.6	7.3
Cattail	20.5	71.6	7.9
Corn cobs	18.5	80.1	1.4
Corn stover	19.2	75.2	5.6
Cotton-gin trash	15.1	67.3	17.6
Macadamia shells	23.7	75.9	0.4
Peach pits	19.9	79.1	1.0
Peanut hulls	21.1	73.0	5.9
Rice hulls	16.7	65.5	17.9
Sudan grass	18.6	72.8	8.6
Sugarcane bagasse	14.9	73.8	11.3
Wheat straw	19.8	71.3	8.9
Pyrolysis chars			
Redwood (421 to 549°C)	67.7	30.0	2.3
Redwood (427 to 941°C)	72.0	23.9	4.1
Oak (438 to 641°C)	59.3	25.8	14.9
Oak (571°C)	55.6	27.1	17.3
MSW and major components			
National average MSW	9.1	65.9	25.0
Newspaper	12.2	86.3	1.5
Paper boxes	12.9	81.7	5.4
Magazine paper	7.3	69.2	23.4
Brown paper	9.8	89.1	1.1
Coals			
Pittsburgh seam coal	55.8	33.9	10.3
Wyoming Elkol coal	51.4	44.4	4.2
Lignite	46.6	43.0	10.4

[a]Adapted from Graboski and Bain (1979) and Jenkins and Ebeling (1985). The data on barks and pyrolysis chars are from Howlett and Gamache (1977). The data on MSW are from Klass and Ghosh (1973). The data on coals are from Bituminous Coal Research (1974). The data on herbaceous biomass are from Jenkins and Ebeling (1985). The data on woods are from Howlett and Gamache (1977) and Jenkins and Ebeling (1985).

examine combustion, pyrolysis, and gasification characteristics. The kinetics of conversion and the conditions for process optimization can also be estimated using TG and DTG. Analysis of selected biomass components by these techniques indicates that pyrolysis is initiated at 150-350°C for hemicelluloses, 275-350°C for celluloses, 250-500°C for lignins, 500-620°C for latex, and 550-900°C for high-molecular-weight resins and oils (cf. Kumar and Pratt, 1996). The combustion of fixed carbon and ash in representative samples of biomass occurs at 900°C according to these studies.

The data presented in Table 8.7 illustrate the pyrolysis characteristics obtained by TG and DTG analysis of several biomass species and parts. This type of data and derived data can be used to project the utility of a given biomass feedstock for thermal conversion. For example, the organic material emitted in the temperature range 320 to 500°C was assumed to be potential tar-forming volatiles (Grover, 1989). Those biomass species having lower emissions of volatiles in this temperature range were judged to be more suitable for conversion to tar-free gases on pyrolysis.

III. PROCESSES

A. HARDWARE

Pyrolysis systems are as varied as combustion systems. The ancient process of making charcoal by the slow pyrolysis of a pile of wood covered with earth is still used today in some developing countries. Pyrolysis times are several days and charcoal is the main product. Before fossil fuels became the preferred feedstocks for chemical production in the early part of the twentieth century, biomass pyrolysis reactors in industrialized countries consisted of various types of ovens and horizontal and vertical steel retorts, essentially all of which were operated in the batch mode. Provision was made for charcoal recovery, pyroligneous acid refining, by-product recovery, and gas recovery and usage. Modern pyrolysis reactor configurations include fixed beds, moving beds, suspended beds, fluidized beds, entrained-feed solids reactors, stationary vertical-shaft reactors, inclined rotating kilns, horizontal shaft kilns, high-temperature (1000 to 3000°C) electrically heated reactors with gas-blanketed walls, single and multihearth reactors, and a host of other designs.

B. COMMERCIAL PYROLYSIS OPERATIONS IN THE EARLY 1900s

The dry distillation of hardwood was commercial technology and was quite common up until the early 1900s (Riegel, 1933). In the older industrial plants,

TABLE 8.7 Proximate Analysis and Pyrolysis Characteristics of Selected Biomass Species[a]

Biomass	Fixed carbon (dry wt %)	Volatile matter (dry wt %)	Ash (dry wt %)	Devolatilization			
				Range (°C)	Max. rate at (°C)	Max. rate (wt %/°C)	Emitted at 320–500°C (wt %)
Bagasse	16.9	75.1	8.0	280-510	385	0.53	54.0
Bamboo dust	15.6	75.3	9.1	240-600	270	0.35	37.8
Cotton stalks	22.4	70.9	6.7	280-520	390	0.70	47.5
Coconut dust	26.8	70.3	2.9	330-600	350	0.71	39.7
Corn cobs	16.2	80.2	3.6	370-710	400	0.49	45.6
Groundnut shell	25.0	68.1	6.9	300-720	505	0.67	49.3
Jute sticks	19.0	75.3	5.7	220-500	390	1.06	57.0
Mustard shells	14.5	70.1	15.4	300-550	370	0.67	42.5
Pigeon pea	14.8	83.5	1.8	290-650	390	0.48	54.0
Pine needles	26.1	72.4	1.5	320-680	410	0.38	44.5
Prickly acacia	22.3	77.0	0.6	270-680	340	0.35	44.1
Prickly sesban stalks	17.0	80.3	2.7	320-650	375	0.45	62.2
Rice husks	19.9	60.6	19.5	340-510	390	0.66	48.0
Sal seed leaves	20.2	60.0	19.7	200-650	440	0.14	22.5
Sal seed husks	28.1	62.5	9.4	340-520	420	0.43	46.3

[a]Adapted from Grover (1989). Thermograms were obtained with powdered samples. The heating rate was 4°C/min in an atmosphere of nitrogen flowing at a rate of 0.3 L/min. The devolatilization range is the temperature at the beginning and end of devolatilization. The volatiles emitted at 320–500°C are the wt % of the original sample.

hardwood logs were placed on steel buggies and pushed into large, horizontal steel retorts. The doors were closed and external heating was supplied with gas or oil, often within a brick enclosure. Openings were provided for removal of volatiles. Heating was rapid for a few hours until the process became exothermic, and then the external heating was reduced and increased again when needed to complete the distillation. The distillation cycle was about 24 h including a 2-h cooling period. The buggies were then cooled for a few days in special airtight cooling chambers. The dry distillation of softwoods to obtain turpentine was carried out in smaller horizontal retorts in which wood chips were placed without buggies. The average product yields per cord of seasoned hardwood from typical commercial pyrolysis processes are about 1025 kg (950 L) of pyroligneous acid containing 7% acetic acid or equivalent, 4% crude methanol and acetone, 9% tar and oils, and 80% water; 454 kg of charcoal; and 212 m^3 of fuel gas having a heating value of 9.3 to 11.2 MJ/m^3 (Lowenheim and Moran, 1975). The pyroligneous acid is allowed to settle to remove insoluble tar and the clear decanted liquor is subjected to extraction or distillation or both to separate acetic acid from methanol and tar. For each tonne of acetic acid produced, about 625 to 800 L of crude methanol is recovered.

Until about the 1930s, when they were displaced by other processes that utilized fossil feedstocks, wood pyrolysis processes were used in industrialized countries for the manufacture of several chemicals and products. One example of this practice is the distillation plant operated by the Ford Motor Company using feedstock of hogged scrapwood from the automobile body plant (Riegel, 1933). A flow schematic for this plant is shown in Fig. 8.3. Vertical, cylindrical, steel retorts 3 m wide by 12 m high with an inside refractory wall 0.46 m thick were used. The hogged wood was dried to a moisture content of 0.5 wt % and consisted of 70% maple, 25% birch, and 5% ash, elm, and oak. Plant capacity was 363 t/day of scrap wood. The retorts were operated continuously for 2-week periods and the heat was supplied entirely by the exothermic pyrolysis reactions or the pyrolysis gas. At startup, the gas was employed to raise the temperature to 540°C, and then external heat was not needed. The average temperatures were 515°C in the center of the retort and 255°C near the bottom. The charcoal was discharged at the bottom of the retorts, cooled, screened, and briquetted. The pyroligneous acid was recovered from the overhead and the pyrolysis gas was used as boiler fuel except for the fuel used on startup of the retorts. The pyroligneous acid was distilled in batch units to remove dissolved tar, and the overhead was then fractionated in other distillation units. The product yields from this plant are shown in Table 8.8. The char, tar, and pitch yields are considerably higher than the yields of chemicals. It is somewhat surprising to note, since wood was a primary feedstock for the manufacture of methanol (wood alcohol) before its displacement

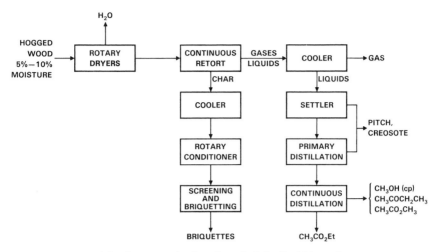

FIGURE 8.3   Continuous hardwood pyrolysis by Ford Motor Company.

TABLE 8.8   Average Product Yields from Commercial Wood
Pyrolysis Plant Operated by Ford Motor Company Using
Badger–Stafford Retorts[a]

Product	Yield per tonne of dry wood
Gas	134.3 m^3 (n)
Char	272 kg
Pitch	29.9 kg
Soluble tar	83.30 L
Ethyl acetate	55.45
Creosote oil	12.30
Methanol	11.80
Ethyl formate	4.81
Methyl acetate	3.58
Methyl acetone	2.47
Ketones	0.86
Allyl alcohol	0.18

[a]Adapted from Riegel (1933). The average heating value of the
dry gas was 11.39 MJ/m^3 (n). The average composition of the
gas in mol % from the retorts was H_2, 2.2; CO, 23.4; CO_2, 37.9;
CH_4, 16.8; C_nH_m, 1.2; O_2, 2.4; N_2, 16.0. The esters were produced
from intermediate acid and product methanol or external ethanol.

by natural gas, that the methanol yield, including that used to esterify the acids formed in the process, is much less than the yields of the other pyrolysis products. In the Ford plant, the acetic acid was converted to esters, since they and not the free acid were needed in other automobile manufacturing operations. However, other companies produced the free acid from the pyroligneous acid by direct solvent extraction. The Brewster process used isopropyl ether as the solvent and the Suida process used a high-boiling wood oil fraction from the pyrolysis plant as the solvent (Riegel, 1933). Eventually, these and other pyrolysis processes were phased out, as they were replaced by synthetic methods based on fossil feedstocks.

C. ADVANCED CHARCOAL PRODUCTION PROCESSES

As already mentioned, the older production methods for conversion of biomass to charcoal are slow processes and the yields are low. Several days are required to complete the process in earthen pits with seasoned wood, and the yields are only about 10 to 15% of the dry wood weight because most of the volatile organics leave the pyrolysis zone before carbonization occurs. But note that the coalification of biomass in nature is truly a long process compared to human-controlled processes. Coal is a product of the gradual, natural decomposition of cellulosic biomass, without free access to air, under the influence of pressure and temperature; it is formed through the successive stages of peat, lignite or brown coal, bituminous or soft coal, and anthracite or hard coal, characterized by increasing carbon content (*cf.* Fieser and Fieser, 1950).

As shown in Tables 8.6 and 8.7, the volatile matter in biomass as measured by the ASTM and TG procedures is much higher than the fixed carbon content, so to significantly increase charcoal yields, the volatiles must be carbonized as well. Closed reactors can be designed to keep the volatiles in the pyrolysis zone for longer periods and increase carbonization. The use of beehive kilns, for example, affords charcoal yields up to 35%, but the process still requires several days for completion (*cf.* Antal *et al.*, 1996). The Ford Motor Company process in Badger-Stafford retorts was performed over 24-h cycles and the charcoal yields were about 27% (Table 8.8).

From a theoretical perspective, pure cellulose contains 44.4 wt % carbon, so the maximum theoretical yield of charcoal, assuming all of the cellulosics can be carbonized, is 44.4 wt %. But with dry wood chars containing about 60 to 70 wt % fixed carbon (*cf.* Table 8.6), the theoretical maximum yields of charcoal including volatile matter and ash from wood feedstocks then correspond to about 65 to 75% by weight of the dry wood. It is evident that if charcoal is the desired product, considerable process improvements should be possible.

A batch process that affords higher charcoal yields with biomass feedstocks over a relatively short reaction time has been developed (Antal *et al.*, 1996). The biomass feedstock, usually logs, wood chips, or nutshells, is maintained under pressure up to about 0.7 MPa at typical temperatures of 450°C for pyrolysis times of 15 min to 2 h. The yields of dry charcoal have ranged from 42 to 62% as shown in Table 8.9. The PDU (process development unit) designed to demonstrate this technology employs a cylindrical steel canister covered by a lid. The canister is charged with biomass that is not predried and is placed within the pressure-tight steel vessel of the PDU, which for demonstration purposes was electrically heated. Startup involves heating the PDU to pyrolysis temperatures and then maintaining the PDU at that temperature. The steam formed from the contained moisture is released as needed to maintain the pressure at 0.7 MPa. Commercial systems will probably use the fuel gas emitted to supply heat. However, the results obtained with the PDU illustrate how the charcoal yields approach the theoretical limits, so the yields of other pyrolysis products may be too low to supply any needed fuel. The results from the PDU suggest that both the moisture in the biomass and the pressure can be manipulated to maximize charcoal yields. The vapors emitted during this batch process are kept in contact with the solid biomass undergoing pyrolysis at the internal pressure of the PDU. These conditions result in increased char and low tar yields at short reaction times compared to those that have been employed in most other processes.

TABLE 8.9 Proximate Analysis of Biomass and Charcoal Yields at Elevated Pressure in Batch Process[a]

Biomass	Moisture (wt %)	Fixed carbon (dry wt %)	Volatile matter (dry wt %)	Ash (dry wt %)	Heating time (min)	Char yield (dry wt %)
Eucalyptus	40.0	75.4	23.3	1.29	114	47.0
Kiawe	21.0	75.9	21.4	2.65	150	47.0
Kukui nut shell	12.5	78.1	20.1	1.79	90	62.1
Leucaena	1.5	82.8	14.4	2.85	15	42.2
Macadamia nut shell	13.5	70.5	28.7	0.81	205	50.8
Palm nut shell	13.4	77.7	18.8	3.48	105	41.6

[a]Adapted from Antal *et al.* (1996).

D. Advanced Pyrolytic Liquids Production Processes

In contrast to the pyrolysis conditions needed to increase charcoal yields, the conditions for increasing the yields of organic liquid products would be expected to involve short heat-up and reaction times and rapid removal and quenching of the organic volatiles before they are carbonized. Further, if the pyrolysis temperature at which maximum devolatilization occurs is chosen, but which is insufficient to gasify the volatiles (i.e., convert them to light fuel gases), more liquid products should be produced. When the reaction temperature is too low, less devolatilization and more char and tar formation occur, and when the temperature is too high, gasification reactions are dominant. Intuitively, pyrolysis at short reaction times and intermediate temperatures would be expected to promote higher organic liquid yields at the expense of other products. Char and gas yields should be low under these conditions. Numerous studies have indeed demonstrated that short-residence-time pyrolysis, or flash pyrolysis, can be performed with biomass feedstocks to maximize liquid yields (*cf.* Bridgwater and Bridge, 1991; Bridgwater and Peacocke, 1995).

In some of the early work on the continuous flash pyrolysis of biomass at atmospheric pressure, it was shown that at optimum temperatures, liquid yields are maximized (Scott and Piskorz, 1983). With entrained flow injection of biomass feedstock of -250 μm to $+105$ μm particle size into a mini-fluidized-bed reactor with sand heat carrier and vapor residence times of 0.44 s, it was found that the maximum yield of liquid products occurs at the optimum temperature, and that yield drops off sharply on both sides of this maximum. Pure cellulose was found to have an optimum temperature for production of liquids at 500°C, whereas the wheat straw and wood species tested had optimum pyrolysis temperatures for maximum liquids at 600°C and 500°C, respectively. The yields of organic liquids were of the order of 55 to 65% of the dry weight of the biomass fed. The liquids contained relatively large quantities of organic acids. Pilot plant tests verified these observations. As research progressed on the conversion of biomass by flash pyrolysis, the optimum conditions for maximum liquid yields were found to be temperatures within the range 400 to 600°C, vapor residence times within the range 0.1 to 2.0 s, particle sizes less than 2 mm and a maximum of 5 mm for wood feeds, and an oxygen-free gaseous atmosphere such as recycled flue gas in the pyrolysis zone (Scott, Piskorz, and Radlein, 1993). Any reactor that can be operated under these conditions and that provides for biomass heating so that the particle temperatures can exceed about 450°C before 10% weight loss occurs can be used as a flash pyrolysis reactor. Such designs include fluidized-bed reactors, circulating fluid beds, transport or entrained flow reactors with or without a solid heat carrier, ablative reactors, and reduced-pressure reactors. Maximum organic liquid yields were projected to be 55 to 65 wt % of woody feedstocks and 40 to 65 wt % of grass feedstocks. The product yields from the flash pyrolysis of softwood (white spruce) and hardwood (poplar) are shown in Table 8.10. The total pyrolytic liquid yields are about the same for each feed, 66.5 and 65.7% of the feed dry weight, but they include several sugars and polysaccharide derivatives that are normally solids at ambient conditions. About 50% of the pyrolytic liquid product is water soluble. The largest fraction is the pyrolytic lignin, the insoluble fraction that remains after water extraction of the pyrolytic liquid. In other research, it has been shown that because of differences in oxidation rates between the anhydrosugar and aromatic lignin-derived pyrolysis products, which are formed in the fast pyrolysis of deionized or prehydrolyzed wood, it is possible to carry out flash pyrolysis with controlled levels of oxygen to selectively oxidize lignins with relatively little effect on the anhydrosugar yields (Piskorz *et al.*, 1995). The recovery of the anydrosugars for use as fermentable sugars or as chemicals is expected to be simplified because their concentrations in the pyrolysis liquid are then significantly increased. The complex nature of the products and the poor selectivity of pyrolysis are evident, but with suitable refining, a wide range of chemicals could be manufactured from the products. Indeed, similar technology has already been commercialized for the production of fuels and chemicals.

A commercial, flash pyrolysis plant (RTP™) built in the United States in 1989 is believed to be the first successful plant in the world based on fast pyrolysis (Graham, Freel, and Bergougnou, 1991). This plant had a capacity of 100 kg/h of particulate hardwood feedstocks and is an upflow design that incorporates complete recirculation of the solid heat carrier in a reactor system capable of operating between 450 and 600°C at a residence time in the range of 0.6 to 1.1 s. The plant is used for the production of boiler fuel and specialty chemicals such as flavorings and natural colorings. The liquid is pourable and pumpable at room temperature and has an HHV ranging from 15 to 19 MJ/kg, including the contained moisture, or approximately the same heating value as the feedstock entering the conversion unit. Typical liquid yields from representative hardwoods at 10-15% moisture content are about 73 wt % of the feedstock, including contained moisture (Graham and Huffman, 1995). In general, the yield increases slightly with an increase in the cellulose content of the feedstock and decreases slightly with an increase in feedstock lignin. However, the energy yield is almost constant since lignin-derived liquids have a higher energy content than cellulose-derived liquids. The liquids are produced as a single phase unlike the heavy tars produced by conventional biomass pyrolysis. Experimentation has shown that by manipulating the vapor condenser operating conditions, the product liquid can be tailored for chemicals and boiler, diesel, or turbine fuel applications without altering the pyrolysis

TABLE 8.10 Flash Pyrolysis Products and Compositions from Softwood and Hardwood[a]

Parameter	White spruce	Poplar
Moisture content, wt %	7.0	3.3
Particle size, μm (max)	1000	590
Temperature, °C	500	497
Apparent residence time, s	0.65	0.48
Product yields, wt % dry feed		
Water	11.6	12.2
Char	12.2	7.7
Gas	7.8	10.8
Pyrolytic liquid	66.5	65.7
Gas composition, wt % dry feed		
H_2	0.02	
CO	3.82	5.34
CO_2	3.37	4.78
CH_4	0.38	0.41
C_2's	0.20	0.19
C_3 +	0.04	0.09
Pyrolytic liquid composition, wt % dry feed		
1,6-Anhydroglucofuranose		2.43
Acetaldehyde		0.02
Acetic acid	3.86	5.43
Acetol	1.24	1.40
Cellobiosan	2.49	1.30
Ethylene glycol	0.89	1.05
Formaldehyde		1.16
Formic acid	7.15	3.09
Fructose + other hexoses	2.27	1.32
Furfural	0.30	
Glucose	0.99	0.41
Glyoxal	2.47	2.18
Hydroxyacetaldehyde	7.67	10.03
Levoglucosan	3.96	3.04
Methanol	1.11	0.12
Methylfurfural	0.05	
Methylglyoxal		0.65
Oligosaccharides		0.70
Water-solubles, above subtotal	34.45	34.33
Pyrolytic lignin	20.6	16.2
Water-soluble losses (phenols, furans, etc.)	11.5	15.2

[a]Adapted from Scott, Piskorz, and Radlein (1993).

process itself. Since 1989, larger commercial plants of capacity up to 70 green t/day of feedstock have been built in North America, demonstration plants have been completed in Europe, and others that range in feedstock capacity from 100 to 350 t/day are in the advanced planning stages (Graham, Freel, and Kravetz, 1996).

In the mid-1990s, the economics of the RTP process for a 100-t/day plant were viable at a product fuel oil price of $5.27/GJ if the biomass feedstocks were available at zero cost. This can occur in several situations, especially with captive sources of waste biomass. The fuel oil price covers all fixed and variable operating costs, the annual capital costs to finance debt (75% of total capital), and an acceptable rate of return on the equity investment (25% of total capital). It is assumed that wood is available in the form of wet chips, and that the drying and grinding equipment and pyrolysis unit are included in the capital cost estimates. Expressed in other terms, the cost of converting 1 dry tonne of feedstock to fuel oil by this process is about $60, which means that the cost of the fuel oil is then about $10.54/GJ if the cost of the biomass feedstock is $60/dry t. For zero-cost fuel oil, a tipping fee of $60/dry t of waste feedstock is required. All of this means that the economics can be favorable in locations where there is an abundance of feedstock at zero or negative cost. In other words, the technology is very site specific, or the cost of competitive products, petroleum fuel oils in this case, must be high enough to justify commercial plants.

An example of one of the first flash pyrolysis processes developed for waste biomass is shown in Fig. 8.4 (U.S. Environmental Protection Agency, 1975; Preston, 1976). In this process, MSW is separated by a sequence of steps to obtain refuse-derived fuel (RDF) and recyclables. The sequence consists of shredding of MSW and air classification to obtain the RDF, magnetic separation of the ferrous metals, screening and froth flotation to recover a glass cullet, and aluminum separation by an aluminum magnet. The RDF is dried in a rotary kiln to about 4 wt % moisture content, and finely divided to a particle size of which 80% is smaller than 14 mesh (1200 μm). The feed, about 0.23 kg of recycled char preheated to 760°C per kilogram of this finely divided material, is rapidly passed through the pyrolysis reactor at atmospheric pressure. The raw product mixture, which consists of product gas and liquid, the char fed to the reactor, and new char formed on pyrolysis, leaves the reactor at about 510°C. Separation of the gas and liquid from the char and rapid quenching to about 80°C yields the liquid fuel. The remaining gas is passed through a series of cleanup steps for in-plant use. Part of the gas is used as an oxygen-free solids transport medium for pyrolysis and part of it as fuel. The raw product yields are about 10 wt % water, 20 wt % char, 30 wt % gas, and 40 wt % liquid fuel. The product char has a heating value of about 20.9 MJ/kg, contains about 30 wt % ash, and is produced at an overall yield

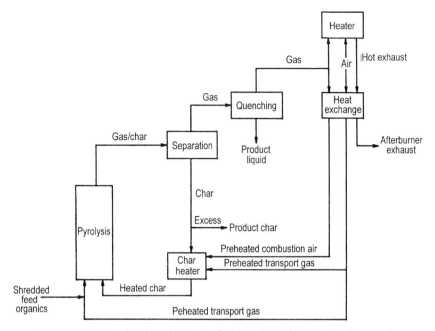

FIGURE 8.4   Liquid fuel production by flash pyrolysis of RDF using char recycle.

of about 7.5 wt % of the dry feed. The corresponding values for the liquid fuel are about 24.4 MJ/kg, 0.2 to 0.4% ash, and 22.5 wt % of dry feed as received (approximately 1 bbl/ton of raw refuse). This product was proposed for use as a heating oil; its properties are compared with those of a typical No. 6 fuel oil in Table 8.11. It is apparent that some major differences exist, but successful combustion trials in a utility boiler with the liquid fuel were performed. A plant designed to process 181 t/day of MSW was built in the United States in 1977, operated for one year, and then after producing several thousand liters of oil, was mothballed because of operating problems in the MSW separation and flash pyrolysis reactor.

Another flash pyrolysis process (GTEFP process) operating at atmospheric pressure was also developed that affords liquid yields in excess of 60% on a dry basis from hardwoods (O'Neil, Kovac, and Gorton, 1990). Yields as high as 70% were projected for commercial plants. The GTEFP process was developed in bench-scale studies and a large-scale PDU in which the particulate feed is entrained in hot combustion gases and pyrolyzed at millisecond residence times and temperatures in the range 500°C. Typical higher heating values of the product oils were 22 MJ/kg.

TABLE 8.11   Typical Properties of No. 6 Fuel Oil and Liquid Fuel from Flash Pyrolysis of Refuse-Derived Fuel[a]

Property	No. 6 fuel oil	Liquid fuel from RDF
Heating value, MJ/kg	42.3	24.6
Density, kg/L	0.98	1.3
Pour point, °C	15–30	32
Flash point, °C	65	56
Viscosity at 87.8°C, SUs	90–250	1150
Pumping temperature, °C	46	71
Atomization temperature, °C	104	116
Analysis, dry wt %		
C	85.7	57.5
H	10.5	7.6
O		33.4
S	0.5–3.5	0.1–0.3
Cl		0.3
N	2.0	0.9
Ash	0.5	0.2–0.4

[a]U.S. Environmental Protection Agency (1975). The pour point, flash point, and viscosity determinations of the liquid fuel from RDF were made on the product containing 14 wt % water as produced.

Short-residence-time pyrolysis of biomass at reduced pressure has been found to improve the yields of liquid products (Roy *et al.*, 1985, 1990). In this research, a large-scale, electrically heated, multiple-hearth PDU afforded pyrolytic oil yields in the range of about 50 wt % of the feedstock with a wide range of wood species. Pyrolysis temperatures in the last hearth were about 450°C. The optimum temperature range was found to be between 350 and 400°C at residence times of about 2 to 30 s (*cf.* Bridgwater and Bridge, 1991). A yield of 60 wt % on a dry, ash-free basis of pyrolytic oil was obtained at an average heating rate of 10°C/min at a total system pressure of 5 to 41 kPa. The process was operated at feed rates of 30 kg/h in a multiple-hearth pilot plant, which was shown to offer advantages in product separation and fractionation because of the primary condensing units attached to each hearth. However, the char and gas yields still comprised about one-third of the products, so it is probable that the liquid yields could be improved further at shorter residence times. A multiple-hearth reactor may not be suitable for biomass pyrolysis at millisecond reaction times.

Still another fast biomass pyrolysis configuration for increasing liquid yields involves what is termed ablative pyrolysis (Diebold *et al.*, 1987, 1990). In this system, biomass particles are entrained tangentially into a vortex tube with a jet of carrier gas at velocities over 100 m/s. This causes the solid particles to be centrifuged to the hot wall of the vortex reactor, where very rapid heat transfer occurs to the surface of the particles. Ablative or surface pyrolysis takes place at high rates essentially independent of the feedstock particle size. This type of conversion process favors chain-cleavage reactions to form oxygenated, organic vapors rather than chars and gases, and is expected to make it possible to design small reactors having high throughput rates. With temperatures of 625°C on the reactor walls, 60 to 70 wt % of the primary vapor products are composed of oxygenated organic compounds and polymer fragments. They condense to form acidic, water-soluble liquids that have nearly the same elemental composition as the feedstock. For softwood feeds, the vortex reactor produces about 58 to 67 wt % of the dry feedstock as primary pyrolysis oils, 10 to 12 wt % as char, 10 to 14 wt % as gases, and 13 to 16 wt % as water. A plant for conversion of 33 t/day of wood wastes has been designed for construction in the Midwest to evaluate this process (Johnson, Tomberlin, and Ayres, 1991; Johnson *et al.,* 1993). Dry wood feedstock (13 mm) is discharged from a hopper and pneumatically transported by recycled pyrolysis gases to a vortex reactor enclosed in a furnace designed to burn pyrolysis gas and natural gas. Natural gas is used for startup. The vapor-rich gases leave the vortex reactor along with the char particles entrained in the gas stream and pass through a cyclone, which is designed to remove 99.5% of the solids. The vapors are condensed at the rate of 11 L/min (dry). This plant is projected to cost $1.5 million, to produce 870 L/h (wet) of oil and 0.21 t/h of char, and to have a net annual revenue before taxes of $194,000, presuming the tipping fee for accepting the waste wood is $11/t, and the oil and char can be sold for $0.11/L and $88/t on the open market. Larger plants having capacities of 90 and 227 t/day are being designed to gain economies of scale.

Other fast biomass pyrolysis techniques for liquid products have been examined wherein a reactive atmosphere is present during conversion to attempt to affect yields and product compositions. Examples are flash pyrolysis in an atmosphere of hydrogen (hydropyrolysis) (Bodle and Wright, 1982; Sundaram, Steinberg, and Fallon, 1984) and methane (methanolysis) (Steinberg, Fallon, and Sundaram, 1983). In hydrogen atmospheres, hydrogenation might be expected to occur, at least to some extent, to yield liquids with lower oxygen and higher hydrogen contents. The experimental data indicated only moderate improvements in product yields and compositions under the test conditions used in an entrained flow pyrolysis reactor. However, substantial changes were observed in methane atmospheres. Flash methanolysis of dried wood particles at residence times of 1 to 2.8 s in an entrained flow pyrolysis reactor at pressures of 138 to 1379 kPa and temperatures between 800 and 1050°C afforded benzene, toluene, and the xylenes (BTX), a heavy oil, ethylene, and carbon monoxide. As much as 12% of the available carbon in the wood feedstock was converted to BTX, 21% to ethylene, and 48% to carbon monoxide at 345 kPa and 1000°C. The maximum heavy oil yield was observed at 345 kPa and 800°C. Data obtained with a methane blanket alone under the same pyrolysis conditions showed that no products were formed. It was concluded from this research that the optimum conditions for maximum production of both BTX and ethylene are a reactor pressure of 345 kPa, a temperature of 1000°C, a residence time of less than 1s, and a methane-to-wood feed ratio of about 5.

E. Summary of Basic Biomass Pyrolysis Methods and Operating Problems

Because of the broad scope of direct biomass pyrolysis, the basic technologies and principal products are tabulated in Table 8.12 to facilitate easy comparison. The conversion conditions and major products shown in this table are typical, but subject to considerable variation. There are several commonalities among the different pyrolysis methods. Pyrolysis time and temperature are clearly the key operating parameters that have the most influence on product yields and distributions. Moderate but optimized temperatures are needed at short residence times to maximize liquid yields, whereas long residence times and

TABLE 8.12   Typical Biomass Pyrolysis Technologies, Conditions, and Major Products[a]

Technology	Residence time	Heating rate	Temperature (°C)	Major Products
Conventional carbonization	Hours–days	Very low	300–500	Charcoal
Pressurized carbonization	15 min–2 h	Medium	450	Charcoal
Conventional pyrolysis	Hours	Low	400–600	Charcoal, liquids, gases
Conventional pyrolysis	5–30 min	Medium	700–900	Charcoal, gases
Flash pyrolysis	0.1–2 s	High	400–650	Liquids
Flash pyrolysis	<1 s	High	650–900	Liquids, gases
Flash pyrolysis	<1 s	Very high	1000–3000	Gases
Vacuum pyrolysis	2–30 s	Medium	350–450	Liquids
Pressurized hydropyrolysis	<10 s	High	<500	Liquids

[a]Adapted from references cited in this chapter and Bridgwater and Bridge (1991).

low temperatures are needed to maximize char yields. Biomass gasification, which can involve pyrolysis at the higher temperatures, is treated in Chapter 9. With few exceptions, the selectivities of specific pyrolysis products are poor. Essentially all of the product liquids produced by direct pyrolysis are highly oxygenated, acidic, generally unstable over time, and contain many compounds, and the product gases are low-energy gaseous mixtures. Some specialty and commodity chemicals can be extracted from the product liquids for market. The oils can often be used directly as fuel for power and steam production. The charcoals can be readily separated from the product mix for sale or captive use, and the product gases can be used as fuel and in some cases as sources of chemicals. Overall, however, most biomass pyrolysis plants will have to deal with complex product separations, waste disposal, and further product refining if pure chemicals and other products are required. The costs of the additional operations will have to be justified based on the markets for the various products that can be manufactured by biomass pyrolysis. Multiproduct slates will dominate most of these plants. This can be quite beneficial from a revenue standpoint when markets fluctuate provided the plant operating conditions can be readily changed to take advantage of the markets for specific products, much like a petroleum refinery.

Up to the 1930s, biomass pyrolysis processes were utilized on a large scale to produce several commodity chemicals from pyrolytic oils. These processes have since been largely replaced by nonpyrolytic processes based on petroleum and natural gas feedstocks. Note that only a few building blocks from fossil feedstocks—ethylene, propylene, butadiene, benzene, toluene, xylene, and synthesis gas—are used to manufacture the vast majority of the organic chemicals sold today (Chapter 13). Since the First Oil Shock in 1973, hundreds and perhaps thousands of research projects have been carried out to perfect and develop biomass pyrolysis technologies for the production of petroleum substitutes. The history of commercializing modern biomass pyrolysis systems in industrialized countries, however, since the First Oil Shock is not very encouraging. In the United States, for example, a few advanced design, biomass pyrolysis plants were built in the 1980s and then closed down, generally because of operating problems and poor economics.

The properties of pyrolytic liquids from biomass such as their high oxygen contents and acidity generally limit their fuel uses to heating oils. The emphasis over the past several years has therefore been to develop methods for upgrading them to more hydrocarbon-like liquids for use as motor fuels and motor fuel components. One approach has focused on catalytic hydrogenation. Hydrogen, which can either be generated from the biomass feed or the conversion products, or be obtained from an independent source, is reacted directly with the pyrolytic liquids or intermediate process streams at elevated pressures and temperatures to yield substitute fuels with higher hydrogen-to-carbon ratios.

In theory, highly oxygenated feedstocks should be capable of reduction to liquid and gaseous fuels at any level between the initial oxidation state of the feed and methane:

$$R(OH)_x + y\ H_2 \rightarrow RH_y(OH)_{x\text{-}y} + y\ H_2O$$
$$R - R' + H_2 \rightarrow RH + R'H.$$

For a cellulosic material containing hydroxyl groups, the reactions might consist of dehydroxylation and depolymerization by hydrogenolysis, during which there is a transition from solid to liquid to gas. In early work to produce hydrocarbon fuels, hydroliquefaction of biomass or wastes was achieved by direct hydrogenation of wood chips on treatment at 10,132 kPa and 340 to 350°C with water and Raney nickel catalyst (Boocock and Mackay, 1980). The wood is completely converted to an oily liquid, methane, and other hydrocarbon gases. Batch reaction times of 4 h give oil yields of about 35 wt % of the feedstock. The oil still contains substantial oxygen, about 12 wt %, but has a heating value of about 37.2 MJ/kg. Distillation yields a major fraction that boils in the same range as diesel fuel and is completely miscible with it.

Catalytic hydrogenation of the pyrolytic liquids at elevated pressures and temperatures and deoxygenation with molecular sieve catalysts that yield hydrocarbon liquids with higher hydrogen-to-carbon ratios than the liquid feedstocks have been studied (cf. Elliott and Baker, 1987; Rajai et al., 1991; Bakhshi, Kaitikaneni, and Adjaye, 1995; Laurent, Maggi, and Delmon, 1995; Horne, Nugranad, and Williams, 1995). Several systems are quite effective for converting pyrolytic liquids to hydrocarbons suitable for use as motor fuels or motor fuel additives. Catalytic hydrogenation is well-developed commercial technology in the petroleum industry and can be applied to pyrolytic oils to obtain partial or complete reduction to hydrocarbons. Deoxygenation with molecular sieve catalysts has the advantage that a source of hydrogen is not needed and BTX and light olefins can be produced by direct passage of the pyrolytic vapors over zeolite catalysts, although conversion conditions must be carefully controlled to avoid formation of undesirable by-products such as coke. High yields of BTX and olefins approaching the theoretical limits can be obtained at high weight hourly space velocities and low steam-to-biomass ratios. This route to olefins appears to offer an economic route to methyl-t-butyl ether (MTBE) and ethyl-t-butyl ether (ETBE) from the light olefins for use as oxygenates in gasolines (Bain et al., 1993).

It would seem that a more practical approach to the upgrading of pyrolytic liquids from biomass is to utilize what is already on hand, namely, the oxygenated product liquids. Instead of conversion to hydrocarbons, which usually requires severe reaction conditions, why not convert the liquids by simple chemistry to other liquids that are suitable for use as motor fuels or additives? Although not directly related to pyrolysis, this approach has been pursued in the development of oxygenated diesel fuels (biodiesel) with natural biomass-derived oils, as will be discussed in Chapter 10. With the possible exception of the lignin components in biomass, the overall thermal efficiencies of converting a highly oxygenated organic feedstock in which the majority of the carbon atoms are bonded to oxygen atoms to other oxygenates should be much more favorable, and lower in cost, than conversion to hydrocarbons. However, there may even be exceptions to this rationale, as will be shown in the following section. Also, there are several opportunities to produce chemical additives suitable for use in modern gasolines as oxygenates by thermochemical processing of biomass (Chapter 13).

IV. MISCELLANEOUS
LIQUEFACTION METHODS

A. Aqueous and Non-aqueous
Non-pyrolytic Conversion

Note that the transformations described here, which take place in an aqueous or non-aqueous liquid medium that may or may not react with the biomass feedstock, are termed nonpyrolytic processes, as compared to pyrolysis processes in which the biomass feedstocks are directly heated.

The direct conversion of cellulosic materials to liquids by heating in aqueous systems has been known for more than 100 years (cf. Ostermann, Bishop, and Rosson, 1980). Pure cellulose is liquefied at 300°C and below at a pressure of 19.3 MPa in less than 1 h with or without added sodium carbonate catalyst up to a concentration of 0.8% in the medium (Molton et al., 1981). A wide range of aliphatic and aromatic alcohols, phenols, hydrocarbons, substituted furans, and alicyclic compounds are formed. The presence or absence of a gaseous atmosphere of carbon monoxide had no effect on the results, which contrasts somewhat to the results obtained from demonstration of a similar conversion system (PERC process) described later. The experimental results support a degradation mechanism that forms acetone, acrolein, and acetoin intermediates, which recondense under alkaline conditions to yield the observed products. Model compound experiments indicated that the ketones and furans are formed by aldol condensations and Michael reactions of carbonyl intermediates. The formation of aromatic hydrocarbons and phenol from these molecules under these conditions appears to involve a variant of the aldol condensation. In other experiments at temperatures of 268 to 407°C, pressures of 21 to 35 MPa, and reaction times of 20 or 60 min, the average liquid yield was 25 wt % of the cellulose converted to acetone-soluble oil (Miller, Molten, and Russell, 1981). About 10 wt % charcoal was produced along with gaseous by-products. The heating of poplar wood chips in water alone at about 330°C in autoclaves affords acetone-soluble liquid oils containing about 20 to 35% oxygen at yields up to about 50 wt % of the feedstock (Boocock et al., 1985, 1987). Little or no char is produced, and the physical breakdown of the chips is believed to occur by water absorption, swelling, and disruption and liquefaction of the matrix, whereupon the absorbed water is regenerated. When poplar chips are used, about one-half the oil is phenolic and one-fourth is phenol itself. The phenol yield is 6.5 wt % on a dry wood basis or 25 wt % based on the lignin content. The relatively high yield of phenol was suppressed under alkaline conditions. It was concluded that the steam formed in the process at the self-generated pressure of about 15.9 MPa was responsible for disruption of the chips. In subsequent experiments with steam injection at 350°C into a downflow reactor, the oil yields from poplar wood chips were in excess of 40%. The oils softened and flowed just above 100°C and their oxygen content was in the low 20% range.

Another interesting catalytic liquefaction method involves the reaction of biomass-water slurries (LBL process) or biomass-recycle oil slurries (PERC process) with sodium carbonate and carbon monoxide gas at elevated temperature and pressure to form heavy liquid fuels. Biomass and the combustible fraction of wastes have been converted at weight yields of 40 to 60% at temperatures of 250 to 425°C and pressures of 10 to 28 MPa. Lower viscosity products are generally obtained at higher reaction temperatures and solid or semisolid products are obtained when the reaction temperature is below 300°C. However, the high nitrogen and oxygen contents and the boiling characteristics and high viscosity range of the liquid products make it difficult to classify them as petroleum substitutes. They would have to be upgraded by other processes. The original PERC process consisted of a sequence of steps: drying and grinding wood chips to a fine powder, mixing the powder with recycled product oil (10% wood powder to 90% recycle oil), blending the mixture with water containing sodium carbonate, and treating the slurry with synthesis gas at about 27,579 kPa and 370°C. The modified LBL process consists of partially hydrolyzing the wood in dilute sulfuric acid and treating the water slurry containing dissolved sugars and about 20% solids with synthesis gas and sodium carbonate at 27,579 kPa and 370°C on a once-through basis. The resulting oil product yield is about 1 bbl/400 kg of chips and is approximately equivalent to No. 6 grade boiler fuel. It contains about 50% phenolics, 18% high-boiling alcohols, 18% hydrocarbons, and 10% water.

The evaluation of pressurized wood-slurry liquefaction by the LBL process of wood-water slurries was performed mainly in small-scale equipment (Ergun, 1981; Davis, 1983). Oils similar to those produced by the PERC process were obtained, but at lower yields. In the late 1970s and early 1980s, the PERC process was evaluated in a PDU (Thigpen and Berry, 1982). The purpose of this work was to demonstrate the direct, continuous, thermochemical liquefaction of biomass. In this process, a synthesis gas mixture was reacted with wood slurried in recycled oil in the presence of 5% sodium carbonate at pressures of 20.7 MPa and temperatures of about 270°C for 1 to 1.5 h. It was suggested by the researchers who developed this process in the laboratory that carbon monoxide reacts with sodium carbonate in the presence of water to form sodium formate, which in turn deoxygenates the biomass feedstock to yield oil.

Study of the mechanism of this complex reduction-liquefaction process led to the suggestion that part of the mechanism involves formate production from carbonate, dehydration of the vicinal hydroxyl groups in the cellulosic feed to carbonyl compounds via enols, reduction of the carbonyl group to an alcohol by formate and water, and regeneration of formate (Appell *et al.*, 1975). The following reactions were suggested:

$$Na_2CO_3 + H_2O + CO \rightarrow 2HCO_2Na + CO_2$$
$$C_6H_{10}O_5 + HCO_2Na \rightarrow C_6H_{10}O_4 + NaHCO_3$$
$$\text{(Wood)} \qquad\qquad\qquad \text{(Oil)}$$
$$NaHCO_3 + CO \rightarrow HCO_2Na + CO_2$$
$$HCO_2Na + H_2O \rightarrow NaHCO_3 + H_2$$
$$H_2 + C_6H_{10}O_5 \rightarrow C_6H_{10}O_4 + H_2O.$$

The approximate stoichiometry of the process developed from the data (Thigpen and Berry, 1982) corresponded to

$$C_6H_{9.84}O_{4.14} + 3.55CO + 2.14H_2 \rightarrow 0.877C_6H_{7.62}O_{0.66} + 3.99CO_2 + 0.22\ CO$$
$$+ 4.36H_2.$$
$$\text{(Wood)} \qquad\qquad\qquad\qquad\qquad \text{(Oil)}$$

In view of the complex nature of the reactants and products, it is likely that a complete understanding of all of the chemical reactions that occur in the PERC process will not be developed unless detailed mechanistic studies are carried out.

In the PDU tests of the PERC process, a crude product oil comparable to No. 6 fuel oil was produced in barrel quantities at yields of about 53 wt % of the feedstock. It had a heating value of up to 34.5 MJ/kg, a specific gravity of 1.1, a viscosity of 0.20 Pa·s at 99°C, and an oxygen content of 12.3%. The distillate from the crude oil had a heating value of 40.4 MJ/kg, a viscosity of 0.01 Pa·s, and an oxygen content of 6.2%. Its characteristics were similar to those of No. 2 fuel oil. These results were obtained from the longest sustained run with Douglas fir; 4988 kg of crude oil were obtained from 9953 kg of feedstock. These data suggest that the PERC process yields what much of the research on the direct liquefaction of biomass has been unable to achieve—a one-step, direct liquefaction process using woody feedstocks that yields a crude product oil similar to a petroleum fuel oil. Unfortunately, operation of the PDU was terminated before several key questions could be answered. Is sodium carbonate necessary? Is synthesis gas necessary, and if so, how much? If one or both of these reactants is eliminated, what is the effect on the crude product oil's composition and yield? Eventually, these questions will be resolved, if and when the process is developed further. But it seems evident from the PDU data that the PERC process is capable of overcoming some of the problems encountered in other direct biomass liquefaction processes.

B. Direct Chemical Liquefaction

One of the more innovative low-temperature, low-pressure, thermochemical techniques of directly liquefying biomass in water involves the use of 57 wt % aqueous hydriodic acid (HI), the azeotrope boiling at 127°C (Douglas and Sabade, 1985). When treated at 127°C with the azeotrope in a stoichiometric excess of 1.6 to 3.8 of the amount required for complete reduction, cellulose is rapidly hydrolyzed and converted to hydrocarbon-like molecules. The yields reach 60 to 70% at reaction times as short as 0.5 min. The laboratory data are consistent with chemistry in which HI acts to form alkyl iodide intermediates that are then converted to hydrocarbons and molecular iodine by further reaction with HI. The stoichiometry developed from the experimental data with cellulose is

$$C_6H_{10}O_5 + 8.28HI \rightarrow C_6H_{9.12}I_{0.18}O_{0.40} + 4.52H_2O + 4.08I_2.$$

Products corresponding to 50% deoxygenation in 1 min, 75% in 30 min, and 92% in 24 h are obtained; charcoal is not formed. Up to 98% of the HI reacted appears as molecular iodine. Ether extraction yields a material that has H:C and I:C ratios of 1.52 and 0.03, and yet there is a 90% reduction in the O:C ratio.

Hydriodic acid is a powerful reducing agent that can even be employed for conversion of benzene to cyclohexane. It is also well known that alcohols can be converted to alkyl iodides on reaction with HI and that alkyl iodides react with HI to form hydrocarbons. Since the experimental data obtained with cellulose shows that most of the HI is ultimately converted to molecular iodine, a cyclic process can be conceptualized in which HI is regenerated. One scheme might use hydrogen sulfide to regenerate HI. Another might use hydrogen. Thus,

$$ROH + HI \rightarrow RI + H_2O$$
$$RI + HI \rightarrow RH + I_2$$
$$(I_2 + H_2S \rightarrow 2HI + S)$$
$$(I_2 + H_2 \rightarrow 2HI).$$

It appears that further research on HI chemistry could lead to processes for direct conversion of biomass to hydrocarbons without the economic penalty associated with operation at high pressure and temperature. The key to the value of such developments resides in the ability to recycle HI. Note that loss of only a small amount of the HI reacted can make the process quite uneconomical, so if it is developed to the point of commercial use, iodine recoveries would have to be substantially improved.

V. COMPARISON OF PYROLYTIC AND NON-PYROLYTIC LIQUIDS

A. PROPERTIES

A few properties of the liquid oils produced by selected flash pyrolysis and the PERC and LBL processes are listed in Table 8.13. It as apparent that there are some basic differences between the two classes of oils. The oils from the flash pyrolysis processes are quite similar, as are the oils from the PERC and LBL processes. But there are major differences in their elemental analyses and

TABLE 8.13   Comparison of Liquids Produced by Pyrolysis and Non-Pyrolytic Thermolysis[a]

Parameter	Atmospheric flash pyrolysis[b]	Ablative flash pyrolysis[c]	PERC Process[d]	LBL Process[e]
Elements, % dry basis				
C	52.1	53.5	78.9	79.2
H	6.2	6.2	8.5	7.8
O	41.4	39.6	12.3	14.4
S			0.06	
Atomic ratio H : C	1.42	1.38	1.28	1.17
Raw product viscosity, Pa·s (at °C)	0.22 (40)	1.3 (30)	0.14 (99)	0.046 (99)
Specific gravity	1.27	1.28	1.11	1.09
Moisture content raw product, wt %	16.6	16.1		
Heating value, MJ/dry kg	22.1	22.3	34.5	33.6
Yield, wt/wt MAF wood	0.53	0.55	0.53	0.25

[a]The data are from Stevens (1994). Some data were not available.
[b]cf. O'Neil, Kovac, and Gorton (1990).
[c]cf. Diebold et al. (1987, 1990).
[d]cf. Thigpen and Berry (1982).
[e]cf. Ergun (1981); Davis (1983).

energy values. The carbon analyses and energy values are much lower and the oxygen analyses are much higher for the pyrolytic oils than for the PERC and LBL oils. This might be expected, since a synthesis gas atmosphere is used to carry out the PERC and LBL processes. The lower yield of oil from the LBL process is apparent when compared to the oil yield from the PERC process.

B. ECONOMIC ANALYSIS OF ATMOSPHERIC FLASH PYROLYSIS AND PRESSURIZED SOLVENT LIQUEFACTION

A detailed comparative economic analysis of the production costs of fuel oil and synthetic gasoline and diesel fuel has been performed for direct liquefaction of biomass (Elliott et al., 1990). To bring this analysis to today's dollars, adjustments should be made to account for inflation and other factors. But since the treatment is a comparative analysis, the differences should remain the same. The basic parameters used for this analysis are listed in Table 8.14.

TABLE 8.14   Parameters Used for Comparative Cost Analysis of Biomass Liquefaction Processes[a]

Parameter	Assumption
Plant capacity	1000 dry t/day of biomass feedstock
Time	September 1987
Place	U.S. Gulf Coast
Currency	U.S. dollars
Annual operating time	8000 h
Labor rate	$20/h including payroll burden
Maintenance labor	1% of fixed capital investment (FCI)
Maintenance materials	3% of FCI
Overhead	2% of FCI
Insurance	2% of FCI
Other fixed operating costs	1% of FCI
Catalyst price	Specified for each process
Feedstock price	$30/t of green wood chips, 50% moisture
Electricity price	$0.065/kWh
Interest rate	10%
Plant life	20 years

[a]Elliott et al. (1990).

Three processing steps were involved in the assessment: liquefaction of wood-chip feedstock to a primary crude oil, catalytic hydrotreatment of the crude oil to a deoxygenated product oil, and refining of the product oil to gasoline or diesel fuel. Atmospheric flash pyrolysis (cf. Scott and Piskorz, 1983; O'Neil, Kovac, and Gorton, 1990) and pressurized solvent liquefaction (cf. Appell et al., 1975; Thigpen and Berry, 1982) were analyzed. Two versions of each process were used—one based on the technology as developed (present technology) and one based on anticipated future improvements (potential technology).

The atmospheric flash pyrolysis process for the present technology case is illustrated in the accompanying flowsheet (Fig. 8.5). One-millimeter particle size wood fibers are rapidly pyrolyzed in a fluidized-bed reactor at 500°C to vapors and char. The condensed vapors form the primary oil product, which contains approximately 39% oxygen on a dry basis. The second and third steps are not shown in Fig. 8.5. In the second step, the primary oil is upgraded in a two-stage catalytic hydrotreatment process using a conventional sulfided cobalt/molybdenum-on-alumina petroleum hydrotreatment catalyst. In the third step, the upgraded product is subjected to distillation, hydrodeoxygenation of the light fraction, hydrocracking of the heavy fraction, catalytic reforming of the gasoline fraction, and steam reforming of the hydrocarbon gas product to produce hydrogen for the process. A gasoline and diesel product slate is produced. For the potential technology case, pyrolysis takes place in a circulating fluidized-bed reactor, which has the advantage of much greater throughput than the present technology case. An advanced three-stage catalytic hydrotreatment is used in the second step and is assumed to yield a high-octane gasoline requiring no further processing other than fractionation.

The present technology case for pressurized solvent liquefaction is illustrated in Fig. 8.6. Wood chips are ground to less than 0.5 mm and mixed with recycled wood-derived oil. A sodium carbonate solution and synthesis gas are added to the slurry prior to preheating. Liquefaction takes place in a tubular, upflow reactor at 350°C, 20.5 MPa, and a 20-min residence time. Gas is flashed from the reactor effluent. A portion of the liquid is recycled. Water is separated from the primary oil and is treated before discharge. The synthesis gas is obtained from a portion of the feedstock, which is gasified in an oxygen-blown gasifier. The product oil is upgraded and refined in a manner similar to the flash pyrolysis oil. The potential case for this process uses an extruder feeder to feed high concentration wood slurries. The oil phase of the slurry consists of recycled vacuum distillate bottoms. Superheated steam is added to the reactor to provide the reactor heat requirement. Sodium carbonate is assumed to be recycled entirely in the distillate bottoms, and no reducing gas is added to the reactor. The liquid product stream is separated into a distillate product and recycled bottoms in a vacuum distillation tower. Catalytic hydrotreatment

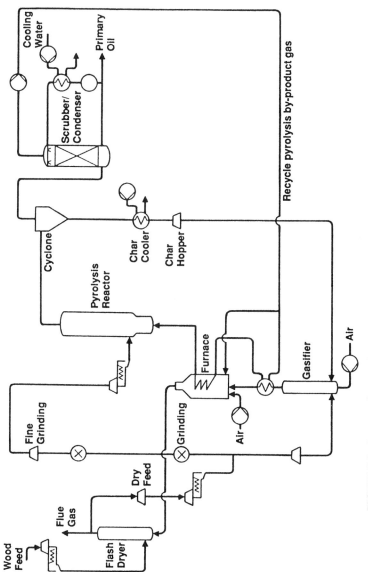

FIGURE 8.5 Process schematic for atmospheric flash pyrolysis of wood chips—present technology.

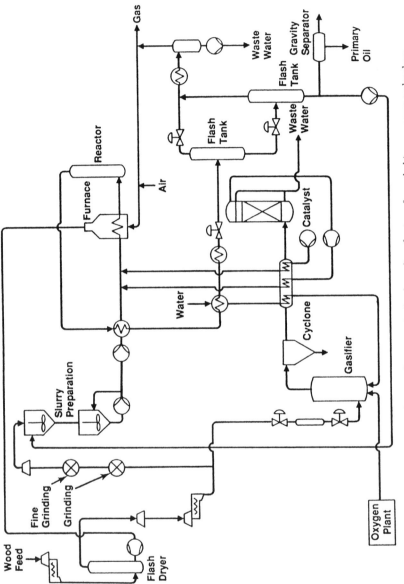

FIGURE 8.6 Process schematic for pressurized solvent liquefaction of wood chips—present technology.

TABLE 8.15 Comparative Economic Analysis of Flash Pyrolysis and Pressurized Solvent Liquefaction[a]

Parameter	Atmospheric flash pyrolysis		Pressurized solvent liquefaction	
	Present (10^6)	Potential (10^6)	Present (10^6)	Potential (10^6)
Fixed capital investment				
Primary liquefaction	38.9	20.6	65.8	37.8
Crude upgrading	36.4	26.8	20.9	20.3
Product finishing	11.4	0.5	11.9	0.6
Total:	86.7	47.9	98.6	58.7
Total capital requirement				
Primary liquefaction	49.8	26.4	84.2	48.4
Crude upgrading	46.6	34.3	26.8	26.0
Product finishing	14.5	0.7	15.3	0.7
Total:	111.0	61.3	126.2	75.1
Fixed operating cost				
Operating labor	6.68	6.43	5.58	4.73
Maintenance labor	0.86	0.49	0.99	0.59
Overhead	1.74	0.96	1.98	1.18
Maintenance materials	2.60	1.44	2.96	1.76
Taxes and insurance	1.74	0.96	1.98	1.18
Others	0.86	0.49	0.99	0.59
Total:	14.48	10.77	14.48	10.03
Variable operating cost				
Feedstock	20.00	20.00	20.00	23.30
Electricity	4.89	3.57	9.72	10.14
Catalyst	0.30	0.02	2.36	0.01
Sludge disposal	0.16	0.02	0.04	0.09
Fuel cost	0.39	0.06	1.32	0.06
Total:	25.74	23.67	33.44	33.60
Capital charges	12.96	7.17	14.75	8.78
Total cost finished product	53.18	41.61	62.67	52.39
Product cost, $/t				
Primary oil	150.00	131.90	423.30	298.10
Finished product	673.80	514.50	815.00	584.70
Energy cost, $/GJ				
Primary oil	9.32	6.91	13.44	12.27
Finished product	16.24	12.99	19.54	14.77
Product energy value, $/GJ				
Primary oil	6.70	6.70	5.20	5.20
Finished product	8.30	8.45	8.15	8.45
Product cost/value ratio				
Primary oil	1.39	1.03	2.58	2.36
Finished product	1.96	1.54	2.40	1.75

[a]Adapted from Elliott et al. (1990). The product energy value is the average U.S. spot market price from 1977 to 1987 for comparable petroleum-based liquid fuel. The average price of U.S. crude oil at the wellhead in 1987 was $2.61/GJ ($15.40/bbl).

is used to upgrade the primary oil as in the present technology case. The upgraded product is assumed to be a high-octane gasoline requiring no further processing other than fractionation.

Detailed estimates were prepared of the capital and operating costs of each of the four processes, which were designed around the results obtained from laboratory, PDU, and pilot-scale tests. A summary of these costs is presented in Table 8.15. The atmospheric flash pyrolysis process is clearly more economical than the pressurized solvent liquefaction process for production of similar products. Using the average U.S. spot market price for fuel oil in the period 1977 to 1987, the ratio of primary oil cost to energy value is about 1.4 at a green wood chip price of $30/t for the present technology case for atmospheric flash pyrolysis. This means that the product is 40% more costly than comparable petroleum fuel. The potential technology case for atmospheric flash pyrolysis is the only process design of the four analyzed that appears capable of producing primary oil product competitive with comparable petroleum fuel. On the basis of sensitivity studies, each process appears to be more sensitive to feedstock cost than to capital cost. Although thermochemical conversion processes are generally capital intensive, the range of capital costs examined had little effect on the final product cost. However, sensitivity analysis of finished product cost and feedstock price showed that when the green wood chip price is $10/t, the ratio is less than 0.9.

It was concluded from this assessment that the most promising process for gasoline production by direct liquefaction of biomass is atmospheric flash pyrolysis. The high-pressure process may have the same future potential, but the uncertainties are much greater.

REFERENCES

Antal, M. J., Jr., and Varhegyi, G. (1995). *Ind. Eng. Chem. Res.* **34** (3), 703.

Antal, M. J., Jr., Croiset, E., Dai, X., DeAlmeida, C., Mok, W. S., Norberg, N., Richard, J., and Majthoub, M. (1996). *Energy & Fuels* **10** (3), 652.

Appell, H. R., Fu, Y. C., Illig, E. G., Steffgen, F. W., and Miller, R. D. (1975). "Conversion of Cellulosic Wastes to Oil," U.S. Bureau of Mines Report Invest. *8013.* Washington, D.C.

Bagrova, R. K., and Kozlov, V. N. (1958). *Trudy Ural'skogo Lesotekhnicheskogo Instituta 12* 97.

Bain, R., Diebold, J., Overend, R., Rejai, B., and Power, A. J. (1993). *In* "Energy from Biomass and Wastes XVI," (D. L. Klass, ed.), p. 753. Institute of Gas Technology, Chicago.

Bakhshi, N. N., Kaitikaneni, S. P. R., and Adjaye, J. D. (1995). *In* "Second Biomass Conference of the Americas: Energy, Environment, Agriculture, and Industry," NREL/CP-200-8098, DE95009230, p. 1089. National Renewable Energy Laboratory, Golden, CO.

Bituminous Coal Research, Inc. (1974). "Gas Generator Research and Development, Phase II. Process and Equipment Development," Report OCR-20-F, PB-235530/3GI. Monroeville, PA.

Bodle, W. W., and Wright, K. A. (1982). *In* "Energy from Biomass and Wastes VI," (D. L. Klass, ed.), p. 1031. Institute of Gas Technology, Chicago.

Boocock, D. G. B., and Mackay, D. (1980). *In* "Energy from Biomass and Wastes IV," (D. L. Klass and J. W. Weatherly III, eds.), p. 765. Institute of Gas Technology, Chicago.

Boocock, D. G. B, Agblevor, F., Holysh, M., Porretta, F., Sherman, K., Kallury, R. K. M. R., and Tidwell, T. T. (1985). *In* "Energy from Biomass and Wastes IX," (D. L. Klass, ed.), p. 1107. Institute of Gas Technology, Chicago.

Boocock, D. G. B, Agblevor, F., Chowdhury, A., Kosiak, L., Porretta, F., and Vasquez, E. (1987). *In* "Energy from Biomass and Wastes X," (D. L. Klass, ed.), p. 749. Institute of Gas Technology, Chicago.

Bradbury, A. G. W., Sakai, Y., and Shafizadeh, F. A. (1979). *J. Appl. Polym. Sci.* **23**, 3271.

Bridgwater, A. V., and Bridge, S. A. (1991). *In* "Biomass Pyrolysis Liquids Upgrading and Utilization," (A. V. Bridgwater and G. Grassi, eds.), p. 11. Elsevier Applied Science, New York.

Bridgwater, A. V., and Peacocke, C. (1995). *In* "Second Biomass Conference of the Americas: Energy, Environment, Agriculture, and Industry," NREL/CP-200-8098, DE95009230, p.1037. National Renewable Energy Laboratory, Golden, CO.

Broido, A. (1976). *In* "Thermal Uses and Properties of Carbohydrates and Lignins," (F. Shafizadeh, K. V. Sarkanen, and D. A. Tillman, eds.), p. 19. Academic Press, New York.

Davis, H. G. (1983). "Direct Liquefaction of Biomass Final Report and Summary of Effort 1977–1983," LBL-16243. Lawrence Berkeley Laboratory, University of California, Berkeley.

Diebold, J. P., and Scahill, J. W. (1988). *In* "Pyrolysis Oils from Biomass, Producing, Analyzing, and Upgrading," (E. Soltes and T. Milne, eds.), ACS Symposium Series 376, p. 31. American Chemical Society, Washington, D.C.

Diebold, J. P., Chum, H. L., Evans, R. J., Milne, T. A., Reed, T. B., and Scahill, J. W. (1987). *In* "Energy from Biomass and Wastes X," (D. L. Klass, ed.), p. 801. Institute of Gas Technology, Chicago.

Diebold, J. P., Scahill, J., Bain, R., Chum, H., Black, S., Milne, T., Evans, R., and Rajai, B. (1990). *In* "Biomass Thermal Processing" (E. Hogan, J. Robert, G. Grassi, and A. V. Bridgwater, eds.), p. 101. CPL Press, Berkshire, UK, October.

Douglas, W. J. M., and Sabade, S. B. (1985). *In* "Bioenergy 84" (H. Egneus and A. Ellegand, eds.), Vol. III, p. 38. Elsevier Applied Science Publishers, London.

Elliott, D. C., and Baker, E. G. (1987). *In* "Energy from Biomass and Wastes X," (D. L. Klass, ed.), p. 765. Institute of Gas Technology, Chicago.

Elliott, D. C., Östman, A., Börje Gevert, S., Beckman, D., Solanantausta, Y., Hörnell, C., and Kjellström, B. (1990). *In* "Energy from Biomass and Wastes XIII," (D. L. Klass, ed.), p. 743. Institute of Gas Technology, Chicago.

Epstein, E., Kostrin, H., and Alpert, J. (1978). *In* "Energy from Biomass and Wastes," (D. L. Klass and W. W. Waterman, eds.), p. 769. Institute of Gas Technology, Chicago.

Ergun, S. (1981). "Review of Biomass Liquefaction Efforts," LBL-13957. Lawrence Berkeley Laboratories, University of California, Berkeley.

Evans, R. J., and Milne, T. A. (1987a). *Energy & Fuels* **1**, 123.

Evans, R. J., and Milne, T. A. (1987b). *Energy & Fuels* **2**, 311.

Evans, R. J., and Milne, T. A. (1988). *In* "Energy from Biomass and Wastes XI," (D. L. Klass, ed.), p. 807. Institute of Gas Technology, Chicago.

Fieser, L. F., and Fieser, M. (1950). *In* "Organic Chemistry," 2nd Ed., p. 567. D.C. Heath and Company, Boston.

Graboski, M., and Bain, R. (1979). *In* "A Survey of Biomass Gasification," Vol II, p. 11-20, Report SERI/TR-33-239. Solar Energy Research Institute, Golden CO, July.

Graham, R. G., and Huffman, D. R. (1995). "Commercial Aspects of Rapid Thermal Processing (RTP™)." Paper presented at Power Production from Biomass II, March 27-28, 1995, Espoo, Finland.

Graham, R. G., Freel, B. A., and Bergougnou, M. A. (1991). *In* "Energy from Biomass and Wastes XIV," (D. L. Klass, ed.), p. 1091. Institute of Gas Technology, Chicago.

Graham, R. G., Freel, B. A., and Kravetz, D. (1996). *In* "Bioenergy '96," Vol. II, p. 590. Southeastern Regional Biomass Energy Program, Tennessee Valley Authority, Muscle Shoals, AL.

Grover, P. D. (1989). "Thermochemical Characterization of Biomass Residues for Gasification," Vol. I. Indian Institute of Technology, Delhi.

Hoffman, D. A., and Fitz, R. A. (1968). *Environ. Sci. Technol.* 2 (11), 1023.

Horne, P. A., Nugranad, N., and Williams, P. T. (1995). *In* "Biomass for Energy, Environment, Agriculture, and Industry," (P. Chartier, A.A.C.M. Beenackers, and G. Grassi, eds.), Vol. 3, p. 1717. Elsevier Science Ltd., Oxford, UK.

Howlett, K., and Gamache, A. (1977). "Forest and Mill Residues as Potential Sources of Biomass," Vol. VI, Final Report, MTR 7347. The MITRE Corporation, McLean, VA.

Jenkins, B. M., and Ebeling, J. M. (1985). *In* "Energy from Biomass and Wastes IX," (D. L. Klass, ed.), p. 371. Institute of Gas Technology, Chicago.

Johnson, D. A., Tomberlin, G. R., and Ayres, W. A. (1991). *In* "Energy from Biomass and Wastes XV," (D.L. Klass, ed.), p. 915. Institute of Gas Technology, Chicago.

Johnson, D. A., Maclean, D., Chum, H. L., and Overend, R. P. (1993). *In* "First Biomass Conference of the Americas: Energy, Environment, Agriculture, and Industry," NREL/CP-200-5768, DE93010050, Vol II, p. 1367. National Renewable Energy Laboratory, Golden, CO.

Johnson, J. L. (1979). "Kinetics of Coal Gasification." John Wiley, New York.

Klass, D. L., and Ghosh, S. (1973). *Chemtech,* 689, November.

Kumar, J. V., and Pratt, B. C. (1996). *American Laboratory,* 15, May.

Lai, W.-C., Krieger-Brockett, B. (1993). *Ind. Eng. Chem. Res.* 32, 2915.

Laurent, E., Maggi, R., and Delmon, B. (1995). *In* "Biomass for Energy, Environment, Agriculture, and Industry," (P. Chartier, A. A. C. M. Beenackers, and G. Grassi, eds.), Vol. 2, p. 1485. Elsevier Science Ltd., Oxford.

Lowenheim, F. A., and Moran, M. K. (1975). "Faith, Keyes, and Clark's Industrial Chemicals," 4th Ed. John Wiley, New York.

McAdams, C. L., and Aquino, J. T. (1994). "Dioxin: Impact on Solid Waste Industry Uncertain," *Waste Age* 25, 11, 103 (November).

Molton, P. M., Miller, R. K., Russell, J. A., and Donovan, J. M. (1981). *In* "Biomass as a Nonfossil Fuel Source," ACS Symposium Series 144, (D. L. Klass, ed.), p. 137. American Chemical Society, Washington, D.C.

Miller, R. K., Molten, P. M., and Russell, J. A. (1981). *In* "Fuels from Biomass and Wastes," (D. L. Klass and G. H. Emert, eds.), p. 451. Ann Arbor Science Publishers, Inc., Ann Arbor, MI.

Nikitin, N. I. *et al.* (1962). "The Chemistry of Cellulose and Wood," (translated in 1966 from Russian by J. Schmorak, Israel Program for Scientific Translations, Jerusalem, Israel). Academy of Sciences of the USSR, Institute of High Molecular Compounds, Moscow-Leningrad.

O'Neil, D. J., Kovac, R. J., and Gorton, C. W. (1990). *In* "Energy from Biomass and Wastes XIII," (D. L. Klass, ed.), p. 829. Institute of Gas Technology, Chicago.

Ostermann, R. D., Bishop, K. A., and Rosson, R. F. (1980). *In* "Energy from Biomass and Wastes IV," (D. L. Klass and J. W. Weatherly III, eds.), p. 645. Institute of Gas Technology, Chicago.

Piskorz, J., Radlein, D., Majerski, P., and Scott, D. (1995). *In* "Second Biomass Conference of the Americas: Energy, Environment, Agriculture, and Industry," NREL/CP-200-8098, DE95009230, p. 1151. National Renewable Energy Laboratory, Golden, CO.

Preston, G. T. (1976). *In* "Clean Fuels from Biomass, Sewage, Urban Refuse, and Agricultural Wastes," (F. Ekman, ed.), p. 89. Institute of Gas Technology, Chicago.

Rajai, B., Evans, R. J., Milne, T. A., Diebold, J. P., and Scahill, J. W. (1991). *In* "Energy from Biomass and Wastes XV," (D. L. Klass, ed.), p. 855. Institute of Gas Technology, Chicago.

Riegel, E. R. (1933). *In* "Industrial Chemistry," Ed. 2, p. 253. The Chemical Catalog Company, Inc., New York.

Roy, C., de Caumia, B., Blanchette, D., Lemieux, R., and Kaliaguine, S. (1985). *In* "Energy from Biomass and Wastes IX," (D. L. Klass, ed.), p. 1085. Institute of Gas Technology, Chicago.

Roy, C., De Caumia, B., Pakdel, H., Plants, P., Blanchette, D., and Labrecque, B. (1990). *In* "Biomass Thermal Processing," (E. Hogan, J. Robert, G. Grassi, and A. V. Bridgwater, eds.), p. 109. CPL Press, Berkshire, U.K.

Scott, D. S., and Piskorz, J. (1983). *In* "Energy from Biomass and Wastes VII," (D. L. Klass and H. H. Elliott, eds.), p. 1123. Institute of Gas Technology, Chicago.

Scott, D. S., Piskorz, J., and Radlein, D. (1993). *In* "Energy from Biomass and Wastes XVI," (D. L. Klass, ed.), p. 797. Institute of Gas Technology, Chicago.

Shafizadeh, F. (1982). *J. Anal. Appl. Pyrol.* 3, 283.

Steinberg, M., Fallon, P. T., Sundaram, M. S. (1983). *In* "Energy from Biomass and Wastes VII," (D. L. Klass and H. H. Elliott, eds.), p. 1171. Institute of Gas Technology, Chicago.

Stevens, D. J. (1994). "Review and Analysis of the 1980-1989 Biomass Thermochemical Conversion Program," NREL/TP-421-7501. National Renewable Energy Laboratory, Golden, CO, September.

Sundaram, M. S., Steinberg, M., and Fallon, P. T. (1984). *In* "Energy from Biomass and Wastes VII," (D. L. Klass and H. H. Elliott, eds.), p. 1395. Institute of Gas Technology, Chicago.

Thigpen, P. L., and Berry, W. L. Jr. (1982). *In* "Energy from Biomass and Wastes VI," (D. L. Klass, ed.), p. 1057. Institute of Gas Technology, Chicago.

U.S. Environmental Protection Agency (1975). "San Diego County Demonstrates Pyrolysis of Solid Wastes," Report SW-80d.2. Washington, D.C.

Zaror, C. A., and Pyle, D. L. (1982). *Proc. Indian Acad. Sci. (Eng. Sci.)* 5 (4), 269.

CHAPTER 9

Thermal Conversion: Gasification

I. INTRODUCTION

In Chapters 7 and 8, the thermal conversion of biomass to energy by combustion and to liquid fuels by pyrolysis and a few nonpyrolytic liquefaction processes was examined. In this chapter, the subject of thermal conversion will be expanded further by addressing biomass gasification. Biomass gasification processes are generally designed to produce low- to medium-energy fuel gases, synthesis gases for the manufacture of chemicals, or hydrogen. More than one million small-scale, airblown gasifiers for wood and biomass-derived charcoal feedstocks were built during World War II to manufacture low-energy gas to power vehicles and to generate steam and electric power. Units were available in many designs. Thousands were mounted on vehicles and many were retrofitted to gas-fired furnaces. Sweden alone had over 70,000 "GENGAS" trucks, buses, and cars in operation in mid-1945 (Swedish Academy of Engineering, 1950). Research continues to develop innovative biomass gasification processes in North America, and considerable research has also been conducted in Europe and Asia. The Swedish automobile manufacturers Volvo and Saab have ongoing programs to develop a standard gasifier design suitable for mass production for vehicles. Much effort has been devoted to the commercialization of biomass gasification technologies in the United States since the early 1970s. A significant number of biomass gasification plants have been built, but many have been closed down and dismantled or mothballed.

There is abundant literature on the thermal gasification of biomass. Information and data carefully chosen from this literature are discussed in this chapter. Information on coal gasification is also included because of its relevancy to the commercialization of biomass gasification; large-scale coal gasifiers have been in commercial operation for several years. This is not the case for most biomass gasifiers. Some of the coal gasification processes are also suitable for biomass feedstocks. Since the conditions required for coal gasification are more severe than those needed for biomass, some coal gasifiers can be operated on biomass or biomass-coal feedstock blends. Indeed, some gasifiers that were originally designed for coal gasification are currently in commercial use with biomass feedstocks.

The pyrolytic gasification of biomass has been interpreted to involve the decomposition of carbohydrates by depolymerization and dehydration followed by steam-carbon and steam-carbon fragment reactions. So the chemistries of coal and biomass gasification are quite similar in terms of the steam-carbon chemistry and are essentially identical after a certain point is reached in the gasification process. Note, however, that biomass is much more reactive than most coals. Biomass contains more volatile matter than coal, and the pyrolytic chars from biomass are more reactive than pyrolytic coal chars.

II. FUNDAMENTALS

A. DEFINITION

Basically, there are three types of biomass gasification processes—pyrolysis, partial oxidation, and reforming. As discussed in Chapter 8, if the temperature is sufficient, the primary products from the pyrolysis of biomass are gases. At high temperatures, charcoal and liquids are either minor products or not present in the product mixture. Partial oxidation processes (direct oxidation, starved-air or starved-oxygen combustion) are those that utilize less than the stoichiometric amounts of oxygen needed for complete combustion, so partially oxidized products are formed. The term "reforming" was originally used to describe the thermal conversion of petroleum fractions to more volatile products of higher octane number, and represented the total effect of many simultaneous reactions, such as cracking, dehydrogenation, and isomerization. Examples are hydroforming, in which the process takes place in the presence of hydrogen, and catalytic reforming. Reforming also refers to the conversion of hydrocarbon gases and vaporized organic compounds to hydrogen-containing gases such as synthesis gas, a mixture of carbon monoxide and hydrogen. Synthesis gas can be produced from natural gas, for example, by such processes as reforming in the presence of steam (steam reforming). For biomass feedstocks, reforming refers to gasification in the presence of another reactant. Examples of biomass gasification by reforming are steam reforming (steam gasification, steam pyrolysis), and steam–oxygen and steam–air reforming. Steam reforming processes involve reactions of biomass and steam and of the secondary products formed from biomass and steam. Steam–oxygen or steam–air gasification of biomass often includes combustion of residual char from the gasifier, of a portion of the product gas, or of a portion of the biomass feedstock to supply heat. The processes can be carried out with or without catalysis.

Under idealized conditions, the primary products of biomass gasification by pyrolysis, partial oxidation, or reforming are essentially the same: The carbon oxides and hydrogen are formed. Methane and light hydrocarbon gases are also formed under certain conditions. Using cellulose as a representative feedstock, examples of some stoichiometries are illustrated by these equations:

Pyrolysis: $C_6H_{10}O_5 \rightarrow 5CO + 5H_2 + C$
Partial oxidation: $C_6H_{10}O_5 + O_2 \rightarrow 5CO + CO_2 + 5H_2$
Steam reforming: $C_6H_{10}O_5 + H_2O \rightarrow 6CO + 6H_2$.

The energy content of the product gas from biomass gasification can be varied. Low-energy gases (3.92 to 11.78 MJ/m^3 (n), 100 to 300 Btu/SCF) are generally formed when there is direct contact of biomass feedstock and air. This is due to dilution of the product gases with nitrogen from air during the gasification process. Medium-energy gases (11.78 to 27.48 MJ/m^3 (n), 300 to 700 Btu/SCF) can be obtained from directly heated biomass gasifiers when oxygen is used, and from indirectly heated biomass gasifiers when air is used and heat transfer occurs via an inert solid medium. Indirect heating of the gasifier eliminates dilution of the product gas with nitrogen in air and keeps it separated from the gasification products. High-energy product gases (27.48 to 39.26 MJ/m^3 (n), 700 to 1000 Btu/SCF) can be formed when the gasification conditions promote the formation of methane and other light hydrocarbons, or processing subsequent to gasification is carried out to increase the concentration of these fuel components in the product gas. Methane is the dominant fuel component in natural gas and has a higher heating value of 39.73 MJ/m^3 (n) (1012 Btu/SCF).

B. STOICHIOMETRIES AND THERMODYNAMICS

Using cellulose as a representative feedstock composition, estimates of the enthalpy changes for some of the primary reactions that take place in biomass gasification systems are shown in Table 9.1. Although many stoichiometries are possible, as alluded to in this table, most of the hypothetical steam gasification reactions listed are endothermic at 300 and 1000 K. If methane is produced, along with the concomitant formation of CO_2, the process becomes progressively more exothermic. The partial oxidation reactions, as expected, are exothermic except at low oxygen levels. The degree of endothermicity and exothermicity of the pyrolysis reactions depends upon the product distributions. As carbon monoxide formation decreases and methane and carbon formation increase, the trend is toward more exothermic processes. It is evident that if fuel gases of higher energy content are desired, the gasification process should be operated to maximize methane and other light hydrocarbon products because their heating values are considerably greater than those of the other fuel components, carbon monoxide and hydrogen, as shown in Table 9.2. As will be shown later, pyrolysis and steam gasification of biomass can be self-sustaining under certain conditions. These types of conversions have each been demonstrated in large facilities.

III. COAL GASIFICATION

Coal gasification is reviewed here to provide a foundation for more detailed discussion of biomass gasification.

A. BRIEF HISTORY

Coal gasification to produce gas for a variety of applications such as fuels, chemicals, and chemical intermediates has been known for many years. The largest application of coal gasification by far has been for manufactured fuel gas production by pyrolytic and partial oxidation processes in which the primary fuel components in the product gas are hydrogen, carbon monoxide, and methane. The first manufactured gas (town gas) plant was built in England in 1812 by London and Westminster Chartered Gas, Light and Coke Company, although the first record of experimental manufactured gas production from coal dates back to seventeenth-century England (cf. Environmental Research and Technology and Koppers Co., 1984; Srivastava, 1993). North America's first manufactured gas plants were built in Baltimore in 1816, in Boston in 1822, and in New York in 1825 (Rhodes, 1974). The early processes involved the carbonization or destructive distillation of bituminous coal at temperatures of 600 to 800°C in small cast-iron retorts to yield "coal gas" (Villaume, 1984). It has been estimated that more than 1500 manufactured gas plants were in operation in the United States during the nineteenth century and the first half

TABLE 9.1 Enthalpies of Selected, Stoichiometric, Cellulose Gasification Reactions[a]

Process	Stoichiometry	Temperature (K)	Enthalpy (kJ)
Pyrolysis	$C_6H_{10}O_5 \rightarrow 5CO + 5H_2 + C$	300	180
		1000	209
	$C_6H_{10}O_5 \rightarrow 5CO + CH_4 + 3H_2$	300	105
		1000	120
	$C_6H_{10}O_5 \rightarrow 4CO + CH_4 + C + 2H_2 + H_2O$	300	-26
		1000	-16
	$C_6H_{10}O_5 \rightarrow 3CO + CO_2 + 2CH_4 + H_2$	300	-142
		1000	-140
	$C_6H_{10}O_5 \rightarrow 3CO + CH_4 + 2C + H_2 + 2H_2O$	300	-158
		1000	-152
	$C_6H_{10}O_5 \rightarrow 2CO + CO_2 + 2CH_4 + C + H_2O$	300	-274
		1000	-276
Partial oxidation	$C_6H_{10}O_5 + 0.5O_2 \rightarrow 6CO + 5H_2$	300	71
		1000	96
	$C_6H_{10}O_5 + O_2 \rightarrow 6CO + 4H_2 + H_2O$	300	-172
		1000	-142
	$C_6H_{10}O_5 + O_2 \rightarrow 5CO + CO_2 + 5H_2$	300	-213
		1000	-180
	$C_6H_{10}O_5 + 1.5O_2 \rightarrow 6CO + 3H_2 + 2H_2O$	300	-414
		1000	-389
	$C_6H_{10}O_5 + 1.5O_2 \rightarrow 4CO + 2CO_2 + 5H_2$	300	-498
		1000	-464
	$C_6H_{10}O_5 + 2O_2 \rightarrow 3CO + 3CO_2 + 5H_2$	300	-778
		1000	-745
Steam gasification	$C_6H_{10}O_5 + H_2O \rightarrow 6CO + 6H_2$	300	310
		1000	322
	$C_6H_{10}O_5 + 2H_2O \rightarrow 5CO + CO_2 + 7H_2$	300	272
		1000	310
	$C_6H_{10}O_5 + 3H_2O \rightarrow 4CO + 2CO_2 + 8H_2$	300	230
		1000	276
	$C_6H_{10}O_5 + 7H_2O \rightarrow 6CO_2 + 12H_2$	300	64
		1000	137
	$C_6H_{10}O_5 + H_2O \rightarrow 4CO + CO_2 + CH_4 + 4H_2$	300	64
		1000	85
	$C_6H_{10}O_5 + H_2O \rightarrow 2CO + 2CO_2 + 2CH_4 + 2H_2$	300	-184
		1000	-175

[a]The standard enthalpies of formation used for the calculations are from Stull, Westrum, and Sinke (1987) and Daubert and Danner (1989). The standard enthalpy of formation of cellulose was calculated from its heat of combustion. The monomeric unit of cellulose is $C_6H_{10}O_5$. The enthalpies are listed in kJ/g-mol of monomeric unit gasified.

TABLE 9.2   Higher Heating Values of Combustibles
Commonly Formed in Gasification Processes

| | Higher heating value | |
Combustible	MJ/m³ (n)	Btu/SCF
Methane	39.73	1012
Ethane	69.18	1762
Propane	98.51	2509
Ethylene	64.43	1641
Propylene	92.42	2354
Benzene	146.1	3722
Carbon monoxide	12.67	322.6
Hydrogen	12.74	324.5

of the twentieth century. The gasification processes used in these plants afforded water gas, producer gas, oil gas, coke oven gas, and blast furnace gas (Liebs, 1985; Remediation Technologies, Inc., 1990).

Natural gas displaced most manufactured gas for municipal distribution in industrialized countries after World War II. In the 1960s and 1970s, interest in developing advanced coal gasification processes was rekindled when it was believed that natural gas reserves would become insufficient in a few years to meet demand. This activity has since declined, but several coal gasification processes developed during this period have been commercialized and are used for production of fuel and synthesis gas.

B. CHEMISTRY

The chemistry of coal gasification is usually depicted to involve the following reactions of carbon, oxygen, and steam (*cf.* Bodle and Schora, 1979). The standard enthalpy change (gram molecules) at 298 K is shown for each reaction.

Gasification:
(1) $C + O_2 \rightarrow CO_2 - 393.5$ kJ
(2) $C + H_2O \rightarrow CO + H_2 + 131.3$ kJ
(3) $C + 2H_2O \rightarrow CO_2 + H_2 + 90.2$ kJ
(4) $C + CO_2 \rightarrow 2CO + 172.4$ kJ

Partial oxidation:
(5) $C + 0.5O_2 \rightarrow CO - 110.5$ kJ

Water gas shift:
(6) $CO + H_2O \rightarrow CO_2 + H_2 - 41.1$ kJ

Methanation:
(7) $2CO + 2H_2 \rightarrow CH_4 + CO_2 - 247.3$ kJ
(8) $CO + 3H_2 \rightarrow CH_4 + H_2O - 206.1$ kJ

(9) $CO_2 + 4H_2 \rightarrow CH_4 + H_2O - 165.0$ kJ
(10) $C + 2H_2 \rightarrow CH_4 - 74.8$ kJ.

In theory, gasification processes can be designed so that the exothermic and endothermic reactions are thermally balanced. For example, consider reactions 2 and 5. The feed rates could be controlled so that the heat released balances the heat requirement. In this hypothetical case, the amount of oxygen required is 0.27 mol/mol of carbon, the amount of steam required is 0.45 mol/mol of carbon, and the oxygen-to-steam molar ratio is 0.6:

$$C + H_2O \rightarrow CO + H_2 + 131.3 \text{ kJ}$$
$$\underline{1.2C + 0.6O_2 \rightarrow 1.2CO - 131.3 \text{ kJ}}$$
$$\text{Net: } 2.2C + H_2O + 0.6O_2 \rightarrow 2.2CO + H_2.$$

Many reactions occur simultaneously in coal gasification systems and it is not possible to control the process precisely as indicated here. But by careful selection of temperature, pressure, reactant and recycle product feed rates, reaction times, and oxygen-steam ratios, it is often possible to maximize certain desired products. When high-energy fuel gas is the desired product, selective utilization of high pressure, low temperature, and recycled hydrogen can result in practically all of the net fuel gas production in the form of methane.

The oxygen-steam ratios required to maintain zero net enthalpy change are given in Table 9.3 for several temperatures and pressures (Parent and Katz, 1948). With increased pressure, the ratio necessary to preserve a zero net enthalpy change diminishes. The decrease is most pronounced at low pressures. The effect of temperature change at constant pressure is also shown in Table 9.3. At lower temperatures, the oxygen-steam ratio doubles for each temperature

TABLE 9.3   Oxygen–Steam Ratios Yielding Equilibrium Products with Zero Net Change in Enthalpy in the Carbon–Oxygen–Steam Reaction[a]

Temperature (K)	Ratio of oxygen to steam (m^3 (n)/kg) at indicated pressure				
	0.1013 MPa	1.0133 MPa	2.0265 MPa	3.0398 MPa	4.0530 MPa
900	3.1	1.1	1.0	0.8	0.7
1000	6.8	2.6	2.0	1.6	1.4
1100	10.9	5.4	4.0	3.2	2.9
1200	11.7	8.8	6.7	6.0	5.3
1300		11.1	9.7	8.7	8.1
1400	12.8	11.9	11.2	10.6	10.3
1500	13.0	12.1	11.9	11.7	11.4

[a]Adapted from Parent and Katz (1948).

interval of 100 K. At higher temperatures, the increase diminishes and finally becomes very small.

The thermodynamic equilibrium compositions and enthalpy changes for the carbon-oxygen-steam system are graphically illustrated at several representative temperatures and pressures in Figs. 9.1 to 9.4 (Parent and Katz, 1948). Increasing pressures tend to lower the equilibrium concentrations of hydrogen and carbon monoxide and increase the methane and carbon dioxide concentrations (Fig. 9.1). Methane and carbon dioxide formation are favored at lower temperatures, and at higher temperatures, carbon monoxide and hydrogen are the dominant equilibrium products (Figs. 9.2 and 9.3). At high temperatures, the reactions occurring in the system are thermodynamically equivalent to reactions 2 and 5. It is also apparent that hydrogen-to-carbon monoxide molar ratios of 1.0 or more are thermodynamically feasible at lower feed ratios of oxygen to steam and low pressure (Fig. 9.4).

Although the utility of thermodynamic data to optimize the operating conditions of a gasification process is of considerable importance, thermodynamics ignore kinetic and catalytic effects and the mechanisms by which processes occur. The data presented here, however, provide valuable guidelines for the design of gasification processes. For coal gasification, the type of coal and reactant contact conditions in the gasifier produce large differences in the raw product gas compositions. In general, the same principles and conclusions apply to biomass gasification. Where experimental conditions are favorable, equilibrium may be approached by prolonged contact of the reactants or by use of catalysts. Where neither of these conditions offers a convenient solution, a compromise between idealized equilibrium and kinetics is necessary.

C. GASIFIER DESIGN AND GASIFICATION

Coal gasifier designs are almost as numerous as the many different types and ranks of coal. The basic configurations, hardware, and operations that have been considered are described here because several of them are applicable to biomass gasification (cf. National Academy of Engineering, 1973, and accompanying references).

Modern coal gasification processes consist of a sequence of operations: coal crushing, grinding, drying, and pretreatment, if necessary; feeding the coal into the gasifier; contacting the coal with the reacting gases for the required time in the gasifier at the required temperature and pressure; removing and separating the solid, liquid, and gaseous products; and treating the products downstream to upgrade them and to stabilize and dispose of solid and liquid wastes, dust, fines, and emissions. A large number of solids-feeding devices have been developed for low-pressure, atmospheric gasifiers. These include

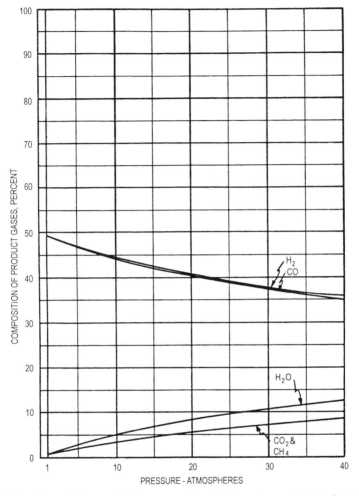

FIGURE 9.1 Change in the equilibrium composition of carbon–steam systems with pressure at 1200 K. From Parent and Katz (1948).

screws and star valves. For gasifiers operating at elevated pressures, lockhoppers and slurry pumping are the two leading solids-feeding devices. Lockhoppers are operated in an intermittent fashion so that coal fills the hopper vessel at atmospheric pressure. The vessel is pressurized with gas; the coal then flows to the gasifier at elevated pressure, and the lockhopper is restored to atmospheric pressure. If one lockhopper is used, the flow of coal to the gasifier is intermittent; two or more can be used for continuous feeding. The ash is

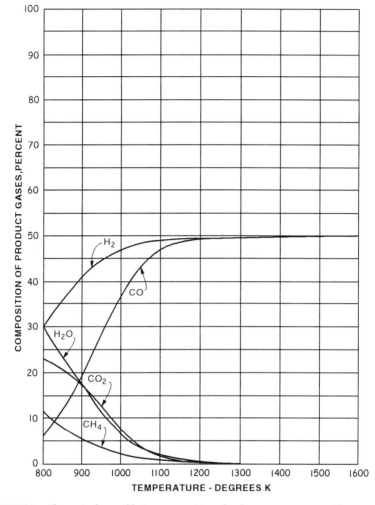

FIGURE 9.2   Change in the equilibrium composition of carbon–steam systems with temperature at a pressure of 1 atm. From Parent and Katz (1948).

withdrawn from the gasifiers as a slurry or by lockhopper. If the ash is molten as in slagging gasifiers, it is ordinarily quenched in water to solidify and break it up before disposal.

Gasifier operating temperatures range from 500 to 1650°C and pressures range from atmospheric to 7.6 MPa. The feedstocks are lump coal or pulverized coal. Processes using moving beds of lump coal can operate at temperatures up to about 980°C if the ash is recovered as a dry solid. Higher temperatures are possible if the ash is removed in a molten state. The methods of contacting

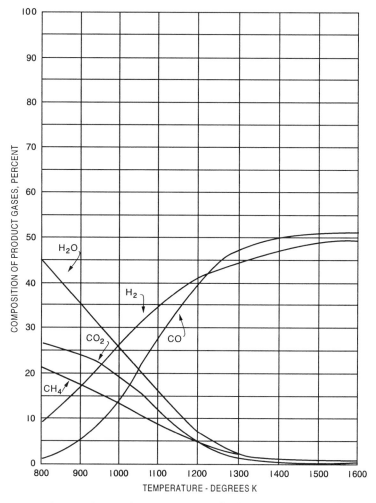

FIGURE 9.3 Change in the equilibrium composition of carbon–steam systems with temperature at a pressure of 20 atm. From Parent and Katz (1948).

the solid coal feed with reactant gases include reactors that contain a descending bed of solids with upflowing gas, a fluidized bed of solids, entrained flow of solids in gas, or molten baths of gasifying media. Modern processes generally utilize fixed-bed reactors operated under nonslagging or slagging conditions, circulating or bubbling fluid-bed reactors with ash recovered from the bed in either a dry or agglomerated form, entrained-flow reactors with pulverized coal suspended in the gas stream wherein gasification is completed before the gas containing the ash leaves the gasifier, or molten bath reactors.

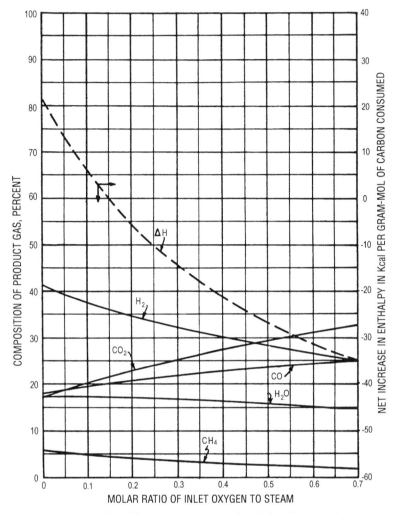

FIGURE 9.4   Variation of equilibrium composition and enthalpy change with oxygen–steam ratio for the carbon–oxygen–steam system at atmospheric pressure and 900 K. From Parent and Katz (1948).

Fixed-bed gasifiers, which are also called moving-bed gasifiers, are usually counterflow systems. Coal is fed at the top of the gasifier and air or oxygen along with steam is generally injected near the bottom. The maximum temperature, normally 930 to 1430°C, occurs at the bottom, and the residence time in the gasifier is 1 to 2 h. The fixed-bed gasifier involves countercurrent flow in which large particles of coal move slowly down the bed and react with gases moving up through the bed. Various processes occur in different zones of the reactor. At the top of the gasifier, the coal is heated and dried while cooling the exiting product gas. The temperatures of the exit gas range from 315°C for high-moisture lignites to 550°C for bituminous coal. As the coal descends through the gasifier, sequential heating, drying, devolatilization, carbonization, and gasification take place. Fixed-bed coal gasifiers are characterized by lower gasification and product gas temperatures, lower oxygen requirements, lower tar and oil production, higher methane content in the product gas, and limited ability to handle caking coals and coal fines.

Fluid-bed gasifiers generally require coal in the 10- to 100-mesh size range, and the maximum bed temperature is determined by the fusion point of the ash, which is usually 815 to 1040°C. Operation below the fusion temperature avoids formation of sticky, molten slag. Fresh coal feed is well mixed with the particles of coal and char already undergoing gasification. Steam and oxygen or air is usually injected near the bottom of the bed. Some unreacted coal and char particles are reduced in size during gasification and are entrained in the hot exit gas. This material is recovered for recycling. The ash is removed at the bottom of the bed and is cooled by heating the incoming feed gas and recycle gas. Fluid-bed gasifiers generally utilize significant recycle of flyash, operate at moderate and constant temperatures, and are limited in their ability to convert high rank coals. Agglomerated ash operation, which can be achieved by incorporation of a hot-ash agglomerating zone in the bottom of the reactor so that the ash particles stick together and grow in size until separated from the unreacted char, improves the ability of the process to gasify high rank and caking coals.

Entrained-flow gasifiers use pulverized coal, about 70% of which is smaller than 200 mesh, and have high feedstock flexibility. The coal particles are entrained in the steam-oxygen feed and the recycled gas stream and gasified at residence times of a few seconds, after which the product gas is separated from the ash. The lower residence times can offer potentially higher through-puts at elevated pressures. Entrained-flow gasifiers can be operated at lower temperatures to maintain the ash as a dry solid, or at temperatures well above the ash fusion point in the slagging mode so that the ash is removed as a molten liquid. Operation at higher temperature results in little or no tars and oils in the product gas.

In molten bath processes, crushed coal is passed with reacting gases into the liquid bath, where gasification occurs. The ash can become part of the liquid bath or can be separated. The media include liquid iron and liquid sodium carbonate.

Low-, medium-, and high-energy gases can be produced in coal gasification processes. The important parameters are essentially the same as those for biomass gasification systems. The higher heating values of the combustible gases commonly formed in coal-derived gases are listed in Table 9.2. As in the case of biomass gasification, the primary combustible components in low-energy product gases are carbon monoxide and hydrogen. In gasifiers where the coal particles are in direct contact with the oxygen-containing gas, nitrogen is a major component in the product gas if air is used as a coreactant instead of oxygen. Medium-energy gases are usually formed with oxygen and contain a higher percentage of combustibles, in addition to hydrogen and carbon monoxide, such as methane. High-energy gases approaching heating values of 39.3 MJ/m³ (n) (1000 Btu/SCF), the approximate higher heating value of pure methane, are produced at lower temperature conditions with oxygen instead of air, to maximize methane concentration. Further processing is necessary to methanate residual carbon monoxide and to separate noncombustible gases to provide a high-energy gas.

D. Product Gas Comparison

A comparison of the heating values and compositions of the raw product gases from selected coal gasification processes is shown in Table 9.4. Some of these processes, a few of which are used for synthesis gas production, have been commercialized. Some have been developed to the point where they might be termed near-commercial, and a few are under development. It is evident that a wide range of gas compositions can be produced by coal gasification. It is also evident that several of the gas compositions and operating conditions can be correlated with thermodynamic principles and the thermodynamics of the carbon-oxygen-steam system. Methane and carbon dioxide yields are generally higher at lower temperatures and higher pressures, as illustrated by the raw gas compositions for the Synthane process, whereas higher temperatures and lower pressures favor carbon monoxide and hydrogen, as illustrated by the raw gas compositions reported for the Koppers–Totzek process. Interestingly, the heating values of the product gases for processes supplied with steam-oxygen coreactants are generally in the same range despite the wide range of operating conditions. The heating values of the product gases from processes supplied with steam-air coreactants are also in the same range, although they are lower than those of the product gases produced by coal-steam-oxygen processes. Arithmetic adjustment of the heating values by deducting nitrogen from the product gases shows that all of them are in the same range.

E. Commercial Processes

The processes listed in Table 9.4 that are reported to be used commercially to supply synthesis gas for methanol production are the Lurgi process, the

Winkler process, the Koppers–Totzek process, and the Texaco process. Downstream adjustment and treatment of the raw product gases is required when these processes are used to supply feedstock or cofeedstock to a typical low-pressure methanol process operating at 220 to 270°C and 5.066 to 10.132 MPa (50 to 100 atm). A few of the operating details of these and other commercial coal gasification processes are presented here.

Dry Ash Lurgi Process

This process is a fixed-bed process that gasifies crushed, dried coal at 620 to 760°C, 2.43 to 3.14 MPa, and residence times of about 1 h. The raw product gas exits the gasifier at 370 to 590°C and contains tar, oil, naphtha, phenols, ammonia, sulfides, and fines. Quenching with oil removes tar and oil. Catalytic shifting and scrubbing of the quenched product gas provides a gas that can be methanated to produce substitute natural gas, or the equivalent of pipeline gas. The process is limited to noncaking coals.

British Gas Lurgi Slagging Process

This process incorporates advancements into Lurgi's dry-ash gasifier that convert the system to a slagging gasifier, reduce the steam requirement to about 15% of that required by the dry-ash gasifier, provide a raw gas with higher carbon monoxide and lower methane, carbon dioxide, and moisture, and improve the capability to use caking coals and a significant amount of fines. The process affords increased gas yields by limiting the net hydrocarbon liquids to naphtha and phenols.

Winkler Process

This process converts crushed coal, oxygen, and steam at 820 to 1000°C and near-atmospheric pressure in a fluid-bed gasifier. After passage of the raw gas through a waste heat recovery section, flyash is removed by cyclones, wet scrubbers, and electrostatic precipitators. Further processing, depending on end use, yields a gas suitable as synthesis gas or pipeline gas.

High-Temperature Winkler Process

This process uses a fluid-bed unit that is especially designed for gasification of brown and hard coals, peat, and biomass. In the case of brown coal, predried feed at 12 wt % moisture is fed along with oxygen and steam to the reactor which operates at 750 to 800°C and 2.53 MPa.

TABLE 9.4 Comparison of Operating Temperatures and Pressures and Typical Product Gas Heating Values and Compositions of Selected Coal Gasification Processes

Process type	Process	Conditions		HHV, MJ/m³ (n)		Raw gas composition, dry mol %[o]					
		(°C)	(MPa)	Steam–O₂	Steam–air	H₂	CO	CH₄	CO₂	N₂	Others
Fixed bed	Lurgi (dry ash)[a]	621–760	2.43–3.14	11.86		40	21	10	28		1
	Lurgi (dry ash)[a]	621–760	2.43–3.14		7.08	25	16	5	14	39	1
	Lurgi (slagging)[b]	1296–1371	2.53	14.25		28	61	8	2	2	
	Ruhr 100[c]	1927	10.13	13.31		32	22	16	28	2	
	Wellman-Galusha[d]	1315	30.40		6.60	15	29	3	3	50	
Fluid bed	Winkler[e]	816–982	0.10–0.61	10.80		42	33	3	21	1	0.3
	Winkler[e]	816–982	0.10–0.61		4.63	13	21	1	7	58	0.2
	Synthane[f]	982	3.55–7.09	15.90		28	17	24	29	0.8	1.3
	CO₂ Acceptor[g]	871	1.01–1.52	17.28		54	17	21	7	0.2	1.4
	U-Gas[h]	1038	2.43		5.89	13	19	5	10	52	0.7
Entrained	Bi-Gas[i]	927–1482	2.03	14.92		24	44	16	14	0.6	1.4
	Koppers-Totzek[j]	1816	0.10	11.78		37	56	0	6	1.1	0.3
	Texaco[k]	1093–1371	2.74	11.78		45	45	1	9		
	Shell Oil Co.	1482	High	11.78		31	67		2		

Molten bath	ATGAS[l]	1427	17.94		10	70	20	13	0.7	0.3
	Molten salt[m]	649–982	12.92		45	34	7		0.4	
	Molten salt[n]	982		5.65	(10)	18	(10)	7	64	1.3

[a]Lurgi Mineraloltechnik Gmbh; commercial.

[b]Lurgi Mineraloltechnik Gmbh, 10-min residence time; commercial.

[c]Ruhrgas AG; demonstrated.

[d]Wellman Engineering Company; 4-h residence time; commercial.

[e]Davy Powergas, Inc.; 30-min residence time; commercial; technology owned by Davy International Corp.

[f]U.S. Bureau of Mines; demonstrated.

[g]Consolidated Coal Company; demonstrated.

[h]IGT; commercial.

[i]Bituminous Coal Research, Inc.; two-stage system with upper stage at 927°C and lower slagging stage at 1482°C; residence time of the order of seconds; demonstrated.

[j]Heinrich Koppers Gmbh; 1-s residence time; commercial.

[k]Texaco, Inc.; commercial.

[l]Applied Technology Corporation; uses molten iron; demonstrated.

[m]M.W. Kellogg Company; uses molten sodium carbonates; demonstrated.

[n]Atomics International; uses molten sodium carbonates; hydrogen and carbon monoxide not differentiated; demonstrated.

[o]Raw gas compositions are rounded figures; raw gas from these processes usually contains small amounts of tars, oils, phenols, ammonia, sulfides, light hydrocarbons, and fines from the ash.

Koppers–Totzek Process

This process can be operated on all types of coal without pretreatment. Dried, pulverized coal and oxygen are converted in a horizontal, entrained-flow gasifier at about 1820°C and near-atmospheric pressure. The raw gas is quenched with water to solidify entrained molten ash, scrubbed to remove entrained solids, and purified to remove hydrogen sulfide and a controlled quantity of carbon dioxide. The resulting product is used as synthesis gas.

Shell Oil Co. Process

This process is a dry feed, entrained-flow, high temperature–high pressure, slagging gasifier that converts a wide variety of coals from lignite to bituminous to a medium-energy gas for combined cycle power generation. The unit operates with pressurized, predried coal, oxygen, and steam at 1500°C and attains carbon conversions above 99%.

Texaco, Inc. Process

This process is a single-stage, pressurized, entrained-flow slagging process that uses a water slurry of ground coal which is pumped along with oxygen to the gasifier. The operating temperature in the gasifier is 1200 to 1500°C. Careful control of the oxygen feed to maintain a reducing atmosphere results in a synthesis gas that is predominantly carbon monoxide and hydrogen.

Destec Energy, Inc. Process

This process is a two-stage, entrained-flow, slagging process for conversion of lignite and subbituminous coal. The preheated water slurry of coal is fed to the first stage where it is mixed with oxygen, the feed rate of which is controlled to maintain the reactor temperature in a specific range, depending on the properties of the coal. The second stage reduces the temperature of the raw product gas to about 1000°C. The coal is almost completely converted to carbon monoxide, carbon dioxide, and hydrogen.

IGT's U-GAS Process

Tampella Corporation is commercializing the U-GAS process, which was developed by the Institute of Gas Technology. Tampella has constructed a 10-MW, integrated U-GAS-combined cycle power plant in Finland that uses coal, peat, and wood wastes as feedstocks. U-GAS incorporates a single-stage, fluid-bed gasifier in which coal reacts with steam and air at 950 to 1090°C at pressures from atmospheric to 3.55 MPa to yield a low-energy gas. Oxygen can be substituted for air, in which case a medium-energy gas is produced.

IV. BIOMASS GASIFICATION

A. INTRODUCTION

The effort to develop and commercialize advanced biomass gasification systems is not nearly as extensive as the effort to develop coal gasification. However, considerable research and pilot plant studies have been carried out since about 1970 on biomass gasification for the production of fuel gases and synthesis gases (cf. Stevens, 1994). Several processes have been commercialized. Basic studies on the effects of various operating conditions and reactor configurations have been performed in the laboratory and at the PDU (process development unit) and pilot scales on pyrolytic, air-blown, oxygen-blown, steam, steam-oxygen, and steam-air gasification, and on hydrogasification. The thermal gasification of biomass in liquid water slurries has also been studied.

The chemistry of biomass gasification is very similar to that of coal gasification in the sense that thermal decomposition of both solids occurs to yield a mixture of essentially the same gases. But as pointed out in the Introduction, biomass is much more reactive than most coals. Biomass contains more volatile matter than coal, gasification occurs under much less severe operating conditions for biomass than for coal feedstocks, and the pyrolytic chars from biomass are more reactive than pyrolytic coal chars. The thermodynamic equilibrium concentrations of specific gases in the mixture depend on the abundance of carbon, hydrogen, and oxygen, the temperature, and the pressure. As in the case of coal feedstocks, increasing pressures tend to lower the equilibrium concentrations of hydrogen and carbon monoxide, and increase the methane and carbon dioxide concentrations. Also, as in the case of coal feedstocks, methane formation is favored at lower temperatures, and carbon monoxide and hydrogen are dominant at high temperatures. Biomass is gasified at lower temperatures than coal because its main constituents, the high-oxygen cellulosics and hemicellulosics, have higher reactivities than the oxygen-deficient, carbonaceous materials in coal. The addition of coreactants to the biomass system, such as oxygen and steam, can result in large changes in reaction rates, product gas compositions and yields, and selectivities as in coal conversion.

Biomass feedstocks contain a high proportion of volatile material, 70 to 90% for wood compared to 30 to 45% for typical coals. A relatively large fraction of most biomass feedstocks can be devolatilized rapidly at low to moderate temperatures, and the organic volatiles can be rapidly converted to gaseous products. The chars formed on pyrolytic gasification of most biomass feedstocks have high reactivity and gasify rapidly. Heat for pyrolysis is usually generated by combusting fuel gas either in a firebox surrounding the reaction chamber or in fire tubes inserted into the reaction chamber. As discussed in Chapter 8, chars, tars and oily liquids, gases, and water vapor are formed in varying amounts, depending particularly on the feedstock composition, heating rate, pyrolysis temperature, and residence time in the reactor. For biomass and waste biomass, steam gasification generally starts at temperatures near 300 to 375°C.

Undesirable emissions and by-products from the thermal gasification of biomass can include particulates, alkali and heavy metals, oils, tars, and aqueous condensates. One of the high-priority research efforts is aimed at the development of hot-gas-cleanup methods that will permit biomass gasification to supply suitable fuel gas for advanced power cycles that employ gas turbines without cooling the gas after it leaves the gasifiers (International Energy Agency, 1991, 1992). It is important to avoid gas turbine blade erosion and corrosion by removing undesirable particulates that may be present. The removal of tars and condensables may also be necessary. Furthermore, utilization of the sensible heat in the product gas improves the overall thermal operating efficiencies. Nonturbine applications of the gas may also be able to take advantage of processes that provide clean, pressurized hot gas, such as certain downstream chemical syntheses and fuel uses. Special filtration and catalytic systems are being developed for hot-gas cleanup. Some of the other research needs that have been identified include versatile feed-handling systems for a wide range of biomass feedstocks; biomass feeding systems for high-pressure gasifiers; determination of the effects of additives, including catalysts for minimizing tar production and materials that capture the contaminants; and suitable ash disposal and wastewater treatment technologies. Research on thermal biomass gasification in North America has tended to concentrate on medium-energy gas production, scale-up of advanced process concepts that have been evaluated at the PDU scale, and the problems that need to be solved to permit large-scale thermal biomass gasifiers to be operated in a reliable fashion for power production, especially for advanced power cycles. Research to develop biomass gasification processes for chemical production via synthesis gas waned in the mid-1980s because of low petroleum and natural gas prices. More attention was given to the subject in the 1990s when the market prices for these fossil fuels began to increase.

Examples of the various types of biomass gasification processes are reviewed in the next few sections before commercial and near-commercial processes are described.

B. GASIFICATION PROCESS VARIATIONS

Pyrolytic Gasification

The primary products of biomass pyrolysis under conventional pyrolysis conditions are gas, oil, char, and water. As the reaction temperature increases, gas yields increase. It is important to note that pyrolysis may involve green or predried biomass, and that product water is formed in both cases. Water is released as the biomass dries in the gasifier and is also a product of the chemical reactions that occur, even with bone-dry biomass. Unless it is rapidly removed from the reactor, this water would be expected to participate in the process along with any added feedwater or steam. As will be shown later, the exothermic heat from the steam gasification of woody biomass under certain conditions appears to be sufficient to eliminate the need for an external heat source or the use of oxygen. Self-sustained steam gasification can effectively be carried out with biomass feedstocks, according to some investigators.

One of the more innovative pyrolytic gasification processes is an indirectly heated, fluid-bed system (*cf.* Alpert *et al.*, 1972; Bailie, 1981; Paisley, Feldmann, and Appelbaum, 1984). This system uses two fluid-bed reactors containing sand as a heat transfer medium. Combustion of char formed in the pyrolysis reactor takes place with air within the combustion reactor. The heat released supplies the energy for pyrolysis of the combustible fraction in the pyrolysis reactor. Heat transfer is accomplished by flow of hot sand from the combustion reactor at 950°C to the pyrolysis reactor at 800°C and return of the sand to the combustion reactor (Fig. 9.5). This configuration separates the combustion

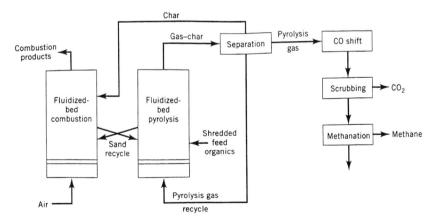

FIGURE 9.5 Methane production by pyrolysis using sand and char recycle in a fluidized two-bed system.

and pyrolysis reactions and keeps the nitrogen in air separated from the pyrolysis gas. It yields a pyrolysis gas that can be upgraded to a high-energy gas (substitute natural gas, SNG) by shifting, scrubbing, and methanating without regard to nitrogen separation. The pyrolysis gas with hybrid poplar feedstocks typically contain about 38 mol % carbon monoxide, 15 mol % carbon dioxide, 15 mol % methane, 26 mol % hydrogen, and 6 mol % C_2' s. This is a medium-energy gas having a higher heating value of about 19.4 MJ/ m^3 (n). The projected gas yields are about 670 m^3 (n) of pyrolysis gas, or about 200 m^3 (n) of methane per dry ton of feed if SNG is produced.

Partial Oxidation

Many thermal conversion processes can be classified as partial oxidation processes in which the biomass is supplied with less than the stoichiometric amount of oxygen needed for complete combustion. Both air and oxygen have been utilized for such systems. When the oxygen is supplied by air, low-energy gases are formed that contain higher concentrations of hydrogen, carbon monoxide, and carbon dioxide than medium-energy gases. When pure oxygen or oxygen-enriched air is used, gases with higher energy values can be obtained. In some partial oxidation processes, the various chemical reactions may occur simultaneously in the same reactor zone. In others, the reactor may be divided into zones: A combustion zone that supplies the heat to promote pyrolysis in a second zone, and perhaps to a third zone for drying, the overall result of which is partial oxidation.

One system (Fig. 9.6) uses a three-zoned vertical shaft reactor furnace (Fisher, Kasbohm, and Rivero, 1976). In this process, coarsely shredded feed is fed to the top of the furnace. As it descends through the first zone, the charge is dried by the ascending hot gases, which are also partially cleaned by the feed. The gas is reduced in temperature from about 315°C to the range of 40 to 200°C. The dried feed then enters the pyrolysis zone, in which the temperature ranges from 315 to 1000°C. The resulting char and ash then descend to the hearth zone, where the char is partially oxidized with pure oxygen. Slagging temperatures near 1650°C occur in this zone, and the resulting molten slag of metal oxides forms a liquid pool at the bottom of the hearth. Continuous withdrawal of the pool and quenching forms a sterile granular frit. The product gas is processed to remove flyash and liquids, which are recycled to the reactor. A typical gas analysis is 40 mol % carbon monoxide, 23 mol % carbon dioxide, 5 mol % methane, 5 mol % C_2's, and 20 mol % hydrogen. This gas has a higher heating value of about 14.5 MJ/m^3 (n).

An example of the gasification of biomass by partial oxidation in which air is supplied without zone separation in the gasifier is the molten salt process (Yosim and Barclay, 1976). In this process, shredded biomass and air are

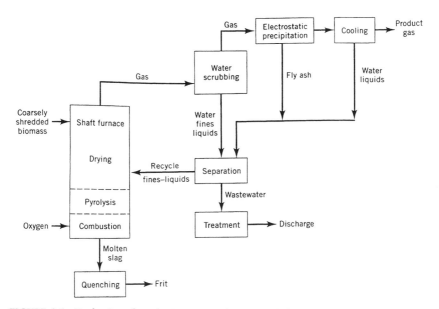

FIGURE 9.6 Production of synthesis gas in a three-zone shaft reactor furnace. From Fisher, Kasbohm, and Rivero (1976).

continuously introduced beneath the surface of a sodium carbonate-containing melt which is maintained at about 1000°C. As the resulting gas passes through the melt, the acid gases are absorbed by the alkaline media and the ash is also retained in the melt. The melt is continuously withdrawn for processing to remove the ash and is then returned to the gasifier. No tars or liquid products are formed in this process. The heating value of the gases produced depends on the amount of air supplied and is essentially independent of the type of feed organics. The greater the deficiency of air needed to achieve complete combustion, the higher the fuel value of the product gas. Thus, with about 20, 50, and 75% of the theoretical air needed for complete oxidation, the respective higher heating values of the gas are about 9.0, 4.3, and 2.2 MJ/m^3 (n).

Many gasifier designs have been offered for the manufacture of producer gas from virgin and waste biomass, and several types of units are still available for purchase. As mentioned in the introduction to this chapter, thousands of producer gasifiers operating on air and wood were used during World War II, particularly in Sweden, to power automobiles, trucks, and buses. The engines needed only slight modification to operate on low-energy producer gas. Although only limited research has been carried out on small-scale producer gasifiers for biomass in recent years, significant design advancements continue to be made even though the gasifiers have been used for more than 100 years. One of the interesting developments is the open-top, stratified, downdraft gasifier in which the feedstock such as wood chips moves downward from the top as it is gasified and air is simultaneously drawn in from the top through successive reaction strata (LaFontaine, 1988; LaFontaine and Reed, 1991). Low-cost, portable gasifiers can be assembled for captive use from ordinary metal cans, garbage containers, and drums that are manually loaded with fuel from the open top. More sophisticated units can of course be manufactured. The open-top biomass gasifier is simple to operate, is inexpensive, and can be close-coupled to a gas engine-generator set without requiring the use of complex gas-cleaning equipment. The system appears to be quite suitable for small- and moderate-scale engine applications from 5 to 5000 HP and portable electric-power generation systems. The gasifier dimensions are sized to deliver gas to the engine based on its fuel-rate requirements, and minimal controls are needed.

A similar, wood-fueled, downdraft gasifier patterned after Swedish reports from the early years of World War II was initially built in the United States in the late 1970s of mild steel. It was used to power an unmodified 1978 Chevrolet Malibu station wagon equipped with a 3.3-L (200-in.^{3}) V-6 engine for a coast-to-coast trip from Jacksonville, Florida, to Los Angeles, California, a distance of about 4300 km (Russel, 1980). Small pine and hardwood blocks of 15 to 25 wt % moisture content were used as fuel throughout the trip. The gasifier was pulled on a small two-wheel trailer behind the vehicle. The system was subsequently driven a total of 8046 km. Examination of the vehicle and all components showed no significant wear or abnormalities. A typical composition of the low-energy fuel gas was reported to be 18 mol % carbon monoxide, 9 mol % carbon dioxide, 1 mol % methane, 17 mol % hydrogen, 45 mol % nitrogen, and 10 mol % water. On a distance traveled basis, about 3.0 to 3.6 kg of wood fuel was estimated to equate to 1 L of gasoline.

Steam Gasification

Steam is also blended with air in some gasification units to promote the overall process via the endothermic steam-carbon reactions to form hydrogen and carbon monoxide. This was common practice at the turn of the last century, when producer gasifiers were employed to manufacture low-energy gas from virgin and waste biomass. The producer gas from these gasifiers generally had heating values around 5.9 MJ/m^3 (n), and the energy yields as gas ranged up to about 70% of the energy contained in the feed.

Study of the steam gasification of biomass in a sequential pyrolysis–steam reforming apparatus has shown that gasification occurs as a two-step process (Antal, 1978). At temperatures in the 300 to 500°C range, volatile compounds are evolved from biomass and some residual char is formed. At about 600°C, the volatile compounds are steam reformed to yield synthesis gases. The condensable tars, oils, and pitches are reduced by the steam reforming reactions to less than 10 wt % of the original feedstock. Table 9.5 is a summary of the steam gasification of pure cellulose that illustrates the effects of temperature and residence time in the steam reformer on product yields. As temperature and residence time are increased, char and tar yields decrease and gas yields increase as expected. A medium-energy gas was produced in these experiments because of the relatively high concentrations of lower molecular weight hydrocarbons in the product gas.

An obvious improvement in the steam gasification of biomass for synthesis gas production is to operate at higher temperatures and to use catalysts to gasify as much of the char and liquid products as possible. Laboratory-scale experiments have been carried out to examine this possibility (Mitchell *et al.*, 1980). Nickel precipitated on silica alumina (1:1) and a mixture of silica alumina and nickel on alumina were evaluated as catalysts for steam gasification at 750°C and 850°C and atmospheric pressure. The results are summarized in Table 9.6. The function of the silica alumina is to crack the hydrocarbon

TABLE 9.5   Sequential Pyrolysis and Steam Reforming of Pure Cellulose in a Close-Coupled Reactor[a]

Gas-phase conditions					
Reactor temperature, °C	500	600	600	600	700
Residence time in reactor, s	9	2	6	10	6
Product yields, wt %					
Gas	53	70	75	80	80
Char	12	11	13	13	13
Tars	35	19	12	7	7
Gas analysis, mol %					
H_2	11	10	10	10	13
CO	40	55	52	55	53
CO_2	42	20	20	16	13
CH_4	2	6	8	8	12
C_2H_4	1	3	4	4	5
C_3H_6	1	1	2	1	1
C_2H_6	1	2	1	2	1
Others	2	3	3	4	2
Gas HHV, MJ/m³ (n)	11.78	19.28	20.34	20.65	19.24
Mass balance, %	64	82	95	85	86
Carbon balance, %	71	69	71	69	88

[a]Antal (1978). The steam superheater was maintained at 350°C, and the pyrolysis reactor was maintained at 500°C. A large excess of steam was passed through the system. The gas yield includes the water of reaction. The carbon balances by improved procedures always exceeded 90%.

TABLE 9.6   Laboratory-Scale Results for Catalyzed Steam Gasification of Wood[a]

Reaction conditions				
Catalyst	Ni:SiAl	Ni:SiAl	Ni on Al	Ni on SiAl
Reactor temperature, °C	750	850	750	850
Wood:catalyst weight ratio	16.1	100	52.5	NA
Steam:wood weight ratio	0.63	1.25	0.71	1.25
Carbon conversion, %				
To gas	73	99.6	77	95
To liquid	Trace	0	Trace	0
To char	27	0.4	23	5
Gas analysis, mol %				
H_2	53.4	56.7	55.9	58.2
CO	28.1	27.9	27.8	28.5
CO_2	15.6	14.9	15.2	13.2
CH_2	2.8	0.5	1.3	0.1
Standard heat of reaction of wood, kJ/kg	490	3101	991	3501
Potential methanol yield, wt % of wood	59	86	64	86

[a]Mitchel et al. (1980). Wood feed rate was 0.3 g/min. All runs were carried out at atmospheric pressure in a single-stage reactor.

intermediates, and the function of the nickel is to promote methane reforming and the hydrogenolysis of higher molecular weight hydrocarbons. It is evident from the data in Table 9.6 that a synthesis gas almost stoichiometric for methanol synthesis can be produced from wood at high yields by catalytic steam gasification in a single-stage reactor at atmospheric pressure. Potential methanol yields over 60 wt % of the wood feedstock were estimated. The advantages of catalytic steam gasification of biomass over steam-oxygen gasification include elimination of the need for an oxygen plant and shift conversion, higher methanol yields for a stand-alone plant, and less carbon dioxide formation. Using the data from the example in Table 9.6 in which the steam-to-wood weight ratio is 0.71, and assuming wood that contains 20 wt % moisture is fed at 100°C with steam at 850°C, the net reactor heat requirement is estimated to be 2800 kJ/kg of dry wood.

The various stoichiometric equations listed in Table 9.1 suggest that synthesis gas mixtures from biomass gasification are generally deficient in hydrogen for methanol synthesis; i.e., the molar ratio of $H_2 : CO$ is less than 2. The use of steam in biomass gasification could conceivably increase hydrogen yields by reaction of residual char, if formed, via the steam-carbon reaction. Steam gasification might also make it possible to use green biomass feedstocks without drying. Under the proper gasification conditions, the use of oxygen or air to meet any heat requirements would be expected to increase the yields of carbon oxides, but an oxygen plant is required in the case of oxygen usage. Gas quality would suffer with air because of nitrogen dilution of the product gases unless air is utilized separately from the gasification process, as already mentioned. However, as just indicated (Mitchell *et al.*, 1980), it has been shown that product gases containing a 2 : 1 molar ratio of hydrogen to carbon monoxide can be produced without use of a separate water gas shift unit:

$$C_6H_{10}O_5 + 3H_2O \rightarrow 4CO + 2CO_2 + 8H_2.$$

Gasification of biomass for methanol synthesis under these conditions would offer several advantages if such processes can be scaled to commercial size.

Commercial methanol synthesis is performed mainly with natural gas feedstocks via synthesis gas. Synthesis gas from biomass gasification could conceivably be used as a cofeedstock in an existing natural gas-to-methanol plant to utilize the excess hydrogen produced on steam reforming natural gas. Examination of a hypothetical hybrid plant has been shown to have significant benefits (Rock, 1982). Typical synthesis gas mixtures from the steam–oxygen gasification of wood and the steam reforming of natural gas are as follows:

> From wood: $2CO + CO_2 + 1.8H_2$
> From natural gas: $5.2/3(2CO + CO_2 + 10H_2)$
> Combined: $5.5CO + 2.7CO_2 + 19.1H_2.$

This combined synthesis gas mixture is stoichiometric for methanol synthesis:

$$5.5CO + 2.7CO_2 + 19.1H_2 \rightarrow 8.2CH_3OH + 2.7H_2O.$$

The stoichiometry for methanol from the unmixed gases is

$$2CO + CO_2 + 1.8H_2 + 0.73H_2O \rightarrow 1.27CH_3OH + 1.73CO_2$$
$$5.2/3(2CO + CO_2 + 10H_2) \rightarrow 5.2CH_3OH + 5.2H_2 + 1.73H_2O.$$

The unmixed synthesis gases produce 6.47 mol of methanol, of which 1.27 mol comes from wood, and the mixed synthesis gases yield 8.20 mol of methanol. In theory, the use of the combined synthesis gases provides 24% more synthesis gas, but methanol production is increased by 58% over that from natural gas alone. Since hydrogen in the purge gas in the reformed natural gas case has been largely consumed in the hybrid case, the total purge gas stream is greatly reduced. This purge gas is normally used as fuel in the reforming furnace and its reduction must be balanced by firing additional natural gas or other fuel for reforming. The use of natural gas and fuel is about 25% lower for the hybrid design than when using natural gas only for the production of the same amount of methanol. In addition, the hybrid version has eliminated the water gas shift and acid gas removal equipment from the wood gasification process alone. This serves to reduce both capital and operating costs associated with wood-derived synthesis gas.

The stoichiometry of this particular hybrid process is approximately as follows:

Wood: $0.5C_6H_{10}O_5 + 1.1O_2 \rightarrow 2CO + CO_2 + 1.8H_2 + 0.7H_2O$
Natural gas: $5.2CH_4 + 6.93H_2O \rightarrow 3.47CO + 1.73CO_2 + 17.33H_2$
Methanol synthesis: $5.47CO + 2.73CO_2 + 19.13H_2 \rightarrow 8.2CH_3OH + 2.73H_2O$
Net: $0.5C_6H_{10}O_5 + 5.2CH_4 + 3.5H_2O + 1.1O_2 \rightarrow 8.2CH_3OH.$

By use of the enthalpy of formation for dry poplar wood of 840.1 kJ/g-mol (361,440 Btu/lb-mol) of cellulosic monomeric unit at 300 K, which is calculated from its measured heat of combustion and the standard enthalpies of formation for the other components, the enthalpy changes for wood gasification (with oxygen) to synthesis gas, the steam reforming of natural gas, and methanol synthesis, are calculated to be −363.8, 1001, and −631.5 kJ, respectively. In theory, the overall enthalpy change is almost zero, 5.7 kJ. Biomass gasification can of course be carried out in several ways, and the gas compositions used for this analysis are idealized. But this type of analysis makes it possible to calculate several parameters of interest. For example, assuming 100% selectivities for intermediates and products, or that no by-products are formed, and that poplar wood and natural gas are accurately represented by ($C_6H_{10}O_5$) and CH_4, the feedstock rates for a 907-t/day (1000-ton/d) methanol plant are estimated to be 0.4 million m^3 (n)/day (288 t/day) of natural gas and 280 t/day of dry wood.

Gasification in Liquid Water

A potential route to synthesis gas from biomass is gasification under conditions in which water is in the liquid or fluid phase at elevated temperature and hydrostatic pressure. Exploratory research has been done in a laboratory-scale, plug-flow reactor with solutions of glucose, the monomeric unit of cellulose, in pure water without addition of any potential catalyst (Klass, Kroenke, and Landahl, 1981; Ng, 1979). Some of the results are summarized in Table 9.7. Gasification experiments carried out below the critical temperature for water (374°C) indicated little or no gasification. At temperatures above 374°C, conversion to a relatively clean synthesis gas began to occur, as shown in this table. Char was not observed. Hydrogen yield and concentration in the product gas and the molar ratio of $H_2 : CO$ exhibited significant increases with increasing temperature. Biomass gasification under these conditions might be expected to offer unique opportunities for homogeneous catalysis at lower capital and operating costs than heterogeneously catalyzed systems.

Heterogeneous catalysts have been found to be effective for the low-temperature, elevated-pressure gasification of 2 to 10% aqueous biomass slurries or solutions that range from dilute organics in wastewater to waste sludges

TABLE 9.7   Noncatalyzed Gasification of Glucose in Water at Above-Critical Pressure and Temperature[a]

Reaction conditions					
Temperature, °C	385	385	500	600	600
Pressure, MPa	27.358	27.358	27.358	27.358	27.358
WHSV	179	98	180	181	90
Residence time in reactor, min	11.1	22.6	10.7	10.4	21.6
Carbon conversion to gas, %	10.5	11.7	18.4	31.1	63.1
Gas analysis, mol %					
H_2	9.6	11.4	23.9	32.3	25.7
CO	33.9	27.2	28.0	11.2	3.6
CO_2	54.2	59.2	42.8	47.5	58.3
CH_4	1.4	1.2	4.0	9.0	12.4
Others	0.8	1.0	1.3		

[a]Klass, Kroenke, and Landahl (1981); Ng (1979). A plug-flow reactor, 0.48 cm ID, was used. The glucose concentration was 3.2 to 3.5 wt %. The WHSV is the weight hourly space velocity in grams of glucose per hour per liter of reactor volume. The critical temperature and pressure of water are 374.1°C and 22.119 MPa.

from food processing (Elliott *et al.*, 1991, 1993). Continuous, fixed-bed catalytic reactor systems have been operated on three scales ranging from 0.03 to 33 L/h. The residence time in the supported metallic catalyst bed is less than 10 min at 360°C and 20,365 kPa at liquid hourly space velocities of 1.8 to 4.6 L of feedstock/L of catalyst/h depending on the feedstock. Aqueous effluents with low residual COD (chemical oxygen demand) and a product gas of medium-energy quality have been produced. Catalysts have been demonstrated to have reasonable stability for up to six weeks. Ruthenium appears to be a more stable catalyst than nickel. The product gas contains 25 to 50 mol % carbon dioxide, 45 to 70 mol % methane, and less than 5 mol % hydrogen with as much as 2 mol % ethane. The by-product water stream carries residual organics and has a COD of 40 to 500 ppm. The medium-energy product gas is produced directly in contrast to medium-energy, gas-phase processes that require either oxygen in place of air or the dual reactor system to keep the nitrogen in air separated from the product.

Hydrogasification

In this process, gasification is carried out in the presence of hydrogen. Most of the research on hydrogasification has targeted methane as the final product. One approach involves the sequential production of synthesis gas and then methanation of the carbon monoxide with hydrogen to yield methane. Another route involves the direct reaction of the feed with hydrogen (Feldmann *et al.*,

1981). In this process (Fig. 9.7), shredded feed is converted with hydrogen-containing gas to a gas containing relatively high methane concentrations in the first-stage reactor. The product char from the first stage is used in a second-stage reactor to generate the hydrogen-rich synthesis gas for the first stage. From experimental results obtained with the first-stage hydrogasifier operated in the free-fall and moving-bed modes at 1.72 MPa and 870°C with pure hydrogen, calculations shown in Table 9.8 were made to estimate the composition and yield of the high-methane gas produced when the first stage is integrated with an entrained-char gasifier as the second stage. Note that although the methane content of the raw product gas is projected to be higher in the moving-bed reactor than in the falling-bed reactor, the gas from the first stage must still be reacted in a shift converter to adjust the H_2/CO ratio, scrubbed to remove CO_2, and methanated to obtain SNG.

Other research shows that internally generated hydrogen for hydroconversion can be obtained in a single-stage, noncatalytic, fluidized bed reactor (Babu, Tran, and Singh, 1980). In this work, hydroconversion was envisaged to occur in a series of steps: nearly instantaneous thermal decomposition of biomass followed by gas-phase hydrogenation of volatile products to yield hydrocarbon

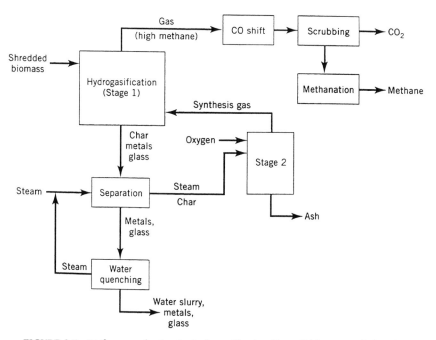

FIGURE 9.7   Methane production by hydrogasification. From Feldmann et al. (1981).

TABLE 9.8 Gas Composition and Yield from Integrated Hydrogasification Process at Stage 1[a]

Product	Free fall	Moving bed
Composition, mol %		
H_2	31.9	13.3
CO	45.9	51.9
CO_2	10.1	16.1
CH_4	10.4	17.2
C_2H_6	1.2	1.1
Benzene	0.5	0.4
Yield, m^3/kg dry feed	1.1	0.95
Fraction of total CH_4 produced in Stage 1 after methanation	0.26	0.52

[a]Feldmann et al. (1981).

gases, hydrogen, carbon oxides, water, and hydrocarbon liquids; rapid conversion of a part of the devolatilized biomass char to methane at appropriate gasification conditions; slow residual biomass char gasification with hydrogen and steam to yield methane, hydrogen, and the carbon oxides; and combustion of residual biomass char, which supplies the energy for the endothermic char gasification reactions. Examination of hydroconversion under a variety of pressure and temperature conditions with woody biomass and hydrogen, steam, and hydrogen-steam mixtures and study of the kinetics of the slower steam-char reactions led to a conceptual process called RENUGAS®, which will be described in more detail later. Biomass is converted in the reactor, which is operated at about 2.2 MPa, 800°C, and residence times of a few minutes with steam-oxygen injection. About 95% carbon conversion is anticipated to produce a medium-energy gas which can be subjected to the shift reaction, scrubbing, and methanation to form SNG. The cold gas thermal efficiencies are estimated to be about 60%. Since this initial work, RENUGAS has been tested at the pilot, PDU, and demonstration scales, and is being commercialized.

Comparative studies on the gasification of wood in the presence of steam and hydrogen have shown that steam gasification proceeds at a much higher rate than hydrogasification (Feldmann et al., 1981). Carbon conversions 30 to 40% higher than those achieved with hydrogen can be achieved with steam at comparable residence times. Steam/wood weight ratios up to 0.45 promoted increased carbon conversion, but had little effect on methane concentration. Other experiments show that potassium carbonate-catalyzed steam gasification of wood in combination with commercial methanation and cracking catalysts can yield gas mixtures containing essentially equal volumes of methane and carbon dioxide at steam/wood weight ratios below 0.25 and atmospheric pressure and temperatures near 700°C (Mudge *et al.*, 1979). Other catalyst combinations produced high yields of product gas containing about 2 : 1 hydrogen/carbon monoxide and little methane at steam/wood weight ratios of about 0.75 and a temperature of 750°C. Typical results for both of these studies are shown in Table 9.9. The steam/wood ratios and the catalysts used can have major effects on the product gas compositions. The composition of the product gas can also be manipulated depending on whether a synthesis gas or a fuel gas is desired.

C. BIOMASS GASIFIER DESIGNS

Basically, biomass gasifiers can be categorized into several reactor design groups: a descending bed of biomass, often referred to as a moving or fixed bed, with countercurrent gas flow (updraft); a descending bed of biomass with cocurrent gas flow (downdraft); a descending bed of biomass with crossflow of gas; a fluidized bed of biomass with rising gas; an entrained-flow circulating bed of biomass; and tumbling beds. Many reactor designs have been evaluated under a broad range of operating conditions. The designs include fixed-bed, moving-bed, suspended-bed, and fluid-bed reactors; entrained-feed solids reac-

TABLE 9.9 Product Gases from Steam Gasification of Wood with and without Catalysts[a]

Parameter	Value			
Gas composition, mol %				
H_2	29	50	0	53
CO	34	17	0	30
CO_2	17	11	48	12
CH_4	15	17	52	4
Reaction conditions				
Primary catalyst	None	Wood ash	K_2CO_3	K_2CO_3
Secondary catalyst	None	None	Ni:SiAl	SiAl
Steam:wood weight ratio	0.24	0.56	0.25	0.75
Reactor temperature, °C	696	762	740	750
Pressure, kPa (gauge)	129	159	0	0
Carbon conversion to gas, %	68	52	68	77
Feed energy in gas, %			76	78
Heating value of gas, MJ/m³ (n)	16.6	17.7	20.6	12.1

[a]Mudge *et al.* (1979) for the K_2CO_3-catalyzed laboratory data with unspecified wood; Feldmann *et al.* (1981) for the other data (PDU) with unspecified hardwood.

tors; stationary vertical-shaft reactors; inclined rotating kilns; horizontal-shaft kilns; high-temperature electrically heated reactors with gas-blanketed walls; single and multihearth reactors; ablative, ultrafast, and flash pyrolysis reactors; and several other designs. There are clearly numerous reactor designs and configurations for biomass gasification, probably more than in the case of coal gasification systems because of the relative ease of thermal biomass conversion.

Fixed-bed, updraft gasifiers are simple to construct and can consist of carbon steel shells equipped with a grate at the bottom fed by a process air manifold, a lockhopper at the top to feed material, and a manifold to remove gas at the top (cf. Miller, 1987). These units are simple to construct and operate and are relatively inexpensive. The gas exiting the gasifier tends to be cool because it has percolated up through the bed and therefore usually contains a fair fraction of lower molecular weight hydrocarbons. Much of the sensible heat has been lost, the feeds are limited to wood chips, and the size is usually not more than 50 million GJ/h. Fixed-bed, downdraft gasifiers consist of two concentric shells. The inner shell holds the material on the grate; the outer shell is used to transport the gas. The gas is drawn out from under the grate through the outer shell to the outside of the system. The gas exits at the combustion zone, and because it is hot, it contains few longer chain hydrocarbons and particulates. The system, however, is more expensive to construct than fixed-bed, updraft gasifiers, and is also limited to sizes up to about 10 to 20 GJ/h and chip feeds. Fluid-bed systems afford more efficient gasification because hot spots are eliminated, diverse feedstocks can be charged, the exit gas has a high sensible heat content, and the gasifiers are capable of scale-up to relatively large sizes. However, the units are more expensive to construct and product gas quality must be carefully monitored because of its higher particulate content.

V. COMMERCIAL AND NEAR-COMMERCIAL BIOMASS GASIFICATION METHODS

A. FEEDSTOCK COMPOSITION IMPACTS

As alluded to in Chapter 8, the ideal biomass feedstock for thermal conversion, whether it be combustion, gasification, or a combination of both, is one that contains low or zero levels of elements such as nitrogen, sulfur, or chlorine, which can form undesirable pollutants and acids that cause corrosion, and no mineral elements that can form inorganic ash and particulates. Ash formation, especially from alkali metals such as potassium and sodium, can lead to fouling of heat exchange surfaces and erosion of turbine blades, in the case of power production systems that use gas turbines, and cause efficiency losses and plant upsets. In addition to undesirable emissions that form acids (SO_x), sulfur can also form compounds that deactivate methanol synthesis catalysts, whereas chlorine can be transformed into toxic chlorinated organic derivatives as well as acids.

Biomass is similar to some coals with respect to total ash content as discussed in Chapter 3, but because of the diversity of biomass, several species and types have relatively low ash and also low sulfur contents. Woody biomass is one of the feedstocks of choice for thermal gasification processes. The ash contents are low compared to those of coal, and the sulfur contents are the lowest of almost all biomass species. Grasses and straws are relatively high in ash content compared to most other terrestrial biomass, and when used as feedstocks for thermal conversion systems, such biomass has been found to cause a few fouling problems.

The high moisture contents of aquatic and marine biomass species make it unlikely that they would be considered as feedstocks for thermal gasification processes. However, a few processes can be performed with aqueous slurries or do not require dry biomass feedstocks as described earlier. As harvested, aquatic and marine biomass species often have moisture contents greater than 90% of the total plant weight. In addition to the relatively high ash contents of herbaceous feedstocks, the nitrogen content is an important factor. Grasses are higher in protein nitrogen than woody feedstocks and can increase nitrogen oxide emissions on gasification.

The compositions of wood compared to those of other potential biomass feedstocks make woody biomass a preferred feedstock for thermal gasification. Although not shown here, most woody biomass species, especially those indigenous to the contiguous United States, are similar in composition. It is important to emphasize that quantitative ash analyses of biomass feedstocks sampled at the plant gate and from storage should be carried out periodically and sometimes continually to provide real-time data needed for process control. There can be large differences in the amounts of specific mineral components in biomass.

A major mechanism of the fouling of heat exchanger surfaces with biomass feedstocks, particularly the straws and herbaceous residues, is the formation in the thermal conversion zone of low-fusion-point alkali metal salt eutectics such as the alkali metal silicates. The problems caused by these salts and the control methods for combustion and thermal gasification systems were discussed in Chapter 8. Several experienced designers of biomass gasifiers and the operators of commercial plants operated on biomass feedstocks have indicated that the problem is usually not severe with gasification systems, but can be with combustion systems. Temperature control to reduce slagging and the formation of molten agglomerates and equipment designs that avoid contact of the internals with hot gases that may contain low-fusion-point particulates are the preferred control methods for minimizing these problems. For biomass gasifiers that are used to supply fuel for gas turbines, the control methods are similar. Some biomass, although high in minerals, may be low in alkali metals. Fouling by sticky particulates is therefore much less with this type of feedstock.

Some gasification process designers claim to have developed proprietary gas processing systems that yield product gases from biomass gasifiers "cleaner than natural gas" using conventional desulfurization processes for sulfur removal and cyclones and proprietary filters to remove ash and char fines. Electrostatic precipitators are not used, and scrubbers are claimed to be optional for some of these systems. These statements are difficult to support without public dissemination of full-scale test results. They are probably true, however, because there are many emissions- and ash-removal systems that have been installed and effectively operated in large-scale commercial biomass combustion plants that meet all requirements. Some of these plants are designed to meet California's stringent South Coast regulations. Much of this experience and technology can be drawn upon to design environmentally clean biomass gasification plants.

Many of the commercial or near-commercial biomass gasification facilities that have been built and operated use green or partially dried feedstocks in which the moisture content of the feedstock to the gasifiers is not specified. The steam-carbon reactions that occur are undoubtedly one of the main reasons for variation in product gas compositions from these systems. Since the carbon content of dry biomass is about 45 wt %, green wood contains about 2.2 kg moisture/kg of carbon. Table 9.10 shows the effects of the moisture content of poplar wood when gasified in an air-blown, downdraft gasifier. As the moisture content of the wood decreases from 34 to 13 wt %, thermal efficiency, product gas heating value, dry gas yield, and the proportion of the combustible components in the dry gas each increase. These data illustrate the importance of specifying feedstock moisture content. Feedstock dryers are essential for some biomass gasification plants depending on the feedstock's moisture content and variation, as well as on the end uses of the product gases.

B. Commercial Biomass Gasification

In the 1970s and early 1980s, about 40 companies worldwide offered to build biomass gasification plants for different applications. Since then, many of the smaller companies and some of the larger ones have gone out of business, discontinued biomass gasification projects, or emphasized established biomass combustion technologies. The problems encountered in first-of-a-kind biomass gasification plants and the low prices of petroleum and natural gas all had an adverse impact on the marketability of biomass gasification technologies. Several of the plants built in North America in the 1970s and 1980s have been

TABLE 9.10 Effects of Moisture Content of Poplar Wood Chips on Product Yield, Gas Composition, and Thermal Efficiency in a Fixed-Bed, Air-Blown, Downdraft Gasifier[a]

Parameter	13 wt %	24 wt %	34 wt %
Input, kg/h	24.0	25.1	25.2
Dry wood equivalent, wt %	40	36	31
Moisture in wood, wt %	6	12	16
Dry air, wt %	54	52	53
Product distribution, wt %			
Dry gas	87	82	76
Tars	3	6	7
Solids	5	3	3
Aqueous condensate	5	9	14
Gas analysis, mol %			
H_2	17.5	16.7	15.1
CO	19.7	16.0	11.9
CO_2	12.7	15.8	17.7
CH_4	3.5	3.2	2.1
C_2H_2	0.3	0	0.1
C_2H_4	1.5	1.4	1.1
C_2H_6	0.1	0.2	0.1
C_3H_8	0.2	0.3	0.2
O_2	1.9	0.9	0.9
N_2	42.7	45.5	50.9
Gas HHV, MJ/m^3 (n)	7.50	6.67	5.22
Thermal efficiency, %	74	68	55

Column header spanning 13/24/34 wt %: Wood moisture content

[a]Adapted from Graham and Huffman (1981). The gasifier was rated at 0.84 GJ/h. The thermal efficiency is (cool gas energy)/(dry wood energy).

shut down, dismantled, or placed on standby. A survey of commercial thermal biomass gasification showed that few gasifiers have been installed in the United States (Miles and Miles, 1989). Most of the units in use are retrofitted to small boilers, dryers, and kilns. The majority of the existing units operate at rates of 0.14 to 1.0 t/h of wood wastes on updraft moving grates. In the United States, many purveyors of biomass gasification technologies have gone out of business or are focusing their marketing activities in other countries or on other conversion technologies, particularly combustion for power generation, in states where combined federal and state incentives make the economic factors attractive. Some existing gasification installations have also been shut down and placed in a standby mode until natural gas prices make biomass gasification competitive again.

Examination of state-of-the-art biomass thermal gasification technology shows that moving-bed gasifiers have been studied and extensively tested

(Babu and Whaley, 1992). Nine atmospheric-pressure updraft gasifiers were commercialized from 1982 to 1986 in Europe and have been successfully operated with wood and peat. Six plants were placed in operation for close-coupled district heating purposes in Finland, while three plants were built in Sweden for district heating and drying wood chips (Kurkela, 1991). In general, the moving-bed systems require close control of feedstock size and moisture content and appropriate means to handle the high tar content of the raw product gases.

The Winkler fluid-bed coal gasifier was successfully scaled up to gasify 25 dry t/h of peat in 1988 by Kemira Oy in Finland (Babu and Whaley, 1992). The product gas was used for the manufacture of ammonia. Major mechanical and process modifications included improvements to the peat lockhopper feeding system, and the control of naphthalene formation by using higher gasifier temperatures and the addition of a benzene scrubber for naphthalene removal. The application of fluid-bed gasifiers to wood and other types of biomass has been commercialized in North America by Omnifuel in Canada and by Southern Electric International, Inc., in Florida, both of which are described later, and Energy Products of Idaho, Inc. The largest and most successful fluid-bed biomass gasification plants to date have been attributed to the Ahlstrom and Gotavarken circulating, fluid-bed gasifiers employed in close-coupled operation with lime kilns in Sweden, Finland, and Portugal (Babu and Whaley, 1992). The gasifiers are about 2 m in diameter and range in height from 15 to 22 m. They are operated at near atmospheric pressure at about 700°C with circulating limestone and are capable of handling mixtures of sawdust, screening residues, and bark. A large-scale circulating fluid-bed gasifier was built in 1992 by Studsvik AB for gasifying RDF (refuse-derived fuel) (Rensfelt, 1991).

These biomass gasifiers are representative low-pressure technologies, which when combined with current state-of-the-art gas-cleanup systems render themselves suitable for close-coupled operation with lime kilns, furnaces, boilers, and probably advanced, combined-cycle power systems. However, from the standpoint of producing methanol, gasification under elevated pressure and temperature is preferred because the equipment size is reduced for the same throughput, the cost of recompression before methanol synthesis should be less, and the noncondensable hydrocarbons and tars are only present in low concentrations in the spent water. The opposing effects of temperature and pressure on C_1-C_4 light hydrocarbon yields can be optimized to afford low yields of these products by careful selection of operating temperature and pressure.

At least five industrial-scale biomass gasifiers were available commercially from U.S. manufacturers in the 1990s. A two-stage stirred-bed gasifier is available from Producers Rice Mill Energy. The company built three gasifiers of 11 to 18 t/day capacity in Malaysia for rice hull feedstocks. Sur-Lite Corporation built small-scale, fluid-bed gasifiers of up to 10 GJ/h capacity for cotton-gin trash in California and Arizona, for rice husks in Indonesia, and for wood and coal in White Horse, Canada. Morbark Industries, Inc., supplies two-stage, starved-air gasification-combustion systems. The units in operation include a 4-GJ/h system for a nursing home in Michigan and a 1-GJ/h system for heating private facilities. Energy Products of Idaho supplies fluid-bed gasifiers to produce low-energy gas. The company has constructed a 57-GJ/h plant in California to fuel a boiler, a 99-GJ/h plant in Missouri to fuel a dryer, and an 85-GJ/h plant in Oregon to generate 5 MW of power from steam. Southern Electric International, a subsidiary of The Southern Company, coordinated the design and construction of a 264-GJ/h, fluid-bed, wood gasification plant in Florida, which has since been dismantled and moved to Georgia. It is described later.

A few representative biomass gasification processes that have been commercialized or that are near commercialization are described here to illustrate some of the details of gasifier designs and the operating results. The biomass pyrolysis plants described in Chapter 8 are not discussed here because the major products are liquids and charcoals, and the by-product gases are used for plant fuel.

Pyrolysis and Partial Oxidation with Air of MSW in a Rotating Kiln

Monsanto Enviro-Chem Systems, Inc., developed an MSW pyrolysis process called the Landgard process through the commercial stage (U.S. Environmental Protection Agency, 1975; Klass, 1982). A full-scale, 1050-t/day plant was built in Maryland and placed in operation in the mid-1970s. The plant was designed to operate for 10 h/day and to accept residential and commercial solid waste typical of U.S. cities. MSW disposal was the primary objective of the plant, not energy recovery. Large household appliances, occasional tires, and similar materials were acceptable feeds; automobiles and industrial wastes were excluded. The process included several operations: shredding of the MSW from storage in 900-HP hammer mills to provide particles small enough (4-cm diameter) to fall through the grates, storage of the shredded MSW which had a heating value of 10.7 MJ/kg, feeding of the shredded MSW to the pyrolysis reactor by twin hydraulic rams, pyrolysis, gas processing, and gas utilization in two waste heat boilers which generated 90,700 kg/h of steam, and processing of the ungasified residue to remove ferrous metals. Pyrolysis took place in a refractory-lined, horizontal, rotary kiln, which was 5.8 m in diameter and 30.5 m long. The kiln was rotated at 2 r/min. The heat required for pyrolysis was provided by burning the MSW with 40% of the theoretical air needed for complete combustion, and supplemental fuel (No. 2 fuel oil) was supplied at a rate of 24.4 L/t of waste. The fuel oil burner was located at the discharge end of the kiln. Pyrolysis gases moved countercurrent to the waste and exited at the feed end of the kiln. The gas temperature was controlled to 650°C, and the residue was kept below 1100°C to prevent slagging. The product gas on a dry basis had a heating value of 4.7 MJ/m³ (n) and consisted of about 6.6 mol % hydrogen, 6.6 mol % carbon monoxide, 11.4 mol % carbon dioxide, 2.8 mol % methane, 1.7 mol % ethylene, 1.6 mol % oxygen, and 69.3 mol % nitrogen. The plant was shut down in January 1981 and was scheduled to be replaced by a direct combustion system. Cost and reliability were cited as the reasons for the change.

Partial Oxidation of MSW with Oxygen in a Slagging, Updraft Gasifier

Union Carbide Corporation developed a partial oxidation process called the PUROX System for converting MSW to fuel gas and an inert slag (Fisher, Kasbohm, and Rivero, 1976). The process was scaled up in the mid 1970s from a pilot plant to a commercial, 181-t/day plant at Union Carbide's facility in West Virginia. A 68-t/day plant was also built in Japan. The plant in West Virginia was operated successfully on MSW. One tonne of refuse required about 0.18 t of oxygen and produced 0.6 t of medium-energy gas with a higher heating value of about 14.5 MJ/m³ (n), 2 t of sterile aggregate residue, and 0.25 t of wastewater. Within the process, 0.03 t of oil was separated in the gas-cleaning train and recycled to the furnace for cracking into additional gas. A typical gas analysis was 40 mol % carbon monoxide, 23 mol % carbon dioxide, 5 mol % methane, 5 mol % C₂+, 26 mol % hydrogen, and 1 mol % nitrogen. The energy balance expressed in terms of percent of the energy in the feedstock was a net 68% recovered in the product gas, 21% lost on conversion, and 11% used for in-plant electric power generation. The 181-t/day plant included front-end shredding and separation equipment for ferrous metal recovery, liquid separation equipment for recycling the condensed oil to the reactor, provision for removal of the slag from the hearth and quenching in a water bath, and treatment of the product gas by water scrubbing and electrostatic precipitation. The reactor was a three-zoned, vertical shaft furnace operating at about 50 cm of water. The RDF was fed to the top of the furnace through a gas seal and oxygen was injected at the bottom. The furnace was maintained essentially full of RDF which continually descends through the reactor. The oxygen in the hearth reacts with char to generate slagging temperatures to melt the glass and metals. Projections indicated that with a 1360-t/day plant composed of 181 to 317-t/day modular units, about 114 million L/year of methanol could be manufactured.

Partial Oxidation of MSW with Air in a Slagging, Updraft Gasifier

In the mid 1970s, Andco Incorporated developed and commercialized a slagging process called the Andco-Torrax System for converting MSW to low-energy gas and an inert glassy aggregate (Davidson and Lucas, 1978; Mark, 1980). Plants ranging in size from 2.4 to 7.5 t/h were installed in Europe, Japan, and the United States. Refuse is charged into the top of the gasifier without prior preparation except to shear or crush bulky items to about one meter or less in the longest dimension. As the refuse descends within the gasifier, it is dried and then pyrolyzed by the hot, oxygen-deficient gases produced in the hearth area. Char from the pyrolysis process and the noncombustible materials continue to descend. Primary combustion air at temperatures of about 1000°C is admitted to the gasifier immediately above the hearth to oxidize the char at temperatures sufficient to slag the inerts. The slag is continuously drained from the gasifier. The low-energy gas, the heating value of which is about 4.7 to 5.9 MJ/m³ (n), from the top of the gasifier is burned in a secondary combustion chamber at slagging temperatures. The slag is collected from this unit also. The heat from the secondary combustion chamber is used for hot water, steam, and power production. It is believed that this process can be used for waste disposal and energy recovery with combined feeds of MSW and tires, sludge, or waste plastics, and for the manufacture of cement. The first Andco-Torrax plant in the United States was built at Disney World in Orlando, Florida. This plant was used for waste paper from restaurants in the theme park. The performance of the gasifier in this plant was felt to be unsatisfactory because "arching" of the feedstock frequently occurred in the upper zone of the gasifier and resulted in feed stoppage. This problem could probably have been eliminated without design changes by densifying the very light waste paper feed.

RDF Gasification in an Atmospheric, Air-Blown, Circulating Fluid-Bed Gasifier

The largest commercial RDF gasification plant in Europe is believed to be the system built in 1992 in Italy that produces low-energy gas at a feed rate of 180 t/day of RDF, which is obtained from 600 t/day of MSW (Barducci et al., 1995, 1996). The hot product gas from the fluid-bed gasifiers is burned in an on-site boiler or is used as industrial fuel. The flue gas from the boiler is cleaned in a three-stage dry scrubber system before being exhausted through the stack. The steam raised in the boiler drives a 6.7-MW condensing steam turbine. Alternatively, the product gas is supplied as fuel to a neighboring cement plant. When the plant is eventually completed by addition of other gasification units, the facility will have the capacity to process 1300 t/day of MSW. The gasification system for this plant was developed by TPS Termiska Processer AB (Morris, 1996). It consists of two circulating fluid-bed gasifiers, each of 54 GJ/h capacity. A downstream cleanup process for the hot product gas is expected to be installed in future plant additions. In this process, the tars in the product gas are catalytically cracked at about 900°C in a dolomite-containing vessel located immediately downstream of the gasifier, and the particulates and alkalis are removed. This is expected to eliminate equipment contamination and filter clogging by tar condensation as the product gas is cooled. The TPS technology is expected to be used for similar projects in The Netherlands and the United Kingdom, and also for the biomass-fueled, integrated gasification, combined-cycle (BIGCC) power plant in Brazil. The BIGCC plant is expected to demonstrate the commercial viability of producing electric power from eucalyptus in an advanced technology plant of 30 MW capacity (see Section V, E).

Wood Gasification in a Pressurized, Air-Blown, Bubbling Fluid-Bed Gasifier

The first commercial, pressurized, air-blown, fluid-bed process for wood feedstocks was developed by Omnifuel Gasification Systems Ltd. and was installed in a plywood mill in Ontario, Canada in 1981 (Bircher, 1982). The unit was an 84-GJ/h gasifier that was supplied with 5.9 t/h of wood chips and wood dust. It operated at about 760°C and 35.5 kPa gauge. The low-energy product gas was used on-site as boiler fuel. Air was introduced at the bottom of the bubbling bed of sand particles and maintained the bed in constant motion as it passed up through the bed. Some of the air caused combustion of feed to maintain the temperature in the desired range, and some reacted with char to yield additional gas. A typical wet gas analysis was 12.3 mol % carbon monoxide, 4.6 mol % methane, 1.6 mol % C_2+ hydrocarbons, 7.8 mol % hydrogen, and 73.7 mol % nitrogen, carbon dioxide, and water. Carbon conversion efficiencies of the order of 99% were obtained, and tar production was very low, of the order of 0.1 to 0.2%. Ash entrained in the product gas was removed by cyclones. Some difficulty was encountered with gas combustion equipment because of the large variation in gas quality and the plant has been shutdown (cf. Klass, 1985). This was attributed to the large range of wood feedstock moisture which varied between 5 and 50 wt %. The heating values of the product gas ranged from 3.1 to 7.9 MJ/m³ (n). Operation with oxygen at 1420 kPa was projected to produce a gas with a heating value of 11.8 to 15.7 MJ/m³ (n).

Wood Gasification in a Low-Pressure, Air-Blown, Bubbling Fluid-Bed Gasifier

In the mid- to late 1980s, a 258-GJ/h, fluid-bed wood gasification plant was built in Florida by Alternate Gas, Inc., for Southern Electric International (Miller, 1987; Makansi, 1987; Bulpitt and Rittenhouse, 1989). Each twin gasifier was 2.44 m in diameter and converted wood chips at the rate of

15.4 t/h into 129 GJ/h of low-energy gas. Hardwood, whole-tree chips, and sawmill residues were the feedstock. Before gasification, the feedstock was predried to 25% moisture in a triple-pass dryer equipped with burners that could burn either product gas or natural gas. About 10 to 20% of the wood charged was combusted in the refractory-lined gasifiers with 25% of the stoichiometric air required to provide the heat needed for gasification, which takes place at 790 to 815°C at 34.5 kPa gauge or less. The product gas was cleaned in two stages of cyclones to remove particulates and was then used as fuel for clay dryers. The gas had a heating value of 5.9 to 7.1 MJ/m^3 (n). The product char after separation from the ash was sold to a charcoal briquette manufacturer. The plant was operated successfully for more than a year and then dismantled and moved to a new location in Georgia by Southern International.

Wood Gasification in an Air-Blown, Crossdraft Gasifier-Combustor

Commercial systems consisting of a close-coupled gasifier and combustor are manufactured by CHIPTEC Wood Energy Systems and are widely used in the Northeast to supply hot water and steam to schools and commercial buildings (Bravakis, 1996). The plants are fueled with wood chips that are conveyed to the refractory-lined gasification chamber by an automated feeder. The fuel can contain up to 45% moisture. An induced draft fan draws air into the crossdraft gasifier, and the resulting low-energy product gas, which is produced under oxygen-deficient conditions, is passed to the combustion chamber. High temperature combustion, a 20:1 turn-down ratio, refractory heat storage, and controlled air allows the gasifier to respond quickly to boiler demand. Gasifier outputs range in size from 0.5 to 10.5 GJ/h, and the corresponding firing rates are 50.8 to 965 kg/h with wood chips having a 35% moisture content. The smaller systems have stationary grates in the combustors, and the larger systems are equipped with moving grates.

Wood Gasification in an Air-Blown Updraft Gasifier

An updraft, wood-chip gasifier was built by Applied Engineering Company in 1980 in Georgia (Jackson, 1982). At that time, the unit was the largest of its kind. It was sized at 26.4 GJ/h, fed with 2.8 t/h of wood chips, and supplied a hospital with steam. A similar unit was built in late 1981 for the Florida Power Corporation. The unit fired one of six boilers in a 30-MW power system. The gasifier was cylindrical in shape, insulated with firebrick, and enclosed in a carbon steel shell. Air was injected at the bottom, and green tree chips having a heating value of about 9.3 to 10.5 MJ/kg and 40 to 50% moisture content were charged at the top. Ash was removed from the bottom. In the design used, oxidation of the wood char occurs at the top of the grate, which is located just above the ash hoppers, and produces temperatures of about 1370°C. Pyrolysis and cracking occur in the middle of the gasifier, and incoming wood is dried by the exiting hot gases. Typical dry gas analyses were 26 to 30 mol % carbon monoxide, 2 to 3 mol % hydrocarbons, 10 to 12 mol % hydrogen, and 58 to 59 mol % carbon dioxide and nitrogen. The heating value of the gas was 5.9 to 6.5 MJ/m³ (n). The gasifiers were operated quite successfully for an extended period of time. It is noteworthy that the carbon monoxide concentration was so high. This may have been caused by the use of green wood with high moisture contents and operation at relatively high temperature in the gasification zone.

Rice Hull Gasification in an Air-Blown Updraft Gasifier

Starting in the early 1980s, PRM Energy Systems, Inc., began to market gasification technology for converting biomass to low-energy fuel gases (Bailey and Bailey, 1996). Several commercial plants based on PRM's air-blown updraft designs for the gasification of rice hulls have been built and operated in Australia, Costa Rica, Malaysia, and the United States. High-silica ash is a salable by-product. An example of this technology is the plant installed in Mississippi in 1995 for the gasification of 300 t/day of rice hulls. The system converts unground rice hulls to fuel gas (121 GJ/h) for an existing boiler-power island which supplies electric power (5 MW capacity) and 6800 kg/h of process steam for parboiling rice. In operation, feedstock is metered into the gasifier by a water-cooled screw conveyor that discharges into the drying and heating zone of the gasifier. The gasifier is a refractory-lined, cylindrical steel shell that is equipped with a fixed grate at the bottom and is mounted in a vertical position. The gasification process is automatically controlled to maintain a preset first-stage gasification zone temperature. Almost all of the ash is removed from the bottom of the gasifier. The low particulate concentration in the product gas makes it possible to direct-fire a boiler without the use of emission control equipment. Total particulate emissions in the boiler exhaust of this plant were determined to be 0.103 kg/GJ.

Biomass Gasification in an Air-Blown Updraft Gasifier

Several small-scale, fixed-bed, updraft gasifiers are operated commercially in Sweden and Finland for the gasification of a wide range of biomass feedstocks, including wood chips, saw mill residues, straw, and RDF (Patel, 1996). The gasifiers are marketed by Carbona Corporation and are refractory-lined shaft furnaces that are fed from the top by a hydraulically operated feeder. The units are equipped with hydraulically rotated mechanical grates at the bottom. Ash sintering is prevented by water vapor contained in the gasification air, and the ash is removed through an ash discharge system installed at the bottom. The moisture content of the feeds can range from 0 to 45%, and the corresponding heating values of the product gas are about 5.5 to 3.8 MJ/m³ (n). The gasifiers can be connected to a hot water or steam boiler depending on whether heat or electric power is desired, or alternatively, the product gas can be used for hot gas generation for kilns and dryers.

Biomass Gasification in a Pressurized, Oxygen-Blown, Stratified, Downdraft Gasifier

The National Renewable Energy Laboratory (Solar Energy Research Institute at that time) designed, built, and operated a 0.9-t/day prototype, downdraft biomass gasifier between 1980 and 1985 (Reed, Levie, and Markson, 1984; Schiefelbein, 1985; Babu and Bain, 1991). In 1985, Syn-Gas, Inc., scaled this process to a 22-t/day plant to develop the concept for the commercial production of methanol. Feedstocks included wood chips, urban wood waste, and densified RDF. Tests in the 22-t/day plant at 870 to 930°C with cedar wood feedstock and oxygen gave 87 to 91% carbon conversions and dry gas analyses of 39 to 45 mol % carbon monoxide, 24 to 30 mol % carbon dioxide, 5 to 6 mol % methane, and 21 to 22 mol % hydrogen; the remainder was C_2-C_3 hydrocarbons. The product gas had a lower heating value (wet) of 8.3 to 9.8 MJ/m³ (n).

Directly Heated, High-Temperature, Steam-Oxygen Fluid-Bed Gasification

The Rheinbraun High-Temperature Winkler process is an outgrowth of the successful operation of two atmospheric Winkler gasifiers operated on lignite feedstocks in Germany from 1956 to 1964 with a combined capacity of 34,000 m³/h of synthesis gas, and subsequent operation of a 1.3-t/h pilot plant beginning in 1978 (Schrader et al., 1984). The process was developed by Rheinische Braunkohlenwerke AG and consists of gasification in a pressurized fluid-bed system supplied with oxygen and steam. Operating pressures and temperatures range up to 1013 kPa and 1100°C. The operating results with lignite at 1013 kPa and 1000°C, and oxygen and steam at 0.36 m³ (n)/kg and 0.41 kg/kg of dry lignite, gave 96% carbon conversion and a combined hydrogen-carbon monoxide yield of 1406 m³ (n)/t. At this steam-to-lignite ratio and an exit gas temperature of 900°C, the raw gas contained about 2 mol % methane. These tests provided the information and data needed to construct a demonstration plant to produce 300 million m³ (n)/year of synthesis gas for methanol synthesis at Rheinbraun's facility. Feedstock tests were conducted for customers worldwide with wood, peat, lignite, and coal feedstocks. Rheinbraun reported that each of these feedstocks is suitable for gasification by their process. Wood, especially, can be converted at high reactor through-put rates.

Directly Heated, Single-Stage, Pressurized, Steam-Oxygen Fluid-Bed Gasification

The RENUGAS process was developed by the Institute of Gas Technology (Evans *et al.*, 1987; Trenka *et al.*, 1991; Trenka, 1996). After tests in a 9.1-t/day PDU, a demonstration plant for 91 t/day of wood or 63 t/d of bagasse feedstock was constructed by the Pacific International Center for High Technology Research in Hawaii at the Hawaiian Commerce and Sugar Company. Bagasse, whole-tree chips, and possibly RDF are being tested in this plant. The gasifier has an inside diameter of about 1.2 m and is fed by a lockhopper and a live-bottom feed hopper. The development work was done in a 0.3-m inside diameter, 9.1-t/day PDU, so the scale-up factor is less than 10. For 92 to 96% carbon conversion, the oxygen requirement ranges from 0.24 to 0.34 kg/kg of wood feed, the dry fuel gas yield ranges from 1 to 1.2 m^3/kg of wood feed, and the heating value of the gas is about 11.8 to 13.5 MJ/m^3 (n). A typical run with whole-tree chips consisted of a feed rate of 321 kg/h with wood containing 9 wt % moisture, 0.69 kg of steam/kg of wood, 0.26 kg of oxygen as air/kg of wood, and gasification at 910°C and 2189 kPa. The heating value of the raw gas on a dry, nitrogen-free basis was 13.6 MJ/m^3 (n) and contained 16 mol % carbon monoxide, 38 mol % carbon dioxide, 17 mol % methane, 1 mol % higher hydrocarbons, and 28 mol % hydrogen. The yield of this gas was 1.04 m^3 (n)/kg of wood (wet). This plant is being operated at pressures up to 2027 kPa over a range of steam/oxygen ratios. The objectives are to demonstrate medium-energy gas production for power generation, hot-gas cleanup, and synthesis gas production. A special system for gas cleanup is being tested in both the 9.1-t/day PDU and the 91-t/day plant (Wiant *et al.*, 1993).

Directly Heated, Pressurized, Steam Gasification Process

This process was developed in the 1970s and early 1980s to the pilot plant stage (0.6 m diameter by 12.2 m, rotating, inclined kiln) by Wright-Malta Corporation (Hooverman and Coffman, 1976; Coffman and Speicher, 1993). Since then, the process has been improved by using a stationary kiln having an internal rotor with vanes. Much of the development work was performed with a stationary kiln that is 0.3 m inside diameter by 3.7 m long. The process is reported to convert as-harvested green wood or any other wet biomass into medium-energy gas of heating value 15.7 to 19.6 MJ/m^3 (n) in the self-pressurized kiln at pressures of about 2027 kPa at 590°C and residence times of about 1 hour. Steam is generated from the moisture in the feedstock and is normally not supplied to the kiln. No air or oxygen is used, and recycled wood ash serves as catalyst. As the biomass moves through the kiln from the cool feed end, it is gradually heated and partially dried, yielding steam. It then undergoes pyrolysis, yielding gas, liquids, tars, and char, all of which move cocurrently down the kiln where they undergo steam gasification and reforming to yield more gas. The inorganic residue is discharged at the hot end, and the hot gas is removed at the cold end after passage through heat-transfer coils in the kiln. The wood decomposition exotherm is reported to be sufficient to sustain the process after initial heat-up by an auxiliary boiler. Work in a small kiln showed that at pressures of 1378 to 2736 kPa and temperatures of 590 to 620°C with sodium carbonate catalyst, any type of green biomass can be gasified to 95 to 98% completion as long as it contains sufficient moisture. Dry gas compositions were about 5 to 10 mol % carbon monoxide, 40 to 50 mol % carbon dioxide, 15 to 22 mol % methane, and 20 to 28 mol % hydrogen. It was estimated that 907 t/day of green biomass at 11.6 MJ/kg would provide an output of 329 t/day of methanol.

Indirectly Heated Steam Gasification

This process, originally called the Pearson-BrightStar Process, was developed by BrightStar Technology, Inc. It consists of the conversion of biomass feedstocks, particularly sawdust and wood chips, by steam gasification in indirectly heated, tubular reactors to afford synthesis gas suitable for methanol production (Smith, Stokes, and Wilkes, 1993) or a medium-energy gas suitable for use in gas turbines (Menville, 1996). A 0.9- to 4.5-t/day pilot plant was operated in Mississippi with sawdust and wood chip feedstocks, but sewage sludge, other biomass feedstocks, and lignites have been tested. The process gasifies partially dried wood at 10 to 15% moisture content with injected steam at a steam-to-carbon ratio of about 2 at low pressure and high temperature to maximize synthesis gas and minimize methane formation. The process is believed by BrightStar to be the first of its kind to utilize externally heated tubular reactors through which the feedstock and steam are passed. Brightstar Synfuels Company, a joint venture of BrightStar and Syn-Fuels Corp., completed construction of a commercial demonstration module in 1996 in Louisiana. This plant requires about 22 dry t/day of wood residue feedstock and has a net energy output of 13.2 GJ/h exclusive of the energy required for reformer firing.

Indirectly Heated, Dual Fluid-Bed, Steam Gasification

A dual fluid-bed process for biomass was developed in the United States in the 1970s and early 1980s by the EEE Corporation (Bailie, 1980, 1981). It was commercialized in Japan by EBARA Corporation. This process, called the Bailie process after its inventor, consists of two circulating fluid-bed reactors that permit the use of air instead of oxygen for conversion of biomass to medium-energy gas. In one bed, feedstock combustion occurs with air to heat the sand bed. The hot sand is circulated to the other reactor where steam gasification of fresh feed and recycled char occurs. The cooled sand is recirculated to the combustion reactor for reheating. This configuration produces a product gas with a heating value of 11.8 MJ/m³ (n) or more. The composition of the gas from the gasifier operated at 650 to 750°C in one of the pilot plants fed with RDF in Japan was 34.7 mol % carbon monoxide, 11.2 mol % carbon dioxide, 12.7 mol % methane, 8.0 mol % other hydrocarbons, 30.0 mol % hydrogen, 2.5 mol % nitrogen, and 0.9 mol % oxygen. The heating value was 17.6 MJ/m³ (n). The pilot plant data indicated that 60% of the carbon resided in the medium-energy gas, 30% was converted to char, and the remaining 10% formed liquid and char. The energy yield as medium-energy gas was between 50 and 60%. The plants operated with RDF feedstocks in Japan were a 36-t/day pilot plant, a 91-t/d demonstration plant, and a 408-t/day commercial plant. These plants have been shut down.

Indirectly Heated, Dual Fluid-Bed, Steam Gasification

This process was developed by Battelle in the 1980s in a dual-bed PDU having a capacity of 20 to 25 t/day (Paisley, Feldmann, and Appelbaum, 1984; Paisley, Litt, and Creamer, 1991). Heat is supplied by recirculating a stream of hot sand between the separate combustion vessel and the gasifier. The PDU used a conventional fluid-bed combustor. In a commercial plant, both the gasifier and the combustor would be operated in the entrained mode to achieve higher throughputs. Tests have been conducted with wood and RDF. The operating ranges of the gasifier in the PDU were 630 to 1015°C at near-atmospheric pressure. The largest gasifier used was 0.25 m inside diameter and had a maximum wood throughput of 1.7 t/h. The heating value of the product gas was 17.7 to 19.6 MJ/m³ (n) and was reported to be independent of the moisture level of the feed. A thermally balanced operation with wood feedstock was achieved at throughputs of 1.5 t/h. Combustor carbon utilization was complete at temperatures above 980°C, and gasifier carbon conversion to gas was 50 to 80% at temperatures above 705°C. Typical nitrogen-free gas compositions were 50.4 mol % carbon monoxide, 9.4 mol % carbon dioxide, 15.5 mol % methane, 7.2 mol % ethane and ethylene, and 17.5 mol % hydrogen. Carbon conversions with RDF were similar to those of wood over a temperature range of 650 to 870°C. The heating values of the product gases were about 21.6 to 23.6 MJ/m³ (n). A commercial plant based on this process has been built to supply fuel gas to a central station power plant in Vermont (Paisley and Farris, 1995; Farris and Weeks, 1996).

Indirectly Heated, Pulse-Enhanced, Fluid-Bed, Steam Gasification

This process was developed by Manufacturing and Technology Conversion International, Inc. (Durai-Swamy, Colamino, and Mansour, 1989; Durai-Swamy et al., 1990). Biomass is reacted with steam in an indirectly heated fluid-bed gasifier at a temperature of 590 to 730°C. This process uses pulse-enhanced, gas-fired, Helmholtz pulse combustors consisting of compact, multiple resonance tubes which serve as the in-bed heat transfer surface. The pulsed heater generates an oscillating flow in the heat transfer tubes that results in turbulent mixing and enhanced heat transfer. Higher heat transfer coefficients than those available in conventional fire-tube configurations were estimated for this process. A medium-energy gas is produced at steam-to-biomass ratios of about 1.0. Based on carbon, the dry gas, char, and tar and oil yields were typically 90%, 4 to 8%, and 1 to 3%, respectively. Dry gas compositions from a wide variety of biomass (wood chips at 20 wt % moisture, pistachio shells and rice hulls at 9 wt % moisture, and recycled waste paper with plastic) ranged from 19 to 24 mol % carbon monoxide, 20 to 28 mol % carbon dioxide, 8 to 12 mol % methane, and 35 to 50 mol % hydrogen. The C_2-C_5 hydrocarbons ranged from a low of about 0.5 mol % to a high of about 6 mol % depending on the feedstock. The higher heating values of the product gas ranged from 12.9 to 15.9 MJ/m^3 (n). This work was conducted in a reactor shell 2.9 m in height; the overall height was 4.6 m including the plenums. The biomass feed rates were about 9 to 13.6 kg/h. Pilot tests in different scales of reactors from 0.2 to 68 t/day with different feedstocks have been carried out. A 15-t/day demonstration unit has been constructed and operated on waste cardboard feedstocks in California, and after relocation to Maryland, the plant was operated on wood chips, straw, and coal (Mansour, Durai-Swamy, and Voelker, 1995). A 109-t/day plant for processing black liquor has been built in North Carolina, and a similar plant has been built in India for processing spent distillery waste. Several cogeneration plants ranging in size from 5 to 50 MW are envisaged for more than 500 sugar mills in India.

C. Apparent Advantages of Steam Gasification

Except during startup, wood pyrolysis is reported to have been carried out commercially in the 1920s and 1930s without an external heat source. For example, the Ford Motor Company's continuous wood pyrolysis plant was operated on hogged hardwood dried to 0.5% moisture content and an external heat source was not needed (Chapter 8). Presuming oxygen is excluded in such processes and that exothermic partial oxidation is not a factor, several exothermic reactions can contribute to the self-sustained pyrolysis of wood— the conversion of carbon monoxide and carbon dioxide to methane or methanol, char formation, and the water gas shift (Table 8.2). Methanation has one of the highest exotherms per unit of carbon converted. These reactions or modifications and combinations of them seem to have occurred in the self-sustained process at a sufficient rate to make the overall process self-sustaining under the operating conditions used by Ford Motor Company.

When applied to biomass feedstocks, few steam gasification systems in which oxygen and air are excluded have been described or operated as autothermal processes since this early work. Wright-Malta Corporation's directly heated, pressurized steam gasification process for the production of medium-energy gas described earlier is one of these (Hooverman and Coffman, 1976; Coffman and Speicher, 1993). An external heat source is needed only during startup, and water is added as a cofeedstock if the biomass feedstock contains insufficient moisture (i.e., less than about 50 wt %). The process was described as follows (Coffman, 1981):

> As the biomass moves through the kiln from the cool feed end, it is gradually heated and first partially dries, yielding steam; then pyrolyzes, yielding gas, liquids, tars, and char. These move co-currently down the kiln. The liquids and tars steam reform, yielding more gas; the char steam-gasifies, yielding still more gas. The hot gas moves back through coils in the auger and kiln wall, giving its heat to the process, and being discharged at the cool end. This regenerative heat and wood decomposition exotherm are sufficient to sustain the process after initial heat-up by an auxiliary boiler. Over-all energy efficiency, raw biomass to clean, dry product gas is estimated to be 88–90%.

As shown in Table 9.1, most of the steam gasification reactions listed are endothermic, but as noted in the discussion of the Wright-Malta process, substantial amounts of carbon dioxide and methane are formed. Many of the gasification reactions that yield these products are exothermic. Char formation and the water gas shift are also exothermic (Table 8.2). Estimated equilibrium gas compositions from the steam gasification of green biomass at different pressures and temperatures shown in Table 9.11 indicate that at the temperature and pressure ranges of the Wright-Malta process, about 2 MPa and 900 K, substantial quantities of carbon dioxide and methane are formed. Calculations show that the process can be exothermic under these conditions. The heat of the exotherm and the sensible heat of the exiting gases, which are passed through tubular heat exchangers in the kiln, and the enthalpy of methanation, which occurs in the kiln and the heat exchangers, apparently drive the process. The total heat released is apparently large enough under Wright-Malta's operating conditions to sustain steam gasification.

TABLE 9.11   Estimated Equilibrium Product Gas Compositions as Function of Pressure and Temperature for the Steam Gasification of Biomass Containing 50.0 wt % Moisture[a]

Pressure (MPa)	Temperature (K)	Gas composition				
		H_2 (mol %)	CO (mol %)	CO_2 (mol %)	CH_4 (mol %)	H_2O (mol %)
0.1013	900	32.5	21.5	25	4	17
	1000	37	45	10.5	1.5	6.5
	1100	38	57	3	0.5	2
	1200	38	60	1	1	1
	1400	38	62	nil	nil	nil
1.0133	900	17	8	33	9	32
	1000	25	22	25	6	21
	1100	31.5	40	13	4	11
	1200	35	53	5	2	5
	1400	38	61	nil	nil	nil
2.0265	900	13	6	34	11	35
	1000	20.5	16.5	28	8	26
	1100	27.5	33	18	5.5	16
	1200	32.5	48	8	3	8
	1400	36.5	60	1	1	1
3.0398	900	11	5	35	12.5	3
	1000	18	14	30	9.5	2
	1100	25	29	20	7	18
	1200	30	45	10.5	4	10
	1400	35	59	2	2	2

[a]Composition of dry biomass assumed to be 44.44 wt % C, 6.22 wt % H, and 49.34 wt % O. Sums of equilibrium gases may not equal 100 because of rounding.

It should be emphasized that many investigators who have specialized in biomass gasification have questioned the validity of the steam gasification of biomass without the application of external heat because only a few autothermal systems have been reported to be operable. It is important to develop additional data to establish whether such systems can be self-sustaining over long periods. If they are, adiabatic, autothermal steam gasification would have several advantages for both medium-energy gas production and synthesis gas production. These include acceptability of a wide range of green biomass feedstocks without pretreatment; lower process energy consumption; direct internal heating of the reactants and therefore more efficient energy utilization; elimination of the need for feedstock dryers, an oxygen plant, and more complex indirectly heated gasifiers and indirectly heated, dual, circulating, fluid-bed gasifiers; and lower overall operating costs because of process simplicity. Another advantage would involve environmental benefits; steam gasification is reported to avoid formation of dioxins and to convert any chlorinated compounds that may be present to salts and clean gas (Mansour, Durai-Swamy, and Voelker, 1995). The disadvantage may be the relatively long solids residence time in the gasifier compared to some of the other processes. This can increase the plant's capital cost for a given throughput rate.

D. PROCESS COSTS

Many economic analyses of biomass gasification for low- and medium-energy gas, synthesis gas, and methanol production have been performed after biomass gasification developments started to increase in the 1970s. The basic approach to many of these analyses is illustrated here by focusing on the manufacture of methanol. For stand-alone methanol plants using biomass feedstocks, the sequence of operations has generally consisted of gasification to a low- or medium-energy gas, steam reforming to essentially all hydrogen and carbon oxides, water gas shift to produce a gas with a molar ratio of hydrogen : carbon monoxide of 2 : 1, acid gas scrubbing to remove carbon dioxide, and methanol synthesis. The gas compositions that would ideally be obtained from each step, using Bailie's indirectly heated steam-gasification process as the source of the synthesis gas, are shown in Table 9.12. Analysis of the cost of synthesis gas production alone, which was reported in the early 1980s for this process (Bailie, 1980, 1981), resulted in a projected capital cost of $22,050/dry t ($20,000/dry ton) of biomass feedstock capacity per day, and a synthesis gas cost of $3.04 to $3.39/GJ ($3.21 to $3.57/MBtu) at a feedstock cost of $31.58/dry t ($28.64 dry ton), or $1.70/GJ ($1.79/MBtu). At that time, the posted prices of natural gas and methanol were $3.16/GJ ($3.00/MBtu) and $10.54/GJ ($10.00/MBtu), or $0.145/L ($0.56/gal). Average capital costs for the steam gasification of biomass in mid-1990 nominal dollars range from about $55,000 to $88,000/dry t ($50,000 to $80,000/dry ton)

TABLE 9.12   Idealized Gas Compositions from Bailie's Indirectly Heated, Steam Gasification Process Applied to Methanol Synthesis from Biomass[a]

Gas	Synthesis gas (mol %)	Reformer gas (mol %)	Water gas (mol %)	Scrubbed gas (mol %)	Methanol synthesis (mol %)
H_2	21.14	50.44	54.60	66.67	
CO	38.88	38.97	27.30	33.33	
CO_2	18.15	10.59	18.10	0.00	
CH_4	15.57	0.00	0.00	0.00	
C_2H_6	6.16	0.00	0.00	0.00	
CH_3OH					100.00

[a]Bailie (1980, 1981)

of biomass feedstock capacity per day depending upon plant size and other factors, so the impact of time on equipment costs is evident. The feedstock and operating costs are also higher.

A plethora of economic projections has appeared in North America on the production of synthesis gas and methanol from biomass since this early work. Governmental regulations regarding motor fuel compositions and the use of oxygenates are undoubtedly responsible in part for this renewed interest. The details of a comparative economic analysis that compared the capital, operating, and methanol production costs of Wright-Malta's steam gasification process, Battelle's steam-air gasification process, IGT's steam-oxygen RENUGAS process, and Shell Oil's coal gasification process as applied to the steam-oxygen gasification of biomass ares summarized in Tables 9.13 and 9.14 (Larson and Katofsky, 1992).

TABLE 9.13   Gasification Processes Used for Economic Analysis of Methanol Production[a]

Process developer:	Wright-Malta	Battelle	IGT	Shell Oil
Gasifier type:	Rotor kiln	Circulating fluid bed	Bubbling fluid bed	Entrained
Gasification process type:	Steam	Steam	Steam–oxygen	Steam–oxygen
Feedstocks				
Type:	Wood	Wood	Wood	Wood
HHV, MJ/dry kg	20.93	20.12	19.12	19.12
Moisture, wt %	45	10	15	11
Feed rate, dry t/d	1650	1650	1650	1650
Steam, t/t dry feed	0	0.314	0.3	0.03
Oxygen, t/t dry feed	0	0	0.3	0.50
Air, t/t dry feed	0	1.46	0	0
Gasification				
Pressure, MPa	1.5	0.101	3.45	2.43
Exit temperature, °C	600	927	982	1045
Exit gas (dry)				
H_2, mol %	20.7	21.1	30.7	33.9
CO, mol %	6.9	46.8	22.2	50.7
CO_2, mol %	37.9	11.3	35.2	14.9
CH_4, mol %	34.5	14.9	12.0	0.2
C_2+, mol %	$(2-3)^b$	6.1	0.4	
Yield, kmol/t dry feed	82.1	58.3	74.7	73.1
Molecular weight, kg/kmol	21.82	21.15	22.25	21.00
Cold gas efficiency, %[c]	79.4	80.7	72.3	80.9

[a]Larson and Katofsky (1992).
[b]This process produces about 2 to 3 mol % C_2+, but is not included in the analysis.
[c]HHV of product gas/Sum of HHVs gasifier (and combustor in Battelle case).

TABLE 9.14   Estimated Capital, Operating, and Production Costs of Methanol from Biomass[a]

Process developer:	Wright-Malta	Battelle	IGT	Shell Oil
Wood feedrate, dry t/day	1650	1650	1650	1650
GJ/h	1439	1383	1315	1315
Methanol production, t/day	1004	705.1	965.9	915.9
GJ/h	949.2	667.0	913.9	866.3
10^3 L/day	1269	891	1220	1157
Capital cost, 10^6				
Installed hardware				
Feed preparation	7.4	18.6	16.4	34.6
Gasifiers	64.0	7.23	29.0	29.0
Oxygen plant	0	0	41.7	59.6
Reformer feed compressor	0	11.0	0	0
Reformer	16.7	15.5	16.7	0
Shift reactors	9.40	9.40	9.40	9.40
Union Carbide Selexol treatment	13.7	14.5	19.4	27.4
Methanol synthesis-purification	48.5	45.1	38.0	43.7
Utilities/auxiliaries	49.9	30.3	42.6	50.9
Subtotal:	200	152	213	254
Contingencies plus:	77	52	71	84
Total working requirement	277	204	284	338
Working capital	20.0	15.2	21.3	25.4
Land	2.30	2.30	2.30	2.30
Operating costs, 10^6/yr				
Variable costs				
Biomass feedstock	22.7	21.8	20.7	20.7
Catalysts & chemicals	1.92	2.88	1.92	1.92
Purchased energy	7.13	0.65	3.08	5.28
Subtotal:	31.7	25.3	25.7	27.9
Fixed costs				
Labor	0.99	1.18	0.99	0.99
Maintenance	5.99	4.55	6.39	7.63
General & direct overhead	4.99	4.25	5.25	6.05
Subtotal:	12.0	9.98	12.6	14.7
Total operating costs	43.7	35.3	38.4	42.6
Levelized costs, $/GJ				
Capital	5.64	4.61	8.33	8.01
Biomass	3.02	3.09	3.82	3.09
Labor & maintenance	1.85	1.82	2.68	2.47
Purchased energy	0.95	0.09	0.57	0.79
Product methanol	11.46	9.61	15.40	14.36
Product methanol, $/L	0.20	0.17	0.27	0.25

[a]Adapted from Larson and Katofsky (1992). All costs are in 1991 U.S. dollars. Capacity factor, 90%.

According to this analysis, the capital, operating, and methanol production costs from a plant supplied with 1,650 dry t/day of wood feedstock ranges from $204 to $338 million, $35.3 to $43.7 million/year, and $0.17 to $0.27/L, respectively. The feedstock cost was assumed to range from $38.19 to $41.88/dry t. At a 90% capacity factor, methanol production ranges from 293 to 417 million L/year depending on the process. Production is highest with the Wright-Malta process and lowest with the Battelle process because a substantial portion of the feedstock is used as fuel to the combustors for the latter process. A generally conservative approach was used for this economic assessment. All unit operations with the exception of biomass gasification were established, commercial technologies when the analysis was performed. The overall cost of methanol is more attractive for the two indirectly heated steam gasification processes (Wright-Malta and Battelle) compared to the methanol cost estimated for the directly heated gasification processes (IGT and Shell). The cost of the oxygen plant is a major contributor to product cost for the directly heated processes. Also, a few of the assumptions made by the analysts appear to disproportionately and adversely affect the cost of methanol from the Wright-Malta process, which when adjusted would provide still lower cost methanol. The utilities and purchased energy costs for this process seem to be excessive because only a small amount of purchased energy would be necessary, as already mentioned in the discussion of the reported autothermal nature of this process. In addition, the requirement for 17 gasification kilns operating in parallel to achieve a target plant capacity of 1650 t/day because of the kilns' low throughput capacity contributed significantly to product cost for the Wright-Malta process. Nevertheless, this type of comparative analysis illustrates the various facets of such economic assessments that should be examined and emphasizes where improvements might be made in the economics of each process.

E. ADVANCED POWER SYSTEMS

Modern, combined-cycle electric power generation systems using gas turbines as the primary generators offer higher thermal efficiencies than conventional steam-turbine systems. Many of the commercial plants in operation today use natural gas-fired, combined-cycle systems in which the hot exhaust from the gas turbines is processed in heat recovery steam generators to afford steam for injection into steam turbines for additional power generation and improved efficiency. Steam injection into the gas turbines along with combustion gases adds further efficiency improvements. Overall thermal efficiencies to electric power are up to twice those of conventional fuel-fired steam turbine systems. Availabilities can be high, the environmental characteristics are excellent, and capital costs are considerably less per unit of electric power capacity compared to the costs of conventional coal-fired plants. One of the largest combined cycle, natural gas-fired plants in the world—a 2,000-MW central station plant in Japan—operates at 95% availabilities (adjusted for mandated inspections).

Integration of coal gasification processes with combined-cycle technologies has opened the way for coal to fuel similar power generation plants at high efficiencies. Integrated gasification-combined cycle (IGCC) systems are in operation throughout the world and have made it possible to resurrect the use of low-cost, high-sulfur fossil fuels for power generation because the gasification process is, in effect, a desulfurization process. Oxygen-blown gasification plants have dominated both commercial and demonstration coal gasification units since as far back as the 1920s. Seventeen commercial plants, having a total of 153 coal gasifiers, are reported to be in commercial operation worldwide (cf. Simbeck and Karp in Swanekamp, 1996). Oxygen is used rather than air in these plants because they produce synthesis gas-based chemicals and premium fuels. In the United States, modern air- and oxygen-blown, fluid-bed gasification processes equipped with hot-gas cleanup systems are being perfected for use with coal feedstocks in IGCC plants. These plants are expected to have good emissions characteristics with one exception—carbon dioxide emissions per unit of fuel will be about the same as those of conventional fossil-fueled power plants. Biomass fuels, because of their relatively short recycling time, would avoid this problem.

It is apparent from the discussion of biomass gasification in this chapter that innovative processes for producing low- and medium-energy fuel gases have been developed for virgin and waste biomass feedstocks and are either about to be or have already been commercialized. These technologies are much improved over conventional, air-blown gasification processes. The availability of suitable fuel gases from modern biomass gasification processes facilitates their coupling with combined cycle power plants in the same manner as fossil-fueled IGCC plants. Biomass-fueled IGCC plants (BIGCC), particularly those having smaller capacities and those used for combined waste disposal and energy recovery, are expected to contribute to the expected 600 GW of new electric generating capacity needed worldwide over the next several years. IGCC plants fueled with both coal and biomass as sequential or combined feedstocks would appear to be a viable alternative because, as already pointed out, some gasification processes are capable of converting both feedstocks. The heat load for conventional Rankine steam-cycle power production using boilers and steam turbines is about 14.8 to 16.9 MJ/kWh; BIGCC technology should have about 25% less heat load and therefore considerably improved economics. The economics of BIGCC systems even as small as 1 to 10 MW in capacity can be very site-specific, but appear to be capable of reasonable rates of return (Craig and Purvis, 1995). Larger biomass integrated-gasification/ steam-injected gas-turbine (BIG/STIG) cogeneration plants are projected to be attractive investments for sugar producers, for example, who can use sugarcane bagasse as fuel (Larson *et al.*, 1991).

A 30-MW power plant fueled with eucalyptus wood from short-rotation energy plantations is planned in Brazil to demonstrate BIGCC technology (Carpentieri, 1993; de Queiroz and do Nascimento, 1993). This plant is projected to operate at an availability of 80% and an overall thermal efficiency of 43% to produce 210 GWh/year from 205,835 m^3/year of wood chips containing 35 wt % moisture. The energy cost is estimated to be $0.45 to $0.65/kWh. The first plant is estimated to have a capital cost of $60 million to $75 million (U.S.); subsequent plants are estimated to cost $39 million to $45 million (U.S.). For sugarcane bagasse, which will be tested as a potential feedstock, the heat rate is estimated to be 8.368 GJ/MWh with fuel consumption at 50 wt % moisture content of 1.021 kg/kWh. It is estimated that the cost of electric power production in 53-MW BIG/STIG plants in Brazil using briquetted sugarcane bagasse is $0.032 to $0.058/kWh (Larson *et al.*, 1991).

Other advanced technologies that are receiving considerable attention include improved designs for combining small biomass gasifiers with motor-generator sets or gas turbines in the multiple-kilowatt range and in the 1 to 5 MW range. Numerous configurations are being developed, although some assessments have ruled out conventional steam turbines because of their relatively low efficiency and high cost at small sizes. Examples of small systems under development include a 1-MW system consisting of a fixed-bed, downdraft gasifier, a gas cleaning system, and a spark-ignited gas engine-generator set; and a 1-MW system consisting of a pressurized fluid-bed gasifier, a hot-gas cleanup system, and a gas turbine (Purvis *et al.*, 1996).

Another advanced technology that can use biomass gasification for power generation employs fuel cells. Fuel cells are devices that electrochemically convert the chemical energy contained in the fuel into direct current electricity and the oxidation products of the fuel. The fuels can be natural gas, and the products can be gases from the gasification of solid fuels, including biomass and derived fuels such as hydrogen and intermediate liquid fuels such as hydrocarbons and ethanol. In one sense, fuel cells are similar to electric batteries, but the fuel and oxidant are continuously supplied from external sources. So, unlike batteries, fuel cells are not consumed or depleted in the process. Also, because fuel cells are not heat engines, they are not Carnot limited and can achieve high fuel-energy-to-electric power conversion efficiencies that can be above 60% based on the energy content of the fuel supplied to the fuel cell. Among the fuel cell configurations, three different types are being developed for power generation by units 100 kW to 25 MW in capacity. They are differentiated by the electrolytes used within the cell—phosphoric acid, molten carbonate, and solid oxide. Some designs such as those that use molten carbonate and solid oxide electrolytes are operated at sufficiently elevated temperatures to be suitable for use in cogeneration applications. A few of these designs are believed to be operable at overall efficiencies as high as 85% based on the energy content of the fuel supplied to the fuel cell. A few small-scale power units using biomass fuels for specialty applications may become available in the next few years, but large-scale fuel-cell power plants are not expected to be available for generating central station power until well into the twenty-first century.

REFERENCES

Alpert, S. B., Ferguson, F. A., Scheeline, H., Marynowski, C., and Louks, B. (1972). "Pyrolysis of Solid Wastes: A Technical and Economic Assessment," NTIS PB 218231/9. Stanford Research Institute, Menlo Park, CA, September.

Antal, M. J., Jr. (1978). In "Energy fromn Biomass and Wastes," (D. L. Klass and W. W. Waterman, eds.), p. 495. Institute of Gas Technology, Chicago.

Babu, S. P., and Bain, R. L. (1991). "A Review of Biomass Gasification in the United States, Solar Energy Research Institute/Syn-Gas, Inc., Program, Task VII, Biomass Conversion, Activity 4, Thermal Gasification, Participating Countries Report, Compiled by Institute of Gas Technology." Institute of Gas Technology, Chicago, December.

Babu, S. P., Tran, D. Q., and Singh, S. P. (1980). In "Energy from Biomass and Wastes IV," (D. L. Klass and J. J. Weatherly, III, eds.), p. 369. Institute of Gas Technology, Chicago.

Babu, S. P., and Whaley, T. P. (1992). Biomass and Bioenergy 2 1, 299.

Bailie, R. C. (1980). In "Energy from Biomass and Wastes IV," (D. L. Klass and J. W. Weatherly, eds.), p. 423. Instituite of Gas Technology, Chicago.

Bailie, R. C. (1981). In "Energy from Biomass and Wastes V," (D. L. Klass and J. W. Weatherly, III, eds.), p. 549. Institute of Gas Technology, Chicago.

Bailey, R. W., and Bailey, R. W., Jr. (1996). In "Bioenergy '96," Vol. I, p. 284. Southeastern Regional Biomass Energy Program, Tennessee Valley Authority, Muscle Shoals, AL.

Barducci, G. L., Daddi, P., Polzinetti, G. C., and Ulivieri, P. (1995). In "Second Biomass Conference of the Americas: Energy, Environment, Agriculture, and Industry," p. 565, NREL/CP-200-8098, DE95009230. National Renewable Energy Laboratory, Golden, CO.

Barducci, G. L., Ulivieri, P., Polzinetti, G. C., Donati, A., and Repetto, F. (1996). In "Bioenergy '96," Vol. I, p. 296. Southeastern Regional Biomass Energy Program, Tennessee Valley Authority, Muscle Shoals, AL.

Bircher, K. G. (1982). In "Energy from Biomass and Wastes V," (D. L. Klass, ed.), p. 707. Institute of Gas Technology, Chicago.

Bodle, W. W., and Schora, F. C. (1979). In "Advances in Coal Utilization Technology," (K. S. Vorres and W. W. Waterman, eds.), p. 11. Institute of Gas Technology, Chicago.

Bravakis, L. T. (1996). "CHIPTEC Wood Energy Systems," South Burlington, VT.

Bulpitt, W. L., and Rittenhouse, O. C. (1989). In "Energy from Biomass and Wastes XII," (D. L. Klass, ed.), p. 753. Institute of Gas Technology, Chicago.

Carpentieri, E. (1993). In "First Biomass Conference of the Americas: Energy, Environment, Agriculture, and Industry," p. 393, NREL/CP-200-5768, DE93010050. National Renewable Energy Laboratory, Golden, CO.

Coffman, J. A. (1981). In "Proceedings of the 13th Biomass Thermochemical Conversion Contractor's Meeting," p. 274, CONF-8110115, PNL-SA-10093. Pacific Northwest Laboratory, Richland, WA.

Coffman, J. A., and Speicher, W. A. (1993). "Draft Final Research & Development Report, Steam Gasification I and Ia," Agreement No. 1039-ERER-ER-88. New York State Energy Research and Development Authority, Albany, NY, December 31, 1993.

Craig, J. D., and Purvis, C. R. (1995). In "Second Biomass Conference of the Americas: Energy, Environment, Agriculture, and Industry," p. 637, NREL/CP-200-8098, DE95009230. National Renewable Energy Laboratory, Golden, CO.

Daubert, T. E., and Danner, R. P. (1989). "Physical and Thermodynamic Properties of Pure Chemicals." Chemisphere Publishing Corporation, NY.

Davidson, P. E., and Lucas, T. W., Jr. (1978). "The Andco-Torrax High Temperature Slagging Pyrolysis System." 175th National Meeting of the American Chemical Society, Anaheim, CA, March 15.

De Queiroz, L. C., and do Nascimento, M. J. M. (1993). In "First Biomass Conference of the Americas: Energy, Environment, Agriculture, and Industry," p. 461, NREL/CP-200-5768, DE93010050. National Renewable Energy Laboratory, Golden, CO.

Durai-Swamy, K., Colamino, J., and Mansour, M. N. (1989). In "Energy from Biomass and Wastes XII," (D.L. Klass, ed.), p. 833. Institute of Gas Technology, Chicago.

Dura--Swamy, K., Warren,, D. W., Aghamohammadi, B., and Mansour, M. N. (1990). In "Energy from Biomass and Wastes XIII," (D. L. Klass, ed.), p. 689. Institute of Gas Technology, Chicago.

Elliott, D. C., Neuenschwander, G. G., Baker, E. G., and Sealock, L. J., Jr. (1991). In "Energy from Biomass and Wastes XV," (D. L. Klass, ed.), p. 1013. Institute of Gas Technology, Chicago.

Elliott, D. C., Sealock, D. C., Phelps, M. R., Neuenschwander, G. G., and Hart, T. R. (1993). In "First Biomass Conference of the Americas: Energy, Environment, Agriculture, and Industry," p. 557, NREL/CP-200-5768, DE93010050. National Renewable Energy Laboratory, Golden, CO.

Environmental Research and Technology and Koppers Co., Inc. (1984). "Handbook on Manufactured Gas Plant Sites." Utility Solid Waste Activities Group, Superfund Committee, Washington, D.C.

Evans, R. J., Knight, R. A., Onischak, M., and Babu, S. P. (1987). In "Energy from Biomass and Wastes X," (D. L. Klass, ed.), p. 677. Institute of Gas Technology, Chicago.

Farris, S. G., and Weeks, S. T. (1996). In "Bioenergy '96," Vol. I, p. 44. Southeastern Regional Biomass Energy Program, Tennessee Valley Authority, Muscle Shoals, AL.

Feldmann, H. F., Choi, P. S., Conkle, H. N., and Chauhan, S. P. (1981). In "Biomass as a Nonfossil Fuel Source," (D. L. Klass, ed.), ACS Symposium Series 144, p. 351. American Chemical Society, Washington, D.C.

Fisher, T. F., Kasbohm, M. L., and Rivero, J. R. (1976). In "Clean Fuels from Biomass, Sewage, Urban Refuse, Agricultural Wastes," (F. Ekman, ed.), p. 447. Institute of Gas Technology, Chicago.

Graham, R. G., and Huffman, D. R. (1981). In "Energy from Biomass and Wastes V," (D. L. Klass, ed.), p. 633. Institute of Gas Technology, Chicago.

Hooverman, R. H., and Coffman, J. A. (1976). In "Clean Fuels from Biomass, Sewage, Urban Refuse, Agricultural Wastes," (F. Ekman, ed.), p. 213. Institute of Gas Technology, Chicago.

International Energy Agency (1991). "Task VII, Biomass Conversion, Activity 4, Thermal Gasification, Summary Report for Hot-Gas Cleanup, Compiled by Institute of Gas Technology." Institute of Gas Technology, Chicago, December.

International Energy Agency (1992). "Task VII, Biomass Conversion, Activity 4, Thermal Gasification, Research Needs for Thermal Gasification of Biomass, Compiled by Studsvik AB Thermal Processes and Institute of Gas Technology." Institute of Gas Technology, Chicago, March.

Jackson, J. F. (1982). In "Energy from Biomass and Wastes VI," (D. L. Klass, ed.), p. 721. Institute of Gas Technology, Chicago.

Klass, D. L. (1982). In "Energy from Biomass and Wastes VI," (D. L. Klass, ed.), p. 1. Institute of Gas Technology, Chicago.

Klass, D. L. (1985). Resources and Conservation 11, 157.

Klass, D. L., Kroenke, I. P., and Landahl, C. D. (1981). "Gasification of Biomass in Liquid Water," unpublished work. Institute of Gas Technology, Chicago, September.

Kurkela, E. (1991). "A Review of Current Activities on Biomass and Peat Gasification in Finland, Task VII, Biomass Conversion, Activity 4, Thermal Gasification, Participating Countries Report, Compiled by Institute of Gas Technology." Institute of Gas Technology, Chicago, December.

LaFontaine, H. (1988). In "Energy from Biomass and Wastes XI," (D. L. Klass, ed.), p. 561. Institute of Gas Technology, Chicago.

LaFontaine, H., and Reed, T. B. (1991). In "Energy from Biomass and Wastes XV," (D. L. Klass, ed.), p. 1023. Institute of Gas Technology, Chicago.

Larson, E. D., Williams, R. H., Ogden, J. M., and Hylton, M. G. (1991). In "Energy from Biomass and Wastes XIV," (D. L. Klass, ed.), p. 781. Institute of Gas Technology, Chicago.

Larson, E. D., and Katofsky, R. E. (1992). "Production of Methanol and Hydrogen from Biomass," PU/CEES Report N0. 271. Princeton University, Princeton, NJ, July.

Liebs, L. H. (1985). In "American Gas Association Operating Section Proceedings," p. 369. American Gas Association, Washington, D.C.

Makansi, J. (1987). *Power* **69**, July.

Mansour, M. N., Durai-Swami, K., and Voelker, G. (1995). In "Second Biomass Conference of the Americas: Energy, Environment, Agriculture, and Industry," p. 543, NREL/CP-200-8098, DE95009230. National Renewable Energy Laboratory, Golden, CO.

Mark, S. D., Jr. (1980). In "Energy from Biomass and Wastes IV," (D. L. Klass and J. W. Weatherly, III, eds.), p. 577. Institute of Gas Technology, Chicago.

Menville, R. L., Jr. (1996). In "Bioenergy '96," Vol. I, p. 289. Southeastern Regional Biomass Energy Program, Tennessee Valley Authority, Muscle Shoals, AL.

Miles, T. R., and Miles, T. R., Jr. (1989). *Biomass* **18**, 157.

Miller, B. (1987). In "Energy from Biomass and Wastes X," (D. L. Klass, ed.), p. 723. Institute of Gas Technology, Chicago.

Mitchell, D. H., Mudge, L. K., Robertus, R. J., Weber, S. L., and Sealock, L. J., Jr. (1980). "Methane/Methanol by Catalytic Gasification of Wood," *CEP*, 53 September.

Morris, M. (1996). In "Bioenergy '96," Vol. I, p. 275. Southeastern Regional Biomass Energy Program, Tennessee Valley Authority, Muscle Shoals, AL.

Mudge, L. K., Sealock, L. J., Jr., Robertus, R. J., Mitchell, D. H., and Weber, S. L. (1979). In "3rd Annual Biomass Energy Systems Conference Proceedings, The National Biomass Program," p. 351, SERI/TP-33-285. Solar Energy Research Institute (NREL), Golden, CO, October.

National Academy of Engineering (1973). "Evaluation of Coal Gasification Technology, Part II, Low- and Intermediate-Btu Fuel Gases," COPAC-7. National Academy of Engineering, Washington, D. C.; Bodle, W. W., and Vyas, K. C. (1975). In "Clean Fuels from Coal II," (F. Ekman, ed.), p. 11. Institute of Gas Technology, Chicago; Knudsen, C. W (1983). In "Handbook of Energy Technology and Economics," (R. A. Myers, ed.), p. 118. John Wiley, NY; Bathie, W. W. (1985). In "Handbook of Energy Systems Engineering, Production and Utilization," (L. C. Wilbur, ed.), p. 1103. John Wiley, NY; Mahagaokar, U., and Krewinghaus, A. B. (1992). In "Kirk-Othmer Encyclopedia of Chemical Technology," Vol. 6, p. 541, 4th. Ed. John Wiley, NY.

Ng, M. L. (1979). "Glucose Gasification in Liquid Water at Sub-Critical Temperature," M.S. Thesis in Gas Engineering, Illinois Institute of Technology, Chicago, May.

Paisley, M. A., and Farris, G. (1995). In "Second Biomass Conference of the Americas: Energy, Environment, Agriculture, and Industry," p. 553, NREL/CP-200-8098, DE95009230. National Renewable Energy Laboratory, Golden, CO.

Paisley, M. A., Feldmann, H. F., and Appelbaum, H. R. (1984). In "Energy from Biomass and Wastes VIII," (D. L. Klass and H. H. Elliott, eds.), p. 675. Institute of Gas Technology, Chicago.

Paisley, M. A., Litt, R. D., and Creamer, K. S. (1991). *In* "Energy from Biomass and Wastes XIV," (D. L. Klass, ed.), p. 737. Institute of Gas Technology, Chicago.

Parent, J. D., and Katz, S. (1948). "Equilibrium Compositions and Enthalpy Changes for the Reactions of Carbon, Oxygen, and Steam," Bulletin No. 2. Institute of Gas Technology, Chicago.

Patel, J. (1996). *In* "Small Scale Gasification of Biomass and MSW." Carbona Corporation, Atlanta, GA.

Purvis, C. R., Cleland, J., Craig, J. D., and Sanders, C. F. (1996). *In* "Bioenergy '96," Vol. II, p. 733. Southeastern Regional Biomass Energy Program, Tennessee Valley Authority, Muscle Shoals, AL.

Reed, T. B., Levie, B., and Markson, M. (1984). "The SERI Oxygen Gasifier-Phase II," SERI/PR/234-2571. Golden, CO, August.

Remediation Technologies, Inc. (1990). "Remediation Alternatives and Costs for the Restoration of MGP Sites," Final Report to Gas Research Institute. Gas Research Institute, Chicago, March.

Rensfelt, E. (1991). "A Review of Biomass Gasification in Sweden, Task VII, Biomass Conversion, Activity 4, Thermal Gasification, Participating Countries Report, Compiled by Institute of Gas Technology." Institute of Gas Technology, Chicago, December.

Rhodes, E. O. (1974). *In* "Bituminous Materials: Asphalts, Tars, and Pitches, Vol. III, Coal Tars and Pitches," (A. J. Hoiberg, ed.), p. 1. Robert F. Krieger Publishing Co., Huntington, NY.

Rock, K. L. (1982). *In* "Energy from Biomass and Wastes VI," (D. L. Klass, ed.), p. 737. Institute of Gas Technology, Chicago.

Russell, B. (1980). *In* "Energy from Biomass and Wastes IV," (D. L. Klass and J. W. Weatherly, III, eds.), p. 819. Institute of Gas Technology, Chicago.

Schiefelbein, G. F. (1985). "Biomass Gasification Research: Recent Results and Future Trends," PNL-SA-12841. Pacific Northwest Laboratory, Richland, WA, February.

Schrader, L., Nitschke, E., Will, H., and Bellin, A. (1984). *In* "Energy from Biomass and Wastes VIII," (D. L. Klass and H. H. Elliott, eds.), p. 747. Institute of Gas Technology, Chicago.

Smith III, A. J., Stokes, C. A., and Wilkes, P. (1993). "The Rebirth of Methanol from Biomass." 1993 World Methanol Conference, Atlanta, GA, November 29–December 1.

Stevens, D. J. (1994). "Review and Analysis of the 1980-1989 Biomass Thermochemical Conversion Program," NREL/TP-421-7501, DE95000287. National Renewable Energy Laboratory, Golden CO.

Stull, D. R., Westrum, E. F, Jr., and Sinke, G. C. (1987). "The Chemical Thermodynamics of Organic Compounds." Robert F. Krieger Publishing Co., Malabar, FL.

Srivastava, V. J. (1993). "Manufactured Gas Plant Sites: Characterization of Wastes and IGT's Innovative Remediation Alternatives." Paper presented at Hazardous and Environmentally Sensitive Waste Management in the Gas Industry, Albuquerque, NM, January.

Swanekamp, R. (1996). *Power* **140** (6), 59.

Swedish Academy of Engineering (1950). "Generator Gas, The Swedish Experience from 1939-1945," (translated from Swedish by the Solar Energy Research Institute in 1979, SP-33-140, Golden, CO, T. B. Reed and D. Jantzen, eds.). Generalstabens Litografiska Anstalts Forlag, Stockholm.

Trenka, A. R., Kinoshita, C. M., Takahashi, P. K., Phillips, V. D., Caldwell, C., Kwok, R., Onischak, M., and Babu, S. P. (1991). *In* "Energy from Biomass and Wastes XV," (D. L. Klass, ed.), p. 1051. Institute of Gas Technology, Chicago.

Trenka, A. R. (1996). *In* "Bioenergy '96," Vol. I, p. 37. Southeastern Regional Biomass Energy Program, Tennessee Valley Authority, Muscle Shoals, AL.

U.S. Environmental Protection Agency (1975). "Baltimore Demonstrates Gas Pyrolysis," First Interim Report SW-75d.i. U.S. Government Printing Office, Washington, D.C.

Villaume, J. F. (1984). "Hazardous and Toxic Wastes: Technology Management and Health Effects" (S. K. Majumdar and E. W. Miller, eds.). The Pennsylvania Academy of Sciences, Easton, PA.

Wiant, B. C., Carty, R. H., Horazak, D. A., and Ruel, R. H. (1993). *In* "First Biomass Conference of the Americas: Energy, Environment, Agriculture, and Industry," p. 571, NREL/CP-200-5768, DE93010050. National Renewable Energy Laboratory, Golden, CO.

Yosim, S. J., and Barclay, K. M. (1976). *In* "Preprints of Papers, 171st National Meeting, American Chemical Society, Division of Fuel Chemistry," Vol. 21, No. 1, p. 73, April 5-9. Washington, D.C.

CHAPTER 10

Natural Biochemical Liquefaction

I. INTRODUCTION

The conversion of solar radiation into chemical energy via photosynthesis results in the growth of woody, herbaceous, and aquatic biomass and the formation of many organic compounds *in situ*, each of which has an intrinsic energy content. The lower the oxygenated state of the fixed carbon in these compounds, the higher the energy content. As discussed in previous chapters, α-cellulose, or cellulose as it is more commonly known, is usually the chief structural element and principal constituent of many biomass species, particularly woody biomass, but is not always the dominant carbohydrate, especially in aquatic species. The lignins and hemicelluloses comprise most of the remaining organic components. In addition, other polymers and a large variety of nonpolymeric organic solids are formed naturally, although not equally, in biomass. Many of these chemicals are or have been used in specialty applications such as pharmaceuticals and industrial formulations. Natural products continue to be discovered, and many have been found to have useful applications. Hundreds of biomass species have also been found to produce low-molecular-weight organic liquids, several of which are used or proposed for use as transportation fuels for vehicles driven by spark- or compression-ignition engines. These liquids are *glycerides* and *terpenes*. The glycerides, which are the primary members of a group of organic compounds called *lipids*, are mainly triglyceride esters of long-chain fatty acids and the triol glycerol. Lipid is a general name for plant and animal products that are structurally esters of higher fatty acids, but certain other oil-soluble, water-insoluble substances are also called lipids. The fatty acids are any of a variety of monobasic acids such as palmitic, stearic, and oleic acids. Among more than 50 fatty acids found in nature, almost all are straight-chain acids containing an even number of carbon atoms. A few biomass species produce esters of fatty alcohols and acids. Certain glycerides are essential components of the human diet and are obtained or derived from animal fats and vegetable oils. In addition to cooking and food uses, many natural glycerides have long been used as lubricants and as raw materials for the manufacture of soaps, detergents, cosmetics, and chemicals. Some are directly useful as motor fuels as formed or can be converted to suitable fuels after relatively simple upgrading using established processes. The derivatives formed on transesterification (alcoholysis) of several natural glycerides with low-molecular-weight alcohols are useful as neat diesel fuels or in diesel fuel blends or as diesel fuel additives. An example is the ester formed on methanolysis of soybean oil.

The term *terpenes* originally designated a mixture of isomeric hydrocarbons of molecular formula $C_{10}H_{16}$ occurring in turpentine obtained from coniferous trees, especially pine trees. Today, the term refers to a large number of naturally occurring hydrocarbons that can be represented as isoprene adducts having the formula $(C_5H_8)_n$, where n is 2 or more, and to an even larger number of derived terpenoids in various states of oxidation and unsaturation. Terpenes are widely distributed in many biomass species and are often found in biomass oils, resins, and balsams. They are classified according to the number of isoprene units contained in the empirical formula: for example, $C_{10}H_{16}$, monoterpenes; $C_{15}H_{24}$, sesquiterpenes; $C_{20}H_{32}$, diterpenes; $C_{30}H_{48}$, triterpenes; and $(C_5H_8)_x$, polyisoprenes. The terpenes thus range from relatively simple hydrocarbons to large polymeric molecules. The lower molecular weight terpenes are usually liquid at room temperature at n = 2 or 3 and are mainly alicyclic structures. Terpenes are monocyclic, bicyclic, tricyclic, etc.; open-chain acyclic terpenes are also known. Examples of terpenes are the dienes limonene (monocyclic monoterpene) and cadinene (bicyclic sesquiterpene), the monoenes α- and β-pinene (bicyclic monoterpenes), the triene myrcene (acyclic monoterpene), the hexaene squalene (acyclic triterpene), and natural rubbers, which are high-molecular-weight polymers of isoprene.

In this chapter, the sources of natural biochemical liquids potentially suitable as motor fuels, their basic properties and conversion chemistry, and their process economics are examined.

II. SOURCES

A. GLYCERIDES

Biodiesel Fuels

Natural glycerides have been investigated as alternative fuels for the compression-ignition engine since the late 1800s. Rudolph Diesel, the inventor of the engine that bears his name, demonstrated a diesel-cycle engine fueled with peanut oil in 1900 at the Paris Exposition. In 1912, Diesel wrote: "The use of vegetable oils may become in the course of time as important as petroleum and the coal tar products of the present time." This obviously has not happened, since the majority of heavy-duty farm and construction vehicles and large trucks are powered by diesel engines that operate on petroleum diesel fuels. Of the two basic types of diesel engines, direct injection, in which the diesel fuel is injected directly into the combustion chamber, and indirect injection, in which the fuel is injected into a precombustion chamber, the direct injection design is more commonly used for the larger vehicles. The direct-injected engines are easier to start and less expensive than the more elaborate but quiet indirect-injected engine (Quick, 1989).

Many natural glycerides can be used with little or no difficulty as diesel fuels for vehicles equipped with indirect-injection engines. Several problems arise when they are used to fuel direct-injection engines. One of problems is caused by the higher viscosity and lower volatility of natural glycerides as compared with the corresponding properties of petroleum diesel fuels (*cf.* Krawczyk, 1996). A comparison of selected properties of No. 2 diesel fuel, soybean and rapeseed oils, their corresponding transesterification products with methanol and ethanol, and the fatty acid makeup of the glyceride oils and esters is shown in Table 10.1.

Performance of the glycerides as a diesel fuel is improved by conversion of the fatty acid moieties of the glycerides to the corresponding methyl or ethyl esters. Fouling problems are significantly reduced, and the viscosities, pour points, and combustion characteristics of the esters in blends with diesel fuel or as neat liquids are superior to those of the natural glycerides. The cetane numbers of the methyl and ethyl esters are about 50 to 65, and they have lower ash, sulfur, and volatilities and higher flash points than conventional diesel fuels. The cetane numbers of the methyl and ethyl esters of the pure fatty acids in the oils have been correlated with the chain length and degree of saturation (Freedman *et al.*, 1990; Clements, 1996; Knothe, Bagby, and Ryan, 1996). For the ethyl esters of the C_{18} fatty acids, stearic, oleic, linoleic, and linolenic acids, the reported cetane numbers are 77, 54, 37, and 27, respectively. The corresponding cetane numbers determined for the free acids

TABLE 10.1 Comparison of Some Typical Properties of Diesel, Soybean Oil, Rapeseed Oil, and Ester Fuels[a]

Property	No. 2 diesel	Soybean oil			Rapeseed oil		
		Oil	Methyl ester	Ethyl ester	Oil	Methyl ester	Ethyl ester
Specific gravity	0.8495	0.92	0.886	0.881	0.91	0.880	0.876
Viscosity at 40°C, mm²/s	2.98	33	3.891	4.493	51	5.65	6.17
Cloud point, °C	-12	-4	3	0		0	-2
Pour point, °C	-23	-12	-3	-3	-21	-15	-10
Flash point, °C	74		188	171		179	124
Boiling point, °C	191		339	357		347	273
Water & sediment, vol %	<0.005		<0.005	<0.005		<0.005	<0.005
Carbon residue, wt %	0.16		0.068	0.071		0.08	0.06
Ash, wt %	0.002		0	0		0.002	0.002
Sulfur, wt %	0.036	0.01	0.012	0.008	0.01	0.012	0.014
Cetane number	49	38	55	53	32	62	65
Copper corrosion	1A		1A	1A		1A	1A
Higher heating value, MJ/kg	45.42	39.3	39.77	39.96	40.17	40.54	40.51
MJ/L	38.58	36.2	35.24	35.20	36.60	35.68	38.00
Fatty acid composition, wt %							
Palmitic (16:0)		9.8	9.9	10.0	1	2.2	2.6
Stearic (18:0)		2.4	3.8	3.8		0.9	0.9
Oleic (18:1)		28.9	19.1	18.9	32	12.6	12.8
Linoleic (18:2)		50.7	55.6	55.7		12.1	11.9
Linolenic (18:3)		6.5	10.2	10.2	15	8	7.7
Eicosenoic (20:1)			0.2	0.2		7.4	7.3
Behenic (22:0)			0.3	0.3		0.7	0.7
Erucic (22:1)			0.0	0.0	50	49.8	49.5
Others		1.7	0.8	0.9	2	6.3	6.6

[a]Adapted from Cruz, Stanfill, and Powaukee (1996); Peterson et al. (1995); Reed (1993); Shay (1993); Stumborg et al. (1993); Auld, Peterson, and Korus (1989). The figures in parentheses after each fatty acid denote the number of carbon atoms and double bonds in the fatty acid. The figures for "Oil" are typical values and are analyses of samples that are not the actual precursors of the esters in this table.

are 62, 46, 31, and 20. These trends would be expected because of the increasing degree of unsaturation from stearic to linolenic acid and the fact that more paraffinic hydrocarbons usually have higher cetane numbers. Similar correlations were found for other esters.

Since the methanol and ethanol transesterification products of a variety of commercial biomass-derived oils, such as soybean, rapeseed, peanut, palm, and sunflower oils, and even waste animal fats have generally been found to be suitable as additives or neat fuels for diesel-powered engines without engine modifications, programs were started, first in several European countries and then the United States, to develop and market what has been termed "biodiesel." Biodiesel is defined as monoalkyl esters of long-chain fatty acids derived from renewable feedstocks, such as vegetable oils and animal fats, for use in compression ignition engines. The emphasis in European biodiesel programs has been on the methyl ester of rapeseed oil, and in the United States, the methyl esters of soybean and rapeseed oils have received the most attention (cf. Krawczyk, 1996). In Europe, the most advanced program is in Austria. Pilot plants, small-scale farm cooperatives, and industrial-scale plants (10,000 to 30,000 t/year capacities) have been built, and fuel standards for rapeseed oil methyl ester have been adopted. In Malaysia, the methyl ester of palm oil is being developed as biodiesel, and in Nicaragua, biodiesel applications of the oil from *Jatropha curcas*, a large shrub or tree native to the American tropics, are under investigation (cf. Jones and Miller, 1991).

Many projects are underway in the United States to complete the U.S. database on the properties and performance of biodiesel. Included in this work are studies on the emissions and performance of engines and vehicles fueled with biodiesel as additives, neat fuels, and fuel blends; the effects of oxygenated additives in biodiesel on cetane number; the effects of interesterification of different vegetable oils followed by transesterification with methanol on biodiesel properties; and the low-temperature flow properties, long-term storage effects, and toxicities of a variety of methyl and ethyl esters. The information being developed in this work continues to support the marketing of biodiesel. Some engine test results and fields trials with diesel vehicles indicate that a few issues must be addressed before biodiesel can be successfully commercialized. In order of importance and frequency of occurrence, these issues concern fuel quality, fuel filter plugging, injector failure, material compatibility, and fuel economy (Schumacher and Van Gerpen, 1996; Schumacher, Howell, and Weber, 1996). It appears that the fledgling biodiesel industry is approaching a stage similar to that of the petroleum industry many years ago, when detailed motor fuel specifications and test procedures were needed to facilitate fuel compatibility and interchangability from different suppliers across the country. The tentative specifications being developed for biodiesel by the American

TABLE 10.3   Annual Commercial Yields of Oilseeds and Seed Oils in the United States[a]

Common name	Species	Seed yield		Seed oil yield			
		Average (kg/ha)	Potential (kg/ha)	Average		Potential	
				(kg/ha)	(L/ha)	(kg/ha)	(L/ha)
Castorbean	Ricinus communis	950	3810	428	449	1504	1590
Chinese tallow tree	Sapium sebiferum	12,553		5548			6270
Cotton	Gossypium hirsutum	887	1910	142	150	343	370
Crambe	Crambe abssinica	1121	2350	392	421	824	940
Corn (high oil)	Zea mays		5940			596	650
Flax	Linum usitatissimum	795	1790	284	309	758	840
Peanut	Arachis hypogaea	2378	5160	754	814	1634	1780
Safflower	Carthamus tinctorius	1676	2470	553	599	888	940
Soybean	Glycine max	1980	3360	354	383	591	650
Sunflower	Helianthus annuus	1325	2470	530	571	986	1030
Winter rape	Brasica napus		2690			1074	1220

[a]Adapted from Lipinsky et al. (1984). Growth is under dry-land conditions except for cotton, which is irrigated. The yield for the Chinese tallow tree is one reported yield equivalent to 6270 L/ha of oil plus tallow and is not an average yield from several sources. It is believed that the yield would be substantially less than this in managed dense stands, but still higher than that of conventional oilseed crops.

The Chinese tallow tree, which has been cultivated and naturalized in the South Atlantic and the Gulf States, deserves special consideration because the yields of potential biodiesel feedstocks are much higher than those of conventional oilseed crops, and because the seeds are not currently harvested for commercial processing or foodstuff usage in the United States. The tree is a member of the Euphorbiaceae family and is native to subtropical China, where it has been cultivated for 14 centuries as a specialty oilseed crop, medicinal plant, and source of vegetable dye, and for uses similar to those of linseed oils (Morgan and Schultz, 1981; Scheld, 1986). Plantations can be established from cuttings, seedlings, or seed. The most convenient and econom-ical method of stand establishment is direct planting of seeds by standard, mechanical, row-crop planters. The tree grows well in saline lands which are marginal for conventional agriculture. Dense plantations of 0.60 to 0.75 m spacing are practical, and coppicing is a feasible system of management. The seed pods yield both a hard vegetable tallow on the outside and a liquid oil on the inside, which together comprise 45 to 50 wt % of the seed. Expressed as weight percentages of the seed, the tallow is typically 25 to 30% of the seed, the oil is 15 to 20%, the hull is about 40%, and the remainder is seed protein and fiber. In some strains, the tallow mainly consists of a single triglyceride, the palmitic-oleic-palmitic triglyceride. The oil is composed largely of glycerides of oleic, linoleic, and linolenic acids and is chemically similar to linseed oil and dehydrated castor oil. Both the tallow and oil have good potential as a source of biodiesel. The annual yields of tallow and oil are also remarkably high, each probably near 14.8 bbl/ha-year from a planting with seeds of about 11,200 kg/ha (10,000 lb/ac).

Certain microalgae represent another source of natural triglycerides. Much research has been done on the culture and compositional characteristics of certain microalgae as a source of "algal oils" (cf. Klass, 1983, 1984, 1985; Brown, 1993). This work has been focused on the growth of the organisms under conditions that can promote glyceride formation. The oils are high in triglycerides and can be transesterified to form biodiesel in the same manner as other natural triglycerides (Hill and Feinberg, 1984). As mentioned in Chapter 4, the production of natural triglycerides from microalgae can sometimes eliminate the high cost of cell harvest and extraction because it may be possible to separate the lipids by simple flotation or extraction from the culture media if they are leaked as extracellular products. As already mentioned, stressed growth conditions can often be used to increase the formation of natural lipids as shown for several microalgae in Table 10.4. The stress was caused by either the use of nutrient-deficient media or the addition of excess salt to nutrient-enriched media. The combination of both nutrient deficiency and salt enrichment appears to enhance lipid formation with *Isochrysis* sp., but to reduce it with *Dunaliella salina*. Interestingly, the free glycerol content can apparently be quite high for *Dunaliella* sp. This suggests either that all of the intracellular glycerol was not esterified, or that the esters were hydrolyzed after formation. *Botryococcus braunii* exhibited relatively high lipid contents under each set of growth conditions, but were the highest, 54.2 dry wt %, under nutrient-deficient growth conditions. Other microalgae have also been found to exhibit similar lipid contents. For example, the lipid content increased from 28 to over 50% of the cell dry weight with increasing nitrogen deficiency for *Nannochloropsis* sp. when grown in media under nitrogen-limited conditions (Tillett and Benemann, 1988). Lipid productivity was demonstrated to have a maximum of 150 mg/L-day at 5 to 6% cell nitrogen and was apparently independent of the light supply and cell density when the initial nitrogen concentration exceeded 25 mg/L. The key to the practical use of triglycerides from microalgae as feedstock for biodiesel production is the rate of triglyceride formation during the growth of microalgae. The yield on conversion of carbon

TABLE 10.4 Proximate Compositions of Microalgae Grown under Different Conditions[a]

Species	Growth conditions	Ash (% dry wt)	Organic component distribution				
			Lipid (% dry wt)	Protein (% dry wt)	Carbohydrate (% dry wt)	Glycerol (% dry wt)	Unknown (% dry wt)
Botryococcus braunii	FW, NE	5.6	44.5	22.0	14.1	0.1	19.3
	FW, ND	7.8	54.2	20.6	14.3	0.1	10.8
	NE, 0.5 molar NaCl	59.6	46.3	15.0	13.3	0.1	25.3
Dunaliella bardawil	ND, 2 molar NaCl	14.7	10.4	9.7	40.4	16.4	23.1
Dunaliella salina	NE, 0.5 molar NaCl	8.6	25.3	29.3	16.3	9.4	19.7
	ND, 0.5 molar NaCl	7.7	9.2	12.5	55.5	4.7	18.1
	NE, 2 molar NaCl	21.7	18.5	35.9	12.5	27.7	5.4
Ankistrodesmus sp.	NE, FW	04.5	24.5	31.1	10.8	0.1	33.5
Isochrysis sp.	NE, 0.5 molar NaCl	12.0	7.1	37.0	11.2	0.1	44.6
	ND, 0.5 molar NaCl	52.0	26.0	23.3	20.5	0.1	30.1
Nanochloris sp.	NE, 1 molar NaCl	65.9	15.3	34.7	15.5	0.1	34.4
	NE, FW	13.6	20.8	33.1	13.2	0.1	32.8
Nitzschia sp.	NE, 1.4 molar Na+	20.4	12.1	16.8	9.2	0.1	61.8

[a]Adapted from Tables 1–9 of National Renewable Energy Laboratory (1983). The culture media are denoted by FW, freshwater; NE, nutrient enriched; ND, nutrient deficient. The analyses were conducted on five different cultures of each species; statistical error analysis showed no standard error of more than 10%.

dioxide to microalgae is important, but high yields are not essential to obtain high triglyceride productivities. They are also a function of the growth rate of the particular microalgae and its triglyceride content. Commercial net productivities of triglycerides from microalgae per unit growth area were projected to be 23 g/m^2-day or 66 t/ha-year several years ago, and improvements through continued research were estimated to raise these figures to 43 g/m^2-day or 124 t/ha-year (Hill and Feinberg, 1984). Although many species of microalgae have since been isolated and characterized, and small outdoor test facilities have been operated to collect data on optimal growth conditions, the sustained production of microalgae and the harvesting of the triglycerides in integrated, large-scale systems have not yet been demonstrated (Brown, 1993). Very high cell densities of the order of 46 million/ml have been achieved in a shallow, outdoor, 50-m^2 raceway in Hawaii for the marine diatom *Phaeodactylum tricornutum*, the lipid content of which was found to be as high as 80% of its dry weight (Glenn, 1982; National Renewable Energy Laboratory, 1982; Shupe, 1982). This and similar projects on the growth of microalgae in fresh and saline water in California and the Southwest have since been terminated. Methods of maintaining optimal levels of nutrients, carbon dioxide, salinity, and temperature for continuous, outdoor microalgal growth, as well as economic methods of recovering triglycerides, must be developed and tested to allow design of large-scale systems.

Research on aquatic biomass by French researchers has resulted in several interesting results with the microalga *B. braunii*. In laboratory studies, this microalga, which under the growth conditions used is reported to have a hydrocarbon content as high as 75% of its dry weight, was reported to have been cultured at hydrocarbon productivities per unit growth area up to 15 g/m^2-day (Casadevall and Largeau, 1982). The dominant hydrocarbons produced are branched-chain olefins. In subsequent work, the French confirmed the high hydrocarbon content of *B. braunii* (Casadevall, 1984; Brenckmann, 1985; Brenckmann *et al.*, 1985; Metzger *et al.*, 1985). Nitrogen limitation was not found to be necessary for high hydrocarbon production, the highest productivities of which were observed during exponential growth. Light intensity did not affect the structure of the hydrocarbons, mainly C_{27}, C_{29}, and C_{31} alkadienes, which were produced in continuously illuminated batch cultures. But adjustment of the light intensity to the proper level gave maximum biomass and hydrocarbon yields. One strain was reported to yield straight-chain alkadienes and trienes as odd-numbered chains from C_{23} to C_{31}, and another strain produces triterpenes of the generic formula C_nH_{2n-10} where n varies from 30 to 37.

The French results with *B. braunii* contrast sharply with the U.S. results and are unexpected because most of the literature on this organism describes lipids and not hydrocarbons as metabolites. The formation of lipids and hydrocarbons together for certain biomass species is not unique, as indicated in this chapter. It is surprising, however, that an organism is reported to produce higher lipid content under stressed growth conditions than that produced under nonstressed growth conditions with no mention of hydrocarbons, and high hydrocarbon production under nonstressed growth conditions with no mention of lipids. It may be that the unknown organic components in *B. braunii* (Table 10.4) are hydrocarbons, but it is unlikely. It is more probable that a major shift in biochemical metabolite formation can occur from one strain to another. In any case, it appears that certain microalgae are capable of high productivities of *lipids or hydrocarbons*. Such organisms could provide significant benefits in processes designed for the manufacture of liquid fuels from biomass.

B. TERPENES

Polyisoprenes

Hydrocarbon production as polyisoprenes in terrestrial biomass by natural biochemical mechanisms is a well-known phenomenon that has been used by industry for many years. Commercial production of pure hydrocarbons as natural rubber, the highly stereospecific polymer *cis*-1,4-polyisoprene, is an established technology. Natural rubber has been reported to occur in hundreds of biomass species including dandelions and goldenrod, but commercial production has been limited to two species: the hevea rubber tree and the perennial desert shrub guayule. Natural rubber has a molecular-weight range between about 500,000 and 2,000,000 and is tapped as a latex from the hevea rubber tree (*Hevea braziliensis*), a member of the family Euphorbiaceae that grows in the tropical climes of South America and southeast Asia. The trees are native to the equatorial, lowland rainforests in the Amazon basin of South America, but have been transplanted to Malaysia and agronomically improved. The trees grow to heights of 15–20 m in 5–7 years and are planted at densities of about 400–500/ha; 200–250 cm/year of precipitation is required for normal growth. The polymer occurs in a cell system between the outer bark of the plant and the cambium, and when an incision is made that penetrates the outer bark, the milky latex oozes out. Mature trees continue to yield latex until they are over 40 years of age. A new incision is generally made on the trunk each day, and about 0.25 kg of latex containing about 30–40 wt % rubber is collected daily. The crude latex rubber contains about 90–95% hydrocarbons, 2.5–3% fatty acids, and sugars, resins, and proteinaceous substances. The annual yield of dry crude rubber from ordinary, mature trees is about 2 kg/tree, but specially selected high-yield trees can yield 9–14 kg/tree or more. The collected latex is strained, diluted with water, and treated with acid to coagulate the colloidal rubber particles. Natural rubbers can be transformed on heating to 280°C into the liquid monoterpene dipentene, the racemic form of limonene. The chemistry of these and similar transformations to potential liquid fuels will be discussed in Section III.

Guayule (*Parthenium argentatum* Gray), a member of the sunflower family, Compositae, that grows in the southwestern United States and in northern Mexico, is another source of natural rubber almost identical with hevea rubber (National Academy of Sciences, 1977). Two-thirds of the rubber is contained in the stems and branches. The remainder is in the roots, so the whole plant must be harvested and extracted. The rubber is contained in single, thin-walled cells in the outer layers, mostly in newly grown tissues. The older cells of the inner xylem and pith produce rubber for several years. There is no rubber in the leaves. In 1910, about 50% of all commercial U.S. rubber was extracted from wild guayule. At that time, the belief that guayule fortunes were soon to be made caused a land boom in the desert areas of the United States and New Mexico. The extensive wild stands, however, could not sustain continued harvesting without replanting, cultivation, or rotational cropping. The wild stands were almost completely devastated. By 1912, many guayule extraction mills were forced to close.

The idea of growing guayule and extracting the rubber latex from the whole plant was tested in full-scale plantations during the rubber shortage in World War II and found to be technically feasible. More than 12,000 ha of guayule were cultivated in California under the Emergency Rubber Project. About 1.4 million tonnes of rubber were produced in two processing facilities. The plants had to be processed within a few days after harvest because of oxidative degradation of the rubber once the plant cells had been exposed to air. Hevea rubber latexes contain antioxidants that preclude such degradation. In the 1940s, guayule strains containing up to 26 dry wt % of rubber were found, and laboratory research indicated that certain amine sprays stimulate guayule's rubber-producing genes to afford plants that contain more than 30% rubber content when harvested (Yokoyama, 1977). However, the strains that were widely cultivated during World War II had been selected before the war and were able to produce only about 20% of the plant's dry weight as rubber after 4 years' growth. The distribution of the components in harvested guayule shrubs is moisture, 45–60 wt %; and expressed in dry weight percentages rubber, 8–26; resins, 5–15; bagasse, 50–55; leaves, 15–20; cork, 1–3; and water solubles, 10–12. It is apparent that the methods for commercially extracting rubber from guayule would have to be considerably more complex than the relatively simple processes used for processing hevea latexes. A process developed for commercial use in Mexico, for example, consisted of the sequential steps of parboiling the shrubs to coagulate the rubber in the latex cells; milling to release the rubber from the cells; separation of the rubber in flotation tanks in which the waterlogged bagasse sinks and the rubber "worms" float and are skimmed off; washing the worms to remove the residual caustic soda used in the milling operations, deresination of the worms with acetone, which is subsequently distilled to obtain the resin and recycle solvent; and purification of the rubber using rubber solvents to remove insolubles.

When guayule is in an active growth phase, it produces little or no rubber, but if the plants are stressed, such as in cool weather or because of reduced moisture supply, biomass growth slows and the photosynthetic products are diverted to rubber production. The rubbers are not metabolized by the plant, even when it is deprived of all carbohydrates and other energy sources, and continue to accumulate for at least 10 years. The resins, which include terpenes, sesquiterpenes, diterpenes, glycerides, and low-molecular-weight polyisoprenes, are found in resin ducts throughout the plant; they constitute 10–15 dry wt % of the plant.

Although wild stands of guayule are remarkably free of disease and insect pests, cultivated plants are susceptible to both. It is not clear that guayule can be economically cultivated in the arid regions of much of its native habitat because the plant may take more than 7 years to develop commercially useful quantities of rubber, presuming natural rubber is the desired product. Guayule plants can survive arid conditions, but if annual rainfall is less than about 36 cm, supplemental irrigation is needed to provide sufficient rubber yields in a reasonable time. In contrast, annual precipitation exceeding about 64 cm can cause excessive biomass growth rather than rubber formation. The introduction of synthetic rubber processes and the unfavorable economics of producing natural rubber in the Southwest from guayule because of the problems alluded to here and the requirement to harvest and extract whole plants basically eliminated guayule rubber from commercial rubber markets. Some reports indicate that the certain areas of the Southwest may provide opportunities for commercial cultivation and rubber production (Texas Agricultural Experiment Station, 1990). This evaluation suggested that guayule must be direct-seeded with varieties containing at least 10% rubber before it can compete for cropland in these areas, and that a viable market is needed for the resin. An option that seems to have been given little consideration is the use of guayule for hydrocarbon liquids production. It is probable that stressed growth conditions would increase both polyisoprene and terpene production, or hydrocarbon production generally. Presuming the polyisoprenes can be converted to terpene liquids, the liquid hydrocarbon yield could then be equivalent to as much as 40 wt % of the dry plant with suitable strains of guayule.

Naval Stores

Terpene extraction from pine trees and other biomass species is also established technology. Naval stores, which are the various resinous substances such as gum turpentines, rosins, and pitches, were obtained from conifers for maintaining wooden ships and as rope coatings for many years. Until recently, the longleaf pine forests in the South constituted the only indigenous U.S. source of naval stores, but the rapid loss of forests due to commercial lumbering operations caused the industry to decline. The pine stumps remaining from lumbering activities were a smaller source of naval stores. The first plant for recovering turpentine and rosin from old stumps was built in 1909. The pine oleoresin industry began to decrease in importance after its peak around that time. Competition from petroleum- and coal-derived substitutes and cheaper products obtained by extraction and steam distillation techniques during pulping operations resulted in oleoresin production of only a small fraction of that in the early 1900s. Research in progress on the stimulation of natural oleoresin formation by chemical injections into trees, and on the benefits of combined timber and oleoresin production in mixed stands, has provided a few leads to improved technology. However, restoration of the naval stores industry as a major source of raw materials for the manufacture of liquid terpenes such as dipentene for conversion to liquid motor fuels seems remote, even though the resulting new markets would stimulate industry growth. The total U.S. and world production of gum, wood, and pulping turpentines was only about 750,000 and 2 million bbl/year in the mid-1990s, each of which is much less than 1 day's U.S. oil consumption. Also, the largest source of turpentine in the United States today, sulfate pulping, is capacity-limited. Major additions are not expected to occur in the next several decades.

Other Herbaceous Biomass

Hundreds of herbaceous biomass species native to North America or that can be grown there have been tested as sources of hydrocarbons as well as glycerides. The objectives of this work have generally been to identify those species that produce potentially useful hydrocarbon fuels, to characterize the yields of hydrocarbons and other components, and to learn what controls the structure and molecular weight of the hydrocarbons within the plant so that genetic manipulation or other biomass modifications might be applied to control these parameters. Some efforts have concentrated on plants that might be grown in arid or semiarid regions of the United States without competition from biomass grown for foodstuffs (cf. Calvin, 1978, 1979, 1985, 1987; Hoffmann, 1983; Johnson and Hinman, 1980; Kingsolver, 1982; McLaughlin and Hoffmann, 1982; McLaughlin, Kingsolver, and Hoffmann, 1983; Adams, 1982; Adams et al. 1983; Adams, Balandrin, and Martineau, 1984; Balandrin, 1984, 1985). Other work has been aimed at perennials adapted to wide areas of North America (cf. Bagby, Buchanan, and Otey, 1981; Buchanan et al., 1978). Table 10.5 lists some of the best mesophytic and xerophytic candidates that have been found to contain natural hydrocarbons and/or triglyceride oils (Klass,

TABLE 10.5 Some Triglyceride- and Hydrocarbon-Producing Biomass Species Potentially Suitable for North America[a]

Family	Genus and species	Common name
Aceraceae	*Acer saccharinum*	Silver maple
Anacardiaceae	*Rhus glabra*	Smooth sumac
Asclepiadiaceae	*Asclepias incarnata*	Swamp milkweed
	sublata	Desert milkweed
	syriaca	Common milkweed
	Cryptostegia grandiflora	Madagascar rubber vine
Buxaceae	*Simmondsia chinensis*	Jojoba
Caprifoliaceae	*Lonicera tartarica*	Red tarterium honeysuckle
	Sambucus canadensis	Common elder
	Symphoricarpos orbiculatus	Corral berry
Companulaceae	*Companula americana*	Tall bellflower
Compositae	*Ambrosia trifida*	Giant ragweed
	Cacalia atriplicifolia	Pale Indian plantain
	Carthamus tinctorius	Safflower
	Chrysathamnus nauseosus	Rabbitbrush
	Cirsium discolor	Field thistle
	Eupathorium altissimum	Tall boneset
	Grindelia aphanactis	Sunflower
	camporum	Sunflower
	squarrosa	Sunflower
	Helianthus annuus	Sunflower
	Parthenium argentatum	Guayule
	Silphium integrifolium	Rosin weed
	laciniatum	Compass plant
	terbinthinaceum	Prairie dock
	Solidago graminifolia	Grass-leaved goldenrod
	leavenworthii	Edison's goldenrod
	rigida	Stiff goldenrod
	Sonchus arvensis	Sow thistle
	Vernonia fasciculata	Ironweed
Cruciferae	*Brassica alba*	White mustard
	napus	Rapeseed
	nigra	Black mustard
Curcurbitaceae	*Cucurbita foetidissima*	Buffalo gourd
Euphorbiaceae	*Euphorbia denta*	
	lathyris	Mole plant, gopher plant
	ulcherima	Poinsettia
	tirucalli	African milk bush
	Ricinus communis	Castor
Gramineae	*Agropyron repens*	Quack grass
	Elymus canadensis	Wild rye
	Phalaris canariensis	Canary grass
	Zea mays	Corn

(*continues*)

TABLE 10.5 (*Continued*)

Family	Genus and species	Common name
Labiatae	*Pycnanthemum incanum*	Western mountain mint
	Teucrium canadensis	American germander
Lauraceae	*Sassafras albidium*	Sassafras
Leguminosae	*Arachis hypogaea*	Peanut
	Copaifera langsdorfii	Copaiba
	multijuga	Soybean
	Glycine max	
Linaceae	*Linum usitatissimum*	Linseed
Malvaceae	*Gossypium hirsutum*	Cotton
Papaveraceae	*Papaver somniferum*	Poppy
Pedaliaceae	*Sesamum indicum*	Sesame
Rhamnaceae	*Ceanothus americanus*	New Jersey tea
Rosaceae	*Prunus americanus*	Wild plum
Phytolaccaceae	*Phytolacea americana*	Pokeweed

[a]Adapted and revised from Klass (1994).

1994). It is apparent that the formation of these substances is not limited to any one family or type of biomass. Interestingly, certain species in the Euphorbiaceae family, which includes *H. braziliensis,* have been reported to form terpene liquids at a yield of 8 dry wt % of the plant and a minimum of 25 bbl/ha-year (10 bbl/ac-year) (Calvin 1978, 1987). As will be shown later, many of these products are suitable as motor fuels or feedstocks for upgrading to motor fuels. "Gasoline plantations" are thus not totally in the realm of science fiction (Maugh, 1976, 1979). It is necessary to emphasize, however, that in order for a liquid hydrocarbon yield of 25 bbl/ha-year to be sustained in the field, the yield of dry biomass per unit growth area must be about 42.1 t/ha-year (18.8 ton/ac-year) when 8 wt % of the dry plant is terpenes.

In extensive research carried out by the U.S. Department of Agriculture over many years on hundreds of indigenous, herbaceous, mesophytic biomass species that grow in the Midwest, the highest yields of "oil" in the best producers ranged from about 1 to 7 wt % of the dry biomass (Buchanan *et al.,* 1978). The oils contain sterols, other free alcohols, free acids, glycerides, and nonglyceride esters, in addition to hydrocarbons. The hydrocarbon fraction contains waxes, isoprene polymers, and terpenoids, and the hydrocarbon yields in the best producers ranged from 0.5 to 1.5 wt % of the dry plant. The highest rated species was common milkweed, *Ascelepias syriaca;* its composition is shown in Table 10.6. The hydrocarbon fraction is essentially all high-molecular-

TABLE 10.6   Composition of Common Milkweed[a]

Component	Amount in whole dry plant (wt %)
Crude protein	11.06
Acetone extract	
Polyphenol fraction	7.2
Oil fraction	4.28
Hydrocarbon fraction	1.39
Ash	9.86
Cellulosics (by difference)	66.21
Total	100.00

[a]Buchanan et al. (1978).

weight polyisoprene, and the oil fraction contains 60.6 wt % saponifiables, presumably esters. The remaining unsaponifiables in the oil fraction should contain the terpenes. Assuming all of the unsaponifiable fraction is terpene, an unlikely possibility since this fraction also contains unsaponifiable resins, sterols, and waxes, the maximum terpene yield is then 1.7 wt % of the whole dry plant. The bulk of the material is clearly cellulosic.

Similar studies on xerophytic biomass species that grow in the Southwest and northwestern Mexico indicate that plants having the greatest potential in these areas are resinous members of the Compositae family including *Grindelia, Chrysothamnus,* and *Xanthocephalum* species (McLaughlin and Hoffmann, 1982; Hoffmann, 1983). Latex-bearing plants, many of which have received widespread attention in recent years, appear to be of secondary importance. The cyclohexane extractables from over 100 of these plants ranged from about 4 to 5.5%.

Figure 10.1 illustrates one of the processing schemes that might be used for separating the various components in hydrocarbon-containing biomass. Acetone extraction removes the polyphenols, glycerides, and sterols, and benzene extraction or extraction with another nonpolar solvent removes the hydrocarbons. If the biomass species in question contained low concentrations of the nonhydrocarbon components, exclusive of the carbohydrate and protein fractions, direct extraction of the hydrocarbons with a nonpolar solvent might be preferred.

Only a few species out of several thousand nonoilseed biomass species that have been screened have been given serious consideration in the United States as a source of liquid hydrocarbon motor fuels. One of the most interesting programs to develop biomass as a source of hydrocarbons was started as a result of reports that the Brazilians recovered liquids suitable for direct use as

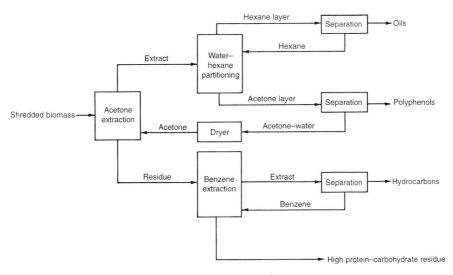

FIGURE 10.1 Example of solvent extraction scheme for separating various components in hydrocarbon-containing biomass.

diesel fuel in trucks without modification from the tropical tree *Copaifera multijuga*, a member of the Leguminosae family, in a manner similar to tapping *H. braziliensis* for natural rubber (*cf.* Calvin, 1978, 1980). The main components of the collected copaiba oil are caryophyllene, bergamotene, and copaene; all are cyclic C_{15} hydrocarbons (Calvin, 1987). Copaiba is the common name of a group of species of the genus *Copaifera* which grow throughout Brazil. A hole is bored horizontally in the trunk into the heartwood of the tree and a bung is placed in the hole. The bung is removed at certain times of the year and the oil flows directly into a container. A single hole in a large tree may yield about 20 to 30 liters of oil in 24 h. The bung is reinserted and 6 months later, another 20 to 30 liters of oil is drained out.

It was estimated that in tree plantation settings, this species should be capable of producing 62 bbl/ha-year (25 bbl/ac-year) of terpene hydrocarbons, so an extensive research program was initiated in the mid-1970's to develop "petroleum plantation agriculture" (Calvin, 1978). Since the tree does not grow in the United States, other species in the Euphorbiaceae family were examined, many of which grow in semiarid areas. It was found that almost every species of the genus *Euphorbia* produces a hydrocarbon latex whose molecular weight is much lower than that of the latex from *H. braziliensis*. The literature indicated the concepts of growing this species in large plantations and of producing motor fuels from such species were not new (Calvin, 1978,

1979). The French harvested 125,000 ha of *Euphorbia resinifera* in 1938 in Morocco and obtained 10,000 L of latex/ha, from which 1700 kg of rubber (benzene extractables) and 2750 kg of gum resin (acetone extractables) were obtained. Three tonnes of hydrocarbons were produced. In addition, the Italians had plans for using various species of *Euphorbia* to produce a material they called "vegetable gasoline" in a refinery at Agodat, but it was not clear what happened to this endeavor. The U.S. work began to focus on the species *Euphorbia lathyris* (gopher plant, mole plant) and *Euphorbia tirucalli* (African milk bush) (Calvin, 1980). Each species can grow in the semiarid western and southwestern parts of the United States where rainfall is about 25 to 50 cm/year, neither requires good soil, and both will grow well in uncultivated rocky areas which are not suitable for food crops. Both also grow upright and are adaptable to mechanical harvesting. *E. lathyris* is an annual, biennial, or even a perennial introduced from the Mediterranean regions, but is now widely naturalized in California and elsewhere in the United States (Sachs *et al.*, 1981). The plant grows from seed to a harvestable height of about 1.2 m in 7 months, is frost hardy, and if not harvested, will survive some winters and grow to seeding maturity the following year. *E. tirucalli* is a perennial propagated from cuttings, grows to a height of about 0.6 m in one year, can reach heights of about 6 m in tropical climes, and is widespread in tropical dry regions (Calvin, 1980; Fukumoto, 1980). This species is sensitive to frost and its range is limited to mild climates in the United States. The origins of *E. tirucalli* are East Africa and Madagascar; it is widespread in India, Java, Okinawa, Sri Lanka, and Taiwan. *E. tirucalli* is actually a shrublike tree and not a herbaceous biomass species; when mature, it might be tapped to collect the latex rather than undergo annual harvesting and processing to recover the product oil. In a plantation in Okinawa planted at a spacing of 0.4 x 0.8 m and a planting density of 30,900/ha, the annual hydrocarbon yield was projected to be about 11 bbl/ha (Fukumoto, 1980).

The initial experimental plantings in southern California showed that *E. lathyris* produced about 8-12% of its dry weight as oil, or not less than about 20 bbl/ha-year with unselected seed and no agronomic experience over a 7-month growing season from February to September (Calvin, 1978, 1980; Nielsen *et al.*, 1977). With proper seed selection, it was felt that oil yields could be raised to about 50 bbl/ha-year (Calvin, 1978), and through genetic and agronomic development, to as high as 65 bbl/ha-year (Johnson and Hinman, 1980). The product oil and other solubles were obtained by harvesting, drying, and extracting the finely powdered plants with selected solvents. The hydrocarbon extract was found to produce the expected suite of products by catalytic cracking at 500°C over Mobil Corporation's zeolite catalyst, and the methanol extract contains hexoses that are all fermentable to ethanol (Calvin, 1980). The product oil from *E. tirucalli* resembles that from *E. lathyris*, and was estimated to be capable of hydrocarbon production at yields of 25–50 bbl/ha-year.

Examination of the literature shows that the solvent extracts from E. *lathyris* and E. *tirucalli* are not "oil" in the sense that they are mainly sesquiterpene hydrocarbons as produced by the tree C. *multijuga*. The compositions of successive acetone and benzene extracts of E. *lathyris*, E. *tirucalli*, and H. *braziliensis* are shown in Table 10.7 (Nielsen *et al.*, 1977). The compositions vary with the plant parts of E. *lathyris* and the glycerides are dominant components in the acetone extract of each species. The respective rubber contents of the latexes from these species are 3, 1, and 87 dry wt % of the latex, and the corresponding sterol contents are 50, 50, and 1 dry wt % of the latex. The rubber fraction from each species is 100% *cis*-polyisoprene. It was concluded that the solvent extracts are complex mixtures of hydrocarbons and other organic compounds. Further study of the compositions of solvent extracts of E. *lathyris* supported the complex nature of the extractables (Nemethy, Otvos, and Calvin, 1981). The dried whole plants were successively extracted with heptane and methanol. The heptane extract is 5% of the dry plant weight and consists of 7% hydrocarbons, 33% fatty acid esters of triterpenoids, 41% tetra- and pentacyclic triterpenoids, ketones, and alcohols, 5% phytosterols and bifunctional compounds, and 15% bifunctional triterpenoids. Methanol extraction of the residue from the heptane extraction, 95% of the dry plant weight, yielded 30% of the dry plant weight, most of which was made up of four simple, water-soluble, fermentable sugars, sucrose, glucose, galactose, and fructose (27% of the dry plant weight) and an ether-soluble fraction (3% of

TABLE 10.7 Compositions of Acetone and Benzene Extractables from Dried, Ground *Euphorbia lathyris*, E. *tirucalli*, and *Hevea braziliensis*[a]

Biomass species	Acetone extract[c]			Benzene extract[c]	
	Glycerides (dry wt %)	Isoprenoids (dry wt %)	Others[d] (dry wt %)	Rubber (dry wt %)	Wax (dry wt %)
E. *lathyris* (leaves)	13.7	2.2	8.3	0.1	0.2
(seeds)	40.0	<0.1	<2.0		
(stems)	1.9	<0.5	2.0		
E. *tirucalli*	2.4	<0.5	2.0	0.07	0.13
E. *tirucalli*[b]	4.4	<0.5	3.4	0.1	0.3
H. *braziliensis*	5.1	<0.5	2.6	1.3	0.2

[a]Adapted from Nielsen *et al.* (1977).
[b]Analysis from University of California, Los Angeles; others from Berkeley.
[c]Successively extracted for 8 h with acetone and then 8 h with benzene.
[d]Other terpenoids.

the dry plant weight). The insoluble residue from the methanol extraction, 65% of the plant dry weight, was bagasse. This work shows that the total amount of hydrocarbons produced by E. *lathyris,* which appears in the heptane extract, is only about 0.4% of the dry plant weight (100[0.05 x 0.07]). Direct acetone extraction of E. *lathyris,* however, yields about 8% of the dry weight of the plant. This fraction is equivalent to the sum of the heptane extract and the ether-soluble fraction in the methanol extract. As will be shown in Section III, most of the acetone extract, which might best be characterized as a terpenoid rosin or hydrocarbon-like products (biocrude), can be converted to substitute petroleum motor fuels (*cf.* Schmalzer *et al.,* 1988).

The projections of liquid hydrocarbon yields from the Brazilian tree C. *multijuga* in plantation settings and the yields of biocrude from E. *lathyris* and E. *tirucalli* have been quite optimistic. However, the main difficulties with the concept of natural hydrocarbon production from biomass are that most of the species that have been tested exhibit low liquid yields compared to the mass of biomass that must be harvested, and the naturally produced liquids are complex mixtures and not pure hydrocarbons (or glycerides). Moreover, the relationship between the minimum hydrocarbon content of dry biomass and biomass yield required to sustain a terpene yield of 25 bbl/ha-year tends to preclude sustainable production at this level. This is perhaps best illustrated by Fig. 10.2. The curve is constructed by assuming the density of terpene hydrocarbons is in the range 0.1347 t/bbl, which is the literature value for

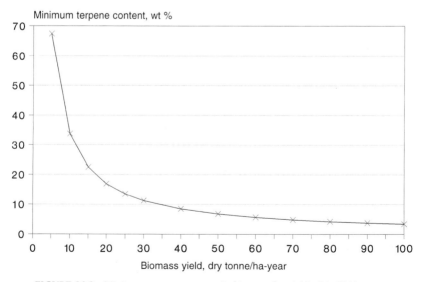

FIGURE 10.2   Minimum terpene content in biomass for yield of 25 bbl/ha-year.

dipentene (density of 0.847 g/cm^3 at 15.5°C). When the biomass yield-terpene content relationships illustrated in Fig. 10.2 are compared with the yield ranges commonly encountered in agriculture, about 5 to 20 dry t/ha-year (2.2 to 8.9 dry ton/ac-year), it is evident that the minimum terpene content of the biomass required to sustain yields of 25 bbl/ha-year ranges from about 17 to 67 wt %, which for most whole-plant extraction systems is very unlikely. For more intensive agricultural practices characterized by high fertilization and irrigation in moderate and semiarid climates, each of which can significantly increase biomass production costs, biomass yields can range up to about 35 dry t/ha-year (15.6 dry ton/ac-year). The minimum biocrude content required at that yield level to sustain liquid yields of 25 bbl/ha-year is still high, about 10 wt % of the dry biomass. There are only a few cases where terrestrial biomass yields are reported to be higher, and most of these species do not produce biocrude. This rather simple explanation of why there has been little success in the search for sustainable biocrude producers at rates of 25 bbl/ha-year has considerable historical support in the literature. Some researchers have suggested that there may be an upper limit for whole-plant hydrocarbons in biomass such as *E. lathyris* that is physiologically determined (Sachs *et al.*, 1981).

One of the ideas to increase the natural production of liquid hydrocarbons in biomass species that can grow in nontropical climates is to transfer the gene that codes for hydrocarbons in *C. multijuga* to *Euphorbia* species (Calvin, 1983). If the functionality of the gene could be retained, relatively pure sesquiterpene liquids might be formed by *E. lathyris* or *E. tirucalli* at high selectivity and sufficient yield to make natural production of liquid hydrocarbon fuels a practical process in the United States. Little research appears to have been done on this concept. Other research has shown that hydroponically grown *E. lathyris* when exposed to increasing levels of salinity caused an increase in the percentage of both hydrocarbons and sugars (Taylor, Skrukrud, and Calvin, 1988). The hydrocarbon fraction, containing mostly triterpenoids, increased by 50% and the sugar fraction, containing mostly sucrose, increased by 88%. This resulted in a shift of available biomass from lignocellulose to sugars and hydrocarbons, although there was little change in the total energy content of the plant and overall growth decreased. A twofold increase in the activity per unit leaf area of the enzyme β-hydroxymethylglutaryl-coenzyme A reductase was also observed with increased salinity. This enzyme is involved in the biosynthesis of triterpenoids, and its response to increased salinity indicates a role in the regulation of natural hydrocarbon productivity. The results suggest that the hydrocarbon content of *E. lathyris* can be changed by manipulation of the growth conditions, so perhaps genetic manipulation could have a similar effect. It is also possible that the stressed growth conditions that occurred at increasing levels of sodium chloride are responsible for reduced biomass yield and higher hydrocarbon yields. Stressed growth has been found to cause similar changes in composition and yield with other biomass species. For example, the phenomenon is well known in the field of polyisoprene rubber production, as already discussed in this chapter, and in the culture of microalgae. Nutrient deprivation and other types of stress are standard means for induction of increased algal production of hydrocarbons and lipids (Shifrin and Chisholm, 1980). Note, however, that in most of the reported cases of increased hydrocarbon content of biomass that produce biocrude under stressed growth conditions, the yield of biomass decreases. So when the objective is to grow and harvest biomass for maximum biocrude, the biocrude production rate per unit biomass growth area is the prime factor, and not necessarily maximum biomass yield. It is doubtful that high biomass yields and high biocrude content will be found in the same biomass species (cf. McLaughlin, Kingsolver, and Hoffmann, 1983).

C. OTHER SOURCES OF SUBSTITUTE DIESEL FUELS

Tall Oils

During the production of wood pulps by the sulfate (kraft) process, a by-product called tall oil is formed by saponification of fatty acid glycerides and esters of the resin acids, particularly from the pulping of pine woods. Soaps are formed as a foam on the black liquor, which itself has an annual worldwide production of about 200 million tonnes and is a major energy resource for the paper industry (Chapter 5). The sodium salts are skimmed from the surface of the black liquor and acidified to yield crude tall oil, which contains about 35% resin acids such as abietic and pimaric acids, 30% fatty acids, and 35% unsaponifiable terpene hydrocarbons and terpenoids such as phytosterols. Crude tall oil can be further separated by vacuum distillation into depitched tall oil and pitch. Depitching refers mainly to removal of the unsaponifiable substances. Catalytic hydroprocessing of tall oil has been shown to yield a high-cetane product suitable as a diesel fuel blending agent, the cetane number of which increases linearly with the addition of the product (Feng, Wong, and Monnier, 1993). The hydroprocessed tall oil consists largely of normal alkanes, over 70% of which are n-heptadecane and n-octadecane. The product has a cetane number of 56. The paraffinic hydrocarbon n-hexadecane (or cetane) is used as the standard in determining the ignition qualities of diesel fuels, and by definition has a cetane number of 100. Commercial No. 2 diesel fuel has a cetane number of about 49. It is estimated that a blend of 25% hydroprocessed tall oil and 75% low-grade diesel fuel stock would have a cetane number over 40. Because of the worldwide distribution of wood pulp manufacturing plants, tall oils could represent a significant source of feedstocks for the manufacture of diesel fuel components. Hydroprocessing will be discussed in more detail in Section IIIA.

Carbohydrate Blends

An interesting alternative to the use of transesterified natural oils and hydrotreated tall oils as substitute diesel fuels is the use of carbohydrates in blends with water and alcohols. These blends are of course not natural oils, but formulations of about 66 wt % glucose, 23 wt % methanol, and 11 wt % water were found to be an effective diesel fuel in experimental demonstrations (Suppes and Wei, 1996). The pentoses and hexoses, as well as polysaccharide mixtures such as the sugar syrups, can be used in place of glucose as long as the carbohydrates are solubilized, presumably by adjusting the water and alcohol content and limiting the average molecular weight of the polysaccharides. Significant cost reductions are projected for such formulations because the fuel costs of the mixtures are estimated to be up to 50% less than the equivalent energy in ethanol. Should this approach to the development of diesel fuel substitutes prove to be practical, the direct use of primary biomass derivatives that are readily obtained from low-grade feedstocks in large quantities (Chapter 11) would be expected to open large fuel markets for biomass fuels. The carbohydrates could be obtained in large quantities with minimal processing.

III. CONVERSION CHEMISTRY

Basically, the chemistry involved in the conversion of natural biomass liquids to conventional motor fuels depends on whether the resulting products are for compression-ignition (diesel-fueled) or spark-ignition (gasoline-fueled) engines. Diesel fuels as already discussed preferably have normal paraffinic structures or properties. One of the reference fuels is n-hexadecane; it has a cetane number of 100. This does not mean, however, that diesel fuels consist entirely of normal paraffinic hydrocarbons having cetane numbers in this range. Many other organic compounds such as various cyclohexane derivatives occur in diesel fuels, and the cetane number ranges from 30 to 55 depending on the grade and additives. In a petroleum refinery, diesel fuel is that fraction which distills after kerosine from a temperature of about 260°C (500°F) up to a maximum of 338°C (640°F) at 90% of the endpoint, and is similar to a gas oil. Gasolines, in contrast, are more complex mixtures of hydrocarbons generally boiling from 32°C (90°F) up to 216°C (420°F) and are blended from a large number of refinery process streams including light virgin naphthas, isomerate, catalytic naphthas, reformate, alkylate, polymer gasoline, hydrocracked naphthas, and C_4-C_5 light hydrocarbons. Isooctane (2,2,4-trimethylpentane) is one of the reference fuels; it has an octane number of 100. Chemically, modern gasolines contain hundreds of organic compounds, the main groups of which are aromatics, branched-chain hydrocarbons, and substituted cycloparaffins. The amounts of aromatic and isomerized paraffinic compounds in the blending components are increased by various refinery processes to increase octane number of the final gasoline blends, and butanes are often added to meet vapor pressure specifications. The Clean Air Act of 1990 requires that reformulated gasoline (RFG) be marketed in certain "nonattainment" areas of the United States to help meet emission standards set by the U.S. Environmental Protection Agency. The benzene content is limited to a maximum of 1.0 vol % of the RFG, and the total aromatic content cannot exceed 25 vol %. RFG must also contain specified amounts of dissolved oxygen to help control emissions. The requirement in the mid-1990s is 2.0 or 2.7 wt % dissolved oxygen, depending upon the area and time of year, and is achieved by blending an oxygenated compound such as ethanol or methyl-t-butyl ether with gasoline or RFG.

It is important to emphasize that all conventional motor fuels—petroleum gasolines and diesel—are produced by complex refining processes. These fuels are mixtures and contain hundreds of organic compounds. None is a single, pure substance. Similarly, almost all biomass-based liquid motor fuels contain numerous organic compounds, but in some cases can consist of relatively few compounds, as in the case of the methyl esters of fatty acids in biodiesel. The exceptions are the lower molecular weight alcohols and a few derivatives that can be used directly as neat motor fuels. They can easily be manufactured from biomass feedstocks as individual compounds.

A. TRIGLYCERIDE FEEDSTOCKS

For biodiesel fuels, the basic idea is to convert natural glycerides to liquid products that have properties closer to those of diesel fuels than those of gasolines. The feedstocks for conversion to biodiesel are usually triglyceride oils from oilseeds. These oils normally contain small amounts of monoglycerides, diglycerides, and free fatty acids. Animal fats and natural non-seed-oil triglycerides are also suitable starting materials.

The first steps in the recovery of oil from oil seeds are to "crush" the seeds and then to separate the oil from the residual seed material (meal). The separation process can involve mechanical presses, solvent extraction, or a combination of both pressing and extraction (cf. Carr, 1989). Mechanical pressing is usually used to separate oil from seeds exceeding 20% oil content. Solvent extraction may be used for materials such as soybeans or press cakes with oil contents less than 20%. Oilseeds and nuts that contain high oil contents are usually processed first by mechanical presses. The crude oil can contain hydratable gums that can be removed by water- or acid-degumming treatments. The degummed oils can then be purified further if desired by conditioning with acid or alkali, bleaching, and other treatments, but can be used for conversion to biodiesel without extensive purification. When triglyceride-bearing biomass species such as microalgae and nonseed oils are utilized as sources of the triglycerides, the separation process is modified accordingly for oil recovery.

The chemistry involved in the conversion is illustrated by the accompanying equation. One equivalent mole of triglyceride is reacted with 3 mol of methanol

$$
\begin{array}{c}
\underset{\substack{| \\ CH_2-OCR_1}}{O} \\
\underset{\substack{| \\ CH-OCR_2}}{O} + 3CH_3OH \xrightarrow{Catalyst} \underset{\substack{| \\ CH_2OH}}{CH_2OH} + R_1CO_2CH_3 + R_2CO_2CH_3 + R_3CO_2CH_3 \\
\underset{CH_2-OCR_3}{O}
\end{array}
$$

Triglyceride

(R_1 - R_3 mainly C_{13} - C_{17})

(or other lower molecular weight alcohol) to yield the products of the trans-esterification reaction—the methyl esters of the fatty acids (biodiesel) in the original triglyceride and glycerol. Variables that affect the yield and purity of the methyl esters include the ratio of the alcohol to triglyceride, the type of catalyst used (acid or alkaline), the temperature, and the purity of the triglyceride (cf. Shay, 1993). At temperatures of 60°C or higher, and stoichiometric ratios of methyl, ethyl, or butyl alcohols to fully refined triglyceride oils, transesterification is complete in 1 h (Freedman, Pryde, and Mounts, 1984). At a 6 to 1 molar ratio of alcohol to triglyceride, a 98% yield of ester is obtained. At the theoretical stoichiometric ratio of 3 to 1, the yield of ester decreases to 82%. The use of crude triglyceride oils instead of the fully refined oils decreases biodiesel yields by 10% or more. Quantitative yields of the methyl and ethyl esters of rapeseed oil can be produced in 1 h at room temperature using 6 to 1 molar ratios of alcohol to oil and 1% sodium hydroxide catalyst (Nye and Southwell, 1983). A typical product yield distribution per tonne of triglyceride feedstock containing 2.5% free fatty acids and 115 kg of methanol is 946 kg of methyl esters, 89 kg of glycerol, and 23 kg of fatty acids (Stage, 1988). A typical flow diagram for the manufacture of methyl esters is shown in Fig. 10.3. Batch processing is the most common method for manufacturing biodiesel. Continuous processes are being developed using techniques such as high-

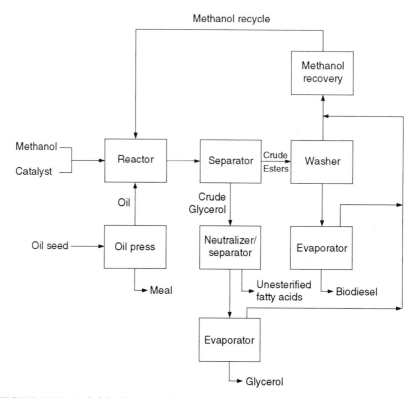

FIGURE 10.3 Methyl biodiesel production by conventional transesterification of seed oil with methanol.

shear-rate mixing (*cf.* Noureddini, Harkey, and Medikonduru, 1996; Anonymous, 1992). Transesterification of triglycerides with lower molecular weight alcohols occurs initially in a two-phase system in which the solubilities of triglyceride and catalyst in the methanol phase, for example, are low. Examination of the chemical and mass transfer characteristics of the reacting systems suggested that a single-phase process can be developed to accelerate production of biodiesel by use of a nonreactive cosolvent such as ethers (Boocock *et al.*, 1996). The addition of 1.27 volumes of cosolvent tetrahydrofuran (THF) per volume of methanol was found to form an oil-rich single phase of a 6:1 methanol–soybean oil mixture. The transesterification process is over 80% complete in 1 min and 95% complete in 20 min. Glycerol separation occurs 4–5 times faster in the presence of the cosolvent. Continuous processes utilizing the cosolvent concept should now be more feasible.

Instead of transesterification, other chemistry can be applied to produce substitute diesel fuels from triglyceride oils as well as animal fats. Catalytic hydroprocessing at elevated temperature and pressure of the oils using conventional refinery processing techniques has been found to convert them to substitute diesel fuels and diesel fuel additives (Stumborg *et al.*, 1993; Wong and Feng, 1995). Typical operating conditions are 300 to 400°C and 2 to 10 MPa. The results are similar to those reported for the hydroprocessing of tall oils discussed in Section II of this chapter (Feng, Wong, and Monnier, 1993). Soybean, canola, and palm oils, for example, are converted under medium-severity, hydroprocessing conditions with cobalt–molybdenum catalysts to products in the diesel-fuel boiling range having cetane numbers of 75 to 100. The respective yields were 77.4%, 70.1 to 81.1%, and 71.1%. The products are straight-chain paraffinic hydrocarbons identical to many of the high-cetane-number hydrocarbons found in conventional diesel fuels, and the process occurs under conditions commonly used in the refinery. The accompanying

$$
\begin{array}{l}
CH_2-\overset{\overset{\displaystyle O}{\|}}{O}C(CH_2)_{12}CH_3 \\
\quad\ \ \overset{\overset{\displaystyle O}{\|}}{} \\
CH-OC(CH_2)_{14}CH_3 \quad + 12H_2 \xrightarrow{\text{Catalyst}} \overset{\overset{\displaystyle CH_3}{|}}{CH_2} + CH_3(CH_2)_{12}CH_3 + CH_3(CH_2)_{14}CH_3 + CH_3(CH_2)_{16}CH_3 + 6H_2O \\
\quad\ \ \overset{\overset{\displaystyle O}{\|}}{} \qquad\qquad\qquad\qquad\ \ \ \overset{|}{CH_3} \\
CH_2-OC(CH_2)_{16}CH_3
\end{array}
$$

equation illustrates the theoretical stoichiometry for a specific triglyceride. Twelve moles of hydrogen are necessary in theory for complete reduction of 1 mol of triglyceride to hydrocarbons. Synthetic diesel yields of 80% of the triglyceride oil feedstocks were routinely achieved after the initial research was completed. Other products produced in the process are propane, carbon dioxide, water, some naphtha, and a small amount of residual material. Since the term biodiesel has been used in the literature to describe the lower molecular weight alkyl esters of natural fatty acids, it may seem inappropriate to use the same term to describe the substitute diesel fuels made from natural triglycerides by catalytic hydroprocessing. The products from hydroprocessing of the oils have been called "super cetane" by some because of their superior properties compared to those of conventional diesel fuel and the methyl and ethyl esters of natural triglycerides. It is evident that the catalytic hydroprocessing of triglycerides can provide a direct route to synthetic diesel fuel components that are fully compatible with conventional diesel fuels without use of methanol or ethanol. This work has also shown that the hydrocarbon products from the process are miscible in all proportions with diesel fuel. They also have a linear effect on the cetane number when blended with petroleum diesel fuel, unlike commercial alkyl nitrate cetane improvers.

It is well known that Mobil Oil Company's MTG process uses shape-selective zeolite catalysts of the ZSM type to promote the conversion of methanol alone to high-grade aromatic gasolines at high selectivities (cf. Yurchak and Wong, 1992). The process was first commercialized in 1985 in New Zealand. Similarly, catalytic conversion of triglyceride oils as well as the monounsaturated ester jojoba oil over shape-selective zeolite catalysts of the ZSM-5 type affords good yields of aromatic-rich gasoline-range liquids (Weisz, Haag, and Rodewald, 1979; Furrer and Bakhshi, 1989, 1990). ZSM-5 catalyst is a crystalline alumino-silicate and is a member of the family of highly siliceous, porous tectosilicates. The high selectivity for methanol conversion results from the fact that the terminal size of the products, mostly aromatics, is limited by the molecular shape selectivity of the catalyst, that is, by the constraining size of the intracrystalline cavities. This is believed to be the reason why ZSM-5 also promotes the conversion of the glycerides, and even terpenes, to a very similar product mixture. Experimental data that illustrate the product distributions with ZSM-5 catalysis are shown in Table 10.8. The behavior of ZSM-5 catalysis contrasts with that of other cracking catalysts for biomass components. The synthetic clay catalysts, for example the nickel-substituted, synthetic mica montmorillonite, and the alumina-pillared, nickel-substituted, synthetic mica montmorillonite, promote the conversion of fatty acids to high-quality gasoline products, but they contain high percentages of branched alkanes and low percentages of aromatics (Olson and Sharma, 1993).

B. Terpenes and Hydrocarbon-like Compounds

A wide range of terpenes and terpenoids can occur in various biomass species. Representative structures are shown in Fig. 10.4. The compounds shown in this figure and related organic structures can be refined and upgraded by processes such as hydrogenation to high-cetane-number liquid fuels. The compounds can also be refined and upgraded for improved combustion characteristics when used as liquid fuels for internal combustion engines. To effectively use terpene liquids as fuels for internal combustion engines, the alicyclic structures can be aromatized by known chemistry to yield p-cymene, for example, as in the case of dipentene conversion to high-octane, substituted aromatics, which are established gasoline components. Alternatively, the alicyclic and acyclic terpenes can be cracked, hydrogenated, and/or isomerized to yield high-octane isoalkanes, which are also established gasoline components. Terpene-derived fuels have low ash and sulfur contents, are miscible with gasolines and diesel fuels, and can have high octane or cetane numbers depending on their structures and the refining procedures used. Fuel components having structures identical to those of the aromatized, cracked, hydrogenated,

TABLE 10.8 Product Distributions on Catalytic Conversion of Methanol, Dipentene, Latex Rubber, and Triglycerides over ZSM-5 Catalyst at Atmospheric Pressure

Feedstock	Conditions		Product distribution					
	Temperature (°C)	WHSV[c]	C_1–C_5 Paraffins (wt %)	C_2–C_5 Olefins (wt %)	C_6–C_9 Aromatics (wt %)	C_6–C_{14} Others (wt %)	Coke (wt %)	Gasoline (%)
Methanol[a]	450	0.67	46	7	43	4	0	48
Latex rubber[a]	482	0.6	7	14	43	10	26	72
Dipentene[a]	482	0.6	5	34	49	12	0	68
Castor oil[a]	500	2.5	19	6	65	7	3	72
Corn oil[a]	450	2.4	28	7	49	13	3	60
Jojoba oil[a]	400	1.3	44	7	42	6	0.7	54
Canola oil[b]	374	6.5	10		55	20	8 (6)	75
Flax oil[b]	373	6.2	10		54	19	0 (17)	73
Linseed oil[b]	371	6.4	8		56	14	0 (22)	70
Mustard oil[b]	374	6.4	10		41	30	7 (12)	71
Rapeseed oil[b]	392	6.4	10		56	23	0 (11)	79
Sunflower oil[b]	372	6.4	10		50	14	13 (14)	64
Soybean oil[b]	371	6.5	8		51	18	9 (14)	69

[a]Adapted from Weisz, Haag, and Rodewald (1979). Hydrogen was passed over the bar charts in the reference. The % gasoline is its approximate percentage in a product consisting of fuel gas (C_1 and C_2), liquefied petroleum gas (C_3 and C_4), gasoline, and light distillate.

[b]Adapted from Furrer and Bakhshi (1989, 1990). A fixed-bed microreactor containing the catalyst was used for these experiments. Hydrogen was not passed over the bed during the process. The % paraffins includes olefins. The % aromatics and % others (liquids) were calculated from data in the reference on the amount of feedstock converted, which ranged from a low of 73% for soybean oil to a high of 86% for canola oil, and the compositions of the liquid products. The figure in brackets under % coke is the percent tars in the total product. The % gasoline is the sum of the % aromatics and % others, which in this reference consists of C_5–C_{12} nonaromatic compounds.

[c]WHSV is the weight of feedstock fed per hour per unit weight of catalyst in the reactor.

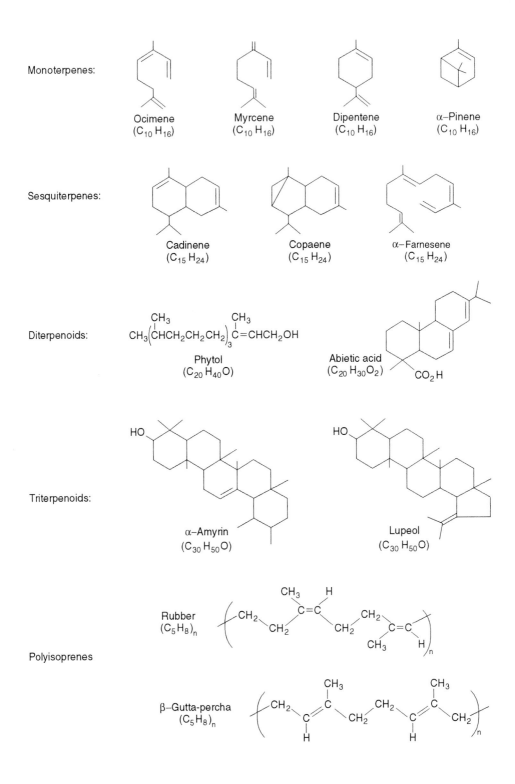

Monoterpenes:

Ocimene
($C_{10}H_{16}$)

Myrcene
($C_{10}H_{16}$)

Dipentene
($C_{10}H_{16}$)

α–Pinene
($C_{10}H_{16}$)

Sesquiterpenes:

Cadinene
($C_{15}H_{24}$)

Copaene
($C_{15}H_{24}$)

α–Farnesene
($C_{15}H_{24}$)

Diterpenoids:

$$CH_3\left(CHCH_2CH_2CH_2\right)_3 \overset{CH_3}{C}=CHCH_2OH$$

Phytol
($C_{20}H_{40}O$)

Abietic acid
($C_{20}H_{30}O_2$)

CO_2H

Triterpenoids:

HO

α–Amyrin
($C_{30}H_{50}O$)

HO

Lupeol
($C_{30}H_{50}O$)

Polyisoprenes

Rubber
(C_5H_8)_n

β–Gutta-percha
(C_5H_8)_n

or isomerized terpene derivatives are found in certain petroleum refinery streams and crude oils.

Polyisoprenes from biomass latexes such as natural rubbers can also be depolymerized and converted to lower molecular weight hydrocarbons by a variety of thermal processes. Zeolite catalysts of the HZSM-5 type have been found to be especially effective as catalysts (cf. Table 10.8). Guayule resin, a by-product from guayule rubber extraction, contains terpenes, fatty acids, low-molecular-weight rubber, high-molecular-weight alcohols, and other products. It is readily converted in the presence of zeolite catalysts to good yields of C_5–C_{10} aromatic and nonaromatic hydrocarbons (Costa et al., 1992). The best results were obtained using a physical mixture of 70 wt % HZSM-5 catalysts (Si/Al, 13) and 30 wt % silica-alumina amorphous matrix (Si/Al, 20), bound with 30 wt % sodium montmorillonite. The light liquid hydrocarbon fraction obtained contains more than 80 wt % C_6-C_{10} aromatics.

IV. PROCESSES AND ECONOMICS

Numerous methods have been tested, demonstrated, and commercialized for the separation and purification of the organic liquids produced by natural biochemical processes. As already discussed, some biomass species produce organic liquids that can be collected over a period of time without harvesting or destroying the plant simply by "tapping" as for copaiba oil, or in the case of microalgae, by skimming the microalgal oil from the water surface after the algal cells are disrupted. Most natural hydrocarbon separation processes, however, involve biomass size reduction, drying, solvent extraction, and separation of the extracted products from the solvent, the recovery and recycling of which is important to minimize costs. Physical separation of the liquids by treatment in extrusion devices or presses is often used for seed-oil separation. Some processes use several solvents for fractionating the different liquid and solid components in the biomass. Depending on the composition and properties of the biomass and the products and purities desired, the separation method may also use thermal and chemical processing techniques in combination with physical processing methods. After separation of the liquids and sometimes co-products, a relatively large amount of residual biomass is left that must

---

FIGURE 10.4   Typical organic compounds in hydrocarbon extracts. For nonpolymeric compounds, a bonded "—" denotes a methyl group at the free end. Similarly, a bonded "=" denotes a methylene group at the free end.

either be processed further to obtain other components of value, disposed of, or used as solid fuel.

A. TERPENES AND BIOCRUDE

A conceptual process sequence to recover biocrude and other products from *E. lathyris* grown in California was analyzed in the early 1980s for economic feasibility (Kohan and Wilhelm, 1980). The process was based on solvent extraction. Since it had been experimentally shown that the entire crude carbohydrate fraction extracted from *E. lathyris* by hot water washing could be directly converted without purification to fermentation ethanol, the sugars were assumed to be produced for sale in this study as a water solution with subsequent conversion by the purchaser to ethanol. Some or all of the bagasse was optionally used on-site for raising steam, and the balance was assumed to be sold as solid fuel. The processing sequence for acetone extraction of field-dried *E. lathyris* (20 wt % moisture) is illustrated in Fig. 10.5. Analysis of the dry feedstock showed that it contained approximately 8 wt % biocrude, 26 wt % hexose sugars, and 7.3 wt % ash. These data were used for the economic analysis. The hypothetical mass and energy balances for the base case of the conceptual process are shown in Table 10.9 for a 907-dry t/day plant (1000 dry ton/day). It is apparent that the sugar and bagasse fractions are the major products. Three cases were assumed for the economic analysis: the optimistic, base, and pessimistic cases. It was assumed that all of the biocrude and sugars are extracted in the process and sold, that all of the bagasse not used as process fuel is sold for $16.10/t, and that the market value of the product sugars is $44.10, $88.20, or $132.30/t for the pessimistic, base, and optimistic cases, respectively. Nonregulated industrial financing was assumed at 100% equity, and the revenue required for biocrude was then calculated for each case. The details of the analysis are shown in Table 10.10. It is evident that at the time of this analysis, when the average price of crude oil at the wellhead in the United States was near its peak during the Second Oil Shock, the estimated market value of the biocrude required for the optimistic case was still about three times that of petroleum. It was concluded that an eightfold and even greater decreases in feedstock costs appear to be necessary to reduce product revenue requirements to levels competitive with petroleum liquid fuels. It was also concluded from sensitivity studies of the revenue required from the sale of biocrude-to-feedstock costs that even if *E. lathyris* containing 10 wt % biocrude and 26 wt % sugars could be successfully cultivated in the central and southern regions of the United States without irrigation at an estimated low feedstock cost of $35.42/ t ($2.05/GJ, $12.10/BOE), the revenue requirements would still be too high for the biocrude. For those

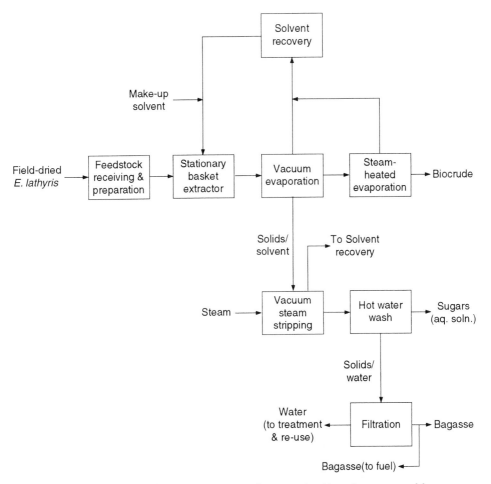

FIGURE 10.5 Conceptual processing sequence for recovering biocrude, sugars, and bagasse from *Euphorbia lathyris.*

conditions, the revenue requirement for the biocrude is reduced to $7.62/GJ or $44.40/BOE, a price that was $5 to $15/bbl higher than that of petroleum crude oil at that time. Adjustment of the estimated costs of this study to today's markets by taking inflation into account would not appear to change the conclusions of this analysis because today's petroleum crude oil prices are still much less in nominal dollars than in 1980.

Field trials to study the growth and compositional characteristics of *E. lathyris* indicated that biocrude would have to sell for $150–$200/bbl, even

TABLE 10.9 Mass and Energy Balances for Recovery of Biocrude, Sugars, and Bagasse by Solvent Extraction of Whole *Euphorbia lathyris*[a]

Process parameter	Mass rate (t/day)	Energy rate (TJ/day)	Energy distribution (%)
Inputs			
Field-dried *E. lathyris*			
Organic matter	841		
Ash	66		
Moisture	227		
Total feedstock	1134	15.8	97.5
Makeup solvent	3	0.1	0.6
Purchased electricity		0.3	1.9
Total:	1137	16.2	100.0
Outputs			
Biocrude	73	2.8	17.3
Sugars (10.7 wt % aqueous solution)	236	4.0	24.7
Bagasse to sales	174	2.6	16.0
Bagasse to process fuel (without ash)	377		
Ash from process fuel	47		
Solvent loss	3	0.1	0.6
Water condensed and treated	227		
Process heat losses		6.7	41.4
Total:	1137	16.2	100.0

[a]Adapted from base case of Kohan and Wilhelm (1980).

with credit for by-products from the extraction process (Sachs *et al.*, 1981). Additional studies supported a cost of biocrude from *E. lathyris* before refining closer to $100/bbl (Hoffmann, 1983; Calvin, 1985).

In other economic analyses of biocrude production, the conversion of residual bagasse to salable co-product electric power was evaluated in an integrated biomass growth-processing system to determine whether the cost of biocrude could be reduced to a practical level (McLaughlin, Kingsolver, and Hoffmann, 1983). Four of the best potential biocrude crops for the Southwest were selected from extensive surveys of native species as model feedstocks. In addition to *E. lathyris*, *Calotropis procera*, *Grindelia camporum*, and *Chrysothamnus paniculatus* were chosen. In this model, biocrude is extracted with nonpolar solvents from the whole plants, and therefore the carbohydrates are in the bagasse fraction. The corresponding biocrude yields are projected to be 8.6, 7.7, 12.7, and 10.8 bbl/ha-year, and the corresponding weight percentages of biocrude in the dry biomass are 8, 5, 15, and 20. The system consists of a 907-dry t/day processing plant and the surrounding acreage required to supply the plant for with feedstock for 300 day/year (272,000 t/year). The processing sequence includes feedstock receiving and storage, feedstock preparation in

TABLE 10.10   Estimated Costs for Recovery of Biocrude, Sugars, and Bagasse from *Euphorbia lathyris* in California at Feed Rate of 907 Dry Tonne per Day[a]

Parameter	Optimistic case	Base case	Pessimistic case
Operating basis			
Annual operating time, h/year	7008	7008	7008
Heating value of biocrude, GJ/t	38.39	38.39	38.39
Value of sugar solution, $/t sugars	132.30	88.20	44.10
Value of bagasse, $/t	16.10	16.10	16.10
Solvent-to-feed solids ratio	1	3	3
Steam, kg/t dry extractor feed	960	2050	2050
Bagasse fuel usage, t/day	200	424	599
Purchased electricity, MW	2.7	3.2	0
Biocrude to sales, t/day	73	73	73
Sugar to sales, t/day	236	236	236
Bagasse to sales, t/day	399	174	0
Capital costs, 10^6			
Feedstock receiving and preparation	2.0	2.5	5.3
Solvent extraction of biocrude	1.2	2.5	3.5
Aqueous extraction of sugars	2.7	4.5	6.6
Utility facilities	3.7	11.4	19.1
General services facilities	1.5	3.1	5.2
Royalties	0.6	1.2	5.0
Land	0.3	0.5	2.0
Interest during construction	0.9	1.9	0.8
Start up expenses	0.3	1.2	3.1
Working capital	1.9	3.1	3.4
Total capital investment	15.1	31.9	54.0
Revenue required, $/GJ biocrude produced			
Feedstock	23.75	29.05	34.32
(Feedstock cost, $/t)	(72.96)	(89.18)	(105.40)
Operating and maintenance	0.98	1.38	1.83
Purchased utilities	1.21	1.72	2.44
Labor costs	1.69	2.75	4.39
Fixed costs	1.14	1.41	1.81
Nonregulated industrial financing	4.31	9.22	15.78
Total revenue required	33.08	45.53	60.57
(Regulated utility financing)	(31.16)	(41.50)	(53.42)
Revenue required, 10^6/year	27.07	37.26	49.57
From sugars	9.17	6.11	3.05
From bagasse	1.87	0.82	0.00
From biocrude	16.03	30.33	46.52
Biocrude cost, $/GJ	19.64	37.06	56.85
$/BOE	115.95	218.80	335.64

[a]Adapted from Kohan and Wilhelm (1980). Nonregulated industrial financing is 100% equity and 15% DCF-ROR; regulated utility financing is 65/35 debt/equity ratio.

hammer and flaking mills, continuous solvent extraction, solvent-biocrude separation by steam-stripping, bagasse drying using waste heat from the power plant, and bagasse combustion in a 40-MW power plant. The electric power is assumed to be sold to the local grid at $0.04/kWh. The annual biocrude revenue is that required to balance the annual capital and operating costs. This model is based on fuel, power, equipment, and materials required for furrow-irrigated agriculture in Arizona. The capital required for agricultural equipment and the conversion plant, and the cost of capital are included in the analysis, the results of which are shown in Table 10.11.

This economic analysis of integrated biomass production–conversion systems indicates that the production cost of biocrude is not competitive. The estimated cost of biocrude from the biomass species selected was not competitive with the high petroleum crude prices at the time of the study. One of the major economic factors that tended to preclude economic feasibility is feedstock cost. In Table 10.11, the cost per dry tonne of feedstock ranges from a low of $62.50 for *C. procera* to a high of $84.93 for *C. paniculatus*. It is apparent that significant reductions in feedstock costs are necessary to effect large reductions in biocrude cost. However, it is perhaps unexpected that as the yield of each of the four species decreases in the order *C. procera, E. lathyris, G. camporum,* and *C. paniculatus,* the cost of biocrude also decreases. The reason for this is that there is a corresponding simultaneous increase in biocrude content of the biomass that is more than sufficient to overcome biomass yield reductions. For example, *C. paniculatus* is grown at the lowest yield, about 35% that of the highest yielding species, *C. procera,* but supplies biocrude at about 70% lower cost. This result emphasizes the importance of feedstock composition when the production of natural products, in this case biocrude, rather than biomass energy itself is desired. The market value of the co-product electric power seems to have little impact on biocrude cost under the conditions assumed for the analysis because the revenue from power sales decreases only slightly as biomass yields decrease. If electric power were the only target product, then maximum biomass yield for conversion to electric power would be the prime objective. In the ideal case, a multiproduct, integrated biomass growth–harvesting–processing system should have sufficient operating flexibility to maximize or minimize the production of individual products at any given time to take advantage of market fluctuations. Of course, the potential changes in feedstock composition could also affect product distributions and would have to be closely monitored.

B. TRIGLYCERIDES

Figure 10.3, shown earlier, illustrates a typical process configuration for the manufacture of methyl biodiesel by transesterification, or methanolysis since

TABLE 10.11 Estimated Costs for Integrated Growth of Four Biomass Species and Recovery of Biocrude and Power Production in Arizona at Feed Rate of 272,000 Dry Tonne per Year[a]

Parameter	Calotropis procera	Euphorbia lathyris	Grindelia camporum	Chrysothamnus paniculatus
Biomass growth				
Habit	Perennial	Annual	Annual	Perennial
Irrigation, cm/ha	150	102	78	46
Land, ha	15,000	21,400	27,300	42,900
Biomass yield, dry t/ha-year	18.14 (20)	12.72 (14)	9.97 (11)	6.34 (7)
Biocrude content, wt %	5	8	15	20
Products				
Biocrude, bbl/ha-year	7.7	8.6	12.7	10.8
GJ/ha-year	41.7	41.8	57.8	49.0
TJ/yr	625	895	1578	2102
Bagasse intermediate, dry t/ha-year	16.9	11.7	8.5	5.5
GJ/ha-year	313.5	216.4	157.1	101.9
TJ/year	4702	4631	4293	4372
Gross electric power, TJ/year	1364	1343	1245	1268
10^6 kWh/year	379.2	373.3	346.1	352.5
Capital and operating costs, 10^6/year				
Feedstock Production	17.0	18.4	19.7	23.1
Solvent extraction	6.0	6.0	6.0	6.0
Combustion	10.0	9.7	8.9	8.4
Total cost	33.0	34.1	34.6	37.5
Revenue required, 10^6/year				
From electricity	15.2	14.9	13.8	14.1
From biocrude	17.8	19.2	21.8	23.4
Biocrude cost, $/GJ	28	21	13	11
$/bbl	154	104	60	51

[a]Adapted from McLaughlin, Kingsolver, and Hoffmann (1983). The water requirements were estimated to be 882 t/dry t of crop for each species. Consumptive water use is approximately 70% of the irrigation water. The model is based on fuel, power, equipment, and materials required for furrow-irrigated agriculture. Biocrude includes the lipids and terpenoids extractable from biomass with nonpolar solvents. Fuel for raising steam is bagasse, which is converted to electric power at a thermal efficiency of 29%. The power is sold for $0.04/kWh. Unstated capital charges are included in the estimates.

glycerol is displaced by methanol. Most biodiesel production systems utilize similar schemes; historically, the process has been conducted in the batch mode, although some effort has been devoted to development of continuous processes (cf. Noureddini, Harkey, and Medikonduru, 1996). Transesterification is promoted by alkaline catalysts such as hydroxide or alkoxides and by acids. Relatively large amounts of alcohol are used to shift the process toward the product monoesters. The excess alcohol is then recovered for recycle. As shown in Fig. 10.3, the products of the process in addition to biodiesel are glycerol and meal if oilseeds are the feedstock.

A detailed economic analysis was performed for methyl biodiesel production plants having processing capacities of 2.72, 4.54, and 21.77 t/day of rapeseed (Gavett, Van Dyne, and Blase, 1993). It was assumed that the feedstock contained 42 wt % oil. The oil content of Brassica oilseeds generally ranges from 38 to 44% of the dry weight. This oil content offers a best-case condition even though the plants are small-scale facilities. Biodiesel production from the facilities is 393,700, 656,200, and 3,244,400 L/year. The results of the analysis are summarized in Table 10.12. The operating conditions and product credits for the analysis are shown in the table. Oilseed presses are used to extract the oil, and established transesterification technology is used to produce the methyl esters. The methyl esters turned out to cost $79.61, $78.81, and $70.71/bbl. The larger facility, which had eight times the capacity of the smaller facility, afforded biodiesel at the lower cost as expected. Sensitivity analysis indicated that a reduction in the annual capital cost to 10% from 21.7% of the investment reduced the ester costs by $2.94 to $3.78/bbl, while a 20% reduction in feedstock cost from $220.50/t to $176.40/t, reduced the cost of the methyl esters about 27% to $57.96, $57.45, and $52.50/bbl, a substantial reduction. However, it was concluded that the production of biodiesel in small-scale plants could not compete with petroleum diesel at the prices that prevailed in the early 1990s. In 1990 and 1991, U.S. farmers paid an average of $36.54 to $37.80/bbl in nominal dollars for bulk diesel fuel delivered to the farm; the price structure was about the same in the mid-1990s. Farmers are exempt from federal road taxes on diesel fuel used in farm production.

The market price of the feed alcohol and the market prices of the by-product glycerol and meal affect the prices of the esters. The relatively low seed-oil yield compared to the total amount of biomass that must be harvested and processed is one of the prime reasons for high triglyceride costs and the high cost of the methyl or ethyl esters from the transesterification process. Since the esters must be competitive with petroleum diesel fuels for large-scale biodiesel markets to emerge, it is unlikely that they will become commercial fuels in the United States in the near term unless triglyceride yields and therefore costs are significantly improved. However, because of the properties

TABLE 10.12 Estimated Costs of Producing Refined Biodiesel from Rapeseed in Small Commercial Plants[a]

Parameter	Feedstock capacity		
	2.72 t/day	4.54 t/day	21.77 t/day
Operating basis			
Oil in rapeseed, wt %	42	42	42
HHV methyl ester, 38.0 MJ/L	38.0	38.0	38.0
Purchased rapeseed, $/t	220.50	220.50	220.50
Purchased methanol, $/L	0.159	0.159	0.159
Residual oil in meal, wt%	15	15	11
Value of glycerol, $/t	73.00	73.00	73.00
Value of meal, $/t	159.86	159.86	159.86
Value of recycle methanol, $/L	0.159	0.159	0.159
Methyl ester to sales, L/year	393,700	656,200	3,244,400
Glycerol to sales, t/year	324.2	540.4	2457
Meal to sales, t/year	347.8	579.6	2770
Capital Cost, $	75,000	115,000	250,000
Capital and Operating Cost, $/year			
Capital cost	16,275	24,955	54,000
Operating cost	234,590	390,190	1,821,880
Sales and administration	25,087	41,514	187,563
Working capital	2875	4758	21,249
Total Cost	278,827	461,417	2,084,692
Revenue Required, $/year			
From glycerol	23,668	39,446	179,341
From recycle methanol	2453	4089	19,627
From meal	55,593	92,655	442,891
From ester	197,113	325,227	1,442,825
Methyl Ester Cost, $/GJ	13.38	13.05	11.71
$/bbl	79.61	78.81	70.71

[a]Adapted from Gavett, Van Dyne, and Blase (1993). The meal is sold at its protein value with soybean meal and no credit is taken for the oil in the meal. Oilseed presses are used to extract the oil and established transesterification technology is used to produce the methyl ester.

of biodiesel–diesel blends, markets for biodiesel as cetane-enhancing additives in petroleum diesel fuels are expected to develop.

An alternative to the integrated processing of oilseeds and seed-oil transesterification is to purchase the oils on the open market and transform them into biodiesel. This eliminates the by-product meal, simplifies the process, and lowers the capital and operating costs. In this case, the principal economic barrier in the United States is the direct purchase cost of $0.53 to $0.79/L ($2 to $3/gal) for triglyceride oils. This is acceptable in certain European regions where heavily taxed diesel fuels sell for $0.53 to $1.06/L, but not in North America (*cf.* Reed, 1993). In the mid-1990s, the wholesale prices of large volumes of natural triglycerides such as palm, soybean, corn, and cottonseed oils ranged from $84 to $126/bbl. The cost of transesterification is about $21/bbl, including glycerol credits, so the cost of biodiesel would still be excessive, about $105 to $147/bbl.

It appears that one of the few approaches to profitable production of biodiesel at petroleum prices under $45/bbl is to utilize the Minor Oilseed Provision of the 1990 U.S. Farm Bill (Gavett, Van Dyne, and Blase, 1993). In this option, the farmer grows rapeseed on land that is removed from production of a crop such as corn, wheat, cotton, and soybeans. The bill permits a farmer to harvest and sell minor oilseed crops grown on this "set-aside" land without losing the program participation payment. In effect, the farmer is paid land rent by the government, but can still produce minor oilseed crops. If allowed by the government, U.S. farmers would grow oilseed crops on set-aside land for subsequent processing and conversion to biodiesel in a farmer cooperative without losing the government subsidy. Also, tax incentives may be legislated, because they are available for other alternative transportation fuels such as ethanol from biomass. Subsidies have been provided for biodiesel in some European countries.

The production of super cetane by direct hydrotreatment of natural triglycerides can, in principle, significantly reduce the operating costs of converting triglycerides to diesel fuel substitutes. The process is less complex than transesterification, as illustrated by the basic operations of the processing sequence shown in Fig. 10.6. Conventional refinery processing is used to produce a

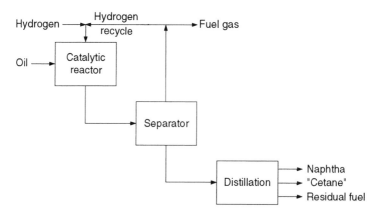

FIGURE 10.6   Catalytic hydroconversion of natural oils, tall oils, and animal fats to cetane enhancers.

paraffinic hydrocarbon product, in which the dominant compounds can be cetane itself and closely related homologues. Super cetane is virtually identical to some of the high-cetane-number normal paraffins in petroleum diesel fuels and is therefore suitable for direct use in diesel fuel blends. The high cetane number of super cetane and its linear effect on the cetane number of any fuel to which it is added are expected to make it valuable as an additive for low-quality middle distillates. A summary of the economic analysis of the production of super cetane from triglycerides and tall oils in which the feedstock is priced at three levels—$67.37, $84.21, and $126.32/t—is shown in Table 10.13 (Stumborg et al., 1993). Although the corresponding costs per barrel are $9.86, $12.32, and $18.48, which are much too low for natural triglycerides, the objective of this study was to determine the sensitivity to feedstock cost for different operating scenarios based on a medium-sized Canadian refinery (5500 m^3/day, 34,600 bbl/day). Feedstock consumption is 23,552 t/year (160,998 bbl/year). The options included a stand-alone, grassroots treating facility (case I), a new conversion unit in an existing refinery or chemical plant (case II), an additional hydroprocessing unit in an existing refinery (case III), and conversion in an existing hydrotreating unit (case IV). The stand-alone facility is not economic because there is insufficient value added to the product and capital expenditures are quite high along with the high overhead costs associated with a small plant. Integrating new units or adding facilities to existing units (cases II and III) shows more promise. These options are limited by feedstocks in the $10/bbl range. Tall oil is the only feedstock that meets this criterion as natural triglycerides cannot be obtained at that price. Case IV is the most tolerant of feedstock price and is viable up to a price of $18.50/bbl. It was concluded from this analysis that use of a waste biomass oil such as tall oil in an existing hydrotreater would be the most likely scenario for commercial exploitation of the technology.

C. COMMERCIAL PROSPECTS FOR NATURAL LIQUID FUELS PRODUCTION

The data presented here indicate that biomass species that contain high levels of hydrocarbon-like compounds and that also grow at high yields per growth area are rare. High biomass yield and hydrocarbon content are unlikely to occur simultaneously in the same species on a continuous basis. This is understandable when considered in terms of the biochemical energy exchanges that occur during biomass growth. When biomass with higher concentrations of the high-energy compounds such as hydrocarbons and triglycerides is formed, more biochemical energy is needed to form reduced carbon compounds than is required to form the conventional structural components which contain

TABLE 10.13 Estimated Simple Rates of Return for Production of Substitute Diesel Fuels and Additives from Triglycerides and Tall Oils in an 80-m³/day Catalytic Hydrotreating Facility[a]

| Case | Capital investment ($10⁶) | Feedstock cost | | Operating cost | | | Net income ($10⁶/year) | Simple ROR (%) |
		Per unit ($/t)	Annual ($10⁶/year)	Ex feed ($10⁶/year)	Total ($10⁶/year)			
I A	17.85	67.37	1.58	2.53	4.11	0.94	5.3
I B	17.85	84.21	1.99	2.53	4.51	0.54	3.0
II A	8.93	84.21	1.99	1.18	3.17	1.88	21.1
II B	8.93	126.32	2.98	1.18	4.16	0.89	10.0
III A	5.64	84.21	1.99	0.92	2.91	2.14	37.9
III B	5.64	126.32	2.98	0.92	3.91	1.14	20.2
IV A	1.01	84.21	1.99	1.68	3.67	1.38	137
IV B	1.01	126.32	2.98	1.68	4.67	0.38	37.6

[a]Adapted from Stumborg et al. (1993). Densities of 0.920 and 0.790 t/m³ are used for feedstock oils and products. Product yield is 80 wt % of the feedstock used. Operating time is 320 days/L, feedstock consumption is 23,552 L/year, production is 23,850,126 L/year, and annual gross revenue at an assumed product sales price of $0.211/L is $5.03 million. All dollar values are converted from Canadian dollars in the original reference to U.S. dollars by a factor of $1.1875 Can./U.S. $, the exchange rate in April 1992. Some of the calculated values may not equal the arithmetic values due to rounding. Case I is a stand-alone, grassroots, hydrotreating facility. Case II is a new conversion unit in an existing refinery or chemical plant. Case III is an additional hydroprocessing unit in an existing refinery. Case IV is conversion in an existing hydrotreating unit.

carbon in a higher oxidation state. The structural components make up the bulk of most biomass. Stressed growth conditions can increase the content of the higher energy compounds within the biomass, but usually at the expense of biomass yield. It appears that the protection of biocrude crops from adverse conditions of water and nutrient stress, in order to avoid a reduction in their growth, may have the ironic effect of impeding formation of the desired biocrude (McLaughlin, Kingsolver, and Hoffmann, 1983). The idea of transferring the gene that codes for terpene hydrocarbon production from the Brazilian tree *Copaifera multijuga* to *Euphorbia* species to try to improve hydrocarbon yields for biomass that grows in the United States would therefore be expected to be difficult to achieve for herbaceous species intended for whole-plant harvesting and extraction. Continuous removal of high-energy compounds from growing biomass should offer more opportunity to produce hydrocarbon-like compounds, because the particular biomass would then be used as a continuous manufacturer of natural liquid fuels.

Excluding oilseed-bearing plants, the data developed from intensive searches to discover plants that produce good yields of nonpolar solvent extractables (lipids, terpenes, waxes) in the United States are not too encouraging. Those species that appear to be promising candidates still need large improvements in yield to make them practical sources of liquid fuels on a sustainable basis. The plants found to afford the highest yields of hexane extractables are generally either latex-bearing plants belonging to the Asclepiadaceae (milkweed), Asteraceae (Compositae), or Euphorbiaceae (spurge) families, or resinous and/or "gummy" Asteraceae plants such as the so-called gumweeds, resinweeds, rosinweeds, tarbushes, and tarweeds (Balandrin, 1985). Much research remains to be done to make it possible to design economically practical biomass production–extraction systems for the production of natural biocrude liquid fuels.

The economic data presented here on the production and conversion of triglycerides to biodiesel show that it is not competitive with petroleum diesel fuels in North America. Commercial markets have developed in certain areas in Europe, but the transportation fuel price structure is considerably different than in North American markets. The production costs of biodiesel are still much higher than those of conventional diesel fuel. Considerable research is in progress to develop improved transesterification processes, but little effort has been expended to develop new cultivars that yield less expensive triglycerides. Barring an oil embargo or a significant increase in diesel fuel prices, research on the development of superior, triglyceride-yielding biomass strains through intensive breeding programs or genetic engineering improvements will have to be carried out before biodiesel can compete in an open energy market. In lieu of significant decreases in biocrude costs, higher value chemical applications of biocrude products would seem to be more attractive until the price of petroleum crude oils increases to the point where biocrudes are competitive.

The largest biomass producers of commodity seed oils—soybean, cotton-seed, groundnut, sunflower, rapeseed, sesame, palm, linseed, and castor—do not appear to be practical sources of diesel fuels without new government mandates or tax incentives. The yields of the triglycerides are too low, and the costs are too high. It is of course possible that the application of modern genetic engineering techniques could be applied to increase the yields of triglycerides and that new higher yielding strains could be developed. This remains to be discovered in long-range research programs. Oilseed crops cannot be a practical source of hydrocarbon fuels either for the foreseeable future because the yields of terpenes, when they are produced at all, are even lower than the triglyceride yields. It appears that even the high triglyceride yields of rapeseed are insufficient to justify use of the transesterification products with methanol or ethanol as fuel at current U.S. petroleum prices (about $20/bbl range in mid-1997).

The Chinese tallow tree, however, clearly has good potential as a source of triglyceride feedstocks for biodiesel and other useful products. Further, the tree exhibits good genetic diversity, which suggests that selection might lead to even higher yields of tallow and oil, and it apparently has no natural enemies in Texas, where it is an introduced species. Although the goal of transferring the gene that codes for terpene hydrocarbon production to a biomass species that grows in the United States has not been achieved yet, it appears that the Chinese tallow tree could be grown commercially now in the Southwest and "tapped" as a large-scale source of substitute hydrocarbon fuels in the form of triglyceride feedstocks for biodiesel or conversion to hydrocarbons. This approach would also avoid whole-plant harvesting. The tree may also be suitable as a source of valuable co-product fiber and chemicals, and possibly wood fuel when the tree is finally harvested. The Chinese tallow tree should be given serious consideration for growth in the United States and elsewhere as a commercial, dedicated energy crop.

The technology for triglyceride production from microalgae has not yet been commercialized. Most of the economic analyses for the production of microalgal liquids reported in the literature indicate they are much too expensive to compete with petroleum fuels. Considerable additional research must be carried out to perfect the process despite the fact that research on microalgal fuel production has been in progress for at least the past two decades.

An economically attractive approach to commercial production of diesel fuel substitutes from natural biomass liquids in North America appears to be the direct conversion of waste biomass oils to super cetane by catalytic hydrotreatment. This technology is expected to be commercialized first in Canada where it was developed. The availability of large amounts of feedstock may initially limit production, but in the near term, this type of process technology, which utilizes conventional refinery processing, is expected to be the entree of natural biomass liquids to diesel fuel markets in North America. Indeed, conventional refinery processes are expected to be the most economic and practical technologies for producing both diesel fuels and motor gasolines from natural biomass liquids. There is little or no technical barrier to the application of refinery processes to these liquid feedstocks, whether they are triglycerides, terpenes, terpenoids, or mixtures.

An innovative option that should be examined further is the use of aqueous carbohydrate blends containing small amounts of alcohol as diesel fuels. These formulations are not natural oils, but if they turn out to be effective fuels, they would be available at relatively low cost from a wide range of biomass feedstocks, including waste biomass.

REFERENCES

Adams, R. P. (1982). In "Energy From Biomass and Wastes VI," (D. L. Klass, ed.), p. 1113. Institute of Gas Technology, Chicago.

Adams, R. P., Balandrin, M. F., Hogge, L., Craig, W., and Price, S. (1983). J. Am. Oil Chem. Soc. 60 (7), 1315, July.

Adams, R. P., Balandrin, M. F., and Martineau, J. R. (1984). Biomass 4, 81.

Anonymous (1992). Chem. Eng. 99 (5), 21.

Auld, D. L., Peterson, C. L., and Korus, R. A. (1989). In "Energy From Biomass and Wastes XII," (D. L. Klass, ed.), p. 1187. Institute of Gas Technology, Chicago.

Bagby, M. O., Buchanan, R. A., and Otey, F. H. (1981). In "Biomass as a Nonfossil Fuel Source," (D. L. Klass, ed.), ACS Symposium Series 144, p. 125. American Chemical Society, Washington, D.C.

Balandrin, M. F. (1984). Science 223 1386; Amer. J. Bot. 71 (No. 5, Pt. 2), 129, May-June; "Screening of Western U.S. Plant Species for Biomass and Energy Sources," Final Report to the U.S. Department of Agriculture, Grant. No. 59-2495-1-6-047-0, December.

Balandrin, M. F. (1985). In "Energy From Biomass and Wastes IX," (D. L. Klass, ed.), p. 1195. Institute of Gas Technology, Chicago.

Boocock, D. G. B., Konar, S. K., Mao, V., and Sidi, H. (1996). In "Second Biomass Conference of the Americas: Energy, Environment, Agriculture, and Industry," NREL/CP-200-8098, DE95009230, p. 961. National Renewable Energy Laboratory, Golden, CO.

Brenckmann, F., Largeau, C., Casadevall, E., and Berkaloff, C. (1985). In "Energy from Biomass 3rd E.C. Conference, Venice, Italy, March 25-29, 1985," (W. Palz, J. Coombs, and D. O. Hall, eds.), p. 717. Elsevier Applied Science Publishers, London.

Brown, L. M. (1993). In "First Biomass Conference of the Americas: Energy, Environment, Agriculture, and Industry," NREL/CP-200-5768, DE03010050, Vol II, p. 902. National Renewable Energy Laboratory, Golden, CO.

Buchanan, R. A., Cull, I. M., Otey, F. H., and Russell, C. R. (1978). Econ. Bot. 32, 131 and 146.

Calvin, M. (1978). Chem. & Eng. News 30, March 20.

Calvin, M. (1979). Energy 4, 851.

Calvin, M. (1980). Die Naturwissenschaften 67, 525.

Calvin, M. (1983). Science 219, 24.

Calvin, M. (1985). *Ann. Proc. Phytochem. Soc. Eur.* **26**, 147.

Calvin, M. (1987). *Bot. J. Linnean Soc.* **94**, 97.

Carr, R. (1989). *In* "Oil Crops of the World," (G. Röbbelen, R. K. Downey, and A. Ashri, eds.), p. 226. McGraw-Hill Publishing Company, New York.

Casadevall, E. (1984). *In* "Solar Energy R&D in the European Community, Energy from Biomass, Capri, Italy, June 7-8, 1983," (W. Palz and D. Pirrwitz, eds.), Series E, Vol. 5, p. 209. D. Reidel Publishing Co., Dordrecht, The Netherlands.

Casadevall, E., and Largeau, C. (1982). *In* "Proceedings of the Workshop on Biomass Pilot Projects and Methanol Production and Algae, Brussels, October 22, 1981," (W. Palz and G. Grassi, eds.), Vol. 2, p. 141. D. Reidl Publishing Co., Dordrecht, The Netherlands.

Clements, L. D. (1996). *In* "Liquid Fuels and Industrial Products from Renewable Resources," (J. S. Cundiff *et al.*, eds.), p. 44. American Society of Agricultural Engineers, St. Joseph, MI.

Costa, E., Aguado, J., Ovejero, G., and Cañizares (1992). *Fuel* **71**, 109.

Cruz, R. O., Stanfill, J., and Powaukee, B. (1996). *In* "Bioenergy '96, Proceedings of the 7th National Bioenergy Conference," Vol. I, p. 364. The Southeastern Regional Biomass Energy Program, Tennessee Valley Authority, Muscle Shoals, AL.

Feng, Y., Wong, A., and Monnier, J. (1993). *In* "First Biomass Conference of the Americas: Energy, Environment, Agriculture, and Industry," NREL/CP-200-5768, DE03010050, Vol II, p. 863. National Renewable Energy Laboratory, Golden, CO.

Freedman, B., Pryde, E. H., and Mounts, T. L. (1984). *J. Am. Oil Chem. Soc.* **61**, 1638.

Freedman, B., Bagby, M. O., Callhan, T. J., and Ryan, T. W., III, (1990). SAE Paper No. 900343, Society of Automotive Engineers, Warrendale, Pennsylvania.

Fukumoto, M. (1980). "The Oil-Bearing Plant Cultivation Under Utilization of Biomass as the Alternative Energy Source," Sekisiu Plastics Co., Ltd., Tokyo.

Furrer, R. M., and Bakhshi, N. N. (1989). *In* "Energy From Biomass and Wastes XII," (D. L. Klass, ed.), p. 1117. Institute of Gas Technology, Chicago.

Furrer, R. M., and Bakhshi, N. N. (1990). *In* "Energy From Biomass and Wastes XIII," (D. L. Klass, ed.), p. 897. Institute of Gas Technology, Chicago.

Gavett, E. E., Van Dyne, D., and Blase, M. (1993). *In* "Energy From Biomass and Wastes XVI," (D. L. Klass, ed.), p. 709. Institute of Gas Technology, Chicago.

Glenn, B. (1982). "Biomass Research Highlights," SERI Technical Information Branch, SERI/SP-281-1742, October. National Renewable Energy Laboratory, Golden, CO.

Hill, A. M., and Feinberg, D. A. (1984). "Fuel Products from Microalgae," SERI/TP-231-2348. National Renewable Energy Laboratory, Golden, CO.

Hoffmann, J. J. (1983). *CRC Critical Reviews in Plant Science* **1** (2), 95.

Johnson, J. D., and Hinman, C. W. (1980). *Science* **208** (4443), 460.

Jones, N., and Miller, J. H. (1991). "*Jatropha curcas*, A Multipurpose Species for Problematic Sites," Land Resources Series No. 1. The World Bank, Washington, D.C.

Kingsolver, B. E. (1982). *Biomass* **2**, 281.

Klass, D. L. (1983). *In* "Energy From Biomass and Wastes VII," (D. L. Klass and H. H. Elliott, eds.), p. 1. Institute of Gas Technology, Chicago.

Klass, D. L. (1984). *In* "Energy From Biomass and Wastes VIII," (D. L. Klass and H. H. Elliott, eds.), p. 1. Institute of Gas Technology, Chicago.

Klass, D. L. (1985). *In* "Energy From Biomass and Wastes IX," (D. L. Klass, ed.), p. 1. Institute of Gas Technology, Chicago.

Klass, D. L. (1994). *In* "Kirk-Othmer Encyclopedia of Chemical Technology," Ed. 4, **12**, p. 16. John Wiley, New York.

Knothe, G., Bagby, M. O., and Ryan, T. W., III (1996). *In* "Liquid Fuels and Industrial Products from Renewable Resources," (J. S. Cundiff *et al.*, eds.), p. 54. American Society of Agricultural Engineers, St. Joseph, MI.

Kohan, S. M., and Wilhelm, D. J. (1980). "Recovery of Hydrocarbon-Like Compounds and Sugars from *Euphorbia Lathyris*," Paper No. 4197. 89th National Meeting, American Institute of Chemical Engineers, Portland, OR, August 17-20.

Krawczyk, T. (1996). *Inform* **7** (8), 800.

Lipinsky, E. S., McClure, T. A., Kresovich, S., Otis, J. L., and Wagner, C. K. (1984). "Fuels and Chemicals from Oilseeds," (E. B. Shultz, Jr. and R. P. Morgan, eds.), AAAS Selected Symposium 91, Chapt. 11, p. 205. Westview Press, Inc., Boulder, CO.

Maugh, T. H., II (1976). *Science* **194** (4260), 46; (1979). *Science* **206** (4417), 436.

McLaughlin, S. P., and Hoffmann, J. J. (1982). *Econ. Bot.* **36**, 323.

McLaughlin, S. P., Kingsolver, B. E., and Hoffmann, J. J. (1983). *Econ. Bot.* **37**, 150.

Metzger, P., Casadevall, E., Coute, A., and Pouet, Y. (1985). In "Energy from Biomass 3rd E.C. Conference, Venice, Italy, March 25-29, 1985," (W. Palz, J. Coombs, and D. O. Hall, eds.), p. 727. Elsevier Applied Science Publishers, London.

Morgan, R. P., and Shultz, E. B., Jr. (1981). *Chem. and Eng. News* **69**, September 7.

National Academy of Sciences (1977). "Guayule: An Alternative Source of Natural Rubber." Washington, D.C.; "Jojoba: Feasibility for Cultivation on Indian Reservations in the Sonoran Desert Region." Washington, D.C.

National Renewable Energy Laboratory (1982). "SERI Biomass Program FY 1982 Annual Report," Draft, SERI/TR-331-1796. Golden, CO.

National Renewable Energy Laboratory (1983). "SERI Biomass Program FY 1983 Annual Report," Draft, SERI/TR-331-2159, p. 45. Golden, CO.

Nemethy, E. K., Otvos, J. W., and Calvin, M. (1981). In "Fuels From Biomass and Wastes," (D. L. Klass and G. H. Emert, eds.), p. 405. Ann Arbor Science Publishers, Inc., Ann Arbor, MI.

Nielsen, P. E., Nishimura, H., Otvos, J. W., and Calvin, M. (1977). *Science* **198** (4320), 942.

Noureddini, H., Harkey, D., and Medikonduru (1996). In "Liquid Fuels and Industrial Products from Renewable Resources," (J. S. Cundiff *et al.*, eds.), p. 83. American Society of Agricultural Engineers, St. Joseph, MI.

Nye, M. J., and Southwell, P. H. (1983). In "Vegetable Oils as Diesel Fuel: Seminar III," (M. O. Bagby and E. H. Pryde, eds.), ARM-NC-28, p. 78. U.S. Department of Agriculture, Peoria, IL.

Olson, E. S., and Sharma, R. K. (1993). In "Energy From Biomass and Wastes XVI," (D. L. Klass, ed.), p. 739. Institute of Gas Technology, Chicago.

Peterson, C., Reece, D., Thompson, J., and Beck, S. (1995). In "Second Biomass Conference of the Americas: Energy, Environment, Agriculture, and Industry," NREL/CP-200-8098, DE95009230, p. 941. National Renewable Energy Laboratory, Golden, CO.

Quick, G. R. (1989). In "Oil Crops of the World," (G. Röbbelen, R. K. Downey, and A. Ashri, eds.), p. 118. McGraw-Hill Publishing Company, New York.

Röbbelen, G., Downey, R. K., and Ashri, A., eds. (1989). "Oil Crops of the World." McGraw-Hill Publishing Company, New York.

Reed, T. B. (1993). In "First Biomass Conference of the Americas: Energy, Environment, Agriculture, and Industry," NREL/CP-200-5768, DE03010050, Vol II, p. 797. National Renewable Energy Laboratory, Golden, CO.

Sachs, R. M., Low, C. B., MacDonald, J. D., Awad, A. R., and Sully, M. J. (1981). *California Agriculture* **29**, July-August.

Scheld, H. W. (1986). In "Energy From Biomass and Wastes X," (D. L. Klass, ed.), p. 177. Institute of Gas Technology, Chicago.

Schmalzer, D. K., Gaines, L. L., Herzenberg, C. L., and Snider, M. A. (1988). "Biocrude Suitability for Petroleum Refineries," ANL/CNSV-69. Argonne National Laboratory, Argonne, IL, June.

Schumacher, L. G., and Van Gerpen, J. H. (1996). In "Liquid Fuels and Industrial Products from Renewable Resources," (J. S. Cundiff *et al.*, eds.), p. 207. American Society of Agricultural Engineers, St. Joseph, MI.

Schumacher, L. G., Howell, S., and Weber, J. A. (1996). In "Liquid Fuels and Industrial Products from Renewable Resources," (J. S. Cundiff et al., eds.), p. 217. American Society of Agricultural Engineers, St. Joseph, MI.

Shay, E. G. (1993). Biomass and Bioenergy 4 (4), 227.

Shifrin, N. S., and Chisholm, S. W. (1980). In "Algae Biomass," (G. Shelef and C. J. Soeder, eds.), p. 177. Elsevier/North Holland, New York.

Shupe, J. W. (1982). Science 216 (11), 1193.

Stage, H. (1988). Fettwissenschaft 90 (1), 28.

Stumborg, M., Soveran, D., Craig, W., Robinson, W., and Ha, K. (1993). In "Energy From Biomass and Wastes XVI," (D. L. Klass, ed.), p. 721. Institute of Gas Technology, Chicago.

Suppes, G. J., and Wei, J. Y. (1996). In "Bioenergy '96, Proceedings of the 7th National Bioenergy Conference," Vol. I, p. 364. The Southeastern Regional Biomass Energy Program, Tennessee Valley Authority, Muscle Shoals, AL.

Taylor, S. E., Skrukrud, C. L., and Calvin, M. (1988). In "Energy From Biomass and Wastes XI," (D. L. Klass, ed.), p. 903. Institute of Gas Technology, Chicago.

Texas Agricultural Experiment Station (1990). "Estimated Cost and Returns for Guayule Production: Texas Trans-Pecos and Winter Garden Regions," MP-1692. The Texas A&M University System, College Station, TX, May.

Tillett, D. M., and Benemann, J. R. (1988). In "Energy From Biomass and Wastes XI," (D. L. Klass, ed.), p. 771. Institute of Gas Technology, Chicago.

Weber, J. A., and Johannes, K. (1996). In "Liquid Fuels and Industrial Products from Renewable Resources," (J. S. Cundiff et al., eds.), p. 350. American Society of Agricultural Engineers, St. Joseph, MI.

Weisz, P. B., Haag, W. O., and Rodewald, P. G. (1979). Science 206 (4414), 57.

Weisz, P. B., and Marshall, J. F. (1979). Science 206 (4414), 24.

Wong, A., and Feng, Y. (1995). In "Second Biomass Conference of the Americas: Energy, Environment, Agriculture, and Industry," NREL/CP-200-8098, DE95009230, p. 902. National Renewable Energy Laboratory, Golden, CO.

Yokoyama, H. (1977). Solar Energy Digest 4, October; Rubber and Plastics News, August 22.

Yurchak, S., and Wong, S. S. (1992). In "Asian Natural Gas—New Markets and Distribution Methods," (D. L. Klass, T. Ohashi, and A. Kutsumi, eds.), p. 593. Institute of Gas Technology, Chicago.

CHAPTER 11

Synthetic Oxygenated Liquid Fuels

I. INTRODUCTION

Many microorganisms effectively serve as catalysts by promoting the biochemical conversion of biomass, particularly its carbohydrate components, to potential and established liquid fuels or liquid fuel precursors. The various enzyme systems generated by the microorganisms are the actual catalysts. Special emphasis is given in this chapter to low- molecular-weight alcohols because they are one of a few classes of liquid motor fuels that can be directly formed on microbial conversion of biomass. Ethanol, for example, is an important blending agent for motor fuels and also a primary microbial conversion product of biomass; it is formed by "fermentation." Fermentation refers to the enzyme-catalyzed, energy-yielding chemical reactions that occur during the breakdown of complex organic substrates, usually but not always under anaerobic conditions, in the presence of certain microorganisms. Fermentation reactions require organic compounds as terminal electron acceptors. Some of these reactions such as the production of beer, wine, and vinegar and the souring of milk, have been known for centuries. Pasteur originally introduced the term fermentation (L. *fermentare*, to boil) to describe microbiological reactions during which gas evolution causes the appearance of boiling (*cf.* Fieser and Fieser, 1950). The microorganisms were originally named "organized ferments" to distinguish them from "unorganized ferments" that had been extracted from plants and animals. Distinction between the two ferments was abandoned when Buchner, in 1897, obtained a cell-free juice from yeast that fermented glucose in exactly the same way as living yeast does. Ethanol has a long history of usage as a liquid fuel for internal combustion engines and is commercially used today by itself (neat fuel) in certain countries, as a motor fuel extender in blends with petroleum fuels, as an additive for octane enhancement, and as a source of dissolved oxygen in modern gasolines.

Several legislative acts and regulations of the U.S. government and the U.S. Environmental Protection Agency (EPA) have resulted in significant expansion of the liquid biofuel industry. The Iran–Iraq war of 1978–1979 led to a highway excise tax exemption in the United States that initially amounted to about $0.16/L ($0.60/gal) of biomass-derived ethanol formulated with gasoline as gasohol, a blend of 10 vol % ethanol and 90 vol % unleaded gasoline. The exemption does not apply to synthetic ethanol manufactured by ethylene hydration. Although the federal exemption has since been reduced to about $0.14/L ($0.54/gal), the availability of federal and additional state tax incentives for biomass-based alcohols and derivatives helped stimulate their development and commercial use as motor fuel components. Also, the U.S. government-mandated use of gasoline-soluble, oxidized organic compounds (oxygenates) in unleaded gasolines to improve their environmental characteristics became effective in November 1992. The mandate has had a large impact on the biofuel industry. The oxygenates are primarily alcohols and ethers.

The oxygenated unleaded gasoline market was initially equivalent to about 30% of total gasoline demand. The U.S. Clean Air Act Amendments of 1990 included a requirement for dissolved oxygen levels in unleaded gasoline of at least 2.7 wt % during the four winter months for 39 so-called carbon monoxide nonattainment areas. The Act also required a minimum of 2.0 wt % dissolved oxygen in reformulated gasoline (RFG) in the nine worst ozone nonattainment areas year-round. The oxygenated RFG market was initially equivalent to about 22% of gasoline demand. Some states have opted to market RFG even though they are not on the nonattainment list. If the area is both a carbon monoxide and an ozone nonattainment area, the RFG must contain at least 2.7 wt % dissolved oxygen during the winter season and 2.0 wt % dissolved oxygen the rest of the year. RFG is limited to a maximum of 1.0 vol % benzene and a total aromatic content of 25 vol %; it has a reduced vapor pressure and cannot contain heavy metal additives.

The raw material for oxygenate manufacture does not have to be biomass, but it is used as feedstock to help supply the growing demand. Since the introduction of RFG with oxygenates to the U.S. motor fuel market in January

1995, significant reductions have occurred in emissions of volatile organic compounds (VOCs) and unburned hydrocarbons, carcinogens, and other criteria pollutants (CO, SO_x, particulates). VOCs react with oxides of nitrogen in the presence of sunlight to form ozone, so RFG also provides considerable protection against ozone-forming emissions. The U.S. Energy Policy Act of 1992 (EPACT) included a strong regulatory program, implemented in the fall of 1996, that requires alternative fuel vehicles in every major U.S. metropolitan market. EPACT sets a national goal of 30% penetration of non-petroleum fuels in the light-duty vehicle market by 2010 and requires that, in sequence, the U.S. government, alternative fuel providers, state and local governments, and private fleets purchase alternative fuel vehicles in percentages increasing over time (Gushee, 1994). Alternative fuels are defined by EPACT to be 85% blends of methanol and ethanol with gasoline (E-85 and M-85), compressed natural gas (CNG), liquefied natural gas (LNG), propane, hydrogen, and any other fuel determined to be substantially not petroleum. All of these incentives and mandates are opening new markets for biomass-derived transportation fuels, especially fermentation liquids.

Because large amounts of ethanol and a few other low-molecular-weight alcohols and derived ethers are commercially marketed in the United States as motor fuel components, a portion of this chapter is devoted to the history of alcohol motor fuel usage and the properties of some of the important oxygenates alone and in gasoline blends (Sections II and III). Several of these oxygenates can be produced from biomass. The use of unicellular microorganisms—yeasts and bacteria—for the production of oxygenates from biomass is then examined (Section IV), and major economic factors are discussed (Section V). Fermentation systems are emphasized. Some microbiological processes that, in theory, can yield suitable liquid motor fuels but have not been perfected yet, and a few thermochemical processes that may ultimately displace microbial systems as the preferred production method for the same biofuels, are also discussed along with a review of economic factors.

II. HISTORY OF ALCOHOL MOTOR FUEL DEVELOPMENT

A. ETHANOL

A review of the historical development of ethanol usage as a motor fuel is in order before discussion of modern production technologies. The history dates back to the beginnings of the internal combustion engine. In an industrial chemistry book published about a century ago (Duncan, 1907), these statements appear:

> One of the most interesting developments of the past decade has been that of
> the internal combustion engine. . . . The question of profitably substituting in
> these engines alcohol for gasoline is one enormously controversial, but out of
> warring testimony there have appeared certain facts that seem unquestionable. . . .
> Alcohol is reproduced in the cycle of the seasons; it is absolutely inexhaustible; it
> is made out of sunshine and air; and its composition does not lessen the value of
> the soil or the energy of the earth. Gasoline, on the contrary, represents a part of
> the stored energy of the earth; it exists only to the extent of about two percent in
> petroleum, and its supply, will in the future inevitably fail.

By far, the vast majority of alcohol usage as motor fuel has occurred with
ethanol. The invention of the four-cycle internal combustion engine in 1877
by Otto and the two-cycle engine in 1879 by Benz involved the testing of
ethanol, other alcohols, and many other organic liquids as potential fuels.
Another factor that played a major role in the development of fuel ethanol
was the passage of laws that permitted the production of tax-free ethanol for
industrial use—England in 1855, The Netherlands in 1865, France in 1872,
Germany in 1879, and then the United States (U.S.I. Chemicals, 1981). Farm
leaders and alcohol distillers lobbied successfully to have the U.S. federal
beverage alcohol tax removed on denatured industrial ethanol during a period
of agricultural price decline in 1906 in the belief that vast new markets for
ethanol fuels would develop (Giebelhaus, 1980). The enactment of the Tax-
Free Industrial and Denatured Alcohol Act of 1906 in the United States,
however, had little effect on the development of ethanol fuel markets because
of the availability of cheap gasoline.

During World War I, various alcohol blends were used by the European
military forces as motor fuels because of gasoline shortages. After the war,
numerous countries other than the United States began a serious effort to
extend their motor fuel supplies by blending ethanol in gasoline (Christensen,
Hixon, and Fulmer, 1934). France passed legislation in 1923 which ultimately
led to compulsory blending of 25 vol % ethanol in high-gravity gasolines to
help reduce agricultural surpluses. A fuel blend called "Monopolin" containing
25 vol % absolute ethanol and 75 vol % gasoline was marketed in Germany
from 1926 to August 1930 in open competition with gasoline. After that time,
compulsory blending to 10 vol % ethanol levels was required for all gasoline
imported or produced within the country. Italian royal decree required that
30 vol % ethanol-70 vol % gasoline blends be used in 1926; this was changed
to 20 vol % ethanol in 1931. A fuel blend called "Motalco" containing 20 vol
% ethanol and 80 vol % gasoline was compulsory in Hungary for all gasolines
over 0.735 specific gravity in 1929. Argentina recommended a fuel containing
30 vol % ethanol and 70 vol % gasoline in 1931, stating that this fuel is superior
to gasoline. The Royal Dutch Shell Company marketed a blend called "Shellkol"
consisting of 15 to 35 vol % absolute ethanol in gasoline in Australia. A law
was passed in Austria in 1931 which compelled the blending of domestic

ethanol up to 25 vol % in gasoline during periods in which the price of ethanol was below that of gasoline. A Brazilian law enacted in 1931 required gasoline importers to use domestic ethanol up to 5 vol % of the gasoline imported. Ethanol–gasoline blends containing 10 to 25 vol % domestic ethanol were required in 1931 in place of gasoline, with the exception of aviation gasoline, and sold in Chile. A fuel blend called "benzolite" containing 55 vol % ethanol, 40 vol % benzene, and 5 vol % kerosine was marketed in China. Latvia required all gasoline, kerosine, benzene, or other liquid fuels for internal combustion engines to contain 25 vol % absolute ethanol. In 1931, a fuel blend called "gasanal" was marketed; it contained ethanol, ethyl ether, and gasoline. Beginning in 1931, Sweden used various blends of ethanol with other organic liquids, including methanol, as motor fuels, but recommended 20 to 30 vol % ethanol in ethanol–gasoline blends after World War I. Czechoslovakian law in 1932 required that all mineral oil fuels contain 20 vol % ethanol. In 1932, Cities Service Company began marketing a fuel blend called "Koolmotor" in the United Kingdom; it contained 10 vol % ethanol, 15 vol % benzene, and 75 vol % gasoline. Yugoslavia required all motor fuels to contain a minimum of 20 vol % ethanol in 1932.

Many of the countries that had enacted laws requiring the blending of ethanol in gasolines suspended them in the late 1930s because of the instability of ethanol supplies (Wilkie and Kolachov, 1942). Germany imported expensive foreign ethanol to meet the legal blending requirements and used 63,500 t of the cheaper synthetic methanol in 1937 to meet the ethanol deficiency. Italy and France suspended all regulations during parts of 1937 and 1938 because of crop failures. It is apparent that the use of ethanol as a motor gasoline extender was widespread in many countries after World War I. This was not the case in the United States.

In the early 1920s, Standard Oil Company (New Jersey) marketed a blend of 20 to 25 vol % absolute ethanol in gasoline in the Baltimore area. The program lasted until 1924, although there seems to be a difference of opinion as to why the venture was terminated. One report indicates that high corn prices in 1924 were the cause (Christensen, Hixon, and Fulmer, 1934); another indicates the reason was storage and transportation difficulties coupled with customer complaints (Giebelhaus, 1980). Standard Oil's customers reportedly encountered clogged fuel lines and carburetors with scale and sediment loosened by the solvent action of the alcohol in the gasoline tank, so it is probable that unfavorable economics and operating problems both caused the demise of the first attempts to market ethanol-gasoline blends in the United States. Subsequent efforts to revive an ethanol fuel program in the late 1920s and early 1930s through federal and state legislation, particularly in the U.S. Corn Belt, failed. The American Petroleum Institute opposed the effort and, together with other groups, was able to block federal tax incentive legislation in 1933 and 1934. However, after Henry Ford and several experts on fermentation ethanol joined hands and organized the First and Second Dearborn Conferences to promote science, industry, and agriculture in 1935 and 1936, a fermentation plant to manufacture 38,000 L/day of anhydrous ethanol specifically for motor fuels was announced for construction in Atchison, Kansas. The plant began operation on March 3, 1937 and manufactured Agrol Fluid, a blend of 78 vol % anhydrous ethanol, 7 vol % other alcohols, and 15 vol % benzene. In 1939, 75 million liters of Agrol Fluid was marketed through more than 2000 service stations in 8 midwestern states (Christensen, Hixon, and Fulmer, 1934; Prebluda and Williams, 1981). The plant closed near the end of 1938 because the cost of fermentation ethanol could not be reduced to a level competitive with that of gasoline. The plant was started again during World War II to manufacture fermentation ethanol for non-vehicular applications.

In the Midwest, American Oil Company began marketing gasohol containing fermentation ethanol on July 1, 1979, becoming the first major petroleum refiner in recent times to market an ethanol-gasoline blend in the United States (cf. Klass, 1980). Several other major oil companies then initiated gasohol marketing programs. Fermentation ethanol was used as a gasoline extender and as an octane enhancer and served to prepare the petroleum and automobile industries for the gradual phase-out of leaded fuels. Gasohol was sold at several thousand retail outlets, and estimates of fuel ethanol production in 1979 ranged from 136 to 227 million L/year (0.86 to 1.43 million bbl/year). At that time, the largest supplier of anhydrous ethanol to the petroleum industry, Archer Daniels Midland Company (ADM), which is located in the U.S. Corn Belt, expanded its production capacity to about 570,000 L/day (1.31million bbl/ year) of 200 proof ethanol from corn. By the mid 1990s, total U.S. fuel ethanol production capacity had increased to over 5.68 billion L/year (35.7 million bbl/year). ADM is still the largest U.S. supplier; it supplied about 38% of the country's demand for fuel ethanol in early 1997 (The Energy Independent, 1997). Gasohol's share of the motor gasoline market in the United States was about 10.8% in the mid-1990s (Bower and Greco, 1997).

The Brazilian fuel ethanol program, called Proalcool, is larger than the U.S. program. It started in the 1930s and was expanded in October 1975 with mandated usage of gasohol (10 vol % ethanol in gasoline) throughout the country (Garnero, 1981). All ethanol-gasoline blends were converted to 20 vol % ethanol shortly thereafter. Ethanol production, primarily from sugarcane, reached a total of 3.41 billion liters in 1980, of which 2.68 billion liters was consumed as vehicular fuel. Neat ethanol-fueled vehicles were introduced to the market in 1979, and by 1982, the percentage of ethanol in blended fuels was raised to 22 vol %. Proalcool was started by government decree and continues today. In 1994, 34.3 million L/day (216,000 bbl/day), or about 12.5 billion L/year (78.8 million bbl/year) of fermentation ethanol was consumed for fuel purposes, accounting for 48.5% of automobile fuel demand (*cf.* Nastari, 1996). In 1995, 4.2 million neat ethanol-fueled automobiles were on the road, accounting for 35% of the total population of passenger vehicles in Brazil.

B. METHANOL

Until the introduction in 1923 to American markets of synthetic methanol manufactured from coal-derived synthesis gas at about one-half the price of wood alcohol, methanol was not given as much attention for motor fuel applications as ethanol (Riegel, 1933). By the early 1930s, when synthetic methanol was well established and had taken over about 75% of the methanol market in the United States, methanol was considered to be a potential alternative fuel for gasoline. But with few exceptions, it was only used as an anti-icing additive, for aircraft injection on take-off, and as a racing fuel where advantage could be taken of the increased power obtainable without regard to economics (Keller, Nakaguchi, and Ware, 1978). The EPA has limited the blending of methanol without cosolvent to a maximum concentration of 0.3 vol % in unleaded gasolines because of phase separation and vapor pressure problems, which will be discussed later.

C. OTHER ALCOHOLS AND ALCOHOL DERIVATIVES

Many oil companies have blended alcohols such as 2-propanol (isopropyl alcohol) and 2-methyl-1-propanol (isobutyl alcohol) in gasolines as anti-icing additives, but not generally as primary fuel components. Also, neat 2-methyl-2-propanol (*t*-butyl alcohol, TBA), the corresponding methyl ether (methyl-*t*-butyl ether, MTBE) and ethyl ether (ethyl-*t*-butyl ether, ETBE), di-isopropyl ether, and *t*-amyl methyl ether (TAME) are marketed in the United States as oxygenates and octane-enhancing additives for gasolines. Arco Chemical Company has marketed blends of methanol and TBA (Oxinol) since 1969 as a gasoline additive to increase octane. TBA also serves as a cosolvent for methanol to improve phase stability of the gasoline blends.

The butanols and their methyl and ethyl ethers have several advantages as oxygenates over methanol and ethanol in gasoline blends. Their energy contents are closer to those of gasoline; the compatibility and miscibility problems with petroleum fuels are nil; excessive vapor pressure and volatility problems do not occur; and they are water tolerant and can be transported in gasoline blends by pipeline without danger of phase separation due to moisture absorption. Fermentation processes (Weizmann process) have been developed for simultaneous production of 1-butanol, 2-propanol, acetone, and ethanol from biomass and are discussed in Section IV. Product mixtures from these fermentations have been employed as "power butanol." Power butanol refers to fuel blends of the same heating value as dry ethanol and is composed of approximately 17% water and 83% combined 1-butanol, acetone, and ethanol in ratios of 7.8:4:1. This mixture is created by blending 80% 1-butanol-20% water, 88% acetone-12% water, and 95% ethanol-5% water mixtures, or the grades of 1-butanol, acetone, and ethanol that can be made by simple distillation. Some of the operating characteristics are different than those of gasolines, but the power output and thermal efficiency data indicate that power butanol and power butanol–gasoline blends function well in standard spark ignition engines with only minor equipment changes (Noon, 1982).

D. ALCOHOL RACING FUELS

As basic knowledge of combustion and engine designs improved after World War I, racing teams began to use alcohol fuel blends formulated with aviation gasoline. Ethanol–benzene–gasoline blends generally ranged from volumetric ratios of about 20:20:60 to 80:10:10 (Powell, 1975). Ethanol, because of its high latent heat of vaporization and low air:fuel ratio requirements compared to those of gasoline, can be used at higher inducted fuel energy densities than gasoline alone to deliver increased power outputs. In the 1930s, methanol's still higher latent heat of vaporization and lower air:fuel ratio requirement provided better performance and became the main power component of many racing fuel blends. Methanol–benzene–gasoline blends were used to establish land and water speed records, and a fuel containing ethanol and gasoline was used to set a world air speed mark. After World War II, engine compression ratios were increased, and the racing community shifted to neat methanol and methanol–nitroparaffin blends. Neat methanol is now the dominant fuel for many racing events.

III. PROPERTIES OF OXYGENATES

Examination of the properties of some of the oxygenates used in the United States is in order before discussing the processes for their manufacture. The efficacy of a particular oxygenate as a neat motor fuel, gasoline or diesel fuel extender, or motor fuel additive depends on its specific properties. The major automobile manufacturers and petroleum refiners and several government agencies and research institutions have carried out large research programs to evaluate oxygenates in spark- and compression-ignition engines to assess their performance, environmental benefits, and compatibility with automotive parts and materials (*cf.* National Renewable Energy Laboratory, 1993). Problems as well as unanticipated benefits and some conflicts between the results of different groups were encountered as the information has developed, despite the long history of alcohol fuel usage. Much of this work has focused on methanol, ethanol, and other oxygenates.

A. COMPARISON OF PROPERTIES

With Gasoline

Table 11.1 presents selected properties of methanol, ethanol, TBA, MTBE, unleaded regular gasoline, typical diesel fuel, and isooctane (2,2,4-trimethyl-pentane). Methanol and ethanol have about 50 and 66% of the volumetric heating value of gasoline, whereas TBA and MTBE have about 80% of the volumetric heating value. All other factors being equal, one would expect that the distance a vehicle can travel per unit volume of these compounds as neat fuels would be less than for gasoline. This is not precisely the case because many complex factors are involved in the performance of an engine-fuel combination.

The differences in the boiling points and latent heats of vaporization between the neat alcohols and gasoline are apparent, although the latent heat of the less polar ether MTBE is close to that of gasoline. The oxygenates listed in Table 11.1 are pure compounds and exhibit specific boiling points, whereas commercial hydrocarbon fuels are mixtures that consist of many paraffinic, aromatic, and naphthenic compounds, and therefore exhibit a boiling range. This difference, coupled with the higher latent heats of vaporization of the alcohols, suggests there may be significant differences in the carburetion of a given engine with a neat alcohol. Also, fuel volatility differences might be expected with oxygenate–gasoline blends compared to the vaporization charac-teristics of neat gasoline, especially because of the interactions that are known for solutions of associated liquids such as alcohols in nonassociated liquids such as hydrocarbons.

Another significant difference in properties is the much lower stoichiometric air/fuel ratios for combustion of methanol and ethanol compared to the ratios for gasoline. Despite the greater flammability range of alcohols in air, this is the reason that alcohol-air mixtures supplied by a gasoline-set carburetor to an engine are too lean for combustion to occur, and the engine will not operate properly. This problem is eliminated in Ford Motor Company's Flexible Fuel Vehicles (FFVs). In addition to some models that can operate on neat methanol or neat ethanol, other FFVs can operate on blends of 85 vol % ethanol (model E-85) or 85 vol % methanol (model M-85), or any lower concentration, including gasoline only. Computerized sensors monitor the amount of alcohol in the

TABLE 11.1 Comparison of Key Properties of Some Oxygenates and Motor Fuels

Property	Methanol	Ethanol	2-Methyl-2-propanol	Methyl 2-methyl-2-propyl ether	Unleaded gasoline	Diesel fuel	Isooctane
Formula	CH_3OH	CH_3CH_2OH	$(CH_3)_3COH$	$(CH_3)_3COCH_3$	C_4–C_{12}	C_8–C_{20}	C_8H_{18}
Molecular weight	32.04	46.07	74.12	88.15	110 avg.	170 avg.	114.23
C, wt %	37.48	52.14	64.81	68.13	85–88	85–88	84.12
H, wt %	12.58	13.13	13.60	13.72	12–15	12–15	15.88
O, wt %	49.94	34.73	21.59	18.15	nil	nil	nil
Density, g/cm³ (20°C/4°C)	0.7914	0.7893	0.7887	0.7405	0.69–0.80	0.82–0.86	0.6919
Atmospheric boiling point, °C	65.0	78.5	82.2	55.2	27–225	240–360	99.238
Latent heat of vaporization, MJ/kg at 20°C	1.177	0.839	0.600	0.321	0.349	0.256	0.314
Latent heat of vaporization, MJ/L at 20°C	0.931	0.662	0.474	0.238	0.251	0.237	0.217
Flash point, °C	11.1	12.8	11.1	−28	−43 to −39	52–96	4
Autoignition point, °C	464	423	478	460	495	260	447
Flammability limits, vol % in air	6.7–36	4.3–19.0	2.4–8.0	1.6–8.4	1.4–76.	1.0–5.0	1.1–6.0
Higher heating value, MJ/kg at 20°C	22.3	29.8	35.55	38.12	47.2	44.9	47.8
Lower heating value, MJ/L at 20°C	15.76	21.09	25.92	26.02	32.16	35.4	33.07
Stoichiometric air/fuel mass ratio	6.45	8.97	11.15	12.50	14.7	14.5	15.07
Stoichimetric air/fuel volumetric ratio	7.16	14.32	28.65	38.18	55	85	59.68
Water solubility, wt % at 20°C	infinite	infinite	infinite	1.4	0.009	nil	nil
Water azeotrope, atm boiling point, °C	none	78.2	79.9	52.2			
Water in azeotrope, wt %		4.4	11.8	3.2			
Research octane number	112	111	113	117	88–98		100
Motor octane number	91	92	110	101	80–88		100
Cetane number	3	8			8–14	40–60	10
Vapor pressure, kPa at 38°C	32	16	12	54	48–103	nil	70

fuel blend and automatically adjust the engine timing and amount of fuel delivered to the engine to provide the best performance. The engine compression ratio is the same as in the gasoline-fueled version because FFVs are designed to operate on gasoline as well. In conventional vehicles, fuel blends containing up to about 20 vol % ethanol generally operate satisfactorily, although there is a blend-leaning effect on delivery of the fuel mixture to the engine equipped with a conventionally adjusted gasoline carburetor. The leaning effect of the blend can often improve fuel mileage if the carburetor is set too rich for gasoline alone, even though the alcohols have a lower energy content. In many late model cars, the air/fuel ratio is automatically controlled by feedback from an oxygen sensor in the exhaust, and the blend-leaning effect does not occur.

The higher octane numbers of the oxygenates compared to those of unleaded regular gasoline in Table 11.1 suggest greater volumetric efficiencies of neat oxygenates than gasoline, provided the engine's compression ratios are high enough to take advantage of the higher octane values. In addition to compression ratio, volumetric efficiency depends on engine design factors such as timing, type of fuel induction system, and breathing capacity, and on fuel parameters such as latent heat of vaporization and the heat received by the fuel charge during its passage through the induction system (Nash and Howes, 1935). For ethanol, and especially methanol, the latent heats of vaporization are sufficiently large that the fuel does not completely evaporate during the suction stroke and continues to evaporate during the compression stroke. For the same amount of fuel evaporated before the inlet valves close, air cooling and the quantity of air drawn into the cylinder are greater. Hence, the temperature of the whole system is lower, and the density of the fuel-air charge is higher, thereby raising the efficiency. It is important to note that the heating value of the fuel only determines fuel consumption for a specific amount of work, not efficiency or power output. Thus, the higher octane values of methanol and ethanol at higher operational compression ratios; the higher latent heats of vaporization and the resulting increases in fuel-air cooling, density, and mass flow; and as will be shown later, the more favorable molar ratio of combustion products to charge with methanol and ethanol all favor greater efficiency of alcohols than hydrocarbon fuels in terms of distance traveled per unit of expended energy. Note that this is *unrelated* to mileage per unit volume of fuel consumed. The greater energy efficiency means that for the same power outputs as gasoline engines, smaller alcohol-fueled engines could be used. The use of alcohols and MTBE as octane-enhancing additives in gasoline blends is apparent from the data in Table 11.1. For 10 vol % blends, the lower volumetric energy content reduces fuel economy in properly adjusted induction systems, as pointed out previously, but not precisely in proportion to the amount of oxygenate in the blend. Losses are usually about 2 to 5% (Gibbs and Gilbert, 1981; Shadis and McCallum, 1980; Stamper, 1979).

Interestingly, several key properties of the methyl ether MTBE and structurally related TBA, both of which are commercially marketed as octane enhancers and oxygenates, are more similar to those of unleaded regular gasoline than those of methanol and ethanol. MTBE is simply the equivalent of the dehydration product of TBA and methanol. Its latent heat of vaporization, heating value, mass and volumetric stoichiometric fuel-air ratios, water solubility, and vapor pressure are closer to the corresponding properties of gasoline. This might be expected for MTBE because a polar hydroxyl group is substituted by a less polar ether linkage, and the molecule is more hydrocarbon-like than the lower alcohols. TBA, although an alcohol, also contains a tertiary butyl group, as does MTBE, and is more hydrocarbon-like than methanol or ethanol. Compared to the lower alcohols, MTBE exerts less blend-leaning effect, less effect on evaporative emissions, and less water-induced phase separation problems in gasoline blends (Greene, 1982). The corresponding ethyl ether of TBA, ETBE, the equivalent of the dehydration product of TBA and ethanol, is closer to gasoline in some of its properties in the same manner as MTBE. It is also commercially marketed as an octane enhancer and oxygenate.

With Diesel Fuel

As for applications of methanol and ethanol as diesel fuel extenders or substitutes, the properties listed in Table 11.1 indicate several difficulties. The cetane numbers, stoichiometric combustion ratios, heating values, ignition temperatures, vaporization characteristics, and boiling points are vastly different from the corresponding properties of diesel. The most direct approach to the use of alcohols as diesel fuels would appear to be the blending of the alcohol with diesel fuel, but only anhydrous ethanol will form solutions. Methanol and lower-proof ethanol are insoluble in diesel fuel. Up to about 30 vol % anhydrous ethanol can be added to diesel fuel without engine modification, but as the percentage is increased, power is reduced, fuel consumption and engine noise increase, and the delay period or the time needed for ignition after the fuel enters the combustion chamber is extended (Barker, Pucholski, and Tholen, 1981; Strait, Boedicker, and Johanson, 1978). Performance with 10 vol % ethanol is about equal to that with diesel fuel. But no performance characteristic is enhanced by ethanol, and contamination by a small amount of moisture causes phase separation. Because of ethanol's low cetane value, severe knock can occur that can be quite destructive because of the high compression pressures involved in diesel engines. The use of additives such as vegetable oils and organic nitrates to improve the cetane numbers of ethanol-diesel fuel blends and neat ethanol, separate injection of the alcohol and diesel fuel into the combustion chamber, formulation of stable alcohol-diesel fuel emulsions, and introduction of the alcohol via direct carburetion in the intake air stream (fumigation) are among the techniques that may eliminate some of the operating problems and permit use of lower molecular weight alcohols as diesel fuels.

B. COMBUSTION CHARACTERISTICS

The stoichiometric equations for complete combustion of the lower molecular weight alcohols, ethers, isooctane, and gasoline are shown in Table 11.2. As pointed out previously, the stoichiometric air-fuel ratios of neat methanol and ethanol are quite low compared to that of gasoline. As the molecular weight increases, the air-fuel ratio increases, but it is still less than the ratio for gasoline up to C_8 alcohols. The amount of air needed for stoichiometric combustion decreases with increasing molecular weight and achieves the level of gasoline (1.72 vol %) for C_8 alcohols. The relatively high volume of air needed for ethanol and methanol to achieve stoichiometric combustion is the reason for the blend-leaning effect. If a spark-ignition engine equipped with a carburetor that is set for delivery of gasoline-air mixtures is converted to methanol-gasoline mixtures without adjustment, the same volume of fuel and air is supplied, but the methanol mixture will be deficient in oxygen. However, as already indicated, the lower stoichiometric air-fuel ratio of methanol and ethanol and their higher latent heats of vaporization than gasoline increase the power outputs that can be obtained from an engine of a given size. The stoichiometric relationships presented in Table 11.2 also show that, with the exception of benzyl alcohol, the molar ratios of products to charge for the alcohols are higher than for gasoline and isooctane. This suggests that one of the reasons for better thermal efficiency of the lower molecular weight alcohols is higher pressures in the combustion chambers and more power output.

The calculated higher heating values of most of the fuels listed in Table 11.2 are listed in Table 11.3. On a heating value per unit mass basis, methanol has the lowest heating value, ethanol is next, and then the values generally increase with molecular weight up to C_7, but are still considerably less than those of gasoline and isooctane. Similarly, on a heating value per unit volume basis, methanol has the lowest heating value, ethanol is next, and then as the molecular weight increases, the volumetric heating value increases, but at a higher rate than the corresponding mass values. The volumetric heating values of the paraffinic C_7 alcohols shown are about the same as those of isooctane, and benzyl and cyclohexyl alcohols are in the gasoline range. This is caused by density differences and the containment of disproportionately more molecules in a unit volume of the liquid of the C_7 alcohols than in the same volume of the other alcohols listed in the table.

TABLE 11.2 Stoichiometric Combustion Air Requirements for Pure Liquid Alcohols, MTBE, ETBE, TAME, Isooctane, and Gasoline[a]

Fuel	Stoichiometry	Product/charge molar ratio	Ratio air/fuel		Fuel in air	
			Mass	Molar	wt %	mol %
Methanol	$CH_3OH + 1.5O_2 + 5.66N_2 \rightarrow CO_2 + 2H_2O + 5.66N_2$	1.061	6.45	7.16	13.43	12.25
Ethanol	$C_2H_5OH + 3O_2 + 11.32N_2 \rightarrow 2CO_2 + 3H_2O + 11.32N_2$	1.065	8.97	14.32	10.03	6.53
Propanols	$C_3H_7OH + 4.5O_2 + 16.98N_2 \rightarrow 3CO_2 + 4H_2O + 16.98N_2$	1.067	10.31	21.48	8.84	4.45
All butanols	$C_4H_9OH + 6O_2 + 22.65N_2 \rightarrow 4CO_2 + 5H_2O + 22.65N_2$	1.067	11.15	28.65	8.23	3.37
All pentanols	$C_5H_{11}OH + 7.5O_2 + 28.31N_2 \rightarrow 5CO_2 + 6H_2O + 28.31N_2$	1.068	11.72	35.81	7.86	2.72
Cyclohexanol	$C_6H_{11}OH + 8.5O_2 + 32.08N_2 \rightarrow 6CO_2 + 6H_2O + 32.08N_2$	1.060	11.69	40.58	7.88	2.40
All hexanols	$C_6H_{13}OH + 9O_2 + 33.97N_2 \rightarrow 6CO_2 + 7H_2O + 33.97N_2$	1.068	12.13	42.97	7.62	2.27
Benzyl alcohol	$C_7H_7OH + 8.5O_2 + 32.08N_2 \rightarrow 7CO_2 + 4H_2O + 32.08N_2$	1.036	10.83	40.58	8.46	2.40
All heptanols	$C_7H_{15}OH + 10.5O_2 + 39.63N_2 \rightarrow 7CO_2 + 8H_2O + 39.63N_2$	1.068	12.45	50.13	7.44	1.96
All octanols	$C_8H_{17}OH + 12O_2 + 45.29N_2 \rightarrow 8CO_2 + 9H_2O + 45.29N_2$	1.069	12.69	57.29	7.30	1.72
MTBE	$(CH_3)_3COCH_3 + 7.5O_2 + 28.31N_2 \rightarrow 5CO_2 + 6H_2O + 28.31N_2$	1.068	11.72	35.81	7.86	2.72
ETBE	$(CH_3)_3COC_2H_5 + 9O_2 + 33.97N_2 \rightarrow 6CO_2 + 7H_2O + 33.97N_2$	1.068	12.13	42.97	7.62	2.27
TAME	$CH_3CH_2C(CH_3)_2OCH_3 + 9O_2 + 33.97N_2 \rightarrow 6CO_2 + 7H_2O + 33.97N_2$	1.068	12.13	42.97	7.62	2.27
Isooctane	$C_8H_{18} + 12.5O_2 + 47.18N_2 \rightarrow 8CO_2 + 9H_2O + 47.18N_2$	1.058	15.07	59.68	7.96	1.65
Gasoline	$C_nH_{2n} + 1.5nO_2 + 5.66nN_2 \rightarrow nCO_2 + nH_2O + 5.66nN_2$	1.051	14.73	57.28	8.13	1.72

[a]Dry air is assumed to contain 20.946 mol % O_2. Gasoline is assumed to have an average molecular formula of C_8H_{16}. MTBE, ETBE, and TAME are methyl 2-methyl-2-propyl ether (methyl t-butyl ether), ethyl 2-methyl-2-propyl ether (ethyl t-butyl ether), and methyl 2,2-dimethyl-1-propyl ether (methyl t-amyl

TABLE 11.3 Higher Heating Values and Heating Values of Stoichiometric Air–Fuel Mixtures of Pure Alcohols, MTBE, Isooctane, and Gasoline

Fuel	Mol. wt.	Higher heating value[a] (MJ/kg)	Higher heating value[a] (MJ/L)	Heating value of air–fuel mixtures[b] (MJ/m³)
Methanol	32.04	22.33	17.70	3.643
Ethanol	46.07	29.77	23.50	3.724
1-Propanol	60.10	33.48	26.93	3.722
2-Propanol	60.10	33.08	25.98	3.678
1-Butanol	74.12	36.07	29.21	3.746
2-Methyl-1-propanol	74.12	36.05	28.88	3.743
2-Methyl-2-propanol	74.12	35.55	28.08	3.692
1-Pentanol	88.15	37.70	30.70	3.758
2-Methyl-2-butanol	88.15	37.27	30.15	3.715
Cyclohexanol	100.16	37.23	35.83	3.736
3-Methy-3-pentanol	102.18	37.99	31.48	3.663
Benzyl alcohol	108.14	34.62	36.07	3.751
1-Heptanol	116.20	39.81	32.72	3.769
3-Ethyl-3-pentanol	116.20	38.91	32.64	3.684
MTBE	88.15	38.12	28.23	3.800
Isooctane	114.23	47.79	33.07	3.745
Gasoline	112.21	47.20	34.0–36.8	3.787

[a]The conditions for combustion are atmospheric pressure, 20°C, and product water in the liquid state. The higher heating value per unit volume is calculated from the density (20°C/4°C) and the higher heating value per unit mass.
[b]Calculated from the heats of combustion at 20°C and the mole percent of fuel in the stoichiometric air–fuel mixture by the following formula:

$$(MJ/kg)(mol \ wt)(mol \ \% \ fuel)(4.1572 \times 10^{-4}).$$

To ascertain any energy-content differences that might exist between equal volumes of fuel-air mixtures, the higher heating values in Table 11.1 and the volume of fuel in the stoichiometric air mixtures in Table 11.2 were converted to heating values per unit volume of stoichiometric air-fuel mixture, as shown in Table 11.3. Despite the many differences in density, heating value, and fuel-air requirement for complete combustion, the heating values of the stoichiometric mixtures for the alcohols, MTBE, isooctane, and gasoline each lie in a very narrow range, about 3.72 MJ/m³ ± 0.08 at 20°C. In theory, a properly aspirated and timed spark-ignition engine would thus be expected to deliver the same power outputs, independent of which fuel is used, with each of the potential fuels listed in Table 11.3. This has been alluded to in the literature on methanol-gasoline blends and gasoline (Goran Svahn, 1979). At suitable air-fuel ratios, the specific energy contents of the fuels had only marginal differences, and the differences in maximum power output were insignificant. But only one alcohol was considered in this work. Furthermore, this argument assumes that complete combustion occurs, and that there are no gross differences between fuels. As already pointed out, there are many differences that cause deviation from ideal behavior.

C. OCTANE NUMBERS

It is apparent that the octane numbers of the neat oxygenates shown in Table 11.1 are higher than those of unleaded gasoline. The octane value of oxygenate-gasoline blends might therefore be expected to increase with increasing oxygenate concentration. Substantial octane benefits can be obtained as the concentration increases, as illustrated in Table 11.4. This table shows that the addition of methanol, ethanol, 1-propanol, and 2-methyl-1-propanol to the base gasoline at concentrations of 5, 10, and 15 wt % in the blend increases research octane number proportionately more than motor octane number, that the blending octane values are higher for the lower octane value base gasoline than the higher octane value base gasoline, and that the octane improvement decreases with increasing molecular weight of the alcohol at equivalent weight percentage additions. Blending value is an octane number calculated from the experimentally determined values by a simple linear equation and corresponds to a hypothetical rating at 100% concentration of the additive. It is a measure of the synergistic octane improvement capability of the additive and is often much higher than the octane number determined with neat additive. The octane-enhancing properties of methanol and ethanol appear to be about the same. Note that the concentration of ethanol required to reach 2.7 wt % dissolved oxygen in an ethanol-gasoline blend is 7.3 vol %, which is significantly less than the amount in U.S. gasohol, 10 vol %. The concentration of dissolved oxygen and ethanol in gasohol is 3.7 wt % and 10.65 wt %.

The blending octane numbers, and a few other parameters that will be discussed later, of a wider range of lower molecular weight alcohols and ethers are listed in Table 11.5. These data show that the octane benefits are broad and can be obtained with a wide range of oxygenates. Many petroleum refiners market regular and premium grades of gasoline that contain several of these oxygenates as octane-improving additives. Table 11.5 also lists the volumetric percentages of each oxygenate required in the oxygenate-gasoline blend to meet the requirement of 2.7 wt % dissolved oxygen in the finished blend. There is about a fourfold increase in the concentration needed from C_1 to C_7 oxygenate to provide this level of dissolved oxygen. It is evident that when

TABLE 11.4 Octane Number of Alcohol–Gasoline Blends[a]

Alcohol	Concentration (wt %)	Gasoline 1		Gasoline 2		Gasoline 1		Gasoline 2	
		RON	MON	RON	MON	BRON	BMON	BRON	BMON
None (base gasoline only)	0	78.5	72.0	90.1	83.5				
Methanol	5	80.9	74.2	92.8	84.0				
	10	84.1	76.4	95.1	84.5	137	111	135	94
	15	87.1	78.0	97.2	85.0				
Ethanol	5	81.6	74.4	92.8	84.2				
	10	83.9	76.7	94.7	84.8	135	115	130	96
	15	86.7	78.7	96.5	85.3				
1-Propanol	5	80.5	73.8	91.9	83.9				
	10	82.8	75.1	93.4	84.2	121	101	119	90
	15	84.5	76.7	94.8	84.5				
2-Methyl-1-propanol	5	80.4	73.5	91.4	83.7				
	10	81.7	74.7	92.9	84.0	113	96	113	88
	15	83.6	75.8	94.0	84.2				

[a]Adapted from Cox (1979). Base gasolines 1 and 2 are the Coordinating Research Council Reference Motor-Fuel Detonation blends 286 and 287, respectively. RON, research octane number; MON, motor octane number; BRON, blending research octane number; BMON, blending motor octane number.

TABLE 11.5 Typical Key Properties of Oxygenates[a]

Oxygenate	Oxygen content (wt %)	Lower heating value at 20°C (MJ/L)	Latent heat of vaporization at 20°C (MJ/L)	Solubility in water (wt %)	Blending RVP at 38°C (kPa)	Blending octane value ([R + M]/2)	For 2.7 wt % O in gasoline (vol %)
MeOH	49.94	15.76	0.931	infinite	214	108	5.1
EtOH	34.73	21.09	0.662	infinite	124	115	7.3
IPA	26.62	24.35	0.529	infinite	97	106	9.6
TBA	21.59	25.92	0.432	infinite	62	100	11.8
IBA	21.59	26.49	0.465	10.0	34	102	11.5
TAA	18.15	27.88	0.440	11.5	41	97	13.6
MTBE	18.15	26.02	0.240	1.4	55	110	14.9
ETBE	15.66	26.99	0.231		28	112	17.3
DIPE	15.66	27.86	0.251		34	105	17.5
TAME	15.66	27.86	0.251	0.6	21	105	16.7
IPTBE	13.77	28.02			17	113	19.3
TAEE	13.77				7	100	19.0

[a]Klass (1984b); Piel (1993); Miller (1993). The oxygenate designations are: MeOH, methanol; EtOH, ethanol; IPA, 2-propanol; TBA, 2-methyl-2-propanol; IBA, 2-methyl-1-propanol; TAA, 2,2-dimethyl-1-propanol; MTBE, methyl 2-methyl-2-propyl ether; ETBE, ethyl 2-methyl-2-propyl ether; DIPE, di-2-propyl ether; IPTBE, 2-propyl 2-methyl-2-propyl ether; TAEE, ethyl 2,2-dimethyl-1-propyl ether; TAME, methyl 2,2-dimethyl-1-propyl ether.

one or more of the compounds shown in Table 11.5 is used to meet the U.S. dissolved oxygen requirements in the gasoline blend, the compound serves the dual purpose of supplying oxygen and increasing octane value. They are more properly called blending agents at the higher concentrations indicated in Table 11.5 rather than additives. Depending on the specific oxygenate used, with the exception of methanol, the percentage contributions to blend volume and energy content at the dissolved oxygen limits specified by the U.S. government can range from about 7 to 19% and up to about 17%. Methanol is an exception because it requires a cosolvent oxygenate, TBA, to eliminate the phase separation and vapor problems described next. Alone, methanol is legally limited to a concentration of only 0.3 vol % in U.S. gasoline blends. Again with the exception of methanol alone, the octane value contributions of the oxygenates can range from about 2 to almost 5 octane numbers.

D. WATER TOLERANCE AND PHASE SEPARATION

Water contamination in an oxygenate-gasoline blend can cause separation into a lower water-rich layer and an upper hydrocarbon-rich layer, and result in engine operating difficulties. The amount of water that the blend can tolerate before separation occurs depends on oxygenate structure, water solubility, and concentration, and on gasoline composition and the effect of cosolvent, if any, in the blend. Early experimental data show that ethanol is much less sensitive to phase separation in benzene than is methanol, and that relatively small amounts of water cause separation in the alcohol concentration ranges permitted today in U.S. gasoline blends (Nash and Howes, 1935). Table 11.6 shows the water tolerance of gasoline blended with methanol and ethanol as a function of alcohol concentration, cosolvent, and aromatic content of the gasoline (Keller, 1979). Water tolerance increases with both methanol- and ethanol-gasoline blends as the temperature and aromatic content of the gasoline increase, and with added cosolvent. But ethanol has a much higher water tolerance in gasoline blends than methanol. For example, at 20°C, the data show that for unleaded gasoline containing 14 vol % aromatics and 10 vol % methanol or ethanol, the ethanol blend can tolerate 0.60 wt % water in the blend before phase separation begins, whereas the methanol blend can only tolerate 0.06 wt %. The low-molecular-weight alcohols have a particularly high affinity for water, as indicated by their water solubilities in Table 11.5, that increases the alcohol's susceptibility to extracting the water under wet storage conditions (Piel, 1993). This limits their use for direct blending at the refinery into gasoline. Elimination of water contamination in gasolines would of course solve the phase separation problem altogether, so it behooves refiners, alcohol blenders, and distributors to keep all equipment as free of water as possible.

TABLE 11.6   Water Tolerance of Alcohol–Gasoline Blends[a]

Alcohol	Concentration (vol %)	Weight percent at indicated temperature					
		−20°C	−10°C	0°C	10°C	15°C	20°C
Methanol	10[b]				0.02	0.04	0.06
	10[c]		0.03	0.07	0.11	0.14	0.16
	10[d]	0.04	0.08	0.12	0.17	0.19	0.23
With cosolvent	10[e]	0.17	0.22	0.27	0.34	0.37	0.43
	15[c]			0.06	0.13	0.17	0.20
	20[e]			0.05	0.20	0.26	
Ethanol	5[c]	0.18	0.19	0.22	0.26	0.28	0.30
	10[b]	0.34	0.40	0.47	0.53	0.57	0.60
	10[c]	0.42	0.47	0.54	0.60	0.64	0.67
	10[d]	0.47	0.52	0.58	0.65	0.68	0.73
With cosolvent	10[f]	0.74	0.82	0.90	0.98	1.30	1.70
	15[c]	0.50	0.80	0.85	1.04	1.09	1.16

[a]Estimated from data in Keller (1979).
[b]In unleaded gasoline containing 14 vol % aromatics.
[c]In unleaded gasoline containing 26 vol % aromatics.
[d]In unleaded gasoline containing 38 vol % aromatics.
[e]In unleaded gasoline containing 26 vol % aromatics and 3.2 vol % 2-methyl-1-propanol.
[f]In unleaded gasoline containing 26 vol % aromatics and 3.2 vol % 1-butanol.

Long-term storage of alcohol–gasoline blends should be avoided to preclude absorption of moisture from air, especially in humid climates. In the United States, ethanol–gasoline blends cannot be transported by pipeline because of the phase separation problem, so ethanol is usually added to the gasoline just before delivery to the service station. Blends of gasoline and the lower polarity ethers, however, can be shipped and distributed without handling restrictions. As shown in Table 11.5, the lower oxygen content of the ethers means that when blended as an oxygenate, they are used at higher concentrations to meet the dissolved oxygen requirement. Most oxygenates with low water solubilities should be able to be blended into gasoline at the refinery to provide dissolved oxygen and octane enhancement. Oxygenates with these characteristics are the butanols and the higher alcohols, as well as the ethers (Piel, 1996).

Neat methanol and ethanol fuels do not present the problem of phase separation since water is soluble in these alcohols in all proportions except at very low temperatures. Interestingly, it is not necessary to use anhydrous neat alcohol fuels in spark-ignition engines. The neat ethanol-fueled automobiles in Brazil operate with 190 proof ethanol (95 vol %), which precludes the energy-consuming step of producing anhydrous ethanol. Indeed, the addition of 10 and 20 wt % water to methanol raises its octane value to about 107 and

112; the corresponding octane values for ethanol are 108 and 115 (Kampen, 1980). Higher thermal efficiencies would be expected for the aqueous alcohols as engine fuels because water has a higher latent heat of vaporization than the alcohols, and the resulting performance should be analogous to the effect of methanol on the performance of methanol-gasoline blends.

E. VAPOR PRESSURE

Methanol and ethanol form azeotropes with many of the hydrocarbons in gasolines. The boiling points and compositions of a few of the constant-boiling mixtures are listed in Table 11.7 (Nash and Howes, 1935). Thus, the boiling points of many of the components are reduced, and at lower temperatures, the vapor pressure of the blend is higher than that of gasoline alone. This is illustrated by the distillation curves shown in Fig. 11.1. The depression of the initial portion of the curves is apparent; the relatively flat portion of the curves for the alcohol blends is sometimes referred to as the alcohol flat. As might be expected from the structures of methanol and ethanol and the data in Table 11.7, the alcohol flat is less depressed with ethanol. The inclusion of ethanol and higher alcohols in the methanol blend also provides less depression.

TABLE 11.7   Constant-Boiling Mixtures of Methanol and Ethanol with Selected Hydrocarbons[a]

Hydrocarbon	Boiling point (°C)	Constant-boiling mixture				
		With methanol			With ethanol	
		Boiling point (°C)	Alcohol in mixture (wt %)		Boiling point (°C)	Alcohol in mixture (wt %)
2-Methylbutane	27.95	24.5	4.0			
2-Methyl-2-butene	37.15	31.75	7.0			
1,5-Hexadiene	60.2	47.05	22.5			
n-Hexane	68.95	50.0	26.4			
Benzene	80.2	58.23	39.55		68.25	32.41
1,3-Cyclohexadiene	80.6	56.38	38.8		66.7	34.0
Cyclohexane	80.75	54.2	37.2		64.9	30.5
Cyclohexene	82.75	55.9	40.0		66.7	35.0
n-Heptane	98.45	60.5	62.0			
Methylcyclohexane	101.8	60.0	70.0		73.0	53.0
Toluene	110.6				76.7	68.0

[a]Nash and Howes (1935). All boiling points are at 101.3 kPa.

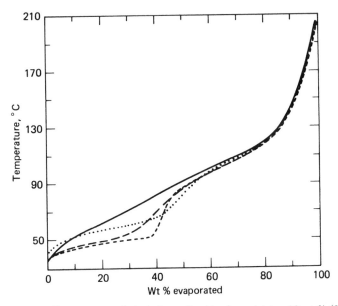

FIGURE 11.1 Distillation curves of alcohol–gasoline blends containing 10 wt % (9.2 vol %) alcohol. (———), base gasoline; (– – – –), methanol; (— — —), 70 wt % methanol–30 wt % C_2–C_4 alcohols; (············), ethanol. From Cox (1979).

The increased front-end volatility from the alcohol flats increases evaporative emissions. At 10 vol % concentrations in gasoline blends, methanol and ethanol have been reported to increase evaporative emissions as much as 130–220% (Stamper, 1980) and 49–62% (Lawrence, 1978), respectively. This can have an adverse effect on the volatility balance of the fuel and also promote vapor lock. It can stop the flow of fuel to the engine and interrupt normal engine operation. To overcome these problems, the compositions of reformulated and modern unleaded gasolines are adjusted by refiners to contain lower boiling fractions, particularly the butanes and pentanes that are used for blending. For gasohol, the EPA has provided a waiver by allowing a 6.9 kPa (1 psi) increase in the vapor pressure over that of gasoline. The starting and driveability problems that occurred in the late 1970s and early 1980s with gasohol have been eliminated, in part by advanced vehicle technology and the use of fuel feedback controls. Methanol-cosolvent oxygenate blends and the other oxygenates are each approved by the EPA for use only up to certain maximum concentrations in the finished gasoline blends as a means of controlling vapor emissions and other characteristics.

IV. PRODUCTION OF OXYGENATES

As will become apparent in what follows, some oxygenates are manufactured by established, commercial, microbial processes using biomass feedstocks. Others can be manufactured by microbial conversion of biomass, but are currently produced using thermochemical conversion methods, usually with fossil feedstocks because of economic or technical factors. A few other oxygenates must be manufactured by thermochemical conversion of fossil feedstocks because suitable microbial processes do not yet exist to produce the oxygenate from biomass. Still others can be produced by a combination of thermochemical and microbial conversion. Microbial conversion systems with biomass feedstocks are emphasized here, but thermochemical methods are briefly reviewed to present a perspective on the options available and what advancements are necessary to perfect suitable processes.

A. METHANOL

As described in Chapter 8, pyrolytic conversion of woody biomass was used to manufacture methanol (wood alcohol) for many years until it was displaced by thermochemical processes that involved gasification to afford intermediate synthesis gas, which is a mixture of hydrogen and carbon oxides, followed by catalytic reduction of the carbon oxides to methanol. In a typical world-scale methanol plant, synthesis gas is produced by steam reforming of natural gas feedstock and shift conversion of CO to obtain the proper molar ratio of H_2 to carbon oxides. The process is strongly endothermic. The hydrogen-producing process is favored by high temperatures and low pressures, and methanol synthesis is favored by low temperatures and high pressures. In a balanced process for natural gas conversion, in which the hydrogen needed is generated from natural gas, the approximate stoichiometries are as follows:

Reforming-shift conversion: $16/3CH_4 + 4/3H_2O + 8/3O_2 \rightarrow 4CO + 4/3CO_2 + 12H_2$
Methanol synthesis: $4CO + 4/3CO_2 + 12H_2 \rightarrow 16/3CH_3OH + 4/3H_2O$
Net: $16/3CH_4 + 8/3O_2 \rightarrow 16/3CH_3OH.$

Preheated natural gas is fed at about 600°C to the reformer and exits at about 880°C and 2.1 MPa. Methanol synthesis is then performed over copper-based catalysts at about 240-270°C and 10.3 MPa. The product gas contains about 5% methanol. By-products are 1-2% dimethyl ether and 0.3-0.5% higher alcohols. Because of equilibrium limitations, conversion of synthesis gas is only a few percent per pass in the catalytic reactor, and the product gas stream after methanol separation at low pressure must be recompressed for recycling with fresh feed.

Synthesis gas can be manufactured from any fossil or biomass feedstock. Biomass feedstocks are not used in industrialized countries for methanol production by pyrolysis or thermal gasification processes because of unfavorable economics. Biomass-based synthesis gas may begin to contribute to at least a portion of methanol demand in the next few decades as natural gas reserves decrease and other fossil feedstocks are unavailable or possibly not desired, as might occur with coal because of environmental issues. Also, combined biomass-natural gas feedstocks can offer substantial benefits in methanol manufacture (Chapter 9). In any case, none of this technology is concerned with microbial conversion. Commercial fermentation processes for methanol have never been developed, although some ethanologenic organisms produce small amounts of methanol by-product. This suggests that there may be a suitable biochemical pathway to methanol if the other pathways can be suppressed.

The thermodynamics of methanol and ethanol production help to explain why methanol is not manufactured by fermentation processes whereas ethanol is. The overall stoichiometries for converting glucose ($C_6H_{12}O_6$) to methanol and ethanol by alcoholic fermentation under physiological conditions (pH 7, 25°C, unit activity in aqueous solution) are

$$C_6H_{12}O_6(aq) + 2H_2O(aq) \rightarrow 4CH_3OH(aq) + 2CO_2(aq)$$
$$C_6H_{12}O_6(aq) \rightarrow 2CH_3CH_2OH(aq) + 2CO_2(aq).$$

The standard enthalpy and Gibbs free energy changes for methanol formation under these conditions are about $+166$ and -99 kJ/mol glucose converted, whereas the corresponding values for the ethanol process are about $+18$ and -235 kJ/mol. The larger decrease in free energy and the almost neutral enthalpy change favor the formation of ethanol over that of methanol. In actual practice, commercial fermentation ethanol forms under slightly exothermic conditions; about 53 kJ is released per gram-mole of ethanol produced.

Every fermentation process must be coupled with the synthesis of adenosine triphosphate (ATP), the universal biochemical energy carrier, from adenosine diphosphate (ADP) and inorganic phosphate. The energy taken up by the cells is used to drive the endothermic synthesis of ATP. Under physiological conditions, this requires between 42 and 50 kJ/mol of ATP formed, but this assumes equilibrium conditions which do not occur naturally in a living cell, so the energy to drive the synthesis of ATP must even be larger, probably near 63 kJ/mole of ATP formed (Thauer, 1976; Thauer, Jungermann, and Decker, 1977). For ethanol synthesis, yeasts of the genus *Saccharomyces* use the Embden–Meyerhof (glycolytic) pathway; the net yield is 2 mol ATP and 2 mol NADH per mole of glucose fermented. Pyruvic acid is the key intermediate. Bacteria of the genus *Zymomonas* use the Entner-Doudoroff pathway; the net yield is 1mol ATP and 2 mol NADH per mole of glucose fermented. For this pathway, pyruvic acid is also the key intermediate. The decrease in free energy of -235 kJ/mol glucose fermented is more than adequate for either pathway. But if the same pathways were available for conversion of glucose to methanol (an unlikely possibility, as discussed later, the decrease in Gibbs free energy of -99 kJ/mol glucose fermented is only slightly more than the amount required by the Entner-Doudoroff pathway, and is considerably less than that required by the Embden–Meyerhof pathway. This is one possible explanation of why fermentation methanol has not yet been developed.

After biochemical conversion of glucose to pyruvic acid intermediate, the next step in ethanol synthesis is nonoxidative decarboxylation and acetaldehyde formation catalyzed by a native decarboxylase, and then acetaldehyde reduction to ethanol catalyzed by a native dehydrogenase.

$$C_6H_{12}O_6 \rightarrow 2CH_3COCO_2H + 4H$$
$$2CH_3COCO_2H \rightarrow 2CH_3CHO + 2CO_2$$
$$2CH_3CHO + 4H \rightarrow 2CH_3CH_2OH.$$

The net reaction is the conversion of 1 mole of glucose to 2 moles each of ethanol and CO_2 with balanced consumption of the reducing power generated during the formation of pyruvic acid. The analogous biochemical conversion of glucose to methanol would involve formation of formaldehyde and its reduction to methanol:

$$C_6H_{12}O_6 \rightarrow 2CH_3COCO_2H + 4H$$
$$2CH_3COCO_2H + 2H_2O \rightarrow 4HCHO + 2CO_2 + 4H$$
$$4HCHO + 8H \rightarrow 4CH_3OH.$$

In this case, although the hypothetical balanced process yields a net 4 moles of methanol and 2 moles of CO_2 per mole of glucose converted, twice the reducing power is needed compared to the ethanol case and cleavage of two additional carbon-carbon bonds is also required. Decarboxylation of pyruvic acid has not been reported to proceed in this manner. These observations, however, do not preclude the possibility of other biochemical pathways and intermediates to fermentation methanol.

Other approaches to the use of microorganisms for methanol production appear to be technically feasible. Methylotrophic organisms are capable of growing nonautotrophically on one-carbon compounds containing a methyl group or on compounds containing two or more unlinked methyl groups. Obligate methylotrophs can grow on methane and methanol, and facultative methylotrophs are in addition capable of growing on other organic compounds such as carboxylic acids and carbohydrates. Most methylotrophs are aerobic organisms. The established biochemical pathway of methane oxidation involves catalysis by methane monooxygenase to afford methanol, followed by sequential reactions catalyzed by dehydrogenases to yield formaldehyde, formic acid, and finally CO_2. These compounds have been detected as exometabolites on microbial conversion of natural gas with methylotrophs, so it has been suggested that leaky mutants of methane utilizers that excrete methanol into the media may offer a microbial route to methanol (Foo, 1978; Foo and Hedén, 1976). But as methanol concentration increases in the media, methane utilization and cell growth are inhibited. Methane utilization by *Pseudomonas* spp. was inhibited at concentrations of only 40 mg/L (Wilkenson *et al.*, 1974). Some blocking agents that shut down methanol oxidation or that remove it from the media appear to be quite effective. The use of iodoacetate with *Methanomonas methanooxidans* allowed 75% of the methane consumed to accumulate as methanol (Brown *et al.*, 1964). Also, it is well known that a large number of mixed microbial populations of acetogenic and methanogenic bacteria are capable of efficient conversion of biomass to approximately equal volumes of methane and CO_2 under anaerobic conditions (see Chapter 12). The overall stoichiometry for the microbial gasification of glucose is

$$C_6H_{12}O_6(aq) \rightarrow 3CH_4(g) + 3CO_2(g).$$

The standard enthalpy and Gibbs free energy changes under physiological conditions are about -131 kJ and -418 kJ/mol glucose converted; methane retains about 95% of the chemical energy contained in the glucose (Klass, 1984a). Thus, the thermodynamic driving force is large and the exothermic energy loss is small. Also, methane fermentation is an established, commercial technology that yields product gases containing 40 to 70 mol % methane.

This information suggests a few conceptual approaches to fermentation methanol that seem to be worth further consideration. One is to employ a staged microbial system in which suitable mixed populations of acetogens and methanogens use biomass under anaerobic conditions in the first stage as a source of carbon and energy to yield CH_4 and CO_2, followed by a second stage of suitable methane-oxidizing bacteria, in which methanol is formed as an exometabolite in the presence of a blocking agent to prevent further oxidation. So a source of oxygen, which could be inorganic chemical oxygen, is necessary in the second stage. The net result is

$$C_6H_{12}O_6(aq) + 1.5O_2(g) \rightarrow 3CH_3OH(aq) + 3CO_2(g).$$

This process is exothermic by about -510 kJ/mol glucose converted and the Gibbs free energy reduction is large, -792 kJ/mol. Since acetate is a key intermediate in methanogenesis, a three-stage system of acetogens, methanogens, and methylotrophs can also be envisaged. There is no need to maintain aseptic conditions in the acetate- and methane-forming stage or stages, although there may be in the methane oxidation stage. Separation of gaseous, water-insoluble methane for delivery to the methanol-forming stage is relatively simple, and the gas is easily sterilized if necessary.

Another conceptual approach to fermentation methanol from biomass is to convert biomass to acetate with suitable acetogenic populations, followed by direct conversion of acetate to methanol. This type of process does not involve intermediate methane and might possibly improve the operational characteristics of the process. The process can be represented by

$$C_6H_{12}O_6(aq) \rightarrow 3CH_3CO_2^- (aq) + 3H^+(aq)$$
$$3CH_3CO_2^- (aq) + 3H^+(aq) + 1.5O_2 \rightarrow 3CH_3OH(aq) + 3CO_2(g).$$

The Gibbs free energy changes of the acetate- and methanol-forming stages are -191 kJ and -601 kJ/mol glucose converted, so the thermodynamic driving force is large for each step. The sum of the free energy changes is -792 kJ, the same as the free energy change of the process shown earlier that involves intermediate methane, because the net result is identical. Suitable microorganisms would have to be found that decarboxylate acetate directly to produce methanol. Alternatively, presuming the availability of a suitable pathway, glucose might be converted to formaldehyde, followed by reduction to methanol:

$$C_6H_{12}O_6(aq) + 2H_2O(aq) \rightarrow 4HCHO(aq) + 2CO_2(g) + 8H$$
$$4HCHO(aq) + 8H \rightarrow 4CH_3OH(aq)$$
$$\text{Net: } C_6H_{12}O_6(aq) + 2H_2O(aq) \rightarrow 4CH_3OH(aq) + 2CO_2(g).$$

The Gibbs free energy changes for the formaldehyde- and methanol-forming stages per mole of glucose converted are 81.9 kJ and -179.4 kJ, or a net reduction of -97.5 kJ, which is not very favorable.

Finally, since methane monooxygenase (MMO) is the only enzyme known that catalyzes the oxidation of methane to methanol, its isolation from suitable methanologenic organisms and use as a catalyst could eliminate the need for blocking agents. Methanol would then be expected to be a final product and not a transitory intermediate in the absence of dehydrogenases. MMO has been isolated from the methanotrophic bacterium *Methylosinus trichosporium* and found to consist of three separate protein components termed hydroxylase (MMOH), reductase (MMOR), and component B (MMOB) (Shu *et al.*, 1997). Kinetic analysis of a single-turnover reaction revealed at least five and probably six intermediates in the catalytic cycle of MMO, among which "Q" is the key methane-oxidizing species. Q has been established to be a diamond core structure that contains two iron atoms, each bound to two oxygen atoms, and a carboxylate bridge that connects the active site to the enzyme. This basic data could lead to development of highly selective, active MMO, possibly in immobilized form on solid supports, or to construction of biomimetic catalysts for methane oxidation.

B. ETHANOL

Commercial Processes

Ethanol is manufactured from a variety of biomass feedstocks by anaerobic fermentation and from ethylene by direct vapor-phase catalytic hydration and sulfation-hydrolysis. The stoichiometries that represent the major processes are as follows:

Biomass fermentation:	$(C_6H_{12}O_6)_n + nH_2O \rightarrow nC_6H_{12}O_6$
	$nC_6H_{12}O_6 \rightarrow 2nCH_3CH_2OH + 2nCO_2$
Direct ethylene hydration:	$CH_2{=}CH_2 + H_2O \rightarrow CH_3CH_2OH$
Indirect ethylene hydration:	$CH_2{=}CH_2 + H_2SO_4 \rightarrow CH_3CH_2OSO_3H$
	$CH_3CH_2OSO_3H + H_2O \rightarrow CH_3CH_2OH + H_2SO_4$
	$CH_3CH_2OSO_3H + CH_2{=}CH_2 \rightarrow (CH_3CH_2O)_2SO_2$
	$2CH_3CH_2OSO_3H \rightarrow (CH_3CH_2O)_2SO_2 + H_2SO_4$
	$(CH_3CH_2O)_2SO_2 + H_2O \rightarrow CH_3CH_2OH + CH_3CH_2OSO_3H.$

Ethanol produced from ethylene feedstocks is called synthetic ethanol. Today, most synthetic ethanol is manufactured by direct vapor-phase catalytic hydration of ethylene rather than by indirect hydration. In a typical process, steam and high-purity ethylene (97%) and recycle ethylene (85%) at a 0.6 molar ratio of water to ethylene are passed over a fixed bed of diatomaceous earth-supported phosphoric acid catalyst at 6.9 MPa and 300°C and a space velocity of 1800 h^{-1}. Conversion per pass is about 4.2%. The product stream is neutralized to remove any entrained acid, unreacted ethylene is separated for recycling, and the dilute aqueous ethanol solution is distilled. The process can be designed to remove impurities in the parts-per-million range to yield high-purity 190- and 200-proof ethanol. The overall yield is about 95% of theory. Co-product diethyl ether and heavy oil by-products are removed during rectification. The more complex ethylene sulfation process is typically carried out in a countercurrent absorber at 1.6-2.4 MPa and 60-70°C containing 96% sulfuric acid followed by hydrolysis at 70°C. The yields based on ethylene converted are about 86-88% ethanol and 5-10% diethyl ether. High purity 190- and 200-proof ethanol can be produced by scrubbing to remove acid, steam distillation to remove ether, and distillation.

Commercial fermentation of biomass-derived sugars is performed with facultative, anaerobic yeasts to yield equimolar amounts of ethanol and carbon dioxide. Yeasts are small, nonmotile, oval cells that reproduce nonsexually by budding. They are classified in all three groups of higher fungi: Ascomycetes, Basidiomycetes, and Fungi Imperfecti. The principal organism for alcoholic fermentation, *Saccharomyces cerevisiae,* is an ascomycetous yeast (bakers' yeast). At a certain stage in its growth, the vegetative cells are transformed into asci, each containing four haploid ascopores. Pairs of germinating ascopores, or the first vegetative cells produced from them, fuse to form diploid vegetative cells.

If simple sugars are present in the biomass feedstock, they can simply be extracted and fermented. If the sugars exist in polymeric or complexed form, they must first be liberated and then extracted and fermented. Ignoring fibrous cellulosics and lignocellulosic biomass feedstocks such as woody biomass for the moment, two basic kinds of biomass are employed as feedstocks (*cf.* Klass, 1984b): sugar crops, and sugar-containing by-products such as sugarcane, sugar beet, sorghum, molasses, and waste sulfite liquors and wood pulping sludges from pulp and paper mills; and starchy crops such as corn and potatoes. Feedstocks such as molasses contain most of the sugars as individual hexoses or disaccharides. The sugars can be directly converted with *S. cerevisiae* without pretreatment. For grains, the cell walls must be disrupted to expose the starch polymers so that they can be hydrolyzed to free, fermentable sugars because the yeasts do not ferment the polymers. The sugar polymers in grain starches contain about 10-20% hot-water-soluble amyloses and 80-90% water-insoluble amylopectins. Both substances yield glucose or maltose on hydrolysis.

The basic steps for conversion of biomass to ethanol in a dry milling plant are shown in Fig. 11.2, and a more detailed conventional schematic for corn is shown in Fig. 11.3 (*cf.* Cole, 1980). Substantial deviation from this scheme is possible. For grains, mechanical grinding is first used to rupture the hull walls and expose the starch polymers. Typically, the grain is pulverized to about 40 mesh in hammermills or other types of grinders, optionally passed through cyclones to remove dust, and suspended in water with agitation. If the fermenters are fed by a wet corn mill instead of a dry milling facility, several by-products are removed in the pretreating steps. This reduces the amount of nonfermentable protein solids and corn oil that pass through the fermenters and appear in the distillers dry grains (DDGs). In wet milling, the crude starch, corn oil, and gluten fractions are obtained through a series of steeping, milling, and separation steps. The products can include gluten meal, gluten feed, corn germ meal, and corn oil. The starch fraction can be processed for sale or converted to syrups, high-fructose corn syrup (HCFS) sweeteners, or ethanol.

In large-scale corn fermentation plants, degermination is often practiced. The floating seed or germ is separated from the surface of the water slurry; the hull, starch, and gluten remain in the mash. The oil-rich germ is dried and pressed to recover corn oil and press cake, which can be sold as cattle feed. Hydrolysis is then initiated with the mash to liberate the sugars. The water slurry is preheated; the pH is adjusted to 7; and the slurry is treated with a commercial enzyme preparation such as a thermophilic bacterial amylase, which is usually made from brewing malts, from controlled germination

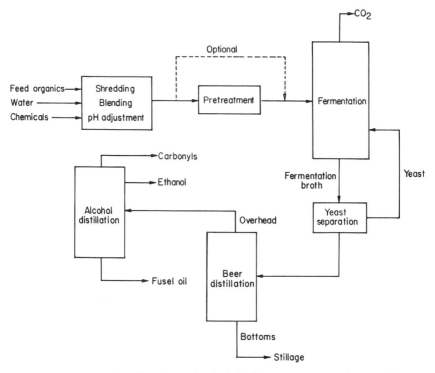

FIGURE 11.2   Ethanol production by alcoholic fermentation. From Klass (1994).

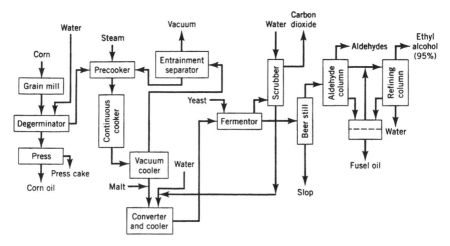

FIGURE 11.3   Flow scheme for manufacture of ethanol from corn. From Klass (1994).

of barley, or from enzymes derived from molds. The mixture is heated to an optimal temperature, usually around 93°C, and "cooked" for an optimal time for the particular enzyme preparation used, usually 15 min. to a few hours. This partial hydrolysis process is called liquefaction. It completes the disruption of the cell walls and breaks the bonds between the sugar polymers to form dextrins, a mixture of oligosaccharides and polysaccharides. The last step in the pretreatment before fermentation is saccharification, or the conversion of the dextrins to free glucose. The mash of dextrins is cooled to the proper temperature and the pH is adjusted to 3 to 5 for the final hydrolysis. This step is catalyzed by addition of a commercial preparation of the enzyme glucoamylase. The mash is held between 60 and 70°C and continuously agitated until hydrolysis is complete. Chemical nitrogen as ammonia, urea, or ammonium salts and other macronutrients such as phosphorus and potassium salts may have to be added, depending on the requirements of the particular strain of yeast used. Sufficient mineral micronutrients are usually supplied in the feed water.

S. cerevisiae contains enzymes that hydrolyze disaccharides to simple sugars and that catalyze the fermentation of four hexoses: glucose, mannose, fructose, and galactose. In addition to the major products ethanol and CO_2, small amounts of by-product glycerol, acetic acid, lactic acid, succinic acid, and fusel oil, which is a mixture of active amyl alcohol and isoamyl alcohols, are formed. Some of the substrate carbon consumed is also converted to small amounts of new yeast cells. About 900 kJ of thermal energy is released per liter of ethanol produced. For grain-derived mashes, the process is generally carried out at pH 3 to 5 and temperatures of 27 to 32°C over a period of about 48 to 72 h, depending on the concentrations of yeast cells and sugars. Higher fermentation temperatures up to about 38°C can be employed to speed up the process, but ethanol losses due to evaporation may become excessive at higher temperatures without ethanol recovery equipment. The concentration of sugar in the mash is adjusted to 22 wt % or less to avoid inhibiting the growth and activity of the yeast cells, the membrane walls of which can be ruptured at higher sugar concentrations. The fermenter effluent, or fermentation broth or beer, contains about 8 to 12 vol % ethanol.

Blackstrap molasses is the residual syrup remaining from the crystallization of cane and sugar beet sugars that cannot be further crystallized without special treatment. The syrupy mixtures contain approximately 55 wt % total free sugars, mainly sucrose, 35 to 45%, and 15 to 20% invert sugar (glucose and sucrose). High-test molasses obtained from the evaporation of raw sugarcane juices contains 70 to 80 wt % total sugars. Fermentation of these feedstocks does not require the extensive pretreatment needed for grain feedstocks, but nutrients such as ammonium sulfate and other salts may have to be added because of deficiencies in the fresh feed. Fermentation is carried out with solutions of about 10 to 20 wt % sugar concentration at pH 4 to 5 and temperatures of 30 to 38°C for about 28 to 48 h. The resulting beer contains about 6 to 10 vol % ethanol.

Potatoes contain about 15 wt % water-insoluble starches surrounded by pulp, so when used as feedstock for fermentation ethanol production, they are usually pretreated with 400-kPa steam for about 1 h to release the starches as a pulp, screened to remove the skins, and hydrolyzed with suitable enzyme preparations to yield a sugar solution ready for fermentation. Pulp and paper mill sludges can contain glucose in as high as 40 to 50 wt % concentrations and a total cellulose plus hemicellulose content of 50 to 75 wt % on a dry weight basis (Kerstetter, Lyford, and Lynd, 1996). They are also suitable feedstocks for fermentation ethanol production after pretreatment, which can be designed around the sugar assays and compositional data developed for each type of sludge.

It is important to emphasize that for moderate- to large-scale fermentation plants, "bank deposits" of suitable yeast strains are necessary to guard against contamination. The capability of propagating active yeast strains adapted to the particular feedstock and conditions used in the process should be maintained to make sufficient inoculum available for fermenter startup after plant upsets and scheduled shutdowns. Also, if yeast cells are separated for recycling as inoculum for fresh feed, the most active cell strata should be selected.

A modern fermentation plant for the manufacture of fuel-grade ethanol includes distillation units to produce anhydrous ethanol from the beer. Units for CO_2 recovery are optional. The first distillation yields an overhead that contains about 55 vol % ethanol and the stillage bottoms, which can be processed and sold as high-protein DDG in the case of grain feedstocks (Weigel, Loy, and Kilmer, 1997). The second distillation yields an ethanol-water azeotrope of 95-96 vol % ethanol (190-192 proof), and the last is usually an azeotropic distillation in older plants that yields anhydrous ethanol. The azeotropic agent is normally benzene, cyclohexane, heptane, or gasoline. These agents remove the remaining water in a ternary azeotrope containing some alcohol, the agent, and water. Diethyl ether is also used as an azeotropic agent to remove the water in a binary azeotrope. These energy-intensive distillations must be carefully performed to maximize plant efficiencies and minimize process steam needs. Many of the modern plants use adsorbents such as molecular sieves and other drying agents to remove small amounts of water.

In the late 1970s and early 1980s, when fermentation ethanol was distributed as a fuel component in the United States after being out of the market for many years, energy consumption in commercial fermentation plants was high. As much as 1.6 times more energy was needed to produce fermentation ethanol than its energy content, 23.6 MJ/L (higher heating value). Since one of the prime objectives of manufacturing fuel ethanol is to conserve and displace petroleum and natural gas, it did not make any sense to use more fossil energy for plant operations than the amount of energy residing in the product ethanol. New, more efficient fermentation methods and techniques for separating ethanol from water solutions were developed. Highly efficient alcohol distillation systems are now used to recover various grades of ethanol with minimum energy consumption (cf. Katzen et al., 1981). Energy consumption has been improved to the point where modern fuel ethanol plants operate on much lower energy inputs than the older plants. Some plants are reported to use as little as 5.6 MJ of steam per liter of ethanol produced and a total energy consumption of 11.1 to 12.5 MJ/L of product ethanol.

The overall efficiency of microbial conversion of the fermentable sugars to ethanol is quite high; over 90% of the fermentable sugars are normally converted to ethanol. Ignoring the small amount of sugars used for cellular growth and maintenance, the theoretical yield of ethanol per mole of hexose fermented is 51.14 wt % of the hexose. Fermentation ethanol and DDG yields per bushel of corn processed in dry milling plants are about 355 L/t of corn (2.5 gal/bu, 85 gal/ton) and 7.5–7.9 kg, respectively. The potential ethanol yields of various biomass feedstocks in conventional fermentation plants are listed in Table 11.8. Interestingly, commercial ethanol yields from sugarcane range from 160 to 187 L/t; this is only about 50% that from corn. The probable cause of this yield difference is the fact that a typical sweet variety of sugarcane, managed for sugar production, consists of approximately 30% soluble fermentable solids and 70% insoluble lignocellulosics (Alexander, 1983). At a typical whole cane yield of 67 t/ha-year in a tropical cane-growing area, the maximum yield of fermentation ethanol is then about 194 L/t of cane. Fermentation ethanol yields for corn wet milling plants are usually about the same or slightly less than those for dry milling plants. The by-product yields per bushel of corn are about 0.45 kg of corn oil, 1.36 kg of gluten meal (60 wt % protein), and 5.90 of gluten feed (21 wt % protein). Carbon dioxide is often recovered from the fermenter off-gas for market.

Fermentation of Lignocellulosics

The use of corn and other grain foodstuffs and animal feeds as fermentation plant feedstocks increases fuel ethanol costs by a substantial amount. The alternative is to use lignocellulosic biomass, the main components of which are crystalline and amorphous cellulose, amorphous hemicellulose, and lignin binder. However, major problems are associated with the production of fermentation ethanol from these materials, such as the difficulty of hydrolyzing cellulosics to maximize glucose yields and the inability of ethanologenic yeasts to ferment the pentose sugars that make up a large portion of the hemicellulose polymers in biomass. Over the last several decades, much effort has been

TABLE 11.8 Potential Fermentation Ethanol Yields from
Various Biomass Feedstocks[a]

Feedstock	Yield (L/t)
Apples	61
Babassu	80
Barley	330
Buckwheat	348
Carrots	41
Cassava	180
Cellulose	259
Cheese whey	23
Corn	355–370
Dates (dry)	330
Figs (dry)	250
Grapes	63
Jerusalem artichoke	83
Molasses	280–288
Oats	265
Peaches	48
Pears	48
Pineapples	65
Plums	45
Potatoes	96
Prunes (dry)	300
Raisins	340
Rice (rough)	332
Rye	329
Sorghum (grain)	332
Sorghum (sweet)	44–86
Sugar beets	88
Sugarcane	160–187
Sweet potatoes	125–143
Wheat	355
Yams	114

[a]The World Bank (1980); Klass (1983); U.S. Department of
Treasury.

directed to perfecting the hydrolysis process for cellulosics and to discovering or genetically engineering other types of microorganisms that can ferment both the C_5 and C_6 sugars. These technologies would make it possible to increase ethanol yields significantly by converting a larger fraction of biomass to ethanol, and to use lower cost biomass feedstocks such as fibrous agricultural crops and residues, woody biomass, bagasse, municipal solid wastes, waste paper, and forest products and lumber industry wastes.

Interest in wood hydrolysis dates back to the early nineteenth century, when it was discovered that cellulose could be dissolved in concentrated sulfuric acid and converted to sugars (Braconnet, 1819). Glucose, the principal sugar produced, could then be readily fermented to ethanol. Wood cellulose and cellulose in most fibrous biomass always occurs together with hemicelluloses and lignins. The association of these materials has been claimed by some to make it difficult to obtain fermentable sugars in high yields by practical processes. The problem persisted for many years. The major factors that affect the accessibility and susceptibility of native cellulose to hydrolysis include its insolubility in water, particle size, extent of lignification, and crystallinity. Cellulose is usually embedded in lignocellulose, an amorphous matrix of hemicellulose and lignin in the cell walls of fibers found in hardwoods and softwoods, cotton, and many other fibrous biomass species. The geometry of the arrangement of the microfibrils in the fiber walls and the strong and weak bonding that has been theorized to exist between the various macromolecules have pronounced effects on the physical properties of the fibers and the accessibility of the sugar polymers. Cellulose itself exists as macromolecules in crystalline and amorphous regions of the fibers. The relative stability of woody biomass to the action of microorganisms has been considered to be due to the fact that cellulose is not in the free state. There is bonding between lignins and cellulose. Many studies have attempted to relate the variability of the properties of cellulosic materials on hydrolysis and the low sugar yields to the complex structures of the lignocellulosics and the relative stability of woody biomass to attack by enzymes and microorganisms. As the lignin content decreases, it was found that cellulose is more accessible and more readily decomposed. Detailed investigations of the hydrolysis of sugar polymers in concentrated and dilute acids at different temperatures and pressures under a variety of batch, staged, and continuous hydrolysis conditions, of the kinetics of hydrolysis, and of the application of enzyme catalysis have led to greatly improved processes that afford near theoretical sugar yields. The lignocellulosic structures and any bonding that may exist between the polymeric components do not present a barrier to development of these improved hydrolytic methods.

The first plant for producing ethanol from cellulose was constructed in South Carolina in 1910; the yield was approximately 83 L/t (20 gal/ton) of sawdust (Fieser and Fieser, 1950). Hydrolysis of sawdust was accomplished with sulfuric acid under pressure. The German Stettin process used in 1918–1919 was based on the same technology. Many process improvements have been made since then. The overall stoichiometry for the hydrolysis and fermentation of pure cellulose,

$$(C_6H_{10}O_5)_n + nH_2O \rightarrow n2CH_3CH_2OH + n2CO_2,$$

corresponds to a theoretical ethanol yield of 56.82 wt % of the cellulose. Presuming the sawdust is dry and contains 50 wt % cellulose, an ethanol yield of 100% from yeast fermentation is about 224 L/t of sawdust.

Luers and Saeman concluded from kinetic studies of the dilute acid hydrolysis of cellulose that the general model representing this process consists of two consecutive first-order reactions: A → B → C (Luers, 1930, 1932; Saeman, 1945, 1981). The first reaction is hydrolysis to yield glucose intermediate and the second reaction involves decomposition of glucose. Saeman found that the hydrolysis reaction is accelerated more than the decomposition reaction by both increased temperature and acid concentration. The predicted general improvement in yield was maintained up to 260°C, and the time to maximum yield at 260°C was 0.45 min. The curves in Fig. 11.4 illustrate the maximum yield of sugar from cellulose at two acid concentrations as a function of temperature. One of the most interesting confirmations of Saeman's kinetic model was reported for pure cellulose

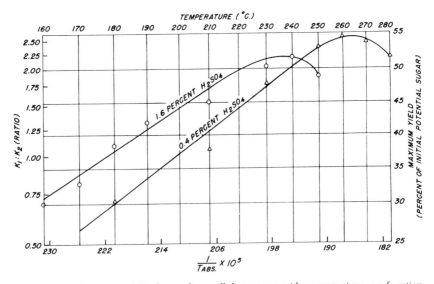

FIGURE 11.4   Maximum yield of sugar from cellulose at two acid concentrations as a function of temperature. From Saeman (1981).

(Grethlein, 1978). A yield of over 50% sugar was obtained using 1% sulfuric acid and a continuous-flow reactor at a residence time of 0.22 min and a temperature of 237°C. It is apparent that the low yields of sugars on acid hydrolysis are attributable to their destruction during hydrolysis. Early attempts at dilute acid hydrolysis by single-stage processing were limited to yields of about 20% fermentable sugars because of the unfavorable ratio of rate constants, k_1 to k_2, of the consecutive reactions. Scholler's pulsed percolation process for hydrolyzing wood chips with dilute acid, which removes sugar as it is formed in the reaction zone, exploits the observation that the first increment of glucose produced is obtained in high yield based on the cellulose consumed. The ethanol yields are 190-200 L/t of oven-dried wood (Faith, 1945; Nikitin, 1962). The yield of glucose is more than twice that obtainable by batch hydrolysis, but is obtained at a concentration of only 4%. The Madison modification of the Scholler process uses a higher temperature, a shorter reaction time, and continuous rather than periodic elution of sugar (Harris and Beglinzer, 1946). The key feature of this process is that wood chips form a rigid porous packed bed which facilitates fast percolation of the acid through the bed. During World War II, 20 plants using the Scholler process were built in Germany. Similar plants were built in China, Korea, Russia, Switzerland, and other countries, and one plant of the Madison type was built in the United States and operated for a short time. The largest application of the percolation process is in Russia, where some 40 plants were built. A few plants that used the Bergius process, which employs hydrochloric acid instead of sulfuric acid, were built in Germany and operated during the war. With the exception of the Russian plants, most of the facilities were shut down after the war because ethanol from lignocellulosics was not economically competitive with synthetic ethanol from ethylene.

The hemicelluloses that occur in association with cellulose are pentosans; they yield pentoses on hydrolysis and are characterized by hemiacetal linkages (Chapter 3). Because of these structural groupings, hemicellulose hydrolyzes at conditions much less severe than those required for cellulose hydrolysis. The repeating units in the polysaccharides are mostly ($C_5H_8O_4$), but some ($C_6H_{10}O_5$) units occur. Hydrolysis yields mostly C_5 pentoses. Xylan, which is made up of xylose units, is the most common polysaccharide in hemicelluloses. Xylose is therefore the dominant pentose in the hydrolysate; arabinose is usually next in abundance, and lyxose and ribose are normally minor pentoses. Minor amounts of mannose, galactose, and the uronic acids also occur in some hemicellulose hydrolysates. Some hemicelluloses contain substantial amounts of glucose residues as well. Basically, the natural, ethanologenic yeasts will not ferment pentoses. So it seems logical to carry out a two-stage hydrolysis in which the hemicellulose is prehydrolyzed and extracted prior to cellulose hydrolysis. The staged use of different concentrations of sulfuric acid and conditions for hemicellulose and cellulose are quite effective (*cf.* Rugg, Armstrong, and Stanton, 1981). For example, cornstalks on prehydrolysis with 4.4% sulfuric acid at 100°C for 50 min. afforded 90% of the theoretical maximum yield of xylose. Subsequent impregnation of the residue with 85% sulfuric acid, dilution to 8%, and hydrolysis at 110°C for 10 min. gave an 89% glucose yield (Sitton *et al.*, 1979). Concentrated hydrochloric acid and mild temperatures (Ackerson, Clausen, and Gaddy, 1991) and even hydrofluoric acid (Ostrovski, Aitken, and Hayes, 1985) afford near theoretical yields of pentoses and hexoses. Hydrolysis with weak acid seems to be superior for hardwoods, which contain larger amounts of the hemicelluloses than softwoods (Harris, Baker, and Zerbe, 1984).

Advanced Hydrolysis

Hundreds of innovative biomass hydrolysis developments have occurred since World War II. Acid hydrolysis processes for biomass have evolved to the point where it is a relatively simple procedure to obtain the monomeric building blocks of the polysaccharides in high yields. The pentoses, hexoses, and lignins can each be isolated by sequential, selective hydrolysis of hemicellulose and cellulose; the residual insolubles are the lignins. The ACOS process, or Acid Catalyzed Organosolv Saccharification process, is an example of an advanced, high-yield process (Paszner and Cho, 1989; Pazner *et al.*, 1993). This process is reported to allow almost quantitative recovery of both sugars and lignin from any wood species in less than 5 min. The ACOS process uses a mixture of 80% or more acetone in water and a catalytic amount of mineral acid at temperatures from 165-230°C, preferably in the continuous single- or two-stage reaction mode to allow control of residence time in the reactor. In a two-stage system, temperature can be controlled in each stage to yield the pentose, hexose, and lignin fractions. The probable reasons the sugars survive the process without destruction seems to be the transformation of the free

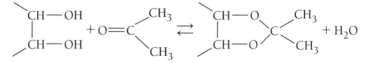

sugars into transitory intermediate ketals at short reaction times, thereby protecting the sugars from further reaction. Ketals are known to be reversibly formed from vicinal diols and acetone under acid catalysis and to be rapidly hydrolyzed under acid conditions at lower temperatures. So if the protected sugars containing one or more ketal groups are rapidly removed from the hydrolysis reactor before they revert to the free sugars, hydrolysis would be expected to occur downstream to liberate the free sugars and regenerate acetone. The experimental data indicate the potential total sugar, ethanol, and lignin yields from the ACOS process shown in Table 11.9. The projected ethanol yield increases are evident for the grain and sugar crop feedstocks because all of the polysaccharides in the nonfermentable fibrous portion of the biomass (straws, stover, rinds, etc.) and the lignocellulosics can be processed to yield near-theoretical amounts of sugars. The sugar yields in the case of grain and sugar crops are doubled or tripled, depending on the type of biomass harvested. Note that the projected ethanol yields in Table 11.9 are high in the ACOS process because they approach the theoretical maximum and include both the pentoses and hexoses. As will be shown later, the pentoses can also be converted to ethanol.

In addition to acid-catalyzed hydrolysis, an intensive effort to develop enzyme-catalyzed hydrolysis of cellulose and hemicellulose in lignocellulosic biomass has been in progress for many years. The required enzymes can be prepared from selected organisms, and modern recombinant DNA methods have also made it possible to prepare cellulases and hemicellulases and other enzyme systems for industrial applications (*cf.* Brennan, 1996). A number of advantages have been attributed to enzyme-catalyzed hydrolysis of lignocellulosics. Corrosion-resistant equipment is not needed. The energy requirements are less because mild hydrolysis conditions are used, whereas high temperatures cause operational and equipment problems and the formation of undesirable degradation products. The sugar yields obtained by enzyme catalysis were believed in the early work to be higher than those obtained by acid catalysis. This is not a significant factor now because of the improvements that have been made for acid catalysis. But there are other advantages in combining enzyme-catalyzed depolymerization and fermentation.

Some of the early work on enzyme-catalyzed hydrolysis of cellulosics used cellulase derived from mutant strains of the fungus *Trichoderma viride* grown on cellulose (Spano, 1976). The filtrate of the enzyme solution was used to treat various types of milled bagasse, corrugated fiberboard, MSW pulps, newspaper, cotton, cellulose pulp, bagasse, and fibrous cotton at pH 4.8 over several hours. Periodic measurements showed that the percent of cellulose saccharification after 48 h ranged up to a high of 78% for fiberboard to a low of 6% for unmilled bagasse. Various pretreatments for newspaper showed that 70% of the sugars were released with pot-milled newspaper on enzyme-catalyzed hydrolysis after 48 h, while all other pretreatments afforded less saccharification. Another enzyme treatment utilized a mutant strain of *T. reesei* grown continuously on specific biomass feedstocks to be hydrolyzed to produce a complete cellulase system (Emert and Katzen, 1981). Enzyme production began on a spore plate with subsequent scale-up to the enzyme production vessel. The interesting feature of this work is that the whole broth from the

TABLE 11.9 Projected Sugar, Ethanol, and Lignin Yields from Grain, Sugar, and Woody Biomass Feedstocks Pretreated by the ACOS Process[a]

Biomass	Plant part	Yield (t/ha-year)	Sugar yield		Ethanol yield		Lignin yield	
			(t/ha-yr)	(t/t)	(L/ha)	(L/t)	(t/ha-year)	(t/t)
Corn	Kernel	14	11.5	0.82	4596	328	5.6[c]	0.40
	Biomass[b]	25	17.5	0.70	9150	366	1.76	0.07
Wheat	Kernel	3	2.2	0.73	1148	383	1.05[c]	0.35
	Biomass[b]	8.3	6.1	0.73	3486	420	1.00	0.12
Sugarcane	Juice[c]	10.0	10.0	1.00	5600	560		
	Biomass[d]	30.4	27.4	0.90	16,128	531	3.67	0.12
Aspen poplar		18	14.8	0.82	7170	398	2.88	0.16
Douglas fir		8	6.1	0.76	3700	463	2.18	0.27
Eucalyptus		48	39.5	0.82	24,100	502	9.6	0.20

[a]Adapted from Paszner and Cho (1989). The yield figures are either in annual units per hectare of growth area or in units per tonne of plant part.
[b]Includes starch.
[c]Sucrose only.
[d]Includes sucrose.
[e]DDGS.

enzyme preparation, which has an optimum hydrolysis temperature of 45 to 50°C, is combined with the yeast and media in the fermenter so that simultaneous saccharification and fermentation (SSF) occurs. The optimum temperature for the combined cellulase and yeast systems is 40°C. The enzyme preparations used converted 90% of the cellulose to glucose in 48 h, and SSF was found to enhance ethanol yields by 25 to 40% because of the removal of products formed during saccharification that inhibit the cellulase system. Glucose and cellobiose are feedback inhibitors of enzymes in the cellulase system, so when glucose is removed as fast as it is formed, there is no product inhibition. A wide variety of potential feedstocks was tested at the pilot plant scale—cotton gin trash, sludges, digester fines and rejects, straw, bagasse, and MSW.

SSF systems have now been widely studied. Optimization of the process and other unit operations such as biomass pretreatment, cellulase production, and xylose fermentation are expected to reduce the cost of ethanol to $0.18/L (cf. Philippidis, 1993). Several biomass pretreatment processes have been developed, including steam explosion, acid-catalyzed steam explosion, organosolv treatment, and ammonia fiber explosion to improve accessibility of the polysaccharides to the enzymes (cf. Wyman, 1993). Studies of a number of substrates have shown that after all of the hemicellulose is converted to sugars, virtually all of the cellulose can be enzymatically hydrolyzed to yield glucose. Highly active enzyme systems have also been developed that function well with alcohol-forming organisms (cf. Wright, 1989). Biological pretreatment of the feedstock with certain enzyme-producing fungi to reduce the need for commercial enzyme preparations is effective in some cases (cf. Akhtar, Kirk, and Blanchette, 1996). Enzymatic cellulose degradation has been modeled from the fungal cellulase system which exhibits endogluconase, exogluconase, and β-glucosidase activity (cf. Recombinant BioCatalysis, Inc., 1996). The endogluconases attack the internal structure, the exoglucanases attack the polymer's terminal groups to generate glucose or cellobiose, and the β-glucosidases catalyze conversion of cellobiose to glucose. The hemicelluloses are composed primarily of xylan, mannan, and glucuronic acid polymers, so balanced hemicellulases contain enzymes (e.g, xylanases and mannanases) that catalyze the hydrolysis of each of these components.

The technical difficulties involved in obtaining sugars and lignins in the free state from lignocellulosic biomass have been addressed by researchers throughout the world and appear to have been solved. The advanced acid- and enzyme-catalyzed hydrolysis methods developed in this effort are efficient and produce high sugar yields under practical conditions from a wide range of virgin and waste biomass feedstocks.

Immobilization

In addition to SSF systems, several other advanced technologies have been incorporated in yeast fermentations. Continuous fermentation with immobilized organisms is one of them. Recycling of yeast in batch operations has many advantages, but the capability of operating in the continuous mode at shorter hydraulic retention times presents opportunities for smaller equipment and lower capital, energy, and labor costs. For example, whole yeast cells of *S. cerevisiae* have been immobilized in calcium alginate beads and used in a large fluid-bed system to ferment molasses and other carbohydrates (Nagashima *et al.*, 1984). Ethanol was continuously produced at a yield, concentration, and productivity over 6 months of stable operation of 95%, 8-10 vol % in the beer, and 20 g/L-h, respectively. The productivity was estimated to be 20 times higher than that of a conventional batch fermentation. Another innovative approach uses ethanologenic organisms immobilized on fiber discs (Clyde, 1982, 1996). The discs are rotated while the sugar solution is passed through the fermenter. With *S. cerevisiae*, cane molasses and saccharified starch were fermented in less than 4 h at conversion efficiencies above 90% and ethanol yields of 49 wt % of the sugar consumed compared to a theoretical yield of 51.14 wt % (Wayman, 1991). With the bacterium *Z. mobilis*, saccharified wheat starch was fermented in 0.9 h at a productivity of 65 g ethanol/L-h, while enzyme-saccharified aspen cellulose was fermented in 20 min. With the yeast *Pichia stipitis* R, coniferous spent sulfite liquor, steam-stripped to remove acetic acid, was fermented in 7 h, compared to 24 h in batch fermentations. Conversion of wood sugars was 96% at yields of 45 to 47 wt % of the sugar consumed, yielding 0.45 to 0.47 g ethanol/g sugar consumed. It is evident that fermentation technology for the production of ethanol can be greatly improved with conventional feedstocks as well as lignocellulosics.

Metabolic Engineering

The next phase in the development of advanced fermentation ethanol processes concerns the conversion to ethanol of all the pentose and hexose sugars released on hydrolysis of lignocellulosics. Traditional bakers' yeast strains promote fermentation of hexoses at high yields, but over long periods of time, and they do not ferment the pentoses. Although some yeasts use both hexose and pentose sugars as sources of carbon and energy and ferment hexoses and xylose, they do not ferment arabinose and the other pentoses. The overall stoichiometry of hemicellulose hydrolysis and pentose fermentation is

$$(C_5H_8O_4)_n + nH_2O \rightarrow nC_5H_{10}O_5$$
$$nC_5H_{10}O_5 + nADP + nP_i \rightarrow 1.667nCH_3CH_2OH + 1.667nCO_2 + nATP.$$

Ignoring the small amount of sugars used by the organisms for growth and maintenance, the theoretical maximum yield of fermentation ethanol from the pentose sugars is 51.14 wt % of the sugars fermented, the same as the theoretical maximum yield from the hexose sugars.

If organisms could be found or metabolically engineered that efficiently ferment both the pentoses and hexoses under practical conditions at high yields and short residence times, fermentation ethanol technology would then have reached another plateau with low-cost lignocellulosic feedstocks. Simultaneous saccharification and fermentation or separate saccharification and fermentation of essentially all the sugars that make up the polysaccharides would each be able to approach the theoretical limit of fermentation ethanol production from the polysaccharides in low-cost lignocellulosic biomass.

As previously mentioned and in the earlier discussion of fermentation methanol, bacteria of the genus *Zymomonas* such as *Z. mobilis* are known to convert hexoses to ethanol at high yields and short residence times. These bacteria are facultative anaerobes that have fermentative capacity and convert only glucose, fructose, and sucrose to equimolar quantities of ethanol and CO_2; the pentoses are not converted. The Entner-Doudoroff pathway is utilized instead of the Embden-Meyerhof pathway, and a net yield of 1 mol of ATP is generated, not 2 mol as in bakers' yeast. But pyruvate is the same key intermediate. In *Z. mobilis*, it is decarboxylated by pyruvate decarboxylase to yield acetaldehyde which is then reduced to ethanol by alcohol dehydrogenase.

Recombinant DNA techniques have provided the methodology needed to combine metabolic pathways from different organisms to effectively "tailor-make" strains that contain the desired traits (*cf.* Lawford and Rousseau, 1993). Thus, if genes encoding the pyruvate-to-ethanol pathway were transferred by recombinant methods into organisms that afford pyruvate from pentoses and hexoses but not ethanol, then essentially all the sugars in lignocellulosics should be fermented. One of the first successful attempts to achieve this goal involved transfer of pyruvate decarboxylase and alcohol dehydrogenase (*pdc* and *adhB*) from ethanologenic *Z. mobilis* to enteric bacteria (Ingram, Conway, and Alterthum, 1991; Ingram, Ohta, and Beall, 1991; Ingram *et al.*, 1997). *Escherichia coli* has the ability to convert *all* of the C_5 and C_6 sugars in biomass to pyruvate, but in contrast to *Z. mobilis*, pyruvate is converted by *E. coli* to organic acids. Recombinant plasmids were constructed which expressed both the *pdc* and *adhB* genes encoding the ethanol pathway. The recombinant strain has the capability of fermenting all of the sugars in biomass to ethanol and CO_2. In the initial constructs, both the native transcriptional promoter and terminator were removed from the *pdc* gene. The remaining *pdc* coding region was ligated upstream from a promoterless *adhB* gene, including the *adhB* transcriptional terminator. Integration of the *Z. mobilis* genes behind the *pfl* promoter, which is a strong promoter that is always active in *E. coli*, led to *E. coli* KO11. In this recombinant strain of *E. coli* and other variations, the natural biochemical pathway is diverted to ethanol. Final concentrations of ethanol with these strains ranged from 40 to 58 g/L at over 90% of the theoretical yield limit. Of more importance, however, is the fact that the ethanol pathway derived from Z. *mobilis* is portable and can be moved to other organisms by the addition of appropriate replicons and promoters. This was done with *Klebsiella oxytoca,* a common organism in pulp and paper waste, and *Erwina,* which causes soft-rot of plant tissue. A large-scale plant for the production of ethanol from lignocellulosic feedstocks using these genetically engineered organisms is planned for operation in Louisiana. This is expected to be the first commercial facility to demonstrate the advantages of recombinant techniques for the production of fermentation ethanol.

A somewhat complementary approach has also been taken in which Z. *mobilis* was metabolically engineered to broaden its range of fermentable substrates to include the C_5 sugar xylose (Zhang *et al.,* 1995). Two operons encoding xylose assimilation and pentose phosphate pathway enzymes were constructed and transferred from E. *coli* into Z. *mobilis* to generate a recombinant strain that efficiently fermented both glucose and xylose. The introduction and expression of the genes encoding xylose isomerase, xylulokinase, transaldolase, and transketolase were necessary for the completion of a functional metabolic pathway that would convert xylose to central intermediates of the Entner-Doudoroff pathway and enable Z. *mobilis* to ferment xylose. The recombinant strain fermented a mixture of glucose and xylose to ethanol at 95% of the theoretical yield within 30 h. The overall stoichiometry of xylose fermentation was illustrated previously and the theoretical ethanol yield is 51.14 wt %. But the net ATP yield from 3 mol xylose is postulated to be 2 mol less than that for conventional xylose fermentation. Because less substrate is used in this pathway for cellular biomass formation with Z. *mobilis,* xylose fermentation is believed by the researchers to be more efficient than in any other microorganism.

A similar genetic manipulation was carried out, except that the genes encoding for the pentose arabinose were isolated from E. *coli* and introduced into Z. *mobilis* (Deanda *et al.,* 1996). The engineered strain grows on arabinose as a sole carbon source and produces ethanol at 98% of the maximum yield, which indicates that arabinose is metabolized almost exclusively to ethanol as the sole fermentation product. Although no diauxic growth pattern was evident, the microorganism preferentially utilizes glucose before arabinose with mixed substrates. The researchers believe that in mixed culture, this strain may be useful with the engineered xylose-fermenting strain for efficient fermentation of the predominant hexose and pentose sugars in lignocellulosic feedstocks. However, since E. *coli* is reportedly capable of generating pyruvic acid intermediate from all the C_5 and C_6 sugars in biomass, recombinant E. *coli* incorporating pyruvate decarboxylase and alcohol dehydrogenase might be expected to ferment pentoses and hexoses at higher yields than mixed cultures of the two recombinant Z. *mobilis* strains, one of whose substrate ranges has been expanded to include xylose, and the other to include arabinose. Xylose and arabinose are normally the dominant pentoses in natural hemicellulose, but the other pentoses present would not be fermented unless further genetic manipulations are carried out. It would also seem to be beneficial to avoid the necessity of using mixed cultures by sequentially engineering the xylose-consuming strain of Z. *mobilis* so that it also metabolizes arabinose and the other pentoses.

Thermochemical Ethanol

It is important to briefly review the various routes to ethanol from synthesis gas,

$$CO + 3H_2 \rightarrow CH_3CH_2OH,$$

because it has a potentially large adverse impact on fermentation ethanol, which is discussed in Section V. The early work targeted the development of heterogeneous catalysts that can be used for direct, gas-phase synthesis of ethanol in the same manner that copper-based catalysts are used for methanol synthesis. The direct hydrogenation of CO to methanol,

$$CO + 2H_2 \rightarrow CH_3OH$$

is established commercial technology. Conversion of synthesis gas to methanol is necessarily low per pass over methanol synthesis catalysts at optimum operating conditions because of equilibrium limitations, but selectivities are high. The Gibbs' free energy changes indicate the thermodynamic driving force below 500 K for CO reduction to ethanol is at least as favorable as the corresponding values for methanol and possibly slightly less favorable at higher temperatures.

During World War II, a very intensive effort was devoted to the development of Fischer-Tropsch chemistry in Germany for converting synthesis gases to motor fuels. Several large-scale experiments that used fused iron catalysts at elevated pressures and temperatures afforded appreciable quantities of water-soluble alcohols, especially ethanol (*cf.* Zorn and Faragher, 1949). The catalysts are obtained by melting mixtures of iron powder and small amounts of TiO_2, MnO, and copper and silicon powders, and then pulverizing and reducing the fused cakes with hydrogen at 650°C. The alcohols comprised about 45% to a high of 60% of the liquid products, about 60-70% of which was ethanol. French investigators later developed catalysts that contain a mixture of Cu and Co oxides, alkali-metal salts, and oxides of Cr, Fe, V, or Mn and that are effective promoters for conversion of synthesis gas to mixed alcohols (The World Bank, 1980). At 250°C and 6.08 MPa, conversion is 35% per pass; selectivity to alcohols is over 95%; and selectivity to two-carbon or normal higher alcohols is greater than 71%. A typical product distribution is 20 wt % methanol, 35 wt % ethanol, 21 wt % 1-propanol, 3 wt % 2-propanol, and 17 wt % 1-butanol. Research done in Japan shows that ethanol is catalytically produced from synthesis gas in reasonably high selectivities at 0.1-5.1 MPa and 150-290°C over Rh, Pt, and Ir carbonyl clusters: for example, $Rh_4(CO)_{12}$, $Pt[Pt_3(CO)_6]_{2-5} \cdot 2N(C_2H_5)_4$, and $Ir_4(CO)_{12}$ impregnated on basic oxides such as MgO, CaO, La_2O_3, Zr_2O_3, and TiO_2 (Kampen, 1980). $Rh_4(CO)_{12}$ on La_2O_3 at 220°C and atmospheric pressure gave a product slate containing 49% ethanol, 19% methanol, 14% methane, 8% carbon dioxide, and 8% other compounds at 23% conversion per pass. These types of catalysts appear to have merit for production of mixed alcohols from synthesis gas.

Research has also been done on the coproduction of fuel-grade methanol, ethanol, and higher alcohols. The C_2-C_4 alcohols are coproduced by catalytic conversion of synthesis gas with modified zinc chromite catalysts in yields up to 30 wt % of the methanol (Brandon et al., 1982; Duhl and Thakker, 1983; Laux, 1977). An increase in temperature, a decrease in space velocity, or a decrease in pressure resulted in an increase in higher alcohol concentration. The presence of CO_2 adversely affected the yields of higher alcohols with increasing space velocity and pressure. Typical gas compositions of 70 mol % H_2, 17 mol % CO, and 13 mol % N_2 at 13.2-28.4 MPa and 300-425°C and gaseous hourly space velocities of 7000-20,000 h^{-1} gave products containing 70-80 wt % methanol, 3-7 wt % ethanol, 2.5-5.5 wt % 1-propanol, 7-23 wt % 2-methyl-1-propanol, trace quantities of 1-butanol and 2-methyl-2-propanol, and small amounts of C_5 alcohols and ethers. This type of product has been termed methyl fuel. The process has been tested at the pilot scale (Paggini and Fattone, 1982). Dow Chemical Company has developed similar one-step processes using molybdenum sulfide catalysts and synthesis gas to manufacture straight-chain terminal alcohols in the C_2-C_4 range; the proportion of methanol can be adjusted from zero to over 90% or to make butanols in significant quantities (Chem. Week, 1984).

Many other research reports could be cited on the direct catalytic conversion of synthesis gases to ethanol-containing mixtures. Virtually all of these reports are similar in that mixed alcohols are formed and ethanol selectivities are not high. However, it is possible to improve ethanol selectivity by direct carbonylation of methanol with synthesis gas (Juran and Porcelli, 1985). Typically, a strong reducing catalyst such as a Co-Ru halide is used in the liquid phase at 27.4-34.5 MPa. Ethanol selectivities are in the 60-80% range; the by-products are acetates and other esters and alcohols. If the reaction is conducted in the liquid phase using a less strongly reducing catalyst, such as a Group VIII metal halide, acetaldehyde can be produced at 80-90% selectivity. Acetaldehyde can then be reduced to ethanol with relatively high overall selectivity. Still another variation on this route to ethanol from synthesis gas is to carbonylate methanol in the liquid phase to form acetic acid instead of acetaldehyde (Winter, 1982, 1986; Juran and Porcelli, 1985). This reaction is reported to proceed in high yield. The acid can either be esterified prior to hydrogenolysis or can be directly reduced. Ethanol is also formed in one step by direct homologation of methanol or by its reaction with synthesis gas in the presence of cobalt octacarbonyl catalysts and iron pentacarbonyl in the presence of a tertiary amine (*Chem. Eng. News*, 1982). Ethanol is produced in high yield in the latter case without formation of water:

$$CH_3OH + 2CO + H_2 \rightarrow CH_3CH_2OH + CO_2.$$

A methanol conversion rate of 14%/h at 220 °C, with ethanol accounting for 72 wt % and methane 28 wt % of the product, was obtained with iron carbonyl-manganese carbonyl catalysts. The ability to synthesize ethanol without co-product water could lead to lower-cost processes that do not require extensive energy-consuming distillations.

Table 11.10 is a summary of the different conversion schemes and conditions for thermochemical ethanol formation discussed here. The primary routes to ethanol and ethanol precursors from synthesis gas are evident. It remains to be determined whether any of these technologies can be developed to produce low-cost thermochemical ethanol.

Thermochemical-Microbial Ethanol

An innovative approach to ethanol production combines thermochemical gasification to generate synthesis gas and its subsequent conversion to ethanol and acetate by the anaerobic spore-former *Clostridium ljungdahlii* (Edgar and Gaddy, 1992). This organism is capable of producing ethanol and acetate from CO and water and/or CO_2 and H_2 in synthesis gas. Under optimal growth conditions, the microorganism produces acetate in preference to ethanol. Conversely, under nongrowth conditions, ethanol is favored.

C. ACETONE-BUTANOL FERMENTATION

Butyric acid-producing bacteria that belong to the genus *Clostridium*, a few members of which are pathogenic such as *C. botulinum*, have been known since the initial work of Pasteur on the microbial conversion of C_6 sugars to simple acids and alcohols (Fieser and Fieser, 1950). Most Clostridia are widely distributed in soil; some fix nitrogen, and many ferment soluble carbohydrates, starch, or pectin with the formation of acetic and butyric acids, CO_2, and H_2. A few also produce acetone, 1-butanol, or both compounds. Two of the most common butyric acid bacteria are *C. butylicum*, which affords mainly acetic acid, butyric acid, 1-butanol, 2-propanol (but no acetone), H_2, and CO_2 from glucose, and *C. acetobutylicum*, which produces mainly acetic acid, butyric acid, 1-butanol, acetone (but no 2-propanol), H_2, CO_2, and small amounts of ethanol from glucose. The product yields from the fermentation of glucose by

TABLE 11.10 Examples of Thermochemical Ethanol Synthesis

Key step	Typical conditions (MPa)	(°C)	Phase	Catalysts	Yields Ethanol (wt %)	By-products (wt %)
Synthesis gas to ethanol	4.1–10.1	240–370	Gas	Cu–Zn–Co transition-metal oxides, alkali	28	CH_3OH, 19; C_3H_7OH-$C_5H_{11}OH$, 23; HCs, 29
Synthesis gas to ethanol	6.1	250	Gas	Cu–Co oxides, alkali-metal oxides	35	CH_3OH, 20; C_3H_7OH 24; C_4H_9OH, 17
Synthesis gas to ethanol	0.1	220	Gas	$Rh(CO)_{12}$ on La_2O_3	49	CH_3OH, 19; CH_4, 14; CO_2, 8; others, 8
Methanol and CO to ethanol		220	Liquid	$Fe(CO)_5$–$Mn_2(CO)_{10}$·R_3N	72	CH_4, 28
Methanol and CO to ethanol	27.6–34.5	200	Liquid	Co–Ru halides	60–80	Higher alcohols, acetates, esters
Methanol to acetaldehyde	24.1	180	Liquid	Group VIII halides	80–90	Higher alcohols, acetates, esters
Methanol to ethanol		220	Liquid	$Fe(CO)_5$–$Co(CO)_8$·R_3N	High	
Acetic acid and H_2 to ethanol	12.4	250	Liquid	Cu–Co–Mn–Mo oxides	High	
Acetate and H_2 to ethanol	2.8	250	Liquid	Cu–Cr oxides	High	

various species of Clostridia are shown in Table 11.11. It is evident that there are large differences in metabolic products and yields among different species, often with complete exclusion of certain products. Acetic acid, butyric acid, CO_2, and H_2 are consistently formed by the five species listed; acetone is formed in only one, *C. acetobutylicum*, and 1-butanol is formed in only two, *C. acetobutylicum* and *C. butylicum*. The yields of each of the products also exhibit considerable variation. It is evident that high yields of 1-butanol occur at the expense of butyric acid.

C. acetobutylicum is a rod-shaped, motile obligate anaerobe that was used to develop commercial acetone-butanol fermentation processes during World War I when there was a great need for acetone solvent in airplane dopes. The process is called the Weizmann process; processes using other organisms were also commercialized. In the Weizmann process, *C. acetobutylicum* was cultured from organisms originally found on corn surfaces and other grains. Other Clostridia species are also used. Sterilized, degermed corn-starch mashes containing about 8.5 wt % corn starch are anaerobically fermented at 37°C over about 36-48 h after addition of the inoculum. The inoculum is grown on sterile corn starch so that it is already adapted to the feedstock. The beer is distilled to remove nonfermented residuals, and the overhead is fractionated to recover the individual liquids. Protein nutrients and buffers to control pH are often added to improve yields. Operation under aseptic conditions is desirable because acetone-butanol fermentations are often upset by contamina-

TABLE 11.11 Moles of Product per 100 Mol of Glucose Fermented by Various Species of Clostridia[a]

Product	C. acetobutylicum	C. butylicum	C. butyricum	C. perfringens	C. tyrobutyricum
Acetic acid	14	17	42	60	28
Acetone	22				
Acetoin	6				
1-Butanol	56	58			
Butyric acid	4	17	76	34	73
Ethanol	7			26	
Lactic acid				33	
2-Propanol		12			
CO_2	221	203	102	176	190
H_2	135	77	235	214	182
C recovered, %	99	96	96	97	91

[a]Adapted from Wood (1961).

tion, especially by bacteriophage. The volumetric yields per bushel of corn (25.5 kg) are approximately 1-butanol, 4.9 L; acetone, 2.4 L; ethanol, 0.7 L; and small amounts of other organic liquids. The corresponding yield ratios by weight are about 70, 25, and 5%. About 30–34 wt % of the hexoses are converted to these liquids. The product gas is about 60 mol % CO_2 and 40 mol % H_2. Other feedstocks such as molasses and beet sugar afford similar results, but other species of Clostridia are often used. For example, molasses feedstock fermented by *C. saccharobutyl acetonicum liquefaciens* affords liquid product yields of approximately 30 wt % of the sugar charged; the weight ratios are 70% 1-butanol, 25% acetone, and 5% ethanol.

Pyruvate is the initial key intermediate in butyric acid fermentations and is formed via the Embden–Meyerhoff pathway. Pyruvate is cleaved to form CO_2, H_2, and acetyl-Coenzyme A (acetyl-CoA), which dimerizes to form acetoacetyl-CoA:

$$CH_3COCO_2H + CoA \rightarrow CH_3COCoA + CO_2 + H_2$$
$$2CH_3COCoA \rightarrow CH_3COCH_2COCoA.$$

This intermediate is subsequently reduced to butyryl-CoA, and the C_4 acid is finally formed by reaction with acetate:

$$CH_3CH_2CH_2COCoA + CH_3CO_2H \rightarrow CH_3CH_2CH_2CO_2H + CH_3COCoA.$$

Acetyl-CoA is regenerated in this process. The overall product yields in moles per mole of glucose converted are approximately 0.5 acetate, 0.75 butyrate, 2 CO_2, and 2 H_2; 2.5 mol ATP are formed. The nonacidic compounds, acetone, 1-butanol, and 2-propanol, are formed by transformation of some of the acetoacetyl-CoA into acetoacetic acid, which is the precursor of acetone and 2-propanol. Some of the butyryl-CoA is the precursor of 1-butanol via intermediate butyraldehyde. Ethanol is formed by reduction of small amounts of acetyl-CoA. The end result of the production of the neutral products by these additional pathways is that the yields of the other products are reduced. The neutral products are in a lower oxidation state than the acidic products and require additional reducing power as NADH to be formed. Some of the product H_2 serves to sustain and provide NADH because higher partial pressures of H_2 during the fermentation promote higher yields of the neutral products, whereas removal of the product H_2 as it is formed has the opposite effect.

Most acetone is manufactured today in the United States by cumene oxidation. It is a co-product with phenol. Acetone is also manufactured by dehydrogenation of 2-propanol, which is made by hydration of propylene. Most 1-butanol is manufactured today by hydrogenation of *n*-butyraldehye, which is obtained by the hydroformylation of propylene (oxo reaction). It is also manufactured by hydrogenation of crotonaldehyde, which is obtained by the successive aldol condensation of acetaldehyde and dehydration of aldol. TBA is manufactured by hydration of isobutylene. This alcohol has not been reported as a fermentation product.

D. AMYL ALCOHOLS

Amyl alcohols occur in eight isomeric forms and have the empirical formula $C_5H_{11}OH$. All are liquids at ambient conditions except 2,2-dimethylpropanol (neopentyl alcohol), which is a solid. Almost all amyl alcohols are manufactured in the United States by the hydroformylation of butylenes. Yeast fermentation processes for ethanol yield small amounts of 4-methyl-1-butanol (isoamyl alcohol) and 2-methyl-1-butanol (active amyl alcohol, *sec*-butyl-carbinol) as fusel oil. However, when the amino acids leucine and isoleucine are added to sugar fermentations by yeast, 87% and 80% yields of 4-methyl-1-butanol and 2-methyl-1-butanol, respectively, are obtained (Fieser and Fieser, 1950). These reactions are not suitable for commercial applications because of cost, but they do indicate the close structural relationship between these C_5 amino acids and the C_5 alcohols. The reactions occur under nitrogen-deficient conditions. If a nitrogen source is readily available, the production of the alcohols is lowered considerably.

V. ECONOMIC FACTORS

Alcohols and ethers comprise the two groups of oxygenates that have been approved by the U.S. government for use in blends with gasolines. Ethanol received the most attention during its development as an additive and fuel extender for gasolines in the 1970s and 1980s, during which there were a number of opposing views regarding economics (*cf.* U.S. Dept. of Agriculture, 1987, 1988). And the net-energy-production potential of modern fermentation ethanol processes is still under discussion (see Chapter 14). Some of the major economic factors that affect production costs of ethanol, the other oxygenates, and the new technologies that appear to have the potential to influence oxygenate markets are examined here. Most of the approved ethers are manufactured by addition of an alcohol to the double bond of an olefin,

$$C = C + ROH \rightarrow CH-COR,$$

and not by a microbial process, although fermentation ethanol usage is increasing for ETBE manufacture. The oxygenates MTBE, thermochemical methanol, and fermentation ethanol are the three largest in terms of annual production, which was about 10.8 billion, 6.5 billion, and 5.7 billion liters, respectively, in the mid-1990s. MTBE is one of the fastest growing oxygenates, averaging about 25% per year in incremental production increases from 1985 to 1995, because of its demand as a gasoline component. Methanol production increased an average of about 7% per year over the same time period because it is used (with isobutylene) to manufacture MTBE and is also used as an oxygenate in RFG with cosolvent. Fermentation ethanol more than doubled in production over the same decade because of its demand as a motor fuel component and for the manufacture of ETBE. Interestingly, synthetic ethanol production from 1985 to 1995 averaged only about 344 million L/year. Note, however, that the consumption of motor gasoline in the United States was about 450 billion L/year in the mid-1990s. This is a good news-bad news market for oxygenates. The good news is that the gasoline market is huge relative to oxygenate production, which ensures continued and growing demand for oxygenates, particularly if the EPA expands oxygenate usage geographically or increases the dissolved oxygen requirement in gasolines. The bad news is that relatively small amounts of motor gasolines were displaced by biomass-derived liquids in the mid-1990s.

A. Embedded Feedstock Costs
and Lignocellulosics

The processes for manufacturing methanol by synthesis gas reduction and ethanol by ethylene hydration and fermentation are very dissimilar and contribute to their cost differentials. The embedded raw-material cost per unit volume of alcohol has been a major cost factor. For example, assuming feedstock costs for the manufacture of methanol, synthetic ethanol, and fermentation ethanol are natural gas at $3.32/GJ ($3.50/10^6 Btu), ethylene at $0.485/kg ($0.22/lb), and corn at $0.098/kg ($2.50/bu), respectively, the corresponding cost of the feedstock at an overall yield of 60% or 100% of the theoretical alcohol yields can be estimated as shown in Table 11.12. In nominal dollars, these feedstock costs are realistic for the mid-1990s and, with the exception of corn, have held up reasonably well for several years. The selling prices of the alcohols correlate with the embedded feedstock costs. This simple analysis ignores the value of by-products, processing differences, and the economies of scale, but it emphasizes one of the major reasons why the cost of methanol is low relative to the cost of synthetic and fermentation ethanol. The embedded feedstock cost has always been low for methanol because of the low cost of natural gas. The data in Table 11.12 also indicate that fermentation ethanol for fuel applications was quite competitive with synthetic ethanol when the data in this table were tabulated in contrast to the market years ago when synthetic ethanol had lower market prices than fermentation ethanol. Other factors also

TABLE 11.12 Estimated Embedded Feedstock Cost in Methanol and Ethanol

Alcohol	Feedstock	Feedstock price	Embedded feedstock cost at 100% yield ($/L)	Embedded feedstock cost at 60% yield ($/L)	Posted price[a] ($/L)	Posted price[a] ($/GJ)	Feedstock in posted price (%)
Methanol	Natural gas	$3.32/GJ ($3.50/10⁶ Btu)	0.07	0.12	0.145	9.20	48–83
Synthetic ethanol	Ethylene	$0.485/kg ($0.22/kb)	0.235	0.39	0.700	33.19	33–56
Fuel ethanol	Corn	$0.098/kg ($2.50/bu)	0.14	0.23	0.357	16.93	39–64

[a]The posted prices are for January 1997.

contribute to the cost differentials. Fuel-grade ethanol is not purified to the degree that synthetic ethanol is, and the cost of corn assumed for this analysis, $2.50/bu, is a low average market price that was as high as $5.00/bu in the mid 1990s when shortages were caused by bad weather. Some of the fermentation plants closed until corn prices returned to levels which permitted profitable operations. Fuel-grade ethanol should be able to compete with synthetic ethanol as long as corn prices remain in the range of $2.50/bu, and the tax incentives available for ethanol from biomass are maintained.

The use of lower cost lignocellulosic feedstocks is expected to permit large reductions in the cost of fermentation ethanol. It is essential, however, that the delivered price of the feedstock be low enough and that it be supplied at sufficient rates to sustain medium-to-large plant operations at a profit. The goal of the U.S. Department of Energy is to achieve fermentation ethanol costs in nominal dollars of about $0.16/L ($0.60/gal) by 2020 with these feedstocks and advanced fermentation processes. Assume that lignocellulosic feedstock contains 65 wt % recoverable, fermentable sugars, which are converted to the maximum yield of fermentation ethanol, 421 L/t (332 kg/t) of dry equivalent feedstock. If the feedstock is delivered to the fermentation plant gate at a cost of $30/t (about $1.62/GJ) on a dry equivalent basis, and the embedded feedstock cost is 50% in the product ethanol, the cost of ethanol, ignoring by-product credits, is $0.14/L and is within the long-term goal of $0.16/L. If this cost goal is to be achieved, it is essential that the delivered price of feedstock in nominal dollars be near $30/dry t or less, presuming the assumptions are correct. Waste biomass, such as municipal solid waste and certain industrial wastes, for which credits can be taken for disposal may offer significant opportunity to use feedstocks at a much lower cost.

A detailed analysis of 22 published investigations of the projected economics of fermentation ethanol production from lignocellulosic feedstocks focused on the influence of plant capacity, capital cost, and overall product yield on ethanol production cost (von Sivers and Zacchi, 1996). Wood was the feedstock and ranged in cost from $22 to $61/dry t. Plant size ranged in capacity from about 40,000 to 700,000 dry t/year of hardwood or softwood. Enzyme-, dilute-acid-, and concentrated-acid-catalyzed hydrolysis methods were employed. The ethanol yields ranged from 48 to 85% of the theory from hexose sugars; the pentoses were ignored. Capital and feedstock costs ranged from 11 to 48% and 14 to 48%, respectively, of ethanol production cost. Both of these percentage ranges represent wide variations. Sensitivity analyses showed that an increase in ethanol yield, a decrease in feedstock cost, or an increase in plant capacity resulted in a decrease in ethanol production cost, as expected. The analysis indicated that the overall ethanol yield is the most important factor, but the variations in ethanol costs estimated from conceptual process designs were large. One of the lowest cost processes in this analysis uses dilute acid hydrolysis of oak wood supplied at $22.00/dry t to a plant having a feedstock capacity of 209,000 t/year (Clausen and Gaddy, 1986). The ethanol yield was 78% of theory, and its production cost was projected to be $0.27/L, 11% and 25% of which were capital and feedstock costs.

B. Tax Incentives

Several federal tax incentives are available in the United States to promote the use of oxygenates from biomass in motor fuels. Some states also provide tax incentives. The incentives encourage production of renewable liquid fuels and have provided a major stimulus to develop the market. In addition, the federal mandates to use oxygenates to reduce emissions of motor gasolines and improve air quality have supplied additional support to develop renewable fuels markets. Table 11.13 lists selected U.S Internal Revenue Code tax incentives for motor fuels and fuel components from biomass. One of the largest incentives is the partial excise tax exemption for fuel alcohol. It is $0.143/L ($0.54/gal) of alcohol in alcohol-motor fuel blends where at least 190-proof alcohol in the blend is from biomass. This exemption effectively applies only to fermentation ethanol. Methanol is excluded because essentially all of it is manufactured from fossil feedstocks. Fermentation processes have not been developed to manufacture methanol from biomass as discussed in Section IV, and thermochemical methanol via synthesis gas from biomass cannot compete with methanol from natural gas. Note that if the other alcohols and ethers were manufactured from biomass by either microbial or thermochemical processes, they would qualify for the tax incentives; it is not necessary that microbial processes be used. If biomass-derived ethanol is used in the manufacture of ETBE from isobutylene, the amount used is eligible for the excise tax exemption. Unless extended by federal legislation, however, the excise tax exemption for alcohols from biomass is scheduled to terminate at the end of 2000.

C. Synthesis Gas Technology vs Fermentation Ethanol

When perfected, synthesis-gas-to-ethanol technology can be expected to have a large impact on fermentation ethanol markets. It is likely that thermochemical ethanol would then be manufactured at production costs in the same range as methanol from synthesis gas, which can be produced by gasification of virtually any fossil or biomass feedstock. Applying the advances that have been made for conversion of lignocellulosic feedstocks via enzymatically catalyzed options, it has been estimated that the production cost of fermentation ethanol

TABLE 11.13   Selected U.S. Federal Tax Incentives for Motor Fuels and Fuel Components from Biomass

Incentive	IRC	Applicable to	What may qualify	Incentive amount[a]
Clean-fuel vehicle income deduction	179A	Qualifying business and non-business vehicles placed in service after June 30, 1993 and before Dec. 31, 2004	Vehicles having OEM or retrofit capability to burn hydrogen or any fuel having combined contents of at least 85% methanol, ethanol, any other alcohol or ether	Limits $50,000 for trucks and vans with GVW more than 26,000 lb, $5000 for trucks and vans with GVW more than 10,000 to 26,000 lb, $2000 for other vehicles; only part of basis applies
Partial excise tax exemption for fuel alcohol	4081	Retail seller and/or end user, depending on seller's choice to pass on or retain all or part of savings to Sept. 30, 2000	Alcohol–motor fuel blends where at least 190-proof alcohol in blend is from biomass	54 cents/gal of qualifying ethanol of at least 190 proof in blends; 60 cents/gal for other qualifying alcohols
Fuel alcohol tax credit	38, 40	Blender only for use as motor fuel in a trade or business, whether produced and used by blender or produced and sold by blender to Dec. 31, 2000	Biomass-based alcohol in motor fuels or neat alcohols	54 or 40 cents/gal of qualifying alcohol of at least 190 proof or 150–190 proof in blends; 60 cents/gal for other qualifying alcohols
Ethyl-t-butyl ether (ETBE) excise tax exemption	4081	Retail seller and/or end user, depending upon seller's choice to pass on or retain part or all of savings	ETBE–fuel blends	54 cents/gal of qualifying ethanol used in producing ETBE in ETBE-gasoline blend
Ethyl-t-butyl ether income tax credit (blender's credit)	38, 40	Blender only for use as motor fuel in a trade or business whether produced and used by blender or produced and sold by blender to Dec. 31, 2000	ETBE–fuel blends	54 cents/gal of qualifying ethanol used in producing ETBE in ETBE–gasoline blend
Small ethanol producer tax credit	40	Small ethanol producers wherein ethanol is sold to another party for blending for trade, business, or retail sale	Plants having production capacity less than 30 million gal/year	10 cents/gal up to 15 million gal of production per year

[a]The incentives are indicated in U.S. units as published.

had been reduced to $0.32/L by the mid-1990s from $1.20/L in 1980 (Wyman, 1995). Other advancements from continuing research are predicted to reduce the production cost to less than $0.18/L so it can compete with gasoline from petroleum crudes at $25/bbl. However, even if lignocellulosic feedstocks and advanced hydrolysis and fermentation technologies make it possible to lower the production cost of fermentation ethanol to the long-term goal of $0.16/L, it is probable that it would not be able to compete with ethanol from synthesis gas in a free market. The production cost of ethanol from synthesis gas is estimated to be in the range of $0.08 to $0.12/L in mid-1990 dollars with biomass feedstocks. With natural gas and similar feedstocks, the cost of methanol from synthesis gas has historically been a small fraction, as low as 15%, of the cost of fermentation ethanol from corn on a volumetric basis.

Another potentially adverse impact on fermentation ethanol markets is presented by the options available for the manufacture of mixed alcohols from synthesis gas. Sufficient experimental data have been accumulated to show how the alcohol yields and distributions can be manipulated and what catalysts and conditions are effective. Some of these data have established the utility of mixed alcohols as motor fuels and motor fuel components.

REFERENCES

Ackerson, M. D., Clausen, E. C., and Gaddy, J. L. (1991). In "Energy from Biomass and Wastes XV," (D. L. Klass, ed.), p. 725. Institute of Gas Technology, Chicago.

Akhtar, M., Kirk, T., and Blanchette, R. (1996). In "International Conference on Biotechnology in the Pulp and Paper Industry: Recent Advances in Applied and Fundamental Research," (E. Srebotnik and K. Messner, eds.), p. 187. Proceedings of the 6th Conference, 1995, Facultas-Universitätsverlag, Vienna.

Alexander, A. G. (1983). In "Energy from Biomass and Wastes VII," (D. L. Klass and H. H. Elliott, eds.), p. 185. Institute of Gas Technology, Chicago.

Barker, L., Pucholski, T., and Tholen, K. (1981). "Use of Ethanol in Diesel Engines," SIM No. 11026, Solar Energy Research Institute [NREL], Golden, Colorado.

Bower, L., and Greco, R. (1997). "Market Trends and American Petroleum Institute Views on Ethanol Use in Motor Vehicles." Paper presented at Biomass Energy Research Association meeting. Washington, D.C., May 1.

Braconnet, H. (1819). Ann. Chem. Phys. 12-2, 172.

Brandon, C. S., Duhl, R. W., Miller, D. R., and Thakker, B. R. (1982). "The Economics of Catalytic Coproduction of Fuel Grade Methanol Containing Higher Alcohols." Paper presented at the 5th International Alcohol Fuel Technology Symposium. Auckland, New Zealand, May 13-18.

Brennan, M. B. (1996). Chem. & Eng. News 74, 31, October 14.

Brown, L. R., Strawinski, R. J., and McCleskey, C. S. (1964). Can. J. Microbiol. 10, 791.

Chem. Eng. News (1982). "Technology: Catalyst Converts Methanol to Dry Ethanol," 60 (38), 41, September 20.

Chem. Week (1984). "Getting Mixed Alcohols from Synthesis Gas," p. 28, November 7.

Christensen, L. M., Hixon, R. M., and Fulmer, E. J. (1934). "Power Alcohol and Farm Relief," Chapter VII. Iowa State College, Ames, IA.

Clausen, E. C., and Gaddy, J. L. (1986). *Biomass Energy Dev.* 3, 551.

Clyde, R. A. (1982). *In* "Energy from Biomass and Wastes VI," (D. L. Klass, ed.), p. 887. Institute of Gas Technology, Chicago.

Clyde, R. A. (1996). *In* "Bioenergy '96," Vol. II, p. 801, Proceedings of the Seventh National Bioenergy Conference. The Southeastern Regional Biomass Energy Program, Muscle Shoals, AL, September 15-20.

Cole, M. S. (1980). *In* "Energy from Biomass and Wastes IV," (D. L. Klass and J. W. Weatherly, III, eds.), p. 659. Institute of Gas Technology, Chicago.

Cox, F. W. (1979). *In* "Proceedings of Third International Symposium on Alcohol Fuels Technology," p. 1, II-22. U.S. Department of Energy, Washington, D.C., May 29-31.

Deanda, K., Zhang, M., Eddy, C., and Picataggio, S. (1996). *App. and Env. Microbiol.* 62 (12) 4465.

Duhl, R. W., and Thakker, B. R. (1983). *In* "Nonpetroleum Vehicular Fuels III," p. 227. Institute of Gas Technology, Chicago.

Duncan, R. K. (1907). *In* "The Chemistry of Commerce," p. 147. Harper & Brothers Publishers, New York.

Edgar, E. C., and Gaddy, J. L. (1992). U.S. Patent 5,173,429.

Emert, G. H., and Katzen, R. (1981). *In* "Biomass as a Nonfossil Fuel Source," (D. L. Klass, ed.), ACS Symposium Series 144, p. 212. American Chemical Society, Washington, DC.

The Energy Independent (1997). "From the Field to the Future, Archer Daniels Midland Company," Vol. 2 (6), 1, March.

Faith, W. L. (1945). *Ind. Eng. Chem.* 27, 9.

Fieser, L. F., and Fieser, M. (1950). "Organic Chemistry," 2nd Ed., Chapter 18, p. 483. D.C. Heath and Company, Boston.

Foo, E. L. (1978). *Proc. Biochem.* 13 (3), 23.

Foo, E. L., Hedén, C. G. (1976). *In* "Microbial Energy Conversion," (H. G. Schlegel and J. Barnea, eds.), Proceedings of the Seminar, p. 267. Göttingen, Germany, October 4-8.

Garnero, M. (1981). "Alcohol Fuels in Brazil," Statement at U.S. Senate Hearing, Subcommittee on Energy, Nuclear Proliferation and Government Processes. Washington, D.C., March 24.

Gibbs, L. M., and Gilbert, B. J. (1981). "Centra Costo County's One-Year Experience with Gasohol," Paper 810440. Society of Automotive Engineers, Detroit, MI, February.

Giebelhaus, A. W. (1980). *Agric. Hist.* 154, 178.

Goran Svahn, L. G. (1979). *In* "Proceedings of Third International Symposium on Alcohol Fuels Technology," p. 1, I-10. U.S. Department of Energy, Washington, D.C., May 29-31.

Greene, M. I. (1982). *Chem. Eng. Prog.* 78 (8), 46.

Grethlein, H. E. (1978). *In* "Fuels From Biomass," (W.W. Shuster, ed.), Vol. I, p. 461, Proceedings Second Annual Symposium. Rensselaer Polytechnic Institute, Troy, NY.

Gushee, D. E. (1994). "Alternative Transportation Fuels: Oil Import and Highway Tax Issues," CRS Issue Brief IB93009. Congressional Research Service, The Library of Congress, Washington, D.C., June 3.

Harris, E. E., and Beglinzer, E. (1946). *Ind. Eng. Chem.* 38, 890.

Harris, J. F., Baker, A. J., and Zerbe, J. I. (1984). *In* "Energy from Biomas and Wastes VIII," (D. L. Klass and H. H. Elliott, eds.), p. 1151. Institute of Gas Technology, Chicago.

Ingram, L. O., Conway, T., and Alterthum, F. (1991). "Ethanol Production by *Escherichia coli* Strains Co-Expressing *Zymomonas* PDA and ADL Genes," U. S. Patent 5,000,000, March 19.

Ingram, L. O., Ohta, K., and Beall, D. S. (1991). *In* "Energy from Biomass and Wastes XIV," (D. L. Klass, ed.), p. 1105. Institute of Gas Technology, Chicago.

Ingram, L. O., Lai, X., Moniruzzaman, M., Wood, B. E., and York, S. W. (1997). *In* "Fuels and Chemicals from Biomass," (B. C. Saha and J. Woodward, eds.), ACS Symposium Series 666. American Chemical Society, Washington, D.C.

Juran, B., and Porcelli, R. V. (1985). *Hydrocarbon Processing* 85, October.

Kampen, W. H. (1980). *Hydrocarbon Processing* **59** (2), 72.

Katzen, R., Ackley, W. R., Moon, G. D., Jr., Messick, J. R., Brush, B. F., and Kaupisch, K. F. (1981). *In* "Fuels From Biomass and Wastes," (D. L. Klass and G. H. Emert, eds.), p. 393. Ann Arbor Science Publishers, Inc., Ann Arbor, MI.

Keller, J. L. (1979). *Hydrocarbon Processing* **58** (5), 127.

Keller, J. L., Nakaguchi, G. M., and Ware, J. C. (1978). "Methanol Fuel Modification for Highway Vehicle Use," Final Report, HCP/W3683-18. U.S. Department of Energy, Washington, D.C., July.

Kerstetter, J. D., Lyford, K., and Lynd, L. (1996). *In* "Liquid Fuels and Industrial Products" (J. S. Cundiff *et al.*, eds.), p. 260. The American Society of Agricultural Engineers, St. Joseph, MI.

Klass, D. L. (1980). *Energy Topics* **1**, April 14.

Klass, D. L. (1983). *Energy Topics* **1**, August 1.

Klass, D. L. (1984a). *Science* **223** (4640), 1021.

Klass, D. L. (1984b). *In* "Kirk-Othmer Encyclopedia of Chemical Technology," Supplemental Volume, 3rd Ed. John Wiley, NY.

Klass, D. L. (1994). *In* "Kirk-Othmer Encyclopedia of Chemical Technology," Vol. 12, p. 16. John Wiley, NY.

Laux, P. G. (1977). "The Catalytic Production and Mechanism of Formation of Methyl Fuel." Paper presented at the International Symposium on Alcohol Fuel Technology. Wolfsburg, Germany, November 21.

Lawford, H. G., and Rousseau, J. D. (1993). *In* "Energy from Biomass and Wastes XVI," (D. L. Klass, ed.), p. 559. Institute of Gas Technology, Chicago.

Lawrence, R. (1978). "Gasohol Test Program, Report 1978," TAEB-79-4A, 79-4A, No. PB-290569. Technology Assessment Evaluation Branch, U.S. Environmental Protection Agency, Ann Arbor, MI, December.

Luers, H. Z. (1930). *Angew. Chem.* **43**, 455.

Luers, H. Z. (1932). *Angew. Chem.* **45**, 369.

Miller, K. D., Jr. (1993). *In* "First Biomass Conference of the Americas: Energy, Environment, Agriculture, and Industry," NREL/CP-200-5768, DE03010050, Vol. II, p. 1159. National Renewable Energy Laboratory, Golden, CO.

Nagashima, M., Azuma, M., Noguchi, S., Inuzuka, K., and Samejima, M. (1984). *Biotechnol. Bioeng.* **26**, 992.

Nash, A. W., and Howes, D. A. (1935). "The Principles of Motor Fuel Preparation and Application," Vols. I and II. John Wiley, NY.

Nastari, P. (1996). *The Energy Independent* **1** (2), 1 [excerpts of address to 16th Congress of the World Energy Council, Tokyo, Japan, October 1996].

National Renewable Energy Laboratory (1993). "Tenth International Symposium on Alcohol Fuels, The Road to Commercialization," Proceedings, Vols. I and II. National Renewable Energy Laboratory, Golden, CO, November 7-10.

Nikitin, N. I. (1962). "The Chemistry of Cellulose and Wood," (translated from Russian by J. Schmorak, Israel Program for Scientific Translations Ltd., Jerusalem, 1966). S. Monson Binder, Jerusalem, Israel.

Noon, R. (1982). *Chem. Tech.* **681**, November.

Ostrovski, C. M., Aitken, J. C., and Hayes, R. D. (1985). *In* "Energy from Biomass and Wastes IX," (D. L. Klass, ed.), p. 895. Institute of Gas Technology, Chicago.

Paggini, A., and Fattone, V. (1982). Paper presented at the Fifth International Fuel Technology Symposium. Auckland, New Zealand, May 13-18.

Pazner, L., and Cho, H. J. (1989). *In* "Energy From Biomass and Wastes XII," (D. L. Klass, ed.), p. 1297. Institute of Gas Technology, Chicago.

Pazner, L., Jeong, C., Quinde, A., and Arardel-Karim, S. (1993). *In* "Energy From Biomass and Wastes XVI," (D. L. Klass, ed.), p. 629. Institute of Gas Technology, Chicago.

Philippidis, G. P. (1993). *In* "Energy From Biomass and Wastes XVI," (D. L. Klass, ed.), p. 545. Institute of Gas Technology, Chicago.

Piel, W. J. (1993). *In* "First Biomass Conference of the Americas: Energy, Environment, Agriculture, and Industry," NREL/CP-200-5768, DE03010050, Vol. II, p. 1116. National Renewable Energy Laboratory, Golden, CO.

Piel, W. J. (1996). "Oxygenates: A Viable Pathway for Non-Petroleum Energy Into Cleaner Transportation Fuels." Paper presented at Biomass Energy Research Association meeting. Washington, D.C., December 12.

Powell, T. (1975). "Racing Experiences with Methanol and Ethanol-Based Motor-Fuel Blends," Paper 750124, Automobile Engineering Congress and Exposition. Society of Automobile Engineers, Detroit, MI, February.

Prebluda, H. J., and Williams, R., Jr. (1981). *In* "Biomass as a Nonfossil Fuel Source," (D. L. Klass, ed.), ACS Symposium Series 144, p. 199. American Chemical Society, Washington, D.C.

Recombinant BioCatalysis, Inc. (1996). "Cellulase CloneZyme Library," CEL-001. Philadelphia, PA.

Riegel, E. R. (1933). "Industrial Chemistry." The Chemical Catalog Co., NY.

Rugg, B., Armstrong, P., and Stanton, R. (1981). *In* "Fuels From Biomass and Wastes," (D. L. Klass and G. H. Emert, eds.), p. 311. Ann Arbor Science Publishers, Inc., Ann Arbor, MI.

Saeman, J. F. (1945). *Ind. Eng. Chem.* **37**, 43.

Saeman, J. F. (1981). *In* "Biomass as a Nonfossil Fuel Source," (D. L. Klass, ed.), ACS Symposium Series 144, p. 185. American Chemical Society, Washington, DC.

Shadis, W. J., and McCallum, P. W. (1980). "A Comparative Assessment of Current Gasohol Fuel Economy Data," Paper 800889. Society of Automotive Engineers, Detroit, MI, August.

Shu, L, Nesheim, J. C., Kauffmann, K., Münck, E., Lipscomb, J. D., and Que, L., Jr. (1997). *Science* 275 (5299), 515.

Sitton, O. C., Foutch, G. L., Book, N. L., and Gaddy, J. L. (1979). *Chem. Eng. Prog.* **52**, December.

Spano, L. A. (1976). *In* "Clean Fuels from Biomass, Sewage, Urban Refuse, and Agricultural Wastes," (F. Ekman, ed.), p. 325. Institute of Gas Technology, Chicago.

Stamper, K. R. (1979). *In* "Proceedings of Third International Symposium on Alcohol Fuels Technology," p. 1, I-9. U.S. Department of Energy, Washington, D.C., May 29-31.

Stamper, K. R. (1980). "Evaporative Emissions from Vehicles Operating on Methanol/Gasoline Blends," Paper No. 801360. Society of Automotive Engineers, Detroit, MI, October.

Strait, J., Boedicker, J. J., and Johanson, K. C. (1978). "Diesel Oil and Ethanol Mixtures for Diesel Powered Farm Tractors," Final Report to State of Minnesota Energy Agency.

Thauer, R. K. (1976). *In* "Microbial Energy Conversion," (H. G. Schlegel and J. Barnea, eds.), Proceedings of the Seminar, p. 201. Göttingen, Germany, October 4-8.

Thauer, R. K., Jungermann, K., and Decker, K. (1977). *Bacteriol. Rev.* **41**, 100.

U.S. Dept. of Agriculture (1987). "Fuel-Ethanol Cost-Effectiveness Study, Final Report, National Advisory Panel on Cost-Effectiveness of Fuel Ethanol Production." Washington, D.C., November.

U.S. Dept. of Agriculture (1988). "Ethanol: Economic and Policy Tradeoffs." Washington, D.C., January.

U.S. Dept. of Treasury. "Alcohol, Tobacco and Firearms Summary Statistics," ATF P 1323.1, Fiscal Year 1976 (4-77); Transitional Quarters Ending September 30, 1976 (8-77); Fiscal Year 1978 (4-81); Fiscal Year 1979 (7-82). Washington, D.C.

U.S.I. Chemicals (1981). "Ethyl Alcohol Handbook," 5th Ed. National Distillers & Chemical Corporation, NY.

von Sivers, M., and Zacchi, S. (1996). *Bioresource Tech.* **56** (2,3), 131.

Wayman, M. (1991). *In* "Energy from Biomass and Wastes XIV," (D. L. Klass, ed.), p. 1145. Institute of Gas Technology, Chicago.

Weigel, J. C., Loy, D., and Kilmer, L. (1997). "Feed Co-Products of the Dry Corn Milling Process;" "Feed Co-Products of the Corn Wet Milling Process." Renewable Fuels Association, Washington, D.C

Wilkenson, T. G., Topiwala, H. H., and Hamer, G. (1974). *Biotechnol. Bioeng.* **16**, 41.

Wilkie, H. F., and Kolachov, P. J. (1942). "Food for Thought." Indiana Farm Bureau, Indianapolis, IN.

Winter, C. L. (1982). *Hydrocarbon Processing* **60** (38), 41.

Winter, C. L. (1986). *Hydrocarbon Processing* **65** (4), 71.

The World Bank (1980). "Alcohol Production from Biomass In the Developing Countries." The World Bank, Washington, D.C.

Wright, J. D. (1989). *In* "Energy from Biomass and Wastes XII," (D. L. Klass, ed.), p. 1247. Institute of Gas Technology, Chicago.

Wood, W. A. (1961). *In* "The Bacteria," (I. C. Gunsalus and R. Y. Stanier, eds.), Vol. 2, p. 59. Academic Press, New York.

Wyman, C. E. (1993). *In* "First Biomass Conference of the Americas: Energy, Environment, Agriculture, and Industry," Vol. II, p. 1010, NREL/CP-200-5768, DE93010050. National Renewable Energy Laboratory, Golden, CO.

Wyman, C. E. (1995). *In* "Second Biomass Conference of the Americas: Energy, Environment, Agriculture, and Industry," p. 966, NREL/CP-200-8098, DE95009230. National Renewable Energy Laboratory, Golden, CO.

Zhang, M., Eddy, C., Deanda, K., Finkelstein, M., and Pictaggio, S. (1995). *Science* **267** (5195), 240.

Zorn, H., and Faragher, W. F. (1949). "The CO-H$_2$ Synthesis at I.G. Farben A.G.," FIAT Final Report No. 1267. Office of the Military Government for Germany, Ludwigshafen, Germany, April 14.

CHAPTER 12

Microbial Conversion: Gasification

I. INTRODUCTION

Certain fermentative microorganisms are capable of converting biomass to methane (CH_4), the dominant fuel component in natural gas, or to molecular hydrogen, which has been proposed as a gaseous fuel for large-scale use. Other microorganisms have the capability of producing hydrogen by performing the chemical equivalent of degrading water into its chemical constituents. The process that yields methane is called methane fermentation, or anaerobic digestion. It takes place in the absence of oxygen, and the microorganisms that perform the process are mixed populations of anaerobic bacteria. Methane fermentation occurs naturally in many ecosystems such as river muds, lake sediments, sewage, marshes, and swamps. It is most conspicuous where plants die and decompose underwater. The water layer acts as a blanket to exclude oxygen and promote the growth of many different anaerobes. Methane fermentation also occurs in the digestive tracts of ruminants. The rumen is supplied with ample quantities of food, is well buffered, has a nearly neutral pH, and is almost free of oxygen. Methane-producing (methanogenic) bacteria develop rapidly and commonly form 100 to 500 L of methane daily per cow.

Three basic methods of generating hydrogen using microorganisms are known. One is fermentation with certain species of heterotrophic anaerobes. Intermediate pyruvic acid is converted to hydrogen and other products. Another method uses photosynthetic organisms to split water (biophotolysis). During normal photosynthesis, including the growth of photosynthetic unicellular biomass, the reducing power generated is always used for reduction of carbon dioxide to carbohydrates and other cellular compounds. But the pathway in some photosynthetic bacteria and microalgae that contain or can synthesize the enzyme hydrogenase can be directed to produce molecular hydrogen. Some of these organisms use fermentation intermediates from biomass as hydrogen donors. The third microbial method uses cell-free chloroplast, ferredoxin, and hydrogenase components extracted from biomass in catalyst formulations for biophotolysis. Dry hydrogen, which has a higher heating value of about 12.7 $MJ/m^3(n)$ (324 Btu/SCF) would seem to be an ideal fuel in many applications because water is the only combustion product. With the exception of small amounts of nitrogen oxides formed when hydrogen is combusted in air, pollutants and partial oxidation products are not formed. However, practical applications of these microbial methods for producing hydrogen have not yet been developed.

In contrast, methane fermentation is used worldwide, either alone or in combination with other processes, for the stabilization and disposal of waste biomass such as domestic, municipal, agricultural, and industrial wastes and wastewaters. During digestion, the amount of organic material, its biological oxygen demand (BOD), and the pathogenic organisms present in the waste are reduced. Many virgin biomass species can also be gasified in the same manner. The gas produced by anaerobic digestion of biomass (biogas) is basically a two-component gas composed of methane and carbon dioxide, although minor amounts of other gases such as hydrogen sulfide and hydrogen may be present. An anaerobic digester (fermenter) operating in a stable mode yields biogas that has a methane content on a dry basis ranging from about 40 to 75 mol %, depending on the operating conditions, and a higher heating value of 15.7 to 29.5 $MJ/m^3(n)$. Dry natural gas and pure methane have higher heating values of about 39.3 $MJ/m^3(n)$ (1000 Btu/SCF). Thus, biogas is a medium-energy gas. Because of these characteristics, biogas obtained by anaerobic digestion of animal manures and human wastes has been used as a fuel for cooking, heating, and lighting for decades in many developing countries. In urban communities, the anaerobic digestion process is often used, frequently in combination with the activated sludge process, to treat municipal sewage (biosolids). Anaerobic digestion is also used for the stabilization and volume reduction of municipal solid waste (MSW) and in industry for the treatment of wastes from meat packing plants, breweries, canneries, and other food processing plants. One of the oldest applications of anaerobic digestion is the stabilization of human wastes in septic tanks.

Wastewater treatment plants in urban communities where municipal biosolids are treated by anaerobic digestion frequently use biogas on-site. Biogas combustion for heat, steam, and electric power generation at municipal wastewater treatment plants is almost universal in many countries. Sanitary landfills are the equivalent of large-scale, batch digesters for MSW and emit biogas (landfill gas or LFG). Sanitary landfills are used worldwide to dispose of solid wastes. LFG is sometimes collected from shallow wells drilled into completed landfills for conversion to electric power or for other applications such as blending with natural gas after removal of carbon dioxide for normal distribution. The recovery of LFG from landfills decreases undesirable methane emissions, eliminates safety hazards that can be caused by migration of LFG in the soil into nearby buildings, and can be a profitable use of what would otherwise be a lost resource. Anaerobic digestion has been used for over 100 years for waste biomass treatment and disposal and as a source of fuel gas. The effort to apply the process to virgin biomass grown specifically for microbial conversion to pipeline-quality gas (substitute natural gas, SNG) is a relatively recent development that started in the early 1970s (Klass, 1974).

Methane fermentation is a multistage process. The complex polymers and compounds in biomass are degraded to lower molecular weight intermediates which are converted to methane and carbon dioxide. Although this representation is an oversimplification of complicated microbiological transformations, it is useful in explaining some of the characteristics of anaerobic digestion such as the effect of pH and acid buildup. The process can be maintained on a large scale for an indefinite period as long as the important fermentation parameters are kept within an acceptable range and fermentable material is available. Almost any kind of biomass feedstock mixture is suitable, with the possible exception of the lignins and keratins, which have low biodegradabilities. But even these materials will undergo microbial gasification over long periods of time. Gasification of a large portion of the biomass can be achieved in the majority of cases.

In this chapter, the early work on methane fermentation, the basic biochemistry and microbiology of the organisms involved, how the process is performed, some of the advancements that have improved the process, and the status of efforts to expand commercial use are discussed. The microbial generation of hydrogen and the factors that have limited its use are also discussed.

II. METHANE FERMENTATION

A. EARLY WORK ON MICROBIAL METHANE

The chemistry of methane, the simplest organic compound known, was first studied by Berthollet in 1786 (*cf.* Roscoe and Schorlemmer, 1878). He analyzed the gas quantitatively, but could not distinguish it from ethylene. This was literally the beginning of thousands of independent investigations of methane conducted over the next 200 years. The biological origin of methane, however, was recognized by Van Helmont, Volta, and Davy long before Berthollet's studies (cf. McCarty, 1982). In 1630, Van Helmont found that flammable gases can be emitted from decaying organic matter. Volta observed in 1776 that there is a direct correlation between the amount of flammable gas emitted and the amount of decaying matter. In 1808, Davy found that during the anaerobic digestion of cattle manure, methane is present in the gas. Direct experimental evidence of the origin of biogas was reported in 1875 (Popoff, 1875). Popoff was able to account for the microbial decomposition of cellulose by the formation of methane. In 1886 and 1887, Hoppe-Seyler found that microorganisms in river muds cause the formation of methane from cellulose and the salts of fatty acids (Hoppe-Seyler, 1886 and 1889). About 20 years later, Omelianski reported that the decomposition of cellulose and the simultaneous formation of methane are caused by bacteria (Omelianski, 1904).

The early work of these investigators established that methane can have a biological origin. Söhngen later substantiated Hoppe-Seyler's observations that fatty acids can yield methane and showed that under certain conditions, hydrogen and carbon dioxide combine in molar ratios of 4:1 to form methane and water (Söhngen, 1906, 1910). After these observations, many researchers studied the microbial formation of methane in relation to such applications as biosolids treatment and the utilization of animal wastes. The position of methane in the carbon cycle was largely determined by observation and analysis of material from river muds and soils. Bacteria that produce methane in an anaerobic environment, from a microbial standpoint, are analogous to those that produce carbon dioxide from methane in an aerobic environment. Anaerobic and aerobic bacteria capable of promoting these reactions were found to exist in many locations throughout the world. Which microbial reactions occur in a given location depend on the organisms and substrates that are present and whether or not a supply of oxygen is available.

Biogas was recognized as a useful fuel gas from this early work. In 1896, biosolids digestion supplied fuel for street lamps in England. In 1897, a waste-disposal tank serving a leper colony in Bombay, India, was equipped with gas collectors and the biogas was used to drive gas engines. In 1925, biogas was found to be satisfactory for general municipal use and was distributed through city mains in Essen, Germany. Millions of low-cost digesters have been operated for many years in China and India on farms and in cooperative village systems to generate biogas from animal manures and human wastes for local use. The idea of microbial gasification of biomass under controlled conditions to produce high-methane fuel gas for useful applications is as old as the early work done to establish the biological origin of biogas. The early work on biogas also suggests that at least some natural gas deposits may possibly have been generated naturally by anaerobic digestion of biomass residues.

To keep biogas and the microbial gasification of biomass in the proper perspective, however, note just a few of the fossil-based methane developments that predate much of the early work done on biogas. Natural gas wells were known in Asia as early as 615. The Chinese reported the transport of natural gas through bamboo tubes for lighting in 900. In 1691, the English researcher Robert Boyle reported that a combustible gas is produced when coal is heated. In 1775, General George Washington described a gas well in West Virginia adjacent to a tract of land granted to him and General Andrew Lewis as a "burning spring." In 1806, the first gas mains laid in a public street were constructed in London. In 1819, the first gas company was formed in France to light the city of Paris. The coal gasification industry was established in the 1800s and was then displaced in many countries by the natural gas industry after World War II as a major part of the international energy economy. The point that will become evident shortly is that anaerobic digestion is a valuable tool for waste biomass treatment and disposal, and biogas is a valuable, renewable fuel that can be recovered and used. But the technology will not begin to assume a large-scale role as an energy resource worldwide until natural gas depletion starts to occur. Instead, small-to-moderate scale applications of methane fermentation for waste treatment and disposal will be the norm.

B. Microbiology of Methanogenic Bacteria

Methanogenic bacteria are unicellular, Gram-variable, strict anaerobes that do not form endospores. Their morphology, structure, and biochemical makeup are quite diverse. More than ten different genera have been described (cf. Zeikus, Kerby, and Krzycki, 1985). All genera have been assigned to the kingdom Archaebacteria, which comprises a group of bacteria typically found in unusual environments, and is distinguished from the rest of the prokaryotes by several criteria, including the number of ribosomal proteins, the lack of muramic acid in the cell walls, membrane lipids that contain isoprenoid side chains bound by ether linkages instead of ester-linked hydrocarbons, and the absence of ribothymine in transfer ribonucleic acid (tRNA). The methanogens have been divided into three groups based on the fingerprinting of their 16S *ribosomal* RNA (rRNA) and the substrates used for growth and methanogenesis (Woese and Fox, 1977; Balch *et al.*, 1979). The methanogens were found to be unexpectedly divergent from other bacteria. A revised taxonomic order was developed based on this work. Group I contains the genera *Methanobacterium* and *Methanobrevibacter;* Group II contains the genus *Methanococcus;* and Group III contains several genera, including *Methanomicrobium, Methanogenium, Methanospirillum,* and *Methanosarcina.* Species classified as *Methanobacterium* are generally rod-shaped organisms that are sometimes curved and that vary in size and arrangement of the cells; the cells may or may not be motile. Species classified in the genus *Methanococcus* are small spherical organisms whose cells occur singly or in irregular masses; some are motile. Methanogens in Group III having large spherical cells that occur in packets and are nonmotile have been classified in the genus *Methanosarcina.* The long, helical, rod-shaped methanogens with polar flagella have been classified in the genus *Methanospirillum* and are also in Group III. The analysis of the 16S rRNA allowed recognition of the archaebacteria as a distinctive group of bacteria that includes the methanogens as well as the halophiles and thermoacido-philes.

Methanogenic bacteria have not been studied as extensively as most other groups of bacteria. Until 1936, all attempts to isolate pure cultures or even to grow colonies on solid media were unsuccessful (Barker, 1936). Taxonomic classification was difficult because mixed cultures were employed in much of the early work. Essentially all of the early work and many recent studies have been carried out with enrichment cultures in which substrates and environmental conditions are chosen to selectively promote the growth of certain microbial species (*cf.* Klass, 1984). By enrichment culture techniques, it is possible to obtain considerable information and data about the morphology of methanogens that have been identified, the conditions that favor their development, and the types of substrates utilized. Other methanogens have been isolated, but remain to be described in more detail before their taxonomic assignment is established. All species that have been studied in pure culture are strictly anaerobic and grow only in the absence of oxygen and in the presence of a suitable reducing agent. Methanogens are much more sensitive to oxygen than most other anaerobes. For this reason, it is much easier to grow methanogenic bacteria in liquid or semisolid media than on the surface of an agar plate. Even in liquid media not fully protected from air, sufficient oxygen may leak into the system to inhibit growth. A roll-tube method (Hungate technique) has been shown to be the most successful method for isolating pure cultures of methanogens (*cf.* Zeikus, 1977).

Several species have been isolated, studied in pure culture, and taxonomi-cally identified and classified. Some of the notable species are *Methanobacterium formicicum, M. bryantii, M. thermoautotrophicum; Methanobrevibacter ruminant-ium, M. arboriphilus, M. smithii; Methanococcus vannielii, M. voltae; Methanomi-crobium mobile; Methanogenium cariaci, M. marisnigri; Methanospirillum hunga-*

tei; and *Methanosarcina barkeri* (*cf.* Zeikus, 1977; Balch *et al.,* 1979; Macario and de Macario, 1982; Zeikus, Kerby, and Krzycki, 1985). Most, if not all, methanogens can use hydrogen and carbon dioxide for methanogenesis and growth. Hydrogen is the electron donor and carbon dioxide is the electron acceptor that is reduced to methane. Thus, most, if not all, methanogens are facultative autotrophs. In addition, some species can use formate for growth and methane production (e.g., *M. vannielii*); others can use methanol, methyl amines, or acetate (e.g., *M. barkeri*). Pure cultures generally grow well in media containing the usual mineral nutrients needed for growth of microorganisms, a reducing agent, and ammonium ion as the nitrogen source. The addition of extracts containing amino acids, growth factors, and other nutritional supplements to synthetic media may not have a beneficial effect, although some species of methanogenic bacteria require complex media for growth (*M. mobile, M. voltae, M. ruminantium,* and *M. smithii*). Several species need large amounts of carbon dioxide because it is used as the primary carbon source. Generally, growth is best in the pH range 6.4 to 7.4. Inhibition may occur at higher pH. But there are exceptions; *M. vannielli* grows best between pH 7 and 9. Despite their diverse morphology, which consists of many different cell shapes and structures, all methanogenic bacteria are unique in that all use simple substrates for energy and growth and all are specialized in their ability to produce methane as a major end product. A few microorganisms that are not classified as methanogens can be induced to produce methane under certain conditions. *Clostridium perfringens,* which normally does not produce methane, can be induced to do so in a peptone-formate medium by addition of a small amount of iodine (Laigret, 1945).

Even though only a few species of methanogenic bacteria are believed to be capable of utilizing acetate as a substrate (McInerney and Bryant, 1981), about 70% of the methane formed in anaerobic biosolids digesters and from lake sediments is derived from the methyl group of acetate (Stadtman and Barker, 1951; Jerris and McCarty, 1965; Smith and Mah, 1965; Cappenburg and Prins, 1974). The carboxyl group yields carbon dioxide. Because of the large number of anaerobes in these systems as well as in other methane fermentation systems, it is probable that there are many yet-to-be identified methanogens that utilize acetate. The relatively simple compounds that serve as carbon and energy sources for methanogenic bacteria are clearly limited, and each methanogenic species is characteristically limited to the use of a few compounds. These compounds are generally supplied as intermediate fermentation products by other anaerobes present in methane fermentation systems. Indeed, it is apparent that several species of fermentative, acetogenic, and methanogenic bacteria are necessary to anaerobically digest the complex substrates in waste and virgin biomass. Mixed cultures are required for complete fermentation. Methane fermentation under sterile conditions is not possible or even desirable in many systems.

C. CHEMISTRY OF MICROBIAL METHANE FORMATION

Some of the early studies of the chemical mechanisms of methane formation by methanogens were carried out with mixed cultures and a single substrate. It was observed that some of the pure substrates can sometimes be converted almost quantitatively to methane and carbon dioxide. The yields of new cellular biomass during methanogenesis are small. The stoichiometries of several of the observed reactions are as follows:

$$CH_3CO_2H \rightarrow CH_4 + CO_2$$
$$4CH_3CH_2CO_2H + 2H_2O \rightarrow 7CH_4 + 5CO_2$$
$$2CH_3CH_2CH_2CO_2H + 2H_2O \rightarrow 5CH_4 + 3CO_2$$
$$2CH_3CH_2OH \rightarrow 3CH_4 + CO_2$$
$$CH_3COCH_3 + H_2O \rightarrow 2CH_4 + CO_2.$$

These equations indicate that the fermentation of acetic acid, propionic acid, butyric acid, ethanol, and acetone all yield the same products, but the ratio of methane to carbon dioxide changes with the oxidation state of the substrate. It is remarkable that the products are independent of substrate structure.

Indirect evidence of the mechanism of methane formation was reported in the early part of the twentieth century (Söhngen, 1910), and in the 1930s (Stephenson and Strickland, 1931, 1933; Fischer, Lieske, and Winzer, 1931, 1932). Söhngen found that enrichment cultures can couple the oxidation of hydrogen with the reduction of carbon dioxide according to:

$$4H_2 + CO_2 \rightarrow CH_4 + 2H_2O$$

These observations were later confirmed with pure cultures. It was found that methane is formed by the reduction of carbon dioxide by hydrogen supplied by the various substrates utilized by the bacteria or, in the case of "*Methanobacillus omelianskii*," by uncombined hydrogen itself (Barker, 1936, 1940, 1941, 1943). *M. omelianskii* turned out to be a mixed culture and will be discussed later.

Much of the early work done on the biochemistry of methane formation supported the position that methane is formed almost exclusively by reduction of carbon dioxide. However, it was shown with methanol and a species of *Methanosarcina* that less than 1% of the methane is derived from carbon dioxide (Schnellen, 1947; Stadtman and Barker, 1949). According to the mechanism proposed by van Neil for catabolism of acetic acid, all of the acid should be oxidized to carbon dioxide, half of which should be reduced to methane (*cf.* Barker, 1936):

$$CH_3CO_2H + 2H_2O \rightarrow 2CO_2 + 4H_2$$
$$CO_2 + 4H_2 \rightarrow CH_4 + 2H_2O$$
$$\text{Net: } CH_3CO_2H \rightarrow CH_4 + CO_2$$

This mechanism was tested by use of ^{14}C-labeled carbon dioxide (Barker, 1943; Buswell and Sollo, 1948; Stadtman and Barker, 1949, 1951; Pine and Barker, 1956; Baresi *et al.*, 1978). Essentially none of the methane was found to be derived from carbon dioxide. Methane is derived entirely from the methyl carbon atoms and carbon dioxide is derived exclusively from carboxyl carbon atoms. Van Neil's mechanism is clearly not valid because the methyl carbon atom is not oxidized to carbon dioxide. Other work has been done to ascertain whether hydrogen atoms are removed during the fermentation of acetic acid, and whether the methyl group is incorporated intact into methane (Pine and Barker, 1954). Water and heavy water were used with deuterated and nondeuterated acetic acid. Acetic acid labeled in the methyl group, when used as the substrate, showed that the isotopic content of acetic acid and methane are the same. Unlabeled acetic acid fermented in the presence of heavy water indicated that about one atom of deuterium per molecule of methane formed is derived from heavy water. It was concluded that the methyl group is transferred from acetic acid to methane as a unit without the loss of attached hydrogen or deuterium atoms.

The fermentation of butyric acid by *Methanobacterium suboxydans* is represented by (Stadtman and Barker, 1951)

$$2CH_3CH_2CH_2CO_2H + 2H_2O + CO_2 \rightarrow 4CH_3CO_2H + CH_4.$$

The oxidation of 2 mol butyric acid to 4 mol acetic acid is coupled with the reduction of 1 mol of carbon dioxide to methane. Tracer experiments showed that 98% of the methane is derived from carbon dioxide. In these examples of methane fermentation involving carbon dioxide reduction, no carbon dioxide is formed in the oxidation of the substrate. The fermentation of propionic acid by *M. propionicum* is more complicated because it involves both carbon dioxide formation and consumption (Stadtman and Barker, 1951):

$$4CH_3CH_2CO_2H + 8H_2O \rightarrow 4CH_3CO_2H + 4CO_2 + 12H_2$$
$$3CO_2 + 12H_2 \rightarrow 3CH_4 + 3CH_4 + 6H_2O$$
$$\text{Net: } 4CH_3CH_2CO_2H + 2H_2O \rightarrow 4CH_3CO_2H + 3CH_4 + CO_2.$$

Tracer experiments with ^{14}C-labeled carbon dioxide or propionic acid indicate that approximately 1 mol of carbon dioxide is formed per mole of propionic acid consumed and that carbon dioxide is the precursor of most of the methane. Tracer experiments were also conducted with propionic acid using enrichment cultures (Buswell *et al.*, 1951). These cultures use carbon dioxide and apparently convert all three carbon atoms of propionic acid to both methane and carbon dioxide in varying amounts.

The simplest chemical mechanism proposed for the conversion of carbon dioxide to methane is a sequential reduction involving formic acid, formaldehyde, and methanol as intermediates. But it has been found that several methanogens cannot use these postulated intermediates as substitutes for carbon dioxide when hydrogen is used as the reductant (Kluyver and Schnellen, 1947). Subsequent experiments with cell-free extracts of methanogens, however, established that methane can be formed from individual one-carbon compounds — CO_2, CH_3OH, HCHO, and HCO_2H. For the three-carbon compounds pyruvic acid (CH_3COCO_2H) and serine ($CH_2OHCHNH_2CO_2H$), the carboxyl carbon atom of pyruvic acid and the hydroxylated carbon atom of serine are converted to methane.

The experimental product distributions and selectivities suggest that methanogenesis is highly efficient for production of methane and carbon dioxide. Thermodynamic calculations support this contention (Chapter 11). Ignoring the small amount of substrate that is used to produce new cells and to provide cellular maintenance energy, the gross stoichiometry of the methane fermentation of glucose can be represented by

$$C_6H_{12}O_6 \text{ (aq)} \rightarrow 3CH_4(g) + 3CO_2(g).$$

The standard Gibbs free energy and enthalpy changes for this conversion under physiological conditions (pH 7, 25°C, unit activities) per mole of glucose fermented are about −418 and −131 kJ, and the mass and energy contents of the methane expressed as fractions of the glucose converted are about 27 and 95%. Thus, the thermodynamic driving force is large; the exothermic energy loss is small; the energy in the glucose is transferred at a higher energy density to a simple gaseous hydrocarbon; methane is easily separated from the aqueous system, and if desired, from the co-product carbon dioxide; methane and carbon dioxide selectivities are high; and the mass of substrate is significantly reduced, which is important if a waste is disposed of or stabilized by anaerobic digestion.

D. FERMENTATIVE AND ACETOGENIC BACTERIA IN METHANE FERMENTATION

Because of the wide variety of complex substrates in biomass, many different bacterial species are necessary to facilitate degradation. The limited number of substrates catabolized by methanogens also requires that other types of organisms be present to implement the overall process. It is apparent that mixed cultures are necessary to convert complex substrates to methane and carbon dioxide. Methane fermentation is a three-stage and possibly a four-stage process that involves, in addition to methanogenic bacteria in the last stage, at least two other groups of organisms that implement the initial stages (McInerney and Bryant, 1981). In the first stage, fermentative bacteria convert the complex polysaccharides, proteins, and lipids in biomass to lower molecular weight fragments and acetate, carbon dioxide, and hydrogen. Another group of bacteria, the obligate, hydrogen-producing acetogenic bacteria, catabolize the longer-chain organic acids, alcohols, and possibly other degradation products formed in the first stage to yield additional acetate, carbon dioxide, and hydrogen. It is probable that some carbon dioxide and hydrogen are also converted to acetate by the acetogens. In the last stage, methanogenic bacteria convert intermediate acetate to methane and carbon dioxide by decarboxylation, and the intermediate carbon dioxide and hydrogen to additional methane. Thus, at least three groups of bacteria are necessary for methane fermentation to proceed—fermentative, acetogenic, and methanogenic bacteria.

The fermentative bacteria found in operating methane fermentations supplied with complex substrates are usually obligate anaerobes in genera such as *Bacteroides, Bifidobacterium, Butyrovibrio, Eubacterium,* and *Lactobacillus.* Many are enteric bacteria, which include the coliform bacteria. The coliform bacteria, classically represented by the pathogen *Escherichia coli* in the genus *Escherichia* and pathogens in the genera *Salmonella* and *Shigella,* are probably the most common fermentative bacteria in methane fermentation because the feedstocks are often biosolids and animal wastes, or the mixed cultures used are derived from active methane fermentations grown on these wastes. The enteric bacteria also include those in the genera *Enterobacter, Serratia, and Proteus,* which occur primarily in soil and water, and plant pathogens of the genus *Erwina.*

The first step in the fermentation of complex substrates by fermentative bacteria is the hydrolysis of polysaccharides to oligosaccharides and monosaccharides, of proteins to peptides and amino acids, of triglycerides to fatty acids and glycerol, and of nucleic acids to heterocyclic nitrogen compounds, ribose, and inorganic phosphate. The sugars are degraded by the Embden–Meyerhof pathway, in the case of fermentative metabolism with enteric bacteria, to intermediate pyruvic acid, which is converted to acetate, fatty acids, carbon dioxide, and hydrogen. At low partial pressures of hydrogen, acetate is favored. At higher partial pressures, propionate, butyrate, ethanol, and lactate are favored, generally in that order (McInerney and Bryant, 1981). There is also a special mode of cleavage of intermediate pyruvic acid to formic acid by enteric bacteria that is not found in other bacterial fermentations (*cf.* Stanier *et al.,* 1986). Some of these bacteria possess the enzyme systems that cleave formic acid by

$$CH_3COCO_2^- + CoA \rightarrow CH_3COCoA + HCO_2^-$$
$$HCO_2^- + H^+ \rightarrow CO_2 + H_2.$$

Members of the genera *Escherichia* and *Enterobacter* contain this enzyme system while those in the genera *Serratia* and *Shigella* and *Salmonella typhi* do not. The amino acids and glycerol are also degraded by the glycolysis pathway to the same products and by other routes. After hydrolysis and glycolysis, some of the fermentation products are suitable substrates for methanogens; others are not.

Further degradation of unsuitable substrates is caused by another group of anaerobes, collectively called acetogenic bacteria. This group is known to exist on the basis of experimental data collected with several cocultures containing one hydrogen-utilizing species such as a methanogen. The acetogens convert the alcohols and higher acids produced on glycolysis to acetate, carbon dioxide, and hydrogen. The isolation of "S" organism from "*Methanobacterium omelianskii*" is the first documented evidence of species of the acetogenic group. Originally, *M. omelianskii* was thought to be a methanogen that catabolized ethanol by (Barker, 1941)

$$2CH_3CH_2OH + CO_2 \rightarrow 2CH_3CO_2H + CH_4.$$

This stoichiometry represented the experimental data and indicated that acetic acid is derived from ethanol and methane is derived from carbon dioxide. To support this interpretation, unlabeled ethanol was incubated with ^{14}C-labeled carbon dioxide; the ^{14}C content of methane was approximately equal to that of the carbon dioxide at the end of the fermentation (Stadtman and Barker, 1949). Later, this result was shown to be caused by the syntrophic association of two strict anaerobes, the unidentified S organism, which converts ethanol to acetate and hydrogen, and a methanogen, which uses the hydrogen to reduce carbon dioxide to methane (Bryant *et al.*, 1967). Neither bacterium alone can grow on ethanol or carbon dioxide, and the growth of S organism is inhibited by the accumulation of hydrogen. Thus, the two organisms have a true symbiotic relationship and are maintained as a mixed culture. The biochemical reactions are

$$2CH_3CH_2OH + 2H_2O \rightarrow 2CH_3CO_2H + 4H_2$$
$$CO_2 + 4H_2 \rightarrow CH_4 + 2H_2O.$$

It has been established that propionate and the longer fatty acids are catabolized by similar syntrophic associations (Boone and Bryant, 1980).

E. BIOCHEMICAL PATHWAYS TO METHANE

From a biochemical standpoint, methanogens contain cofactors not found in other bacteria, including the carriers of the carbon dioxide during its reduction to methane—methanopterin, methanofuran, and CoM (*cf.* Stanier *et al.*, 1986). Other factors are believed to function as hydrogen carriers during these reductions. The oxidation of hydrogen is presumed to occur on the outside of the cytoplasmic membrane, while carbon dioxide (or bicarbonate) reduction occurs inside the cell. The net result is the equivalent of transporting two protons out of the cell for each hydrogen oxidized. Because so many different bacteria are involved in methanogenesis and the fact that most of the methane is derived from acetate and not carbon dioxide, multiple pathways are undoubtedly involved, even for carbon dioxide reduction.

The biochemical pathway by which carbon dioxide is incorporated as cellular carbon and the mechanism of coupling methanogenesis to ATP synthesis has not been established with certainty, although knowledge of this and related pathways is about to expand greatly (see Section II,F). Classical electron transport pathways are probably not operative, but metabolic pathways have been proposed for methanogenic bacteria that synthesize methane during growth on single-carbon substrates and hydrogen (*cf.* Zeikus, Kerby, and Krzycki, 1985). All methanogens seem to be universally capable of using hydrogen as an electron donor and carbon dioxide as an electron acceptor. Many are also capable of using sulfur, sulfate, and nitrate as electron acceptors. Most methane fermentation systems reduce sulfate and other sulfur compounds that may be present to hydrogen sulfide, which forms insoluble sulfides with heavy metals in the fermentation broth. Methanogenesis ceases when methanogens of Groups I and II are grown in the presence of sulfur, whereas those in Group III continue methanogenesis simultaneously with sulfur reduction (*cf.* Stanier *et al.*, 1986).

Several factors make the biochemical pathways for methane production from pure substrates difficult to elucidate. One is that some of the early work was carried out with mixed cultures, so the experimental data may be questionable. Another is that the observations made with one pure methanogen do not necessarily apply to another. The third is that complex substrates complicate matters further. Although the basic carbon flows to products in acetogenic and methanogenic bacteria are predictable, a better understanding of the exact biochemistries (enzymes, coenzymes, electron carriers, and their cellular localization) is needed even for single-carbon substrates to test the various proposed models for carbon and electron flow and energy conservation during growth (Zeikus, Kerby, and Krzycki, 1985).

The catabolism of ethanol by acetogenic S organism to form acetate is inhibited by hydrogen and proceeds at good growth rates only when a hydrogen utilizer is present. This can be explained by use of the standard Gibbs free energy changes for the dominant reactions of the major groups of bacteria in methane fermentation (Table 12.1). Ethanol conversion to acetic acid by acetogens has a slightly positive free energy change, so coupling of this reaction

TABLE 12.1   Estimated Gibbs Free Energy Changes of Selected Biological Reactions in Methane Fermentation under Physiological Conditions[a]

Reaction	$\Delta G^{\circ\prime}$ (kJ)
Fermentative bacteria	
$(C_6H_{10}O_5) + H_2O \rightarrow C_6H_{12}O_6$	-18
$C_6H_{12}O_6 + 6H_2O \rightarrow 6CO_2 + 12H_2$	-26
$C_6H_{12}O_6 \rightarrow 2CH_3COCO_2^- + 2H^+ + 2H_2$	-112
$C_6H_{12}O_6 + 2H_2O \rightarrow CH_3CH_2CO_2^- + H^+ + 3CO_2 + 5H_2$	-192
$C_6H_{12}O_6 \rightarrow CH_3CH_2CH_2CO_2^- + H^+ + 2CO_2 + 2H_2$	-264
Acetogenic bacteria	
$CH_3CH_2CO_2^- + H^+ + 2H_2O \rightarrow CH_3CO_2^- + H^+ + CO_2 + 3H_2$	$+72$
$CH_3CH_2CH_2CO_2^- + H^+ + 2H_2O \rightarrow 2CH_3CO_2^- + 2H^+ + 2H_2$	$+48$
$CH_3CH_2OH + H_2O \rightarrow CH_3CO_2^- + H^+ + 2H_2$	$+10$
$HO_2C(CH_2)_2CH(NH_2)CO_2^- + 3H_2O \rightarrow 2CH_3CO_2^- + HCO_3^- + H^+ + H_2 + NH_4^+$	-34
$CH_3COCO_2^- + H_2O \rightarrow CH_3CO_2^- + CO_2 + H_2$	-52
$HOCH_2CH(OH)CH_2OH + 2H_2O \rightarrow CH_3CO_2^- + HCO_3^- + 2H^+ + 3H_2$	-73
$2H_2NCH_2CO_2^- + 2H^+ + 4H_2O \rightarrow CH_3CO_2^- + 2HCO_3^- + H^+ + 2H_2 + 2NH_4^+$	-83
$2CO_2 + 4H_2 \rightarrow CH_3CO_2^- + H^+ + 2H_2O$	-95
$2HCO_3^- + 4H_2 + H^+ \rightarrow CH_3CO_2^- + 4H_2O$	-105
$C_6H_{12}O_6 + 4H_2O \rightarrow 2CH_3CO_2^- + 2HCO_3^- + 4H^+ + 4H_2$	-206
$C_6H_{12}O_6 + 2H_2O \rightarrow 2CH_3CO_2^- + 2H^+ + 2CO_2 + 4H_2$	-216
$C_6H_{12}O_6 \rightarrow 3CH_3CO_2^- + 3H^+$	-311
Methanogenic bacteria	
$CH_3CO_2^- + H^+ \rightarrow CH_4 + CO_2$	-36
$CH_3OH + H_2 \rightarrow CH_4 + H_2O$	-113
$CO_2 + 4H_2 \rightarrow CH_4 + 2H_2O$	-131
$HCO_2^- + H^+ + 3H_2 \rightarrow CH_4 + 2H_2O$	-134
$HCO_3^- + H^+ + 4H_2 \rightarrow CH_4 + 3H_2O$	-136
Inorganic reducing bacteria	
$S + H_2 \rightarrow HS^- + H^+$	-28
$SO_4^{2-} + H^+ + 4H_2 \rightarrow HS^- + 4H_2O$	-152
$SO_3^{2-} + 2H^+ + 3H_2 \rightarrow H_2S + 3H_2O$	-173
$NO_3^- + 2H^+ + 4H_2 \rightarrow NH_4^+ + 3H_2O$	-600
$2NO_2^- + 2H^+ + 3H_2 \rightarrow N_2 + 4H_2O$	-794
$2NO_3^- + 2H^+ + 5H_2 \rightarrow N_2 + 6H_2O$	-1121

[a]The free energy changes were calculated from the standard Gibbs free energies of formation in Thauer, Jungermann, and Decker (1977) or are from the reference. The conditions are 25°C, pH 7, and aqueous solutions at unit activity where possible. Methane, H_2, and CO_2 are in the gaseous state. Cellulose is assumed to have the same standard free energy of formation per unit of glucose as glycogen, and the hydrolysate is assumed to be α-D-glucose.

with a methanogenic reaction that has a strongly negative free energy change is thermodynamically favorable. Other trends can also be perceived from the free energy changes. The thermodynamic driving force for a few of the acetogenic reactions in which the higher acids are converted to acetate are positive, whereas that for direct conversion of glucose to acetate is strongly negative. For fermentative bacteria, the free energy changes listed in Table 12.1 are negative, but cellulose hydrolysis is the least favorable reaction.

Complete conversion of glucose to carbon dioxide and hydrogen in dark fermentations, while having a slightly negative free energy change, has not been considered an efficient process for the production of hydrogen, since every fermentation reaction must be coupled with the synthesis of ATP from adenosine diphosphate and inorganic phosphate (Chapter 11). Under physiological conditions, ATP synthesis requires about 42 to 50 kJ/mol ATP formed; the glucose-to-hydrogen reaction supplies only about 26 kJ/mol glucose converted. Note that each of the methanogenic reactions in Table 12.1 exhibits a negative free energy change. Methanation of carbon dioxide and bicarbonate is more favored than direct conversion of acetate, which produces about 70% of the methane in methane fermentation. Thermodynamic data are very useful for making predictions and explaining methane fermentation, but judgment should be exercised in interpreting them.

Information accumulated from the examination of pure compounds and natural products as substrates for methane fermentation and the characterization of anaerobic organisms indicate that the scheme shown in Fig. 12.1

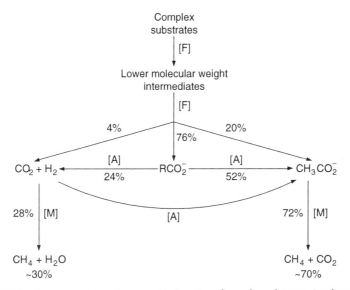

FIGURE 12.1 Routes to methane by anaerobic digestion of complex substrates. Bracketed letters denote mixed cultures of bacteria; F, fermentative; A, acetogenic; M, methanogenic. From Zehnder, Ingvorsen, and Marti (1982).

accounts for the actions of the three major groups of bacteria in the process and the sources of methane and carbon dioxide (Zehnder, Ingvorsen, and Marti, 1982). The fermentative bacteria accomplish hydrolysis and conversion of the complex substrates to intermediates in yields of about 4% carbon dioxide and hydrogen, 20% acetate, and 76% intermediate higher acids and other lower molecular weight compounds. The acetogenic bacteria convert about one-third of the higher acids and lower molecular weight compounds to additional carbon dioxide and hydrogen, and two-thirds to additional acetate. About 70% of the methane and carbon dioxide is produced by methanogenic bacteria from acetate, and 30% is produced from carbon dioxide and hydrogen. Direct observation of operating methane fermentation systems are in accord with this scheme. For example, when a steady-state fermentation is upset by an undesirable change in environmental conditions or an operating parameter that reduces gas production, the pH decreases while the volatile acids in the fermentation broth and carbon dioxide evolution increase. This, as well as several other features of methane fermentation, can be predicted with this model.

The phasic or stepwise nature of methane fermentation suggested by Fig. 12.1 is also supported by many observations of the behavior of individual substrates. For example, when pure glucose was fermented in the batch mode with an inoculum from an active sewage sludge digester, almost all the glucose was assimilated in the first 30 h of fermentation; the product gas during this period contained 70 to 100% carbon dioxide (Ghosh and Klass, 1978). No methane was detected for the first 6 h. Most of it was collected after about 95% of the glucose had been consumed. The gas production data indicate that methanogenic bacteria function at a much lower rate than the fermentative and acetogenic bacteria, which rapidly catabolize glucose. Other observations that support stepwise methane fermentation have been made with *Macrocystis pyrifera* (giant brown kelp), a complex substrate (Klass and Ghosh, 1977). As shown in Fig. 12.2, some denitrification occurred in the first few hours, as indicated by the nitrogen peak at a concentration of about 70 mol % of the product gas. Maxima in the concentrations of hydrogen and carbon dioxide in the biogas were observed at 13 h (about 28 mol %) and 103 h (about 89 mol %) during the early portion of the process when methane production was low. Methane production rapidly increased from about 100 to 300 h. Its concentration reached a plateau of about 85 to 89 mol % while carbon dioxide concentration rapidly decreased and stabilized at about 10 to 15 mol %. The production of hydrogen and nitrogen fell to zero during this period.

The microbial transformations and stages in anaerobic digestion are supported by experimental data accumulated over many years. The overall schematic of the process shown in Fig. 12.3 is perhaps the simplest chemical representation of the hydrolysis, acid-formation, and methane-formation stages. As discussed in Section IV, this information led to the development of

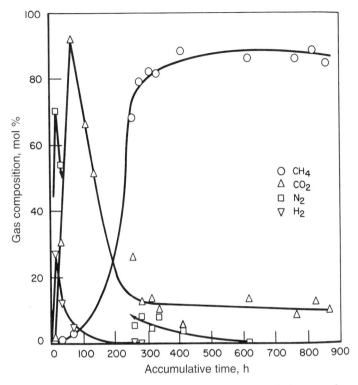

FIGURE 12.2 Variation of gas composition with time for 36-day batch digestion of *Macrocystis pyrifera*. Conditions: 2 L liquid volume in 2-L fermenter, continuous mixing, pH 6.8 to 7.2 with NaOH additions, mixed inoculum, chopped kelp passed through a 0.95-cm screen, 1.41 wt % volatile solids in charge. Results: Maximum gas production rate as volume (n) per liquid volume per day: CH_4, 0.294; CO_2, 0.188; H_2, 0.079; time to maximum gas production: CH_4, 294 h; CO_2, 103 h; H_2, 12.6 h; energy in gas as percentage of substrate energy: 20.4 CH_4, 0.45 H_2; volatile solids reduction: 20.3 %. From Klass and Ghosh (1977).

what has been called two-phase methane fermentation or digestion in which methanogenesis is physically separated from hydrolysis and acid formation. This resulted in significant improvements in process performance that can easily be obtained at low cost.

F. Genome Sequence of Methanogens and Gene Identification

The first complete genome sequence for a member of the kingdom Archaebacteria, *Methanococcus jannaschii*, has been determined (Bult *et al.*, 1996). *M.*

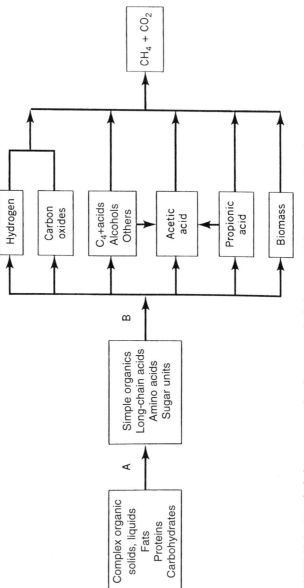

FIGURE 12.3 Microbial phases in anaerobic digestion: A, hydrolysis; B, acidification; C, methane fermentation.

jannaschii is a strict anaerobic methanogen that was isolated from the sea floor at the base of a 2600-m-deep "white smoker" chimney on the East Pacific Rise (Jones *et al.*, 1983). It grows at pressures of up to 200 atm and over a temperature range of 48 to 94°C; the optimum is near 85°C. Such organisms are often called hyperthermophiles, as opposed to superthermophiles and extremophiles, which grow at still higher temperatures and are also known. M. *jannaschii* grows on carbon dioxide and hydrogen and on formate as substrates, but appears to lack the genes to use methanol and acetate. All of the known enzymes and enzyme complexes associated with methanogenesis have been identified in this organism, the sequence and order of which are believed to be typical of methanogens (DiMarco *et al.*, 1990). The organism also contains the genes necessary to fix nitrogen. They have also been identified. Determination of the complete genome sequence for M. *jannaschii* is the first for an autotrophic methanogen and provides, along with the other genome sequences that have been completed, the opportunity to compare biochemical pathways among the Archae, Eukaryotes, and Prokaryotes.

This investigation and others have shown that of the pathways that fix carbon dioxide, the Ljungdahl–Wood pathway is used by methanogens to fix carbon (Wood, Ragsdale, and Pezacka, 1986). This pathway consists of a noncyclic, reductive acetyl coenzyme A-carbon monoxide hydrogenase pathway, which is facilitated by a carbon monoxide dehydrogenase complex (Blaat, 1994). The complete Ljungdahl–Wood pathway, encoded in the M. *jannaschii* genome, depends on the methyl carbon in methanogenesis, but methanogenesis can occur independently of carbon fixation (Bult *et al.*, 1996).

The data compiled in identifying the genes and gene sequence in M. *jannaschii* are quite astounding, because this effort and others that will most certainly follow will advance knowledge to the point where biochemical pathways and microbial performance can be analyzed in a quantitative manner. The entire biochemical mechanism will be elucidated. Meanwhile, until this occurs, particularly with the basic groups of microorganisms that make up methane fermentation systems, a semiqualitative approach is necessary to develop and improve the technology.

III. MICROBIAL HYDROGEN

A. Hydrogen-Producing Microorganisms

In the 1960s, hydrogen-producing microorganisms were categorized into four groups: strict heterotrophic anaerobes, facultative heterotrophic anaerobes that do not contain cytochromes as electron carriers, heterotrophic anaerobes that contain cytochromes as electron carriers, and photosynthetic microorganisms

(Gray and Gest, 1965). This classification system is still valid today for most microorganisms. Electron sources for the strict heterotrophic anaerobes are usually two- or three-carbon intermediates such as pyruvic acid, ethanol, and acetaldehyde. *Clostridium butylicum* and *Methanomonas aerogenes* are examples of bacteria in this group. Members of the second group use formic acid as the electron donor. Examples are *Escherichia coli* and *Bacillus macerans*. *Desulfovibrio desulfuricans* is one of the few members of the third group. This microorganism uses sulfate as a terminal oxidant for energy-yielding, cytochrome-linked anaerobic oxidations (e.g., of lactate), but certain strains can liberate molecular hydrogen from pyruvate and formate when sulfate is absent.the photosynthetic microorganisms classified in the fourth group consist of several algae and bacteria. They are described in more detail in Part C.

B. Hydrogen Fermentation and Cell-Free Enzyme Catalysts

The Embden–Meyerhof pathway via pyruvic acid is the normal route to fermentation products from biomass carbohydrates (Chapter 11). Enteric bacteria appear to have the unique capability of converting pyruvic acid directly to formic acid (Section II,D):

$$CH_3COCO_2^- + CoA \rightarrow CH_3COCoA + HCO_2^-$$

So formic acid can be a major end product in sugar fermentations. Since some of the enteric bacteria also contain the enzyme system formic "hydrogenlyase," which degrades formic acid to equimolar quantities of hydrogen and carbon dioxide, hydrogen can be a major end product. This enzyme system consists of at least two enzymes, a soluble formate dehydrogenase and a particulate hydrogenase. The most frequent mode of sugar breakdown by enteric bacteria, however, results in the formation of several products, principally ethanol and mixed acetic, lactic, succinic, and formic acids, or hydrogen if formic hydrogenlyase is present. *E. coli*, for example, is an effective enteric organism for this fermentation, but the selectivity is poor for individual products, including hydrogen. The product distribution in moles per 100 mol of glucose fermented is acetic acid, 36; formic acid, 2.5; lactic acid, 79; succinic acid, 11; ethanol, 50; hydrogen, 75; and carbon dioxide, 88 (*cf.* Stanier *et al.*, 1986). Such fermentations are characteristic of the genera *Escherichia, Photobacterium, Proteus, Salmonella, Shigella, Yersinia,* and *Vibrio,* and they occur in some *Aeromonas* species. Some enteric bacteria produce an additional major product, 2,3-butanediol. The selectivities are still poor for individual products. This fermentation is characteristic of enteric bacteria in the genera *Enterobacter* and *Serratia,* most species of *Erwina,* and some *Aeromonas* and *Photobacterium* species. Other bacteria that produce hydrogen as one of the end products in sugar fermentations are certain Gram-positive, spore-forming members of the genera *Bacillus* and *Clostridium*. These bacteria do not require formic hydrogen-lyase and are able to produce hydrogen and carbon dioxide directly from pyruvic acid, but again, hydrogen selectivity is poor. For example, the main products of sugar fermentation by *B. macerans* are ethanol, acetone, acetate, formate, carbon dioxide, and hydrogen. The main fermentation products of *B. polymyxa* are 2,3-butanediol, ethanol, carbon dioxide, and hydrogen. Multiple fermentation products, including hydrogen, are also produced by various species of Clostridia (*cf.* Table 11.11). The use of fermentative organisms to produce molecular hydrogen would appear to be a very inefficient use of biomass because hydrogen is usually a minor product and multiple products are formed.

In some cases, the selectivity for hydrogen can be relatively high because only a few organic products are formed. This is especially true when cell-free enzyme preparations are employed as catalysts for the conversion of specific biomass components. The energy yield as hydrogen is still small because most of the feedstock energy resides in the organic products. To illustrate, a cell-free, glucose dehydrogenase-hydrogenase system extracted from appropriate bacteria was found to be an effective catalyst at near-ambient conditions for the conversion of glucose to hydrogen and gluconic acid, the only organic product (Woodward *et al.*, 1996). However, the energy yield as hydrogen is only about 8%. Cellulose can be substituted for glucose and converted to gluconic acid and hydrogen in a single stage if cellulase is added to promote the hydrolysis step.

As indicated in the previous section, methane fermentation can be separated into the acid and methane phases, and molecular hydrogen is produced at least to some extent in the acid phase. Under certain conditions, substantial quantities can be formed. The Gibbs free energy changes for several microbial conversions of acetogenic bacteria are quite favorable for hydrogen production, as shown in Table 12.1. Consequently, the possibility of producing molecular hydrogen as an energy product in the acid phase of methane fermentation is of interest. Complete conversion of glucose to carbon dioxide and hydrogen in dark fermentations is not regarded as an efficient process for hydrogen production because of the energy available for ATP synthesis. However, many of the transformations in which partial conversion of biomass to organic intermediates occurs yield relatively large amounts of co-product hydrogen. The problem, of course, is that if the biosynthesis of methane is the objective, any hydrogen that is withdrawn from the process reduces methane yield and the overall transfer of energy from biomass to methane. The stoichiometries show that if the acid phase of methane fermentation is carried out to maximize the yield of molecular hydrogen as an end product, a large amount of substrate-derived organic intermediates such as the acids will remain in the medium.

Nevertheless, molecular hydrogen can be recovered as a product or co-product of the anaerobic fermentation of biomass under batch or continuous anaerobic fermentation conditions. The conversion of waste pea shells illustrates the batch mode of producing hydrogen (Kalia and Joshi, 1995). A pea shell slurry, 1 wt % total solids, was inoculated with an enriched, mixed culture of acidogens from cattle manure and incubated for 2 days, and then inoculated with an enriched, mixed culture of hydrogen producers, also from cattle manure. Incubation at ambient temperatures with pH control and periodic flushing with nitrogen over a 6-day period, when gas evolution stopped, provided a total biogas yield of 362 L/kg of volatile solids (VS, organics) reduced. The gas consisted of 119 L (33 mol %) of hydrogen and 8 to 12 mol % of hydrogen sulfide. The remainder was carbon dioxide. Assuming the higher heating value of dry pea shells is about 18.5 MJ/kg, the energy yield as hydrogen is about 8% in these studies. In other studies of hydrogen production by continuous anaerobic fermentation of pure glucose, the hydrogen content of the biogas is as high as about 75 mol %, but the yield is low because of co-product organic acids and the dilution rates necessary to produce high-hydrogen content gases (*cf.* Vavilin, Rytow, and Lokshina, 1995).

C. Photosynthetic Microorganisms

In the 1940s, it was discovered that intact cells of certain unicellular, photosynthetic microalgae (Gaffron and Rubin, 1942) and Gram-negative eubacteria (Gest and Kamen, 1949) are capable of generating molecular hydrogen by biophotolysis. Atmospheric oxygen must be excluded from the organisms' environment so that the highly oxygen-sensitive enzyme hydrogenase, present in the organisms or synthesized by the cells under anaerobic conditions, can catalyze the formation of hydrogen. Since oxygen is generated at the same time, it must be eliminated or reduced during the process if the formation of hydrogen and oxygen occurs together in the same reaction zone. Eventually, it became apparent that the ability to produce molecular hydrogen is widespread over different taxonomic and physiological types of photosynthetic microorganisms. A key element in the microbial generation of molecular hydrogen with photosynthetic microorganisms is that the reducing power is supplied by organic compounds in the biomass feedstock, the biomass fermentation products, or by inorganic compounds, or is synthesized by cellular biomass. Except under special conditions, the biophotolysis of water does not occur in such a way that all the energy needed to convert water to its components is from light.

To understand how photosynthetic organisms produce molecular hydrogen, consider the mechanism of the process. The splitting of water to hydrogen and oxygen using the radiant energy of visible light and the photosynthetic apparatus of green plants and certain bacteria and algae involves the oxidation of water to liberate molecular oxygen and electrons. The electrons are raised from the level of the water-oxygen couple, $+0.8$ V, to 0.0 V by Photosystem II (Chapter 3), and then undergo the equivalent of a partial oxidation process in dark reactions to a positive potential of $+0.4$ V. Photosystem I then raises the potential of the electrons to as high as -0.7 V. In normal oxygenic photosynthesis, the reducing power of these electrons is transferred by a series of reactions to the oxidized form of nicotinamide adenine dinucleotide, or its phosphate, to afford the reduced form, which in turn participates in the reduction of carbon dioxide to organic material. In the biophotolysis of water, these electrons are diverted from carbon dioxide fixation to the enzyme hydrogenase, which catalyzes the reduction of protons to hydrogen. Thus, the overall net chemistry is simply the photolysis of water to hydrogen and oxygen:

$$H_2O \rightarrow 0.5O_2 + 2H^+ + 2e^-$$
$$2H^+ + 2e^- \rightarrow H_2$$
$$\text{Net: } H_2O \rightarrow 0.5O_2 + H_2.$$

Note that the standard Gibbs free energy and enthalpy changes under physiological conditions for biophotolysis (or photolysis) are about 237 and 242 kJ per mol water reacted. Also note that water is the source of electrons that reduce the pyridine nucleotides, and both photosystem II and Photosystem I are required for this process to occur.

Microorganisms that use water as an electron donor—the eukaryotic, chlorophyll-containing algae and the prokaryotic cyanobacteria (formerly called blue-green algae)—perform the equivalent of oxygenic or plant photosynthesis. The cyanobacteria contain a pigment system similar to that of the photosynthetic eukaryotes and are the only microorganisms capable of oxygenic photosynthesis and atmospheric nitrogen fixation. Photosystem II produces oxygen by photooxidation of water that occurs by a mechanism called cyclic photophosphorylation in which the only product is ATP, and Photosystem I couples the production of ATP and oxygen evolution with the reduction of the nonheme iron proteins, ferredoxins. Reduced ferredoxins then reduce the pyridine nucleotides in the dark reactions. Other photosynthetic microorganisms—the prokaryotic green and purple eubacteria—carry out anoxygenic photosynthesis under anaerobic conditions and generate reducing power from electron donors other than water. Anoxygenic phototrophs use either organic compounds or reduced inorganic compounds as electron donors, and the known photosynthetic eubacteria can fix carbon dioxide. Reducing power is generated by what is called reverse electron transport, as in certain chemoautotrophs that derive their energy from the oxidation of inorganic compounds, or from organic compounds by other routes. Photosystem II is absent in the green and purple eubacteria, but Photosystem I is present, and each group of microorganisms is also capable of fixing atmospheric nitrogen.

Under certain anaerobic conditions, molecular hydrogen can be generated by some of the photosynthetic organisms in these groups. For example, it was found that anaerobically adapted green algae will simultaneously produce hydrogen and oxygen for over 16 h when illuminated with visible light and an inert gas is used to sweep out the gaseous products to maintain a low partial pressure of oxygen (Greenbaum, 1981). At the end of the 16-h period, the algae are still viable and can be rejuvenated by exposure to a cycle of normal aerobic photosynthesis. Then another period of hydrogen and oxygen production can follow, and the system can be recycled. However, the rate of hydrogen production is too low to be practical. In a somewhat similar manner, a mutant (B4) of the unicellular green alga *Chlamydomonas reinhardi,* which lacks Photosystem I, grown on acetate was found to photoautotrophically fix carbon dioxide and simultaneously generate hydrogen and oxygen in an inert atmosphere using only Photosystem II (Greenbaum *et al.,* 1995). This work indicates that Photosystem II alone can reduce ferredoxins to generate reduced pyridine nucleotides and that Photosystem I is not necessary for autotrophic photosynthesis. Other studies show that mutants of *Chlamydomonas* without Photosystem I can grow photoautotrophically with oxygen evolution using atmospheric carbon dioxide as the sole carbon source (Lee *et al.,* 1996). These experiments are the equivalent of water photolysis, at least during the initial illumination stage. Under anaerobic conditions in which hydrogenase is activated, photoevolution of hydrogen occurred at the beginning of illumination. But at steady state, hydrogen evolution approached zero and reduction of carbon dioxide by the Calvin–Benson cycle became the exclusive sink for reductant generated by photolysis of water. There may be other as yet unknown electron transfer routes that can promote the process. The investigators believe these observations may lead to methods of doubling the thermodynamic efficiency of converting light into chemical energy, because a single photon rather than two span the potential difference between water oxidation and carbon dioxide or proton reduction. These observations also appear to be in conflict with the traditional rationale and Z scheme (Chapter 3) used to explain how the potential differences in conventional aerobic photosynthesis are overcome.

Nitrogen-fixing, heterocystous cyanobacteria such as *Anabaena cylindrica* can also liberate hydrogen and oxygen simultaneously under anaerobic, nitrogen-deficient conditions (Benemann and Weare, 1974). *A. cylindrica* does not form heterocysts when grown with a chemical nitrogen source, but both nitrogenase synthesis and heterocyst formation occur when cultures are deprived of chemical nitrogen. Under this condition, photosynthesis occurs in the nonheterocystic vegetative cells with the fixation of carbon dioxide and the formation of oxygen and cellular biomass. The carbohydrates produced are believed to be supplied to the heterocysts, which then convert them to a reductant to fix atmospheric nitrogen. Photosystem II is not present in the heterocysts. When atmospheric nitrogen is absent, the reducing power is used by the heterocysts to convert protons to molecular hydrogen. Careful regulation of gas flows and nutrient and cell concentrations resulted in about 2 to 3% conversion of the incident fluorescent light energy to hydrogen energy (Benemann, 1978). Under nitrogen-deficient conditions, the enzyme nitrogenase apparently acts like the enzyme hydrogenase, or hydrogenase is synthesized in the stressed cells.

Several of the purple bacteria can produce molecular hydrogen photosynthetically under anaerobic conditions when nitrogenase activity is induced in an inert, nitrogen-free atmosphere, and a suitable organic or inorganic electron donor is supplied. Under these conditions, purple bacteria can anaerobically oxidize acetate, for example, to carbon dioxide and hydrogen:

$$CH_3CO_2H + 2H_2O \rightarrow 2CO_2 + 4H_2.$$

Green sulfur bacteria under anaerobic conditions and in a nitrogen-free atmosphere might also be expected to generate molecular hydrogen, but none is known that can grow photoheterotrophically. The green nonsulfur bacterium *Chloroflexus* might be expected to produce hydrogen, too, because it appears to derive organic nutrients from cyanobacteria in natural hot springs and is also chemoheterotrophic.

Hydrogen production was first demonstrated in cell-free, chloroplast-ferredoxin-hydrogenase preparations (Arnon, Mitsui, and Paneque, 1961). In addition, synthetic chloroplast membranes have been proposed for the generation of hydrogen in which quantum acts take place on opposite sides of the membrane. On one side, hydrogen evolution occurs; oxygen is generated on the other side. Partial experimental support for this concept was obtained by Calvin and co-workers (*cf.* Calvin, 1978, 1979) and others. Several sensitizers and catalysts are still needed to complete the cycle. When developed, the system is expected to mimic natural photosynthesis in green plants except that the electrons are diverted from carbon dioxide fixation to hydrogen formation. Calvin suggested that it might be possible to totally reduce carbon dioxide to hydrocarbons with such systems. Continuously operable *in vitro* systems in which activity can be sustained remain to be constructed.

IV. ANAEROBIC DIGESTION
SYSTEM CHARACTERISTICS

A. CONVENTIONAL SYSTEMS

In its simplest configuration, anaerobic digestion is carried out at the proper fermentation conditions in a closed digester to eliminate oxygen inhibition. A schematic representation of the anaerobic digestion of municipal biosolids and the typical distribution of the components within the digester are shown in Fig. 12.4. The process is carried out in the batch, semicontinuous, or continuous operating mode. In the latter two modes, the digesters are intermittently or continuously supplied with an aqueous slurry of the feedstock, and an equal amount of fermenter broth is withdrawn. In batch systems, steady-state conditions cannot be achieved because the components and compositions within the digester are constantly changing. In the semicontinuous or continuous modes, methane fermentation can take place in the steady state as the organisms grow at the maximum rate permitted by the inflow of substrate and nutrients. For optimum methane recovery and fuel quality, the batch mode is not ordinarily used, except for small-scale systems, because the gas composition varies with time and the equipment costs are usually higher than those of continuous systems for the same throughput rates.

The important operating parameters are the composition, physical form, and energy content of the substrate; the inoculum source and activity; the feeding frequency and rate of nutrient and substrate addition to the digester; the hydraulic and solids retention times (HRT and SRT) within the digester; the pH, temperature, and mixing rate within the digester; the gas removal rate; and the amount and type of recycling. Numerous studies have been conducted on how these parameters affect methane production rate and yield, substrate reduction, volatile acid formation, gas composition, energy recovery, and steady-state operation. Reactor configuration and design also influence the performance of the process.

So-called standard or low-rate digestion is often utilized for wastewater stabilization (Fig. 12.5). It is normally carried out with intermittent feeding and withdrawal at mesophilic fermentation temperatures of 30 to 40°C or thermophilic fermentation temperatures of 50 to 60°C, total retention times of 30 to 60 days, and loading rates per unit of digester capacity of about 0.5 to 1.5 kg VS/m³-day. Stratification within the digesters usually occurs resulting in layers of digesting biosolids, stabilized biosolids, and a supernatant, which often has a scum layer. High-rate digestion is conducted in a similar manner, or with continuous feeding and withdrawal, and mixing is used to provide homogeneity. The retention times are about 20 days or less. Under these conditions, loading rates can be increased to about 1.6 to 6.4 kg VS/m³-day.

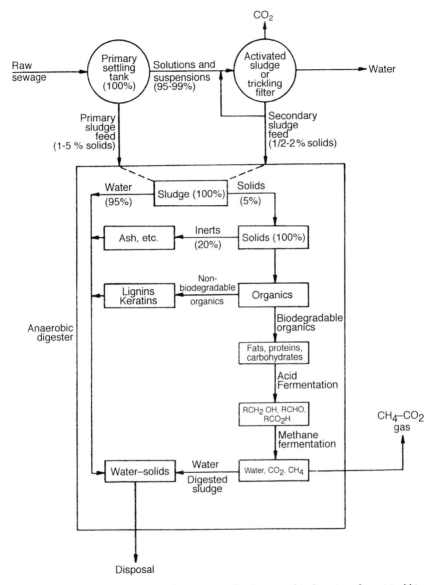

FIGURE 12.4 Typical distribution of components for the anaerobic digestion of municipal biosolids.

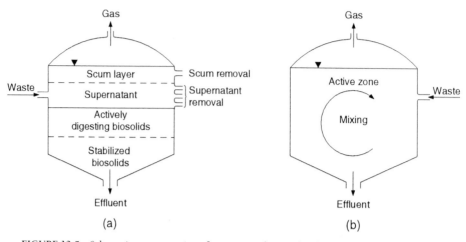

FIGURE 12.5 Schematic representation of conventional anaerobic digesters. (a) Low-rate digestion: Intermittent feeding and withdrawal, heated at 30 to 40°C (mesophilic) or 50 to 60°C (thermophilic), 30- to 60-day retention times, loading rates of 0.5 to 1.5 kg VS/m³-day. (b) High-rate digestion: Continuous or intermittent feeding and withdrawal, mixing, heated at mesophilic or thermophilic temperatures, retention times of 20 days or less, loading rates of 1.6 to 6.4 kg VS/m³-day.

Many modifications of these processes are used. An example is the anaerobic contact process in which digested biosolids are separated for recycling from the digester effluent. Generally, in a well-operated, balanced digester, the organic material fed to the digester will yield biogas at about 0.8 to 1.1 m³/kg of volatile solids destroyed. Many wastewater treatment plants in the United States maximize the usage of biogas as a fuel. Table 12.2 summarizes the biogas utilization program at the West-Southwest Treatment Plant of the Metropolitan Sanitary District of Greater Chicago. Although the economic data for low-pressure steam production are from the 1970s, they illustrate the relative differences and show that the payout can be quite favorable. About 60% of the biogas is used for digester heating at this facility.

Numerous waste and virgin biomass types and species have been evaluated as substrates for methane fermentation. Although there are major differences in energy content, moisture, volatile solids, and ash contents among the various materials, the biogas production parameters, volatile solids reductions, and energy recovery efficiencies as methane span a relatively narrow range under high-rate balanced digestion conditions. This is illustrated by the data in Table 12.3, which show composition and digestion performance for each of several substrates under similar high-rate conditions. This kind of performance is somewhat unexpected because most (but not all) of the organic components

TABLE 12.2 Biogas Production and Utilization from
Municipal High-Rate Digestion of Activated Biosolids at
Mesophilic Temperatures[a]

Plant or biogas parameter	Amount of parameter
Plant description	
Digester size	9462 m^3
Number of digesters	12
Design capacity	272 t/day
Activated sludge feed	3.3% total solids
Average biogas production	72,500 m^3 (n)/day
Average methane production	47,300 m^3 (n)/day
Biogas utilization	
For digester heating	45,600 m^3 (n)/day
For low-pressure steam	26,900 m^3 (n)/day
Biogas utilization economics	
Capital cost for steam plant	$830,000
Annual operating cost	$150,000
Annual natural gas cost for steam	$569,400
Approximate payout for biogas use	2 years

[a]Rimkus, Ryan, and Michuda (1979).

are present in each substrate. Compositional differences would be expected to control methane fermentation and lead to dissimilar biogas production rates and methane yields per unit of volatile solids added to the digester.

The major organic components in most waste biomass and terrestrial biomass species are proteins, celluloses, hemicelluloses, and lignins. Terrestrial biomass species usually conform to this general rule (Chapter 3). Marine biomass species contain lesser amounts of celluloses and hemicelluloses and larger amounts of proteins, reduced monosaccharides such as mannitol, and other carbohydrate polymers such as algin. These characteristics and the amounts of the individual components converted under high-rate, mesophilic, methane fermentation conditions are illustrated by the data in Table 12.4 for a few of the substrates from Table 12.3. Lignin conversion is small as expected because of its polyaromatic structure and the resulting low biodegradability. On the basis of these and other data, the general order of decreasing biodegradability is monosaccharides (glucose, mannitol, etc.), hemicelluloses, algins, cellulose, proteins (crude), and lignins. Several factors should be kept in mind when considering data of this type. The first is that the reduction in concentration of a particular component does not necessarily mean that the material was gasified. The organic structures in the substrate may have been modified during fermentation and therefore not detected in the analysis of the

TABLE 12.3 Comparison of Biomass Compositions and Methane Fermentation Under High-Rate Mesophilic Conditions[a]

Parameter	Primary biosolids	Primary activated biosolids	RDF–biosolids blend	Biomass-waste blend	Coastal Bermuda grass	Kentucky bluegrass	Giant brown kelp	Water hyacinth
Carbon, dry wt%	43.7	41.8	42.1	43.1	47.1	46.2	26.0	41.0
Nitrogen, dry wt %	4.02	4.32	1.91	1.64	1.96	4.3	2.55	1.96
Phosphorus, dry wt %	0.59	1.30	0.81	0.43	0.24	0.48	0.48	0.46
Ash, dry wt % of total solids	26.5	23.5	8.4	17.2	5.05	10.5	45.8	22.7
VS, dry wt % of total solids	73.5	76.5	91.6	82.8	95.0	89.8	54.2	77.3
HHV, MJ/kg (dry)	19.86	18.31	17.20	20.92	19.04	19.19	10.26	16.02
C/N ratio	10.9	9.7	22.0	26.3	24.0	10.7	10.2	20.9
C/P ratio	74.1	32.2	52.0	100	196		54.2	89.1
Biogas production rate, volume (n)/liquid volume-day	0.74	0.84	0.59	0.52	0.56	0.52	0.62	0.47
Methane in biogas, mol %	68.5	65.5	60.0	62.0	55.9	60.4	58.4	62.8
Methane yield, m³ (n)/kg VS added	0.313	0.327	0.210	0.201	0.208	0.150	0.229	0.185
VS reduction, %	41.5	49.0	36.7	33.3	37.5	25.1	43.7	29.8
Feedstock energy in gas, %	46.2	54.4	39.7	38.3	41.2	27.6	49.1	35.7

[a]Klass (1980). Methane fermentation conditions were daily feeding, continuous mixing, 35°C, pH 6.7 to 7.2, 12-day retention time, 1.6 kg VS/m³-day, except for kelp, which was 2.1 kg VS/m³-day. All biomass was 1.2 mm or less in particle size. The blends are described in detail in the original references. VS is volatile solids (organic matter). HHV is higher heating value.

TABLE 12.4 Comparison of Conversion of Organic Components under High-Rate Mesophilic Conditions[a]

Component	Coastal Bermuda grass		Giant brown kelp		Biomass-waste blend	
	% of VS	% converted	% of VS	% converted	% of VS	% converted
Crude protein	12.3		29.3	8	12.0	24
Cellulose	31.7	65	8.9	8	44.6	32
Hemicellulose	40.2	67			37.8	86
Lignin	4.1	9			5.5	0
Mannitol			34.5	71		
Algin			26.2	85		

[a]Klass (1980). Same fermentation conditions as in Table 12.3. All values are percentages by weight.

residual solids, or they may be detected in another fraction. The second factor is that a particular component in one substrate may not have precisely the same molecular structure as the corresponding fraction in another substrate, even though the analytical results are the same. As a result, the fractions identified as the same component in two substrates may have different degradabilities. This is supported by the cellulose data, which indicate that fractions identified as cellulose in different substrates have different degradabilities. Cellulose exists in complexed form in biomass that contains components such as lignin (Chapter 11). In this state, it is less accessible and has a lower degradability than free cellulose. Thus, cellulose conversion in methane fermentation systems might be expected to vary greatly depending on biomass species and maturity. Another factor concerns protein conversion, which was estimated in Table 12.4 by crude protein analysis (Kjeldahl nitrogen value times 6.25). Amino acid assays are necessary to determine true protein degradability, which is often high. It is evident from the data in Table 12.4 that if the degradabilities of certain biomass components in a conventional high-rate digestion system are low, pretreatment to enhance hydrolysis and provide higher concentrations of biodegradable intermediates such as sugars should increase gasification and methane yields. The use of appropriate advanced methane fermentation systems might also increase these parameters.

For high-rate digestion where all of the basic steps—hydrolysis, fermentation, acidogenesis, and methanogenesis—take place simultaneously in the same vessel in the presence of each bacterial group, one of these steps might intuitively be thought of as rate-limiting. Considerable experimental work has been done by many research groups to examine the kinetics of methane fermentation, and many reports of empirical observations, particularly with single substrates, have led to proposals regarding methane fermentation kinetics. For example, as the SRT is reduced from 20 days to about 2 days, the volatile acids in the digester increase and the methanogens tend to be washed out in the digester effluent. This type of evidence led many to conclude that conversion of the intermediate acids to methane by methanogenic bacteria limits the rate of the overall process. However, in fermentations with a complex substrate containing large amounts of cellulosics, cellulose hydrolysis might be rate-limiting.

Some investigators felt that these observations also supported transfer of the gaseous products to the gas phase as the rate-limiting step, and concluded that the design specifications for faster methane fermentations might include vigorous agitation, low pressure, and elevated temperature (Finney and Evans, 1975). However, with the exception of methane fermentation at thermophilic temperatures, which increase the methane production rate because of higher reaction rates, it has been known for many years that rapid, continuous agitation of anaerobic digesters is not necessary, and in some cases is even harmful (Stafford, 1982). Reduced pressure also provides little or no benefit (Hashimoto, 1982).

B. TWO-PHASE DIGESTION

Consideration of the requirements of mixed microbial groups in the anaerobic digestion process and the apparent rate limitation of methanogenesis led to proposals to physically separate the acid- and methane-forming phases of methane fermentation to take advantage of the stepwise nature of the process. The optimum environment for each group of organisms might then be maintained and the kinetics of the overall process improved. This appeared to offer improvements over conventional high-rate methane fermentation, where the environmental parameters are chosen to satisfy the requirements of the limiting microbial population. The techniques suggested for separating the acid- and methane-forming phases included selective inhibition of the methanogens in the acid-phase digester by manipulation of kinetic factors, addition of chemical inhibitors, and balancing of redox potentials (Borchardt, 1967); selective diffusion of the acids from the acid-phase digester through permeable membranes to the methane-phase digester (Schaumburg and Kirsch, 1966; Hammer and Borchardt, 1969; Borchardt, 1971); kinetic control by adjusting dilution rates to preclude growth of methanogens in the acid-phase digester (Pohland and Ghosh, 1971; Anonymous, 1971); and others (cf. Klass, 1984). Kinetic control is the simplest technique in concept and is likely to present the least operational difficulty. Kinetic control and acid- and methane-phase separation were first demonstrated with a soluble substrate, glucose (Pohland and Ghosh, 1971), and then with a particulate substrate, activated biosolids (Ghosh, Conrad, and Klass, 1975).

An example of the determination of the kinetic constants of the separate phases—acidogenesis (not acetogenesis) of pure cellulose, pure glucose, and activated biosolids is shown in Table 12.5. Comparison of the maximum specific growth rates (μ_{max}) shows that acid-phase fermentation of glucose is the fastest of the reaction steps studied. The other reactions in order of decreasing rate are acid-phase conversion of activated biosolids and cellulose and methane-phase conversion of acetic acid. In an overall process supplied with hydrolyzable cellulosics, methanogenesis is rate-limiting, assuming that acetate is the main intermediate in the methane-phase reactor. The saturation constants (K_S) provide information on the effects of substrate concentration on reaction rate. The low K_S for glucose means that high acidification rates can be achieved at low concentrations, while the very high K_S for cellulose and activated biosolids means that much higher concentrations of these substrates would be needed to reach conversion rates comparable to those of glucose. The K_S of acetate in the methane-phase reactor is much larger than that of glucose, but still much less than those of the insoluble substrates studied (cellulose and activated biosolids). Theoretical substrate conversion rates per unit reactor volume were estimated in this work from

$$R = \frac{S_o\,(\mu_{max}\theta - 1) - K_S}{\theta(\mu_{max}\theta - 1)}$$

where R is the substrate converted per liquid volume at hydraulic retention time θ, and S_o is the substrate concentration in the feed. For each substrate, plots of θ vs R yield a family of curves with maxima whose positions depend on S_o. The plots can be used to estimate the optimum θ to achieve maximum feed conversion in the shortest time at the lowest digester volume. At 30 to

TABLE 12.5 Comparison of Kinetic Constants of Mesophilic Acidogenic Fermentation of Glucose, Cellulose, and Activated Biosolids and Mesophilic Methanogenic Fermentation of Acetic Acid[a]

Kinetic constant	Acidogenesis of			Methanogenesis of acetate
	cellulose	glucose	activated biosolids	
μ_{max}, day^{-1}	1.7	7.2	3.84	0.49
g_{min}, h	9.8	2.3	4.3	33.9
K_S, g/L	36.8	0.4	26.0	4.2

[a]Ghosh and Klass (1978); Ghosh, Conrad, and Klass (1975). Kinetic constants are: μ_{max}, maximum specific growth rate; g_{min}, minimum generation time; and K_S, saturation constant or substrate concentration at which the specific growth rate is $\frac{1}{2}\mu_{max}$.

37°C, application of this equation indicated that optimum conversion of glucose can be achieved at θ's of 4 h and 4 days in the acid- and methane-phase reactors. For cellulosics and activated biosolids, the corresponding θ's are much higher, about 1 to 2 days for acid-phase digestion and 5 to 8 days for methane-phase digestion.

Laboratory data for acid- and methane-phase digestion of activated biosolids in Table 12.6 illustrate the course of two-phase digestion by kinetic control (Ghosh and Klass, 1978). The acid-phase unit was operated at a short retention time and a high loading rate. Methane yield and production rate were very low in the acid phase. The low pH and short retention time in this unit precluded growth of methanogenic bacteria. The methane-phase unit was operated on the liquid effluent from the acid phase at about half the retention time of a high-rate unit. Methane production rates were high, and the methane concentration in the gas from the methane-phase reactor was about 10 to 15% higher than that from a high-rate unit. When applied to a hypothetical commercial plant for two-phase digestion of municipal biosolids at a daily feed rate of 231 t of dry biosolids, the operating parameters of the two-phase system indicate that the estimated capital costs are about 60% less than those of a high-rate plant at the same throughput, mainly because the digester volume requirements are about one-third those of the high-rate plant. The total digester volume for the high-rate plant was 76,455 m^3 for eight digesters, and the

TABLE 12.6   Two-Phase Digestion of Municipal Biosolids[a]

Parameter	Acid phase	Methane phase
Temperature, °C	37	37
pH controlled	No	No
pH	5.7–5.9	7.0–7.4
Retention time, day	0.5–1	6.5
Loading rate, kg VS/m^3-day	24–43.2	
Methane production rate, volume (n)/liquid volume-day	0.006–0.6	4.4–8.4
Methane in gas, mol %	19–44	61–78
Methane yield, m^3 (n)/kg VS gasified	0.006–0.07	0.50–0.76
Effluent volatile acids, mg/L as acetic acid	3700–5100	100–150

[a]Ghosh and Klass (1978). The feed was 90% activated biosolids and 10% primary biosolids from the Metropolitan Sanitary District of Greater Chicago. The vessel sequence was a 10-L complete-mix, stirred tank reactor (CSTR) acid-phase digester, an effluent storage vessel, and a 10-L CSTR methane-phase digester. The higher heating value of the dry charge was 26.0 to 27.9 MJ/kg.

corresponding total volume needed for the projected two-phase plant was 27,014 m³, which consisted of 3851 m³ for four acid-phase digesters and 23,163 m³ for four methane-phase digesters. Studies of the conversion of glucose in experimental reactors that were identical, except that one was used for conventional digestion and two were used for two-phase digestion, showed that at the maximum specific loading, the biogas production rate and chemical oxygen demand (COD) turnover rate were about four times higher for the two-phase system than for the conventional system (Cohen *et al.*, 1980). A schematic of a two-phase, anaerobic digestion design is shown in Fig. 12.6 (Ghosh, Conrad, and Klass, 1975).

Results of a laboratory comparison of high-rate and two-phase digestion of an industrial waste are shown in Table 12.7 (Ghosh and Henry, 1981). Two-phase digestion facilitated conversion at much shorter retention times with more concentrated feed. When these data were applied to a hypothetical commercial plant supplied with waste solids, the digester volume required for the two-phase system was about one-third that of a high-rate system with the same throughput. Also, the net production of methane, after the biogas needed for plant fuel is withdrawn, was 73% more than that of the high-rate plant. The increase in net methane production is possible because less process fuel is needed for the two-phase plant due to the higher loading of volatile solids in the feed slurry. Less liquid is heated to maintain the process temperature.

The relationship of the acid and methane phases to the scheme in Fig. 12.1 raises several interesting questions, such as where the acetogenic bacteria are located and whether methane is formed exclusively from acetate or from both carbon dioxide reduction and acetate in the methane-phase reactor. The experimental data useful in answering these questions include the observations that for a glucose-fed, two-phase digestion system, more than 96% of the products from the acid-phase reactor were carbon dioxide, hydrogen, acetate,

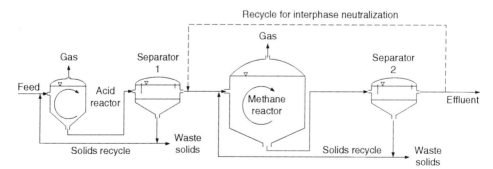

FIGURE 12.6 Schematic of two-phase digestion model. From Ghosh, Conrad, and Klass (1975).

TABLE 12.7   Comparison of High-Rate and Two-Phase Digestion of Soft-Drink Waste at 35°C[a]

Parameter	High-rate digestion	Two-phase digestion
Loading rate, kg VS/m³-day	0.64	4.8
Retention time, day	15	7.4
Biogas production rate, volume (n)/liquid volume-day	0.39	2.74
Biogas composition		
Methane, mol %	61	63
Hydrogen, mol %	0	3
Methane yield, m³ (n)kg VS added	0.37	0.37
Volatile solids reduction, %	72	64
COD reduction, %	84	96
Feed energy in gas, %	76	75

[a]Ghosh and Henry (1981). The feed was obtained from a modern soft-drink canning plant. The high-rate digestion was conducted in a 7-L CSTR. The vessel sequence for the two-phase system was a 2.5-L CSTR and a 5.5-L upflow anaerobic filter. The higher heating value of the charge was 19.8 MJ/kg.

and butyrate on complete assimilation of the glucose (Cohen et al., 1979). The acid-phase gas represented about 12% of the influent COD and contained approximately equimolar amounts of carbon dioxide and hydrogen. No methane was detected. Butyrate was present in the acid-phase effluent at about three times the concentration of acetate. About 98% of the organic substances fed to the methane-phase reactor were converted to a small amount of cellular biomass and product gas containing 84 mol % methane and 16 mol % carbon dioxide. These data support the view that the small amount of hydrogen from the acid-phase reactor is derived mainly from fermentative bacteria because of the unfavorable thermodynamics of acetogenesis without coupling to a methanogenic reaction, which would have yielded methane; that acetogenic bacteria are present in the methane-phase reactor and convert butyrate to acetate and hydrogen, which is rapidly converted to methane since no hydrogen is detected in the product gas from the methane-phase reactor; and that a good portion of the methane is derived from carbon dioxide reduction, because the methane concentration is much higher than 50 mol % in the product gas and butyrate is the main carbon source in the acid-phase effluent. The observations of high butyrate concentrations in the acid-phase effluent and hydrogen in the acid-phase gas are also in accord with evidence that high hydrogen partial pressures promote the formation of higher fatty acids (McInerney and Bryant, 1981). The high rates and loadings of acid-phase digestion would be expected to lead to rapid generation of reduced nicotinamide adenine dinucleotide, which reduces carbon dioxide to methane when coupled to a methanogenic reaction. This makes the transfer of hydrogen, which is often called interspecies hydrogen transfer, thermodynamically favorable. But since there are few or no methanogenic bacteria in the acid-phase reactor, the reducing power of reduced nicotineamide adenine dinucleotide is transferred through fermentative pathways to yield higher fatty acids and other products (lactate, ethanol). Thus, for glucose conversion by two-phase digestion, it seems reasonable to conclude that the acid-phase reactor contains fermentative bacteria as the dominant organisms, and the methane-phase reactor contains both acetogens and methanogens as dominant species. This probably also applies to other methane fermentations because essentially the same gas compositions are obtained from other two-phase digestion systems that are supplied with complex substrates such as municipal biosolids and industrial wastes.

V. COMMERCIAL DEVELOPMENT

A. METHANE FERMENTATION

Hundreds of projects have been carried out since the 1970s to develop improved anaerobic digestion processes for commercial use. A major goal of much of this effort has been to optimize digester designs and the operating conditions to efficiently treat various solid and liquid wastes so that discharge and disposal requirements are satisfied. Anaerobic digestion processes for municipal wastewaters, MSW, farm and agricultural wastes, the waste streams from food processors and beverage manufacturers, and wastes from other types of industrial facilities such as pulp and paper manufacturers have received special attention. Biogas recovery has usually been a minor factor in most of the process improvements, but methane (and carbon dioxide) yield and selectivity range from good to excellent. Methane production correlates with volatile solids reduction, so digester performance improvements generally produce more biogas. Biogas is always a valuable product of the digestion process. The carbon dioxide from many installations can be recovered and sold, and the digested solids are often marketable as fertilizer, animal bedding, and sometimes animal feed. The digested solids have higher nutrient values per unit mass than the undigested feed solids and are effectively slow-nitrogen-release fertilizers because the readily biodegradable material has already been gasified.

Typical anaerobic digestion facilities can contain hydrolysis units, digesters, gas cleanup and dehydration units, and liquid effluent treatment units, as illustrated in Fig. 12.7. Many improvements have been incorporated into the ancillary unit operations as well as the methane fermentation process itself.

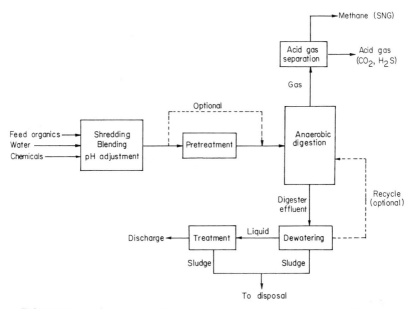

FIGURE 12.7   Schematic of methane production by anaerobic digestion of biomass.

The process improvements have included digester operation at high solids loadings; use of immobilized bacteria on solid supports; addition of materials such as activated carbon, flyash, enzymes, lactobacillus cultures, and growth factors to digesters; chemical, physical, and electrical (arc discharges) pretreatment of the influent substrate or posttreatment of the residual digested solids to increase biodegradability before recycling or forwardcycling; integration of anaerobic digestion with thermal gasification of ungasified residuals from the digester; and innovative digester designs such as plug-flow, fixed-film packed-bed, fixed-film fluidized-bed, upflow sludge blanket, and baffle-flow digesters that permit longer SRTs than HRTs. Most of the digesters based on these designs can be used with either high-rate or two-phase digestion. Integration of advanced digester designs with unit operations such as feed pretreatment for substrates that are resistant to enzyme-catalyzed hydrolysis and posttreatment of the residual digested solids have led to practical systems that maximize biogas production and minimize liquid effluent and residual solids disposal problems.

   In the early 1980s, there was a large increase in the number of commercial anaerobic digestion plants constructed in the United States, particularly advanced technology systems for industrial wastes (cf. Klass, 1983). Several of the new digesters were of the fixed-film or upflow sludge blanket types. Full-scale upflow sludge blanket units ranging in size from 13,000 to 66,000 kg COD/day were installed for the treatment of beer, potato starch, and potato cooking wastes, and a large downflow packed-bed system with a capacity of about 120,000 kg COD/day was installed for treatment of rum still bottoms. The COD reductions for these plants are about 80 to 85%.

In the late 1970s, four commercial and six pilot two-phase digestion plants were built and used for several types of high-COD industrial liquid wastes in Belgium and Germany (Ghosh *et al.*, 1982). The pilot plants had capacities ranging from 45 to 180 kg COD/day, and consisted of completely mixed, acid-phase digesters followed by upflow sludge blanket, methane-phase digesters. They were operated in the mesophilic temperature range and achieved COD reductions up to 90%. Three of the plants processed beet sugar wastes, 2 processed distillery wastes, and one processed citric acid wastes. The first full-scale, two-phase plant was built in 1980 in Belgium for the stabilization of liquid wastes generated during flax retting and had a capacity of 350 kg COD/day; 87% COD reductions were obtained. The gas production rate was about 4.0 vol (n)/vol-day, and the yield was about 0.40 m^3(n)/kg of COD added. Three other full-scale, two-phase plants were installed in Germany to stabilize wastes from beet sugar, starch-to-glucose, and potato chip factories. Their capacities were 15,000, 20,000, and 32,000 kg of COD/day, respectively. Since then, other two-phase anaerobic digestion facilities have been built, including several in the United States. One of the first, commercial, two-phase digestion facilities in the United States uses the ACIMET® process for treatment of municipal biosolids. It was installed in a 45-million L/day wastewater treatment plant in Woodridge, Illinois (DuPage Group, 1993). Steady-state operations were achieved over long periods of time and demonstrated 30% higher rates of biosolids conversion, the elimination of foaming problems, a significant reduction of sulfide in the biogas, and a 30% increase in methane compared to conventional anaerobic digestion. The process also consistently demonstrated a reduction in pathogen content in the effluent of less than 10,000 CFU (colony-forming units). Other systems using this process have been installed for treatment of municipal biosolids. Among the first two-phase commercial plants for anaerobic digestion of industrial wastes in the United States are facilities for digestion of industrial wastewaters from the manufacture of jam and jelly, corn-based ethanol, bakers' yeast, and beer (Sax and Lanting, 1991). These plants use upflow sludge blanket digesters for acid-phase digestion.

One of the interesting features of the two-phase digestion process is that it can be retrofitted to an existing anaerobic digestion plant to effectively double its treatment capacity at a considerably reduced capital cost compared to the cost of a grassroots plant of the same additional incremental capacity. The U.S. Environmental Protection Agency's projection of new anaerobic digestion capacity needed in the United States for municipal wastewater treatment in the next two decades is estimated to have a capital cost of $20 billion or more. The retrofitting of two-phase digestion capability to existing plants for the purpose of increasing plant capacity has the potential of reducing this cost by a very large amount.

North America and Europe will probably never make such widespread use of anaerobic digestion for methane production as China and India. There were 5.8 million digesters in China in 1978, most of them in Szechwan province (4.3 million) where they provided cooking fuel for some 20 million people (Smil, 1979). China's long-range plan at that time called for 70 million digesters by 1985. Similar usage patterns in the United States would require rather complete automation and unattended operation at low cost, factors that are not compatible for small-scale systems. But small-scale digestion systems are increasingly used in the United States, and commercial turn-key plants and package systems are available. The prime operating objective of most of the U.S. systems is waste stabilization and disposal. The demand for municipal anaerobic digestion capacity has increased almost exponentially with urban population growth. Many municipal wastewater treatment plants are now operating anaerobic digestion units to maximize biogas recovery for in-plant use (cf. Haug, Moore, and Harrison, 1995). Commercial use of anaerobic digestion on farms has shown modest growth in the United States (cf. Lusk, 1994). Although many farm-scale systems have been shut down, several manure-fueled, high-rate systems are in operation, and a few manure-fueled, plug-flow systems have been built for electric power generation. Farm-scale, anaerobic lagoon systems covered with gas-impermeable polymeric membranes for liquid and solid animal waste stabilization and biogas recovery have been operated for several years and have exhibited excellent performance with little or no evidence of groundwater pollution. The biogas is used on-site. The largest manure-to-methane plant in the United States was placed in operation in 1977 in Guyman, Oklahoma, on a large cattle feedlot for the production of pipeline-quality gas (39.7 MJ/m^3(n)) at a rate of 16 million m^3/year for transmission to Chicago (Meckert, 1978). CSTR digesters were operated under mesophilic conditions. Delivery of pipeline-quality gas began on April 1, 1978, to Natural Gas Pipeline Co.; the co-products were fertilizer and cattle feed. Because of operating problems, especially during the winter months, plant availability was eventually reduced. Gas delivery was then terminated because SNG could not be produced at a competitive price, and only the co-products were marketed. Because of fuel costs in Central Europe, the interest in farm-scale and community biogas plants for animal manures is much greater (Köberle, 1995). Approximately 470 plants were operating in the mid-1990s. Most of them are small or medium-sized, farm-scale plants that use 1 to 20 m^3/day of feedstock. Larger plants that use more than 20 m^3/day are located in Germany, Denmark, Italy, and The Netherlands. Some 200 plants, including nine large farm-scale plants, are operated in Germany. More than 100 different plant designs are used in Europe. Two of the primary farm-scale and community plant designs were developed in Germany: horizontal flowthrough digesters using steel tanks, often previously used as gasoline tanks, and vertical steel digesters.

The anaerobic digestion of the combustible fraction of MSW (refuse-derived fuel, RDF) was used commercially in the United States from the 1930s to the 1950s in several cities. The cities of Richmond, Indianapolis, and Marion in the state of Indiana found that the anaerobic digestion of garbage was satisfactory and a suitable disposal method (cf. Bloodgood, 1936; Backmeyer, 1947; Ross, 1954). Other U.S. cities (Lansing, Michigan, New York, and Washington, D.C.) also used anaerobic digestion for garbage disposal (cf. Wyllie, 1940; Taylor, 1941). The disposal of RDF by anaerobic digestion has since been terminated in almost all U.S. cities, although considerable research was carried out to develop modern, integrated RDF recovery-digestion systems. An example of this research is the effort made to apply modern high-rate digestion to RDF for combined disposal and methane production in the mid-1980s. A proof-of-concept plant was built and operated from 1978 to 1985 in Pompano Beach, Florida (cf. Isaacson et al., 1988; Mooij and Pfeffer, 1986). This plant was sized to process 90 t/day of RDF, but several operating and design problems precluded operation at design capacity. The plant was eventually shut down in 1985 and dismantled. A full-scale RDF digestion plant was constructed in Dinwiddie, Virginia, in 1994, operated for about 1 year, and then shut down because of financial problems. This plant was sized to accept 20,000 t/year of MSW at the plant gate and operated under mesophilic conditions with CSTR digesters. In contrast to the apparent lack of interest in the anaerobic digestion of MSW in the United States, 15 full-scale plants were in operation in Europe and 20 more were being planned or under construction in the mid-1990s (International Energy Agency, 1994).

LFG is the only MSW-derived biogas currently (late 1990s) produced in the United States on a commercial scale. Sanitary landfills for MSW disposal are batch analogues of high-solids, anaerobic digesters and produce biogas of about the same composition. LFG in the form of medium-heating value gas is recovered and either is upgraded to pipeline-quality gas by removal of carbon dioxide and small amounts of hydrogen sulfide and other minor contaminants or is used directly as fuel gas. Of the approximately 20,000 MSW landfills in the United States, including approximately 5000 active landfills that are in the process of being filled, LFG recovery systems are installed in about 140. The LFG is used for conversion to electricity via internal combustion engines or turbines, for upgrading to pipeline gas for injection into natural gas pipelines or distribution systems, and for direct use after minimal processing for boilers, industrial processes, and greenhouses (cf. Nichols, 1996).

The U.S. Environmental Protection Agency has estimated that LFG could be used to cost-effectively recover biogas from more than 700 U.S. landfills. The avoidance of methane emissions from these landfills would correspond to removal of 12 million cars from the total vehicle population, according to this agency. Until recently, LFG emissions to the atmosphere were not subject to federal regulation. This situation changed on March 12, 1996, with the promulgation of New Source Performance Standards and Emissions Guidelines for landfills under Title 1 of the Clean Air Act. A landfill must now install LFG control systems to meet these requirements if it satisfies these three criteria: It received waste after November 8, 1987, it has 2.5 million t or greater of permitted design capacity, and the emissions of nonmethane organic compounds are greater than 50 t/year. These regulations appear to have two opposing effects on LFG as a commercial energy resource. LFG recovery systems are now required on more of the existing landfills, which will therefore result in increased LFG usage. But the additional cost of control systems for new, large-scale landfills will tend to reduce the number of systems constructed because of negative effects on landfill profitability in some areas, and hence LFG recovery in some of these areas might be expected to decrease.

Total biogas production from commercial usage of anaerobic digestion in the United States, including biogas from waste stabilization and wastewater treatment, is a very small amount of methane when compared with U.S. natural gas consumption (Chapter 2). It is not possible to supply more than a small fraction of natural gas consumption with biogas unless a major, long-term effort is made to expand anaerobic digestion capacity. Such a program will not be undertaken until there are strong signs that natural gas depletion is starting to cause nationwide shortages (Chapter 1). The effort made in the United States in the 1970s and 1980s to develop large-scale supplies of SNG from dedicated energy crops and biogas is exemplified by the Ocean Energy and Food Farm Project undertaken jointly by the American Gas Association, the Gas Research Institute, the U.S. Department of Energy, and the U.S. Navy. The availability of the projected amounts of SNG from giant brown kelp, other dedicated biomass energy crops, and possibly other feedstocks including coal and other sources of carbon oxides would make it possible for the U.S. natural gas industry to continue to utilize its vast transmission and distribution systems without modification to deliver methane to its customers when natural gas depletion begins. Large-scale transmission in existing pipelines is not considered to be practical with hydrogen instead of methane because of large-scale leakage problems, the associated hazards, and the costs of modification.

Estimates of the amounts of anchored giant brown kelp (Fig. 12.8) harvested off the California Coast indicated an energy yield of 7 quad/year (Szetela et al., 1974), or about 32% of natural gas consumption (21.7 quad) at that time.

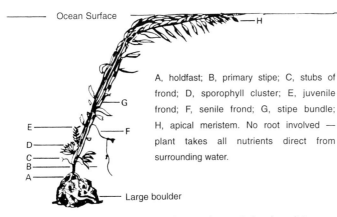

FIGURE 12.8   Diagram of young adult giant brown kelp plant (*Macrocystis pyrifera*). From Leese (1976).

Other estimates indicated that the anaerobic digestion of anchored giant brown kelp grown in 24,500-ha (100,000-ac) kelp farms could produce 0.42 to 1.27 billion m^3/year (15 to 45 billion ft^3/year) of methane at a cost of $2.85 to $8.54/MJ ($3 to $9/million Btu) (Wilcox and Leese, 1976). Another estimate indicated that a square ocean area 266 to 539 km (165 to 335 miles) on each edge could supply all U.S. natural gas consumption (540 billion cubic meters) at a cost of $2.85 to $5.69/GJ (Sharer and Flowers, 1979). Still other estimates indicated that 10% of natural gas demand could be supplied at a kelp yield of 112 dry t/ha-year (50 dry ton/ac-year) from an ocean area of 14,200 km^2 (5500 mi^2) (Bryce, 1978). The idea of such a farm was considered to be manageable when considered in terms of the vast expanse of open ocean area available (Figs. 12.9 and 12.10). In addition, the fertilizing nutrients required to sustain optimal growth rates are available in the ocean starting at depths of about 90 m and can be upwelled to the growing kelp at relatively low cost. The research schedule called for construction of a commercial 24,500-ha demonstration farm by 1992. Extensive work was done on the design and installation of a small, floating, modular farm platform for the growth of anchored kelp (*cf.* Barcelona *et al.*, 1979; North, Gerard, and Kuwabara, 1981) and on the anaerobic digestion of kelp (*cf.* Klass, Ghosh, and Chynoweth, 1979). Unfortunately, winter storms destroyed the platform and the project was terminated. The technical feasibility of the concept of large-scale SNG production from virgin biomass was partially supported by this work. This is an example of technology that is "on the shelf" and available for commercial development as natural gas shortages develop with the onset of depletion.

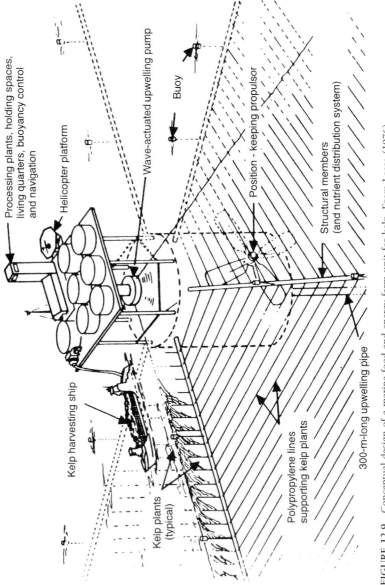

Processing plants, holding spaces,
living quarters, buoyancy control
and navigation

Helicopter platform

Wave-actuated upwelling pump

Buoy

Position - keeping propulsor

Structural members
(and nutrient distribution system)

Kelp harvesting ship

Kelp plants
(typical)

Polypropylene lines
supporting kelp plants

300-m-long upwelling pipe

FIGURE 12.9   Conceptual design of a marine food and energy farm unit for kelp. From Leese (1976).

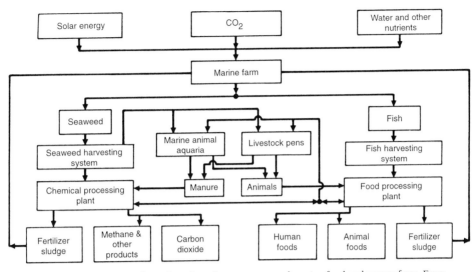

FIGURE 12.10 Feedstocks and products from a conceptual marine food and energy farm. From Leese (1976).

B. MICROBIAL HYDROGEN

Unlike methane fermentation, hydrogen fermentation is characterized by low yields and selectivities. The selectivity for hydrogen is insufficient to justify the development of large-scale production methods unless major improvements can be made. Genetic manipulation would probably be the preferred approach to incorporate the characteristics into the organisms that are needed. But as noted here, the complete conversion of glucose to carbon dioxide and hydrogen in dark fermentations is not an efficient process. The overall difficulty with the production of molecular hydrogen as a primary product from biomass by use of fermentative microorganisms is that the yield of hydrogen is too low. The cabability of performing these processes at near-ambient conditions appears attractive compared to thermal gasification processes. But unless hydrogen yields can be significantly increased, the technology has little practical value. An alternative that may be suitable for practical applications is a process that produces molecular hydrogen along with a high-value organic derivative, both in reasonably high yields, so that the recovery of each is justified. The same rationale also applies to the use of cell-free enzyme preparations from fermentative organisms as catalysts to perform the equivalent of thermochemical, catalytic conversion of biomass components to hydrogen. Catalytic activity would of course have to be sustained.

The research discussed here supports the view that the biophotolysis of water can be performed with live, photosynthetic microorganisms that contain both Photosystems I and II or either Photosystem I or II. Although the technical feasibility of hydrogen production has been established, several technical barriers tend to preclude the development of practical processes. The hydrogenase system is highly sensitive to oxygen and requires anaerobic conditions to maintain activity. This is not easily achieved when oxygen is a co-product. Intermittant or continuous replacement of inert atmospheres or the use of oxygen-scavenging agents to keep the partial pressure of oxygen as low as possible would incur additional processing cost. When hydrogen and oxygen are produced simultaneously in the same reactor, the additional costs of gas separation and dealing with an explosive mixture, which under ideal conditions corresponds to the stoichiometric mixture of these gases for water formation, must also be considered. The physical separation of hydrogen and oxygen production in staged reactors or the separation of hydrogen and oxygen production with time in the same reactor might improve the process, but a practical biophotolysis system has yet to be demonstrated. The efficiency of photosynthetic hydrogen formation is low. The upper limit has been estimated to be 3% for well-controlled systems. This limits the capital cost of useful systems to low-cost materials and designs. For the present, microbial hydrogen production is in the laboratory stage of development. Continued research may provide the knowledge needed for practical applications of the technology, especially if it can be developed to the point where biological systems can emulate or at least approach pure photolysis.

Synthetic chloroplasts and possibly cell-free preparations that can be used over long periods of time as catalysts for the photolysis of water without loss in activity appear to offer the best opportunity for microbial or pseudo-microbial hydrogen production. Many problems remain to be solved before this technology is perfected. It does seem to have the most promise when compared with microbial methods, because it has the potential of providing a solar method of splitting water into its components by the equivalent of pure photolysis. Note that numerous heterogeneous and some homogeneous catalysts, composed of a variety of chemical compounds, many of which are stable and commonly available, have been researched over the years for both water photolysis and thermal water splitting. Some of these catalysts are effective, though none has yet been demonstrated in a practical system that can supply low-cost hydrogen at a reasonable scale. The availability of low-cost hydrogen from a strictly catalytic water-splitting technology would of course make an infinite amount of hydrogen available from a common resource for a host of energy and other applications.

Perhaps the greatest potential of catalytically active synthetic chloroplasts and cell-free preparations is the accompanying knowledge that would evolve to permit development of catalysts that facilitate extracellular, photosynthetic conversion of ambient carbon dioxide to low-cost hydrocarbons under comparatively mild, controlled conditions. The availability of such catalysts for which activity can be sustained could conceivably eliminate all future shortages of high-energy-density liquid and gaseous hydrocarbon fuels and feedstocks, make hydrocarbon fuels renewable at short recycle times in the same context that biomass energy is renewable, and reduce fossil fuel consumption. There does not appear to be any insurmountable technical barrier that precludes such developments.

REFERENCES

Anonymoous (1971). In "Biological Waste Treatment," (R. P. Canale, ed.), p. 85. Interscience Publishers, New York.

Arnon, D., Mitsui, A., and Paneque, A. (1961). Science 134 (3488), 1425.

Backmeyer, D. P. (1947). Sewage Works 19 1, 48.

Balch, W. E., Fox, G. E., Magrum, L. J., Woese, C. R., and Wolfe, R. S. (1979). Microbiol. Rev. 43, 260.

Barcelona, M., Gerard, V., Kuwabara, J., Lieberman, S., Manley, S., and North, W. J. (1979). In "3rd Annual Biomass Energy Systems Conference Proceedings, The National Biomass Program, June 5-7, 1979," p. 33, SERI/TP-33-285. National Renewable Energy Laboratory, Golden, CO.

Baresi, L., Mah, R. A., Ward, D. M., and Kaplan, I. R. (1978). Appl. and Env. Microbiol. 36 (1), 186.

Barker, H. A. (1936). Arch. Mikrobiol. 7, 404.

Barker, H. A. (1940). Leeuwenhoek 6, 201.

Barker, H. A. (1941). J. Biol. Chem. 137, 153.

Barker, H. A. (1943). Proc. Nat. Acad. Sci. U.S. 29, 184.

Benemann, J. R. (1978). In "Energy from Biomass and Wastes," (D. L. Klass and W. W. Waterman, eds.), p. 557. Institte of Gas Technology, Chicago.

Benemann, J. R., and Weare, N. M. (1974). Science 184 (4133), 174.

Blaat, M. (1994). Antonie van Leeuwenhoek 66, 286.

Bloodgood, D. E. (1936). Sewage Works 8, 3.

Boone, D. R., and Bryant, M. P. (1980). Appl. Environ. Microbiol. 40, 626.

Borchardt, J. A. (1967). Pro. 3rd Int. Conf. Water Pollut. Res. 1, 309.

Borchardt, J. A. (1971). Adv. Chem. Ser. 105, 108.

Bryant, M. P., Wolin, E. A., Wolin, M. J., and Wolfe, R. S. (1967). Arch. Mikrobiol. 59, 20.

Bryce, A. J. (1978). In "Energy from Biomass and Wastes," (D. L. Klass and W. W. Waterman, eds.), p. 353. Institute of Gas Technology, Chicago.

Bult, C. J. et al. (1996). Science 273 (5278), 1058.

Buswell, A. M., and Sollo, F. W. (1948). J. Am. Chem. Soc. 70, 1778.

Buswell, A. M., Fina, L., Mueller, H., and Yahiro, A. (1951). J. Am. Chem. Soc. 73, 1809.

Calvin, M. (1978). "Fermentation and Hydrocarbons." Presented at II Latin-American Botanical Congress, Brasilia, Brazil, January 22-27, Lawrence Berkeley Laboratory Paper 7270. University of California, Berkeley, CA.

Calvin, M. (1979). Energy 4, 851.

Cappenberg, T. E. and Prins, R. A. (1974). Antonie van Leeuwenhoek J. Microbiol. Serol. 40, 457.

Cohen, A., Breure, A. M., van Andel, J. G., van Deursen, A. (1980). Water Res. 14, 1439.

DiMarco, A. A., Bobik, T. A., and Wolfe, R. S. (1990). Annu. Rev. Biochem. 59, 355.

DuPage Group (1993). "News Release: DuPage Group Announces Commercial Availability of the ACIMET(R) Process for the Treatment of Municipal Sludge." Institute of Gas Technology, Chicago.

Finney, C. D., and Evans II, R. S. (1975). *Science* **190** (4219), 1088.

Fischer, F, Lieske, R., and Winzer, K. (1931). *Brennstoffchemie* **12**, 193.

Fischer, F, Lieske, R., and Winzer, K. (1931). *Biochem. Z.* **24**, 52.

Gaffron, H., and Rubin, J. (1942). *J. Gen. Physiol.* **26**, 219.

Gest, H., and Kamen, M. D. (1949). *J. Bacteriol.* **58**, 239; *Science* **109** (2840), 558.

Ghosh, S., and Henry, M. P. (1981). Paper presented at the 36th Annual Purdue Industrial Waste Conference. Lafayette, IN, May 12-14.

Ghosh, S., and Klass, D. L. (1978). *Process Biochem.* **13** (4), 15.

Ghosh, S., Conrad, J. R., and Klass, D. L. (1975). *J. Water Pollut. Control Fed.* **45**, 591.

Ghosh, S., Ombregt, J. P., DeProost, V. H., and Pipyn, P. (1982). *In* "Energy from Biomass and Wastes VI," (D. L. Klass, ed.), p. 323. Institute of Gas Technology, Chicago.

Gray, C. T., and Gest, H. (1965). *Science* **148** (3667), 186.

Greenbaum, E. (1981). *In* "Energy from Biomass and Wastes V," (D. L. Klass and J. W. Weatherly, III, eds.), p. 480. Institute of Gas Technology, Chicago.

Greenbaum, E., Lee, J. W., Tevault, C. V., Blankinship, S. L., and Mets, L. J. (1995). *Nature* **376**, 438.

Hammer, M. S., and Borchardt, J. A. (1969). *Proc. Am. Soc. Civ. Eng.* **95**, 907.

Hashimoto, A. G. (1982). *Biotechnol. Bioeng.* **24**, 9.

Haug, R. T., Moore, G. L., and Harrison, D. S. (1995). *In* "Second Biomass Conference of the Americas: Energy, Environment, Agriculture, and Industry," p. 734, NREL/CP-200-8098, DE95009230. National Renewable Energy Laboratory, Golden, CO.

Hoppe-Seyler, F. (1886). *Z. Physiol. Chem.* **10**, 201.

Hoppe-Seyler, F. (1989). *Z. Physiol. Chem.* **11**, 561.

International Energy Agency (1994). "Biogas from Municipal Solid Waste," IEA Bioenergy Agreement, Task XI: Conversion of MSW Feedstock to Energy, Activity 4: Anaerobic Digestion of MSW, 1994. Minister of Energy/Danish Energy Agency, Copenhagen, Denmark, July.

Isaacson, R., Pfeffer, J., Mooij, P., Geselbracht, J. (1988). *In* "Energy from Biomass and Wastes XI," (D. L. Klass, ed.), p. 1123. Institute of Gas Technology, Chicago.

Jerris, J. S., and McCarty, P. L. (1965). *J. Water Pollut. Control Fed.* **37**, 178.

Jones, W. J., Leigh, J. A., Mayer, F., Woese, C. R., and Wolfe, R. S. (1983). *Arch. Microbiol.* **136**, 254.

Kalia, V. C., and Joshi, A. P. (1995). *Bioresource Technology* **53** (2), 165.

Klass, D. L. (1974). *Chem. Technol.* **3**, 689.

Klass, D. L. (1980). *In* "Proceedings Bio-Energy '80," p. 143. World Congress, Atlanta, GA, April 21-24.

Klass, D. L. (1983). *In* "Energy from Biomass and Wastes VII," (D. L. Klass and H. H. Elliott, eds.), p. 1. Institute of Gas Technology, Chicago.

Klass, D. L. (1984). *Science* **223** (4640), 1021.

Klass, D. L., and Ghosh, S. (1977). *In* "Clean Fuels from Biomass and Wastes," (W. W. Waterman, ed.), p. 323. Institute of Gas Technology, Chicago.

Klass, D. L., Ghosh, S., and Chynoweth, D. P. (1979). *Proc. Biochem.* **14** (4), 18.

Kluyver, A. J., and Schnellen, C. G. T. P. (1947). *Arch. Biochem.* **14**, 57.

Köberle, E. (1995). *In* "Second Biomass Conference of the Americas: Energy, Environment, Agriculture, and Industry" p. 753, NREL/CP-200-8098, DE95009230. National Renewable Energy Laboratory, Golden, CO.

Laigret, J. C. R. (1945). *Acad. Sci.* **221**, 359.

Lee, J. W., Tevault, C. V., Owens, T. G., and Greenbaum, E. (1996). *Science* **273** (5273), 364.

Leese, T. M. (1976). *In* "Clean Fuels from Biomass, Sewage, Urban Refuse, and Agricultural Wastes," (F. Ekman, ed.), p. 253. Institute of Gas Technology, Chicago.

493

Lusk, P. (1994). "Methane Recovery from Animal Manures: A Current Opportunities Casebook," NREL/TP-421-7577, DE95004003. National Renewable Energy Laboratory, Golden, CO, December.

Macario, A. J. L., and de Macario, E. C. (1982). *Immunology Today* **3** (10), 279.

McCarty, P. L. (1982). In "Anaerobic Digestion 1981," p. 3, Proceedings of the 2nd. Symposium on Anaerobic Digestion. Travemunde, Germany, September 6-11.

McInerney, M. J., and Bryant, M. P. (1981). In "Conversion Processes for Energy and Fuels," (S. Sofer and O. R. Zaborsky, eds.), p. 277. Plenum, New York.

Meckert, G. W., Jr. (1978). In "Energy from Biomass and Wastes," (D. L. Klass and W. W. Waterman, eds.), p. 431. Institute of Gas Technology, Chicago.

Mooij, H. P., and Pfeffer, J. T. (1986). "RefCoM - Proof-of-Concept Experiment," Final Report 1978-1985, GRI-86/0134. Gas Research Institute, Chicago.

Nichols, M. (1996). *Waste Age* **27** (8), 89.

North, W. J., Gerard, V. A., and Kuwabara, J. S. (1981). In "Biomass as a Nonfossil Fuel Source," (D. L. Klass, ed.), ACS Symposium Series 144, p. 77. American Chemical Society, Washington, D.C.

Omelianski, W. (1904). *Centr. Bakt.* **11**, 369.

Pine, M. J., and Barker, H. A. (1954). In "Proceedings of the 54th General Meeting, Society of American Bacteriologists," p. 98.

Pine, M. J., and Barker, H. A. (1956). *Appl. Microbiol.* **14**, 368.

Pohland, F. G., and Ghosh, S. (1971). *Environ. Lett.* **1**, 255.

Popoff, L. (1875). *Chem. Zentralbl.* **6**, 470.

Rimkus, R. R., Ryan, J. M., and Michuda, A. (1979). Paper presented at the American Society of Civil Engineers Conference. Atlanta, GA, October 22-29.

Roscoe, H. E., and Schorlemmer, C. (1878). "A Treatise on Chemistry." D. Appelton, New York.

Ross, W. E. (1954). *Sewage Ind. Wastes* **26** 2, 140.

Sax, R. I., and Lanting, J. (1991). In "Energy from Biomass and Wastes XIV," (D. L Klass, ed.), p. 891. Institute of Gas Technology, Chicago.

Schaumburg, F. D., and Kirsch, E. J. (1966). *Appl. Microbiol.* **14**, 761.

Schnellen, C. G. T. P. (1947). "Dissertation, Tech." University of Delft, The Netherlands.

Sharer, J. C., and Flowers, A. (1979). *Grid* **2**, (1).

Smil, V. (1979). *Dev. Dig.* **XVII** (3), 25.

Smith, P. H., and Mah, R. A. (1965). *Appl. Microbiol.* **14**, 368.

Söhngen, N. L. (1906). *Centr. Bakt.* **15**, 513.

Söhngen, N. L. (1910). *Rec. Trav. Chim. Pays-Bas.* **29**, 238.

Stadtman, T. C., and Barker, H. A. (1949). *Arch. Biochem.* **21**, 256.

Stadtman, T. C., and Barker, H. A. (1951). *J. Bacteriol.* **61**, 81; *ibid, 62*, 269.

Stafford, D. A. (1982). *Biomass* **2**, 43.

Stanier, R. Y., Ingraham, J. L., Wheelis, M. L., and Painter, P. R. (1986). "The Microbial World," 5th Ed. Prentice-Hall, Englewood Cliffs, NJ.

Stephenson, M., and Strickland, L. H. (1931). *Biochem. J.* **25**, 205.

Stephenson, M., and Strickland, L. H. (1933). *Biochem. J.* **27**, 1517.

Szetela, E. J., Krascella, N. L., Blecher, W. A., and Christopher, G. L. (1974). In "Proceedings of Division of Fuel Chemistry," 168th National ACS Meeting, Atlantic City, NJ, September 8-13, 1974, p. 172. American Chemical Society, Washington, D.C.

Taylor, H. (1941). *Eng. News Rec.* **127**, 441.

Thauer, R. K., Jungermann, K., and Decker, K. (1977). *Bacteriol. Rev.* **41**, 100.

Vavilin, V. A., Rytow, S. V., and Lokshina, L. Ya. (1995). *Bioresource Technology* **54** (2), 171.

Wilcox, H. A., and Leese, T. M. (1976). *Hydrocarbon Processing,* **86**, April.

Woese, C. R., and Fox, G. E. (1977). *Proc. Natl. Acad. Sci. USA* **74**, 5088.

Wood, H. G., Ragsdale, S. W., and Pezacka, E. (1986). *Trends Biochem. Sci.* **11**, 14.

Woodward, J., Mattingly, S. M., Danson, M. J., Hough, D. W., Ward, N., and Adams, M. W. W. (1996). *Nature Biotechnology* **14**, 879.

Wyllie, G. F. (1940). *Sewage Works J.* **12** 4, 760.

Zehnder, A. J. B., Ingvorsen, K., and Marti, T. (1982). *In* "Anaerobic Digestion 1981," (D. E. Hughes *et al.*, eds.), p. 45. Elsevier, Amsterdam.

Zeikus, J. G. (1977). *Bacteriol. Rev.* **41**, 514.

Zeikus, J. G., Kerby, R., and Krzycki, J. A. (1985). *Science* **227** (4691), 1167.

CHAPTER 13

Organic Commodity Chemicals from Biomass

I. INTRODUCTION

Biomass is utilized worldwide as a source of many naturally occurring and some synthetic specialty chemicals and cellulosic and starchy polymers. High-value, low-volume products, including many flavorings, drugs, fragrances, dyes, oils, waxes, tannins, resins, gums, rubbers, pesticides, and specialty polymers, are commercially extracted from or produced by conversion of biomass feedstocks. However, biomass conversion to commodity chemicals, which includes the vast majority of commercial organic chemicals, polymers, and plastics, is used to only a limited extent. This was not the case up to the early 1900s. Chars, methanol, acetic acid, acetone, and several pyroligneous chemicals were manufactured by pyrolysis of hardwoods (Chapter 8). The naval stores industry relied upon softwoods as sources of turpentines, terpenes, rosins, pitches, and tars (Chapter 10). The fermentation of sugars and starches supplied large amounts of ethanol, acetone, butanol, and other organic chemicals (Chapter 11).

Biomass was the primary source of organic chemicals up to the mid- to late 1800s when the fossil fuel era began, and was then gradually displaced by fossil raw materials as the preferred feedstock for most organic commodities. Aromatic chemicals began to be manufactured in commercial quantities as a by-product of coal coking and pyrolysis processes in the late 1800s. The production of liquid hydrocarbon fuels and organic chemicals by the destructive hydrogenation of coal (Bergius process) began in Germany during World War I. The petrochemical industry started in 1917 when propylene in cracked refinery streams was used to manufacture isopropyl alcohol by direct hydration. In the 1920s, aliphatic chemicals in the form of alcohols, glycols, aldehydes, ketones, chlorinated hydrocarbons, esters, and ethers were beginning to be produced in quantity by synthesis from hydrocarbon feedstocks supplied by the petroleum and natural gas industries. Methanol from synthesis gas replaced wood alcohol from wood pyrolysis in the 1920s and early 1930s. And commercial use of Fischer–Tropsch chemistry for the catalytic conversion of synthesis gas to aliphatic hydrocarbon fuels and chemicals began in 1936, also in Germany. The petroleum industry then entered the field of aromatics production on a large scale because of the unprecedented demand for toluene needed for the manufacture of trinitrotoluene at the outbreak of World War II in 1939. In the 1940s and 1950s, the BTX fraction of catalytic reformate—benzene, toluene, and xylenes—derived from naphthas in refineries became the largest commercial source of aromatic chemicals.

The accompanying development of a wide variety of thermal conversion processes for the manufacture of commodity-type organic chemicals resulted in the displacement of processes that used biomass feedstocks for the manufacture of the same chemicals. The reasons why this occurred are complex. An obvious one is that the petroleum and chemical industries had invested heavily in research to develop highly efficient thermal conversion processes for hydrocarbon feedstocks. These processes afforded the desired products in high yields and selectivities at low costs. This effort resulted in the need for only a few building blocks—synthesis gas, the light olefins ethylene, propylene, and butadiene, and BTX—to manufacture about 90% of the synthetic organic chemicals that are marketed today. The building blocks are readily obtained at low cost from petroleum, natural gas, and in some cases coal. Construction of world-scale plants in large petrochemical and chemical facilities became a reality with the widespread installation of long-distance pipelines for transmission of petroleum and natural gas.

It was not possible to justify the manufacture of most commodity-type organic chemicals from biomass after World War II, even though technically, all of them can be manufactured from biomass feedstocks. But the chemical industry is expected to slowly revert to biomass as petroleum and natural gas costs increase. Biomass is also expected to continue to be the preferred feedstock where its structural characteristics either facilitate certain processes or are desired in the end product. Examples include some chemicals in which oxygenated carbon atoms or functional groups that are present in biomass are retained. Biomass is also preferred and is sometimes required as feedstock for the manufacture of synthetic chemicals that are used in foodstuffs.

Environmental issues and regulations are beginning to adversely affect the manufacture of chemicals from fossil feedstocks. On an annual basis, the U.S. chemical industry produces almost 2 billion tonnes of waste, most of which is wastewater, and generates about 90% of all U.S. industrial hazardous wastes and more than 40% of all U.S. industrial toxic wastes. The 1990 Clean Air Act Amendments set standards for industrial sources of 148 chemical pollutants that are required to be met by 2003 in the United States. Additional regulations are expected that will expand the existing requirements and include more of the smaller companies. The federal environmental regulations will continue to add significantly to chemical production costs well beyond 2003, particularly to the costs of organic chemicals from fossil feedstocks. Environmentally benign "green feedstocks" have started to attract investors. Advanced biomass conversion processes, both thermal and microbial, that provide an opportunity to supply commodity chemicals at costs that are potentially competitive with the costs of the same chemicals from fossil feedstocks are being developed or already exist. It appears that it will be only a matter of time before petroleum and natural gas prices start to increase because of localized shortages. The effort to develop energy-efficient, high-yielding biomass conversion processes indicates that the technology will be perfected and available when needed. In some countries, an organic chemicals industry based on renewable biomass feedstocks has already been established.

In this chapter, organic chemicals in the commodity category and the potential of biomass to replace fossil feedstocks for the manufacture of these chemicals are discussed. Some of the low-volume specialty chemicals and new products that are or can be manufactured from biomass are also examined.

II. COMMERCIAL ORGANIC CHEMICALS AND PRODUCTS

A. INDUSTRY CHARACTERISTICS

The U. S. chemical industry is the world's largest producer of chemicals (American Chemical Society *et al.,* 1996). In 1995, total shipments were valued at $367 billion, or about 24% of the worldwide market, which was valued at $1.3 trillion. Japan, Germany, and France ranked next in total production. At that time, the United States was second in chemical exports, about $60.5 billion, or 14% of the total export market. The U.S. chemical industry had the largest trade surplus ($20.4 billion) of all U.S. industry sectors and was the largest manufacturing sector in terms of value added. Overall, the chemical industry is the third largest manufacturing sector in the United States. It represents approximately 10% of all U.S. manufacturing. In the mid-1990s, the industry employed about 850,000 people.

The U.S. chemical industry is a large user of energy in the form of feedstock and process energy. It used about 5.1 quad of energy in 1991, approximately one-half of which was attributed to feedstock consumption; the remainder was attributed to fuel and power needs (U.S. Dept. of Energy, 1995). The overall efficiency of energy use by the chemical industry has improved by 51% since 1974. This is due to development of improved technology and energy management and a general shift toward the production of less energy-intensive products. Simultaneously, the proportion of energy used for feedstocks in the chemical industry has grown from 39% in 1970 to about 50%. Much of this increase is attributed to the growth in the use of polymeric materials. Organic chemicals; plastics, resins, and rubbers; fertilizers; and inorganic chemicals consume about 50, 15, 12%, and 10% of the total energy used, respectively. According to the Synthetic Organic Chemical Manufacturers Association (SOCMA), which represents the U.S. batch and custom chemical industry, the synthetic organic chemicals industry produces 95% of the 50,000 chemicals manufactured in the United States.

It is obvious that the chemical industry is a major economic force in the United States, and that the organic chemical and organic product sectors are dominant components.

B. U.S. Production of Organic Chemicals

One hundred major organic commodity chemicals that are manufactured in the United States are listed in Table 13.1. This listing includes 32 of the organic chemicals in the American Chemical Society's top-50 tabulation (the remaining 18 chemicals in the ACS listing are inorganic chemicals) and fermentation ethanol, excluding alcoholic beverage production. Many of the remaining organic chemicals in Table 13.1 are in the top 100 when ranked by production, but note that several others are manufactured for which production data are not available. The percent annual change in production and the estimated process energy and feedstock energy consumed for many of the chemicals, and the estimated market value for all 100 chemicals are included in this table. The feedstock usually employed in the manufacture of each chemical is also listed. Before these data are discussed, a few general comments are in order.

Traditional synthesis processes are used to manufacture almost all the chemicals listed in Table 13.1. And although it is not apparent from the tabulated data, most of these processes utilize homogeneous or heterogeneous catalysis

TABLE 13.1 One Hundred Major Organic Chemicals Manufactured in the United States Ranked by Production[a]

No.	Product	Annual production (10^9 kg)	Average percent annual change 1994–95	Average percent annual change 1985–95	Usual organic feedstock	Energy consumed per product mass Process (MJ/kg)	Energy consumed per product mass Feed (MJ/kg)	Average market Price ($/kg)	Average market Annual value ($10^9)
1.	Ethylene	21.30	5	5	Natural gas liquids	23.3	58.1	0.55	11.74
2.	Propylene	11.65	7	6	Natural gas liquids	16.3	56.8	0.42	4.89
3.	MTBE	7.99	29	25	C_4H_9/methanol	13.0	50.5	0.29	2.32
4.	Ethylene dichloride	7.83	3	4	C_2H_4	4.0	42.2	0.37	2.90
5.	Benzene	7.24	5	5	BTX	2.3	81.2	0.33	2.39
6.	Urea	7.07	−2	2	$CO_2/(NH_3)$	2.8	30.6	0.20	1.41
7.	Vinyl chloride	6.79	8	5	C_2H_4	17.4	53.4	0.45	3.06
8.	Ethylbenzene	6.19	27	6	C_2H_4/benzene	2.3	66.5	0.56	3.47
9.	Styrene	5.16	1	4	Ethylbenzene	12.8	74.3	0.64	3.30
10.	Methanol, synthetic	5.12	−7	8	$CO/(H_2)$	11.6	25.8	0.21	1.08
11.	Mixed xylenes	4.25	3	6	BTX	10.2	43.0	0.31	1.32
12.	Ethanol, fermentation	3.90		17	Corn	18.8	31.0	0.42	1.64
13.	Formaldehyde (37 wt %)	3.68	−1	4	Methanol	0.0	43.9	0.24	0.88
14.	Terephthalic acid[b]	3.61	5	2	p-Xylene	12.8	57.8	0.88	3.18
15.	Ethylene oxide	3.46	5	3	C_2H_4	0.4	68.1	0.99	3.43
16.	Toluene[c]	3.05	0	3	BTX	10.2	43.0	0.31	0.95
17.	p-Xylene	2.88	1	3	BTX	22.3	59.1	0.46	1.32
18.	Cumene	2.55	8	5	C_6H_6/benzene	2.3	66.2	0.51	1.30
19.	Ethylene glycol	2.37	−14	2	Ethylene oxide	13.9	49.6	0.37	0.88
20.	Acetic acid, synthetic	2.12	18	5	Methanol/(CO)	10.7	27.7	0.79	1.67
21.	Phenol, synthetic[d]	1.89	6	4	Cumene	15.1	99.3	0.90	1.70
22.	Propylene oxide	1.81	8	5	Propylene	16.3	58.4	1.41	2.55

(continues)

TABLE 13.1 (Continued)

No.	Product	Annual production (10⁹ kg)	Average percent annual change		Usual organic feedstock	Energy consumed per product mass		Average market	
			1994–95	1985–95		Process (MJ/kg)	Feed (MJ/kg)	Price ($/kg)	Annual value ($10⁹)
23.	1,3-Butadiene[c]	1.67	9	5	Butanes/enes	51.1	58.4	0.49	0.82
24.	Carbon black	1.51	2	3	Residual oil	27.9	93.0	0.64	0.97
25.	Isobutylene	1.47	5	12	Butanes/enes	9.3	85.3	0.68	1.00
26.	Acrylonitrile	1.45	6	3	Propylene	6.0	78.8	1.17	1.70
27.	Vinyl acetate	1.31	-5	3	C_2H_4/HOAc	15.6	59.2	0.97	1.27
28.	Acetone	1.25	4	4	Cumene	15.1	99.3	0.86	1.08
29.	Butyraldehyde	1.22	22	8	Propylene/(CO)			0.95	1.16
30.	Cyclohexane	0.969	9	3	Benzene	2.3	58.1	0.44	0.43
31.	Adipic acid	0.816	0	2	Cyclohexane	29.1	84.0	1.62	1.32
32.	Nitrobenzene	0.748	15	6	Benzene			0.73	0.55
33.	Bisphenol A	0.736	-5	6	Acetone/phenol			2.07	1.52
34.	Caprolactam	0.714	4	4	Cyclohexanone	7.9	54.5	2.05	1.46
35.	Acrylic acid	0.698	7		Propylene			1.92	1.34
36.	n-Butanol	0.677	1	8	Propylene/(CO)			1.10	1.49
37.	Isopropyl alcohol	0.646	-2	1	Propylene			0.71	0.46
38.	Aniline	0.631	10	7	Nitrobenzene			1.08	0.66
39.	Methyl methacrylate	0.622	4	5	Acetone/(HCN)			1.32	0.82
40.	Cyclohexanone	0.501			Phenol			1.61	0.81
41.	Methyl chloride	0.483	7	10	Methane			0.85	0.41
42.	o-Xylene	0.465	12	4	BTX			0.42	0.20
43.	Propylene glycol	0.462	9	7	Propylene oxide			1.43	0.66
44.	Phthalic anhydride	0.451	4	2	Naphthalene			0.84	0.38

45.	Acetone cyanohydrin	0.410			Acetone/(HCN)			0.77	0.32
46.	Toluene diisocyanates	0.395	2.7		Toluene			2.09	0.83
47.	Dodecylbenzene	0.386			C_6H_6/dodecene			1.25	0.48
48.	Ethanol amines^f	0.369	8	4	Ethylene oxide			1.26	0.46
49.	Diethylene glycol	0.354	2		Ethylene oxide			0.71	0.25
50.	Carbon tetrachloride	0.344			Methane			0.79	0.27
51.	2-Ethyl-1-hexanol	0.337	1	3	C_3H_6/CO or RCHO			1.23	0.42
52.	Ethanol, synthetic	0.284	−3	0	Ethylene	4.7	31.4	0.42	0.12
53.	Isoprene	0.281			HCs/turpentines			0.68	0.19
54.	1,4-Butanediol	0.266			C_3H_6/CO or THF			1.24	0.33
55.	Methyl ethyl ketone	0.264	−3	1	2-Butanol			1.01	0.27
56.	Ligninsulfonic acid salt	0.258			Sulfite liquor			0.36	0.09
57.	Chloroform	0.256			Methane			0.87	0.22
58.	Maleic anhydride	0.251	16	3	Benzene			0.93	0.23
59.	C_{12}-benzenesulfonate Na	0.217			C_6H_6/dodecene			1.34	0.29
60.	1-Butene	0.209	4.5		Ethylene			0.57	0.12
61.	$C_{3,12}$-alkylphenols	0.204			Phenol/olefin			1.83	0.37
62.	Glycerol	0.190	2		C_3H_6 or fats/oils			2.03	0.39
63.	Methylene chloride	0.183			Methane			0.95	0.17
64.	Toluene-2,4-diamine	0.175			Toulene			1.98	0.35
65.	Methyl chloroform	0.152			Vinyl chloride			1.50	0.23
66.	Tallow acids, Na salt	0.151			Tallow			0.88	0.13
67.	C_{16-18}-acids/esters/salts	0.150			Seed/plant oils			1.10	0.17
68.	Ethyl acetate	0.148			Ethanol/HOAc			1.26	0.19
69.	Dicyclopentadiene/cyclo.	0.146	6		HC cracking			0.62	0.09
70.	Sorbitol (70 wt %)	0.145			Dextrose			0.53	0.08
71.	1-Dodecene	0.140			Ethylene			0.41	0.06

(continues)

TABLE 13.1 (*Continued*)

No.	Product	Annual production (10^9 kg)	Average percent annual change		Usual organic feedstock	Energy consumed per product mass		Average market	
			1994–95	1985–95		Process (MJ/kg)	Feed (MJ/kg)	Price ($/kg)	Annual value (10^9)
72.	Chlorodifluoromethane	0.139			CCl$_4$/(HF)			2.35	0.33
73.	Ethyleneamines	0.136	4		Ethylene oxide			3.04	0.41
74.	Citric acid	0.136			Molasses			1.87	0.25
75.	Butyl acetate	0.134	5.5		Butanol/HOAc			1.19	0.16
76.	C$_{12}$-benzenesulfonic acid	0.131			C$_6$H$_6$/dodecene			0.85	0.11
77.	C$_{12-18}$ alcohols	0.127			Ethylene			2.00	0.25
78.	Perchloroethylene	0.118		−9	Propane			0.71	0.08
79.	Dioctyl phthalate	0.117			Anhydride/ROH			0.93	0.11
80.	Melamine	0.116	5		Urea			1.32	0.15
81.	Tetrahydrofuran	0.114			Butanediol or furan			2.93	0.33
82.	Chlorobenzene	0.113	0.4		Benzene			1.21	0.14
83.	Lysine	0.113			Corn			6.06	0.69
84.	Propionic acid	0.107	5		Ethylene/(CO)			0.90	0.10
85.	Naphthalene	0.107	−0.2		Naphthas			0.86	0.09
86.	Benzoic acid	0.104	5		Toluene			1.39	0.14
87.	Pentaerythritol	0.090			CH$_3$CHO/HCHO			2.49	0.22
88.	Neopentyl glycol	0.087			C$_3$-CHO/HCHO			1.65	0.14
89.	Ethyl chloride	0.069			Ethylene			0.72	0.05
90.	Butyrolactone	0.063			Butanediol or THF			3.44	0.22
91.	Methyl isobutyl ketone	0.063			Acetone			1.64	0.10

No.	Chemical					
92.	Dichlorodifluoromethane	0.058		$CCl_4/(HF)$	4.51	0.26
93.	Isobutyl alcohol	0.055		Propylene/(CO)	1.10	0.06
94.	Acrylamide	0.054	1	Acrylonitrile	1.76	0.10
95.	Cresols	0.052	1	Coal tars, resids	1.32	0.07
96.	Triethylene glycol	0.050	2	Ethylene oxide	0.93	0.05
97.	Propyl acetate	0.041		Propanol/HOAc	1.30	0.05
98.	Hexamethylenetetramine	0.036		$HCHO/NH_3$	1.58	0.06
99.	p-Dichlorobenzene	0.036		Benzene	1.52	0.05
100.	Monosodium glutamate	0.032		Sugars	1.65	0.05

[a]Adapted from Kirschner (1996), *Chem. Mkt. Reporter* (1997), United States International Trade Commission (1995), Tonkovich and Gerber (1995). The production data for which 10-year averages are indicated pertain to 1995 and the organic chemicals produced in the largest quantities. The data for monosodium glutamate are for 1995 (SRI Consulting). The remaining production data are for 1994, except for fermentation ethanol and lysine, which are for 1996. Fermentation ethanol does not include ethanol for beverages. With the exception of fermentation ethanol, the energy consumption data are from Lipinsky and Ingham (1994). Note that production pertains to quantities that are not internally used by a company, and that the production of many chemicals is not publicly available.

[b]Includes acid and dimethyl ester.

[c]Tar distillers and coke-oven operators not included. All grades; includes material used in blending motor fuel.

[d]Does not include data for coke and gas-retort ovens.

[e]Rubber grade.

[f]Mono-, di-, and triethanolamines.

to convert a relatively limited number of feedstocks from petroleum or natural gas, or primary derivatives of these feedstocks. It is perhaps unexpected that most of the organic chemicals listed in Table 13.1 are produced by thermochemical processes that were discovered and perfected many years ago, from about 1915 to 1965. The advances incorporated into each process since then have resulted primarily from improved process controls, reactor designs, and unit operations such as distillation, extraction, evaporation, filtration, pumping, and heat transfer management through the application of modern chemical engineering principles. Although catalysts have been greatly improved for many of these processes, the chemistry of the basic process has not changed much for most individual organic commodities.

It is evident that microbial processes play a minor role in the manufacture of these chemicals. This is because the large-scale thermochemical conversion of hydrocarbon feedstocks to organic chemicals can be carried out rapidly in an energy-efficient manner at low cost. Microbial conversion of the same feedstocks to the same products is not technically feasible in most cases. Many thermochemical conversion plants are so-called world-scale facilities, such as several of the methanol plants that can produce from 1500 to 3000 t/day of product. The economies of scale are often significant. On an annual basis, microbial processes currently account for commercial production of about 13.6 billion kilograms of mostly specialty chemicals, including carboxylic and amino acids, antibiotics, industrial and food enzymes, pharmaceuticals, and agricultural chemicals (American Chemical Society et al., 1996). On a mass basis, this is a small fraction of total organic chemical production. Fermentation ethanol is one of the exceptions. Because of its large-volume use as an oxygenate and octane enhancer in unleaded gasolines, it ranks number 12 in U.S. production as shown in Table 13.1. The largest fermentation plant in the United States has the capacity to manufacture about 600 t/day of ethanol. Microbial products, including fermentation ethanol for gasolines, account for about $10 billion in annual sales at the bulk level. Synthetic ethanol produced by ethylene hydration ranked number 50 in production even though it is a direct replacement for fermentation ethanol in gasolines. U.S. federal regulations, however, favor fermentation ethanol usage in unleaded gasolines by providing an excise tax exemption for blenders of ethanol and gasolines (Chapter 11). Synthetic ethanol is not eligible for this exemption because it is not manufactured from renewable feedstocks. Ethylene is obtained from petroleum and natural gas.

It is apparent from the data in Table 13.1 that most organic chemical commodities are extracted from hydrocarbon mixtures, are derived from hydrocarbons, or are the basic raw materials for the manufacture of other products. As the ranking decreases, many of the high-volume chemicals are used as feedstocks for lower ranked organic chemicals. Ethylene and propylene are the top two organic chemicals. They are used as feedstocks for many of the lower ranked chemicals and as monomers for the manufacture of large quantities of polymers. Methyl-t-butyl ether (MTBE) ranks third in production, is the highest ranking oxygenated organic chemical, and has exhibited disproportionately large increases in annual production compared to the other commodities in this table. MTBE is used as an oxygenate and octane enhancer in unleaded gasolines in the same manner as fermentation ethanol (Chapter 11). Urea is the second highest ranking oxygenated compound, but has shown little growth over the last several years. It competes with ammonia as a nitrogen source in agriculture and is also used as a monomer in thermosetting urea-formaldehyde resins. Methanol from synthesis gas is the third ranking oxygenate, one of the largest uses of which is in the manufacture of MTBE. So it is unexpected that the percentage change in annual consumption of methanol is substantially less than that of MTBE. Annual production of about 8 billion kilograms of MTBE requires about 2.9 billion kilograms of methanol at theoretical MTBE yields. Approximately 32%, or for the data in Table 13.1, 1.6 billion kilograms of methanol production, is used for MTBE (Hunt, 1993). The difference is accounted for by large methanol imports (Miller, 1993; Peaff, 1994).

The energy consumption data tabulated for several of the top-ranked organic chemicals illustrate how process and feedstock energy consumption vary per mass unit of product. The quantities of feedstocks required at 100% yield were generally calculated from the stoichiometric equations, the heats of combustion of the reactants, and sometimes the energy invested in obtaining the reactants (Lipinsky and Ingham, 1994). For example, the feedstock energy value used for p-xylene in terephthalic acid production includes the heat of combustion of p-xylene and the energy invested in obtaining it from mixed xylenes. This approach is somewhat misleading because a significant portion of the energy content of the reactants resides in the organic product, so the feedstock energy is not literally consumed. A somewhat more detailed analysis of the energy consumption data of the type shown in Table 13.1 can help pinpoint whether process improvements are feasible and where improvements might be made. For example, a modern wet-milling plant for fermentation ethanol that uses cogeneration on-site to supply power and steam consumes about 18.8 MJ of process energy per kilogram of ethanol from corn, as shown in Table 13.1. This is made up of 8.96 MJ of thermal energy and 6.19 MJ (0.564 kWh) of electrical energy equivalent per liter of product ethanol, assuming molecular sieve distillation (Shapouri, Duffield, and Graboski, 1996). The process energy consumption is considerably less than that of the older fermentation ethanol processes, but it is still slightly more than 60% of the energy content of the ethanol, which has a higher heat of combustion of 29.8 MJ/kg. On the other hand, the actual yield of ethanol is about 7.5 kg/bu (0.294 kg/kg, 9.46 L/bu) of corn. At an ethanol yield of 100% of the theoretical amount from glucose equivalent, assuming glucose is the key reactant and ignoring the incorporation of water into the key reactant during hydrolysis of the polysaccharides, the stoichiometry requires that one bushel of corn (25.5 kg) yields, in addition to 7.5 kg (29.4 wt %) of ethanol, 7.2 kg (28.2 wt %) of CO_2 and 10.8 kg (42.4 wt %) of other material as by-product. Feedstock energy consumption then corresponds to 31.2 MJ/kg of ethanol produced (at an HHV for glucose of 15.6 MJ/kg). This means that the total process and feedstock energy consumption in a modern wet-milling plant is about 50 MJ/kg of ethanol produced. Any effort made to improve the energy conservation characteristics of the process should therefore be directed to reducing process energy consumption rather than to increasing the yield of ethanol via conversion of by-product, presuming the theoretical yield of ethanol is consistently obtained from sugar fermentation, or to development of high-value by-product markets. This of course is what is done in a modern, multiproduct fermentation ethanol plant (Chapter 11).

The estimated annual value for each of the organic chemicals shown in Table 13.1 was calculated as the product of annual production and average market price. Generally, the higher ranked chemicals have higher annual values; most of them are more than $1 billion annually. Based on average plant capacities, the capital investment needed per kilogram of annual production capacity for the higher ranked chemicals is generally in the range $0.25 to $0.60 (Lipinsky and Ingham, 1994). Chemical syntheses performed in refineries or large petrochemical facilities tend to have lower capital intensities, as expected.

C. U.S. PRODUCTION OF POLYMERS AND COPOLYMERS

A tabulation similar to Table 13.1, but listing the major polymers and copolymers manufactured in the United States, is shown in Table 13.2. Many of the polymers and copolymers in this table and their applications were developed during and after World War II, and most are manufactured from the higher ranked organic commodities listed in Table 13.1 by traditional thermochemical polymerization. Petroleum and natural gas are the primary raw materials. With few exceptions, the polymeric products from these feedstocks have annual values of more than $1 billion. It is evident that consumer demand for plastics and rubbers is very large and that the market for polymers and copolymers has exhibited good growth.

In contrast to the polymeric products derived from fossil fuels, the production of synthetic polymers from biomass—the rayons and cellulose acetates—is quite small despite the fact that cellulose is one of the world's most abundant raw materials. The rayons are any of a variety of regenerated celluloses manufactured via intermediates such as alkali salts and cellulose xanthates. The process via xanthate intermediates was introduced in 1892 and is still used for the production of fabrics and films (cellophane). Wood pulps containing 80-90% cellulose are suitable feedstocks, since the hemicelluloses are soluble in alkali and are removed during production of alkali cellulose. The cellulose acetate derivatives were known before 1914 when they were used to manufacture small quantities of synthetic fibers and plastics. During World War I, large quantities were manufactured to replace cellulose nitrate coatings, and then partially acetylated derivatives were made in large quantities in the 1940s for the manufacture of acetate rayons. Today, several types of esters and ethers are synthesized from cellulose for use in the manufacture of fibers, fabrics, coatings, films, and magnetic tapes. As illustrated by the differences in production, the polymers made from fossil feedstocks apparently have characteristics that far outweigh the advantages of the polymers that can be synthesized from cellulosic feedstocks.

Presuming market prices are competitive, one route to development of larger markets for biomass in polymeric products is to convert biomass feedstocks to the same monomers that are used for synthesis of the large-volume polymers and copolymers from fossil-derived monomers. An alternative is to discover and develop natural and synthetic "biopolymers" that have superior or unique properties. Each of these approaches is discussed in the next section.

III. BIOMASS AS FEEDSTOCK FOR SYNTHETIC ORGANIC CHEMICALS

A. INTRODUCTION

Biomass can serve as a source of large-volume organic chemicals that retain the basic structural characteristics of biomass and either are or have the potential of becoming commodity chemicals. Dextrose (D-glucose) from the hydrolysis of starch or cellulose, high-fructose corn syrup from the enzymatic isomerization of dextrose hydrolysate, and D-xylose from the hydrolysis of the hemicelluloses are commodity products. Biomass feedstocks present several opportunities for varying product distributions as a function of market demand and other factors, or in which an optimum mix of products is chosen based on feedstock characteristics. The system depicted in Fig. 13.1 illustrates how such a plant might be configured (*cf.* Goldstein, 1976). Mild acid hydrolysis of hemicellulose in wood, for example, affords a solution that is predominantly either xylose or mannose, depending on the type of wood, and a cellulose-lignin residue. Acid treatment of this residue yields a glucose solution, which can be combined with mannose for alcoholic fermentation, and a lignin residue. Phenols can be made from this residue by hydrogenation, and 2-furancarboxyaldehyde

TABLE 13.2   Major Polymers and Copolymers Manufactured in the United States[a]

						Average market	
	Annual	Percent annual change		Energy consumed			Annual
	production			Process		Price	value
Category and name	(10^9 kg)	1994–95	1985–95	(MJ/kg)	Feed	($/kg)	(10^9)
Thermosetting resins							
Phenol, other tar-acid	1.45	−1	2	16.3	82.0	1.68	2.44
Urea	0.82	−5	4	9.3	56.8	0.99	0.81
Unsaturated polyester	0.72	7	3	18.6	83.7	1.35	0.97
Epoxy	0.29	5	5				
Melamine	0.13	−3	4	9.3	56.8	1.35	0.18
Thermoplastic resins							
Polyethylene	5.84	2	4	9.3	83.0	0.99	5.78
Polypropylene[b]	5.08	1	5	10.5	74.6	0.97	4.93
Styrene polymers							
Polystyrene	2.58	−3	3	9.3	87.9	1.19	3.07
Styrene–butadiene	1.38	−10	4				
ABS	0.68			13.9	93.9	2.03	1.38
PVC and copolymers[c]	5.58	5	6	11.6	73.9	1.04	5.80
Polyesters	1.72	18	15	27.9	78.1	1.54	2.65
Polyurethanes[d]	1.71			7.0	74.4	2.20	3.76
Polyamides, nylon type	0.46	8	10	18.6	73.0	2.82	1.30
Synthetic rubber							
Styrene–butadiene	0.88	4	2			1.41	1.24
Polybutadiene	0.53	5	3			1.71	0.91
Others[e]	0.43	5	4				
Ethylene-propylene	0.27	2	4			3.37	0.91
Nitrile	0.08	0	4			2.80	0.22
Polychloroprene	0.07	−8	−2			4.17	0.29
Synthetic fibers							
Polyester	1.76	1	2	27.9	78.1	1.54	2.65
Nylon	1.23	−1	1	18.6	96.0	2.93	3.60
Olefin[f]	1.10	0	7			1.17	1.30
Acrylic[g]	0.20	−2	−4			2.23	0.45

TABLE 13.2 (*Continued*)

Category and name	Annual production (10^9 kg)	Percent annual change		Energy consumed		Average market	
		1994–95	1985–95	Process (MJ/kg)	Feed (MJ/kg)	Price ($/kg)	Annual value (10^9)
Cellulosics							
Rayon	0.13	6	−2				
Acetate[h]	0.10	−8	0.1			4.67	0.47

[a]Adapted from Kirschner (1996), *Chem. Mkt. Reporter* (1997), *Plastics Tech.* (1997). With one exception, the production data figures are for 1995. Polyurethane production is for 1994. The energy consumption data are from Lipinsky and Wesson (1995). The unit cost data for the synthetic rubbers are from the *Chem. Mkt. Reporter* staff and are for mid-1997.

[b]1995 figure includes Canadian production.

[c]Beginning with 1994, includes Canadian production.

[d]Includes only foams, the largest-volume category. Does not include thermosetting polyurethane resins made from triisocyanates.

[e]Includes neoprene; butyl, polyisoprene; chlorosulfonated polyethylene; polyisobutylene; and acrylo, fluoro, and silicone elastomers.

[f]Includes olefin film, olefin fiber, and spun-bonded polypropylene.

[g]Includes modacrylic.

[h]Includes diacetate and triacetate.

(furfural) can be made from xylose by strong acid treatment. The commodity chemicals obtained from the primary biomass feedstocks, the C_5 and C_6 sugars and the lignins, in this system design are ethanol, furfural, and phenols.

Other compounds that can be produced directly from biomass in good yields, but which do not retain the basic structural characteristics of biomass, are also classified as commodity chemicals. Examples are acetic acid, methane, and synthesis gas. They are not manufactured in large volumes from biomass because fossil fuels are the preferred feedstocks in commercial production systems. Technically, biomass can serve as a feedstock for production of the entire range of commodity organic chemicals presently manufactured from fossil fuels. The various routes to large-volume chemicals from biomass will be examined later. Consider first some of the existing biomass-based chemicals, most of which are specialty chemicals that are manufactured for commercial markets.

Naturally occurring organic chemicals have long been extracted from a variety of biomass species. Alkaloids, steroids, vitamins, amino acids, proteins, enzymes, oils, gums, resins, waxes, and other complex organic chemicals are isolated and marketed as specialty products or used for conversion to other specialty products. Thousands of examples exist—the glycoside digitonin used in heart medicines from the purple foxglove plant, the narcotic morphine from the opium poppy plant, the antimalarial quinine from tropical trees of the

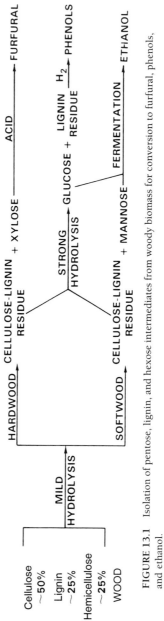

FIGURE 13.1 Isolation of pentose, lignin, and hexose intermediates from woody biomass for conversion to furfural, phenols, and ethanol.

Cinchona species, the chemotherapeutic agent taxol from Pacific yew tree members in the genus *Taxus,* the algins and alginic acids from the ocean kelp *Macrocystis pyrifera,* the high-erucic-acid canola oils from rapeseed, the lesquerella oils that are high in hydroxy-unsaturated fatty acids from cruciferous *Lesquerella* species, the B vitamins from various biomass species, the alkaloid scopolamine from the belladona plant, the enzyme cellulase in certain bacteria that catalyzes the hydrolysis of cellulose to glucose, etc. For all practical purposes, the list is endless. All of these specialty products are organic or organometallic compounds and most are normally marketed as high-value, specialty products. Their structures are complex; only a few are semisynthetic or synthetic products. The retention of chirality by natural organic chemicals used as pharmaceuticals, dietary supplements, and biological agents is beneficial because physiological activity depends on it. The majority of complex organic chemicals that occur naturally in biomass will therefore continue to be obtained by extraction from suitable species except when the cost of synthesis from biomass feedstocks is acceptable and competitive. Some examples of complex synthetic organic chemicals from biomass feedstocks will be cited later.

Chemicals from wood, or silvichemicals, are in a class by themselves. Many silvichemicals are commercially available. Natural polyisoprene rubbers from trees that are members of the genus Hevea continue to be harvested and marketed. Turpentines supply fragrances, perfume ingredients, aroma chemicals, flavorings, and a large number of terpene intermediates for the synthesis of high-value chemicals such as vitamins. Synthetic terpene resins prepared by cationic polymerization of terpene hydrocarbons are used in hot-melt and pressure-sensitive adhesives. Wood pulps are sources of specialty sugars used in drugs, sweeteners, plasticizers, and surfactants. In addition to the commercial regenerated celluloses described previously, wood pulps are used for production of cellulose esters such as the acetates and butyrates for fibers, films, and moldings, and for production of ethers such as ethylcellulose, carboxymethylcellulose, and hydroxyethylcellulose for use as gums and thickening agents. The alkaline pulping liquors are usually combusted to recover the pulping chemicals (and energy) for recycle, but the alkali lignins are also separated for use in formulating resins, rubbers, and emulsions. Dimethyl sulfide, dimethyl sulfoxide, and dimethyl sulfone are recovered from the volatile fraction of kraft black liquors from sulfate pulping. The sulfonated lignins in sulfite pulping liquors are a source of semisynthetic vanillin and are separated for use in formulating adhesives, binders, dispersants, tanning agents, and similar products. The alkaline pulping liquors yield fatty and resin acid salts that, on acidification, afford tall-oil fatty acids for conversion to dimer acids, polyamide resins, and other products that are used in inks, resins, adhesives, and surface coatings. Phenolic acids and waxes extracted from various barks are used in a variety of resin applications.

The primary building blocks and the main processing methods for production of the more important biomass-derived chemicals are tabulated in Fig. 13.2. All of the chemicals are either commercially manufactured by the indicated routes now or were manufactured in the past. Secondary processing of the derived chemicals in this figure would make it possible to manufacture a large percentage of the chemicals presently manufactured from fossil raw materials. Obviously, many different routes to organic chemicals from biomass feedstocks exist because of the chemistry of the polysaccharides and the substituted aromatic compounds in biomass. In a sense, biomass refining is analogous to petroleum refining. Petroleum is a mixture of aliphatic, alicyclic, and aromatic hydrocarbons that are rearranged and converted to higher value products on refining. The compounds present in biomass are subjected to other rearrangement processes, and sometimes processes that are the same as those used for petroleum, on conversion to higher value chemicals.

In the United States and most industrialized countries, the practical value of the commercial production of an organic chemical from biomass depends strongly on the availability and price of the same chemical produced from petroleum or natural gas, and occasionally coal. As will be shown shortly, there is no technical barrier that precludes production of commodity chemicals from biomass feedstocks. And it will become evident that many of the more complex organic chemicals are best synthesized from biomass feedstocks or can be extracted from appropriate biomass species. These chemicals can also be synthesized from hydrocarbon feedstocks, but the costs are often prohibitive. In the midst of the fossil fuel era, the less complex, commodity organic chemicals are preferentially manufactured from petroleum or natural gas, whereas complex specialty chemicals are derived from biomass. But commodity organic chemicals are open to entry by biomass feedstocks if they can provide economic advantages. Note that many of the routes described here to commodity chemicals from biomass were in commercial use in the past, are still in commercial use, have recently been commercialized, or have been developed and are available for commercial use.

B. THERMOCHEMICAL SYNTHESIS OF BUILDING BLOCKS

Much of the technology discussed in previous chapters can be applied to the synthesis of organic chemicals from biomass. The pyrolysis of wood described in Chapter 8 illustrates how several commodity chemicals were manufactured in the 1920s and 1930s by the Ford Motor Company. The yields and selectivities

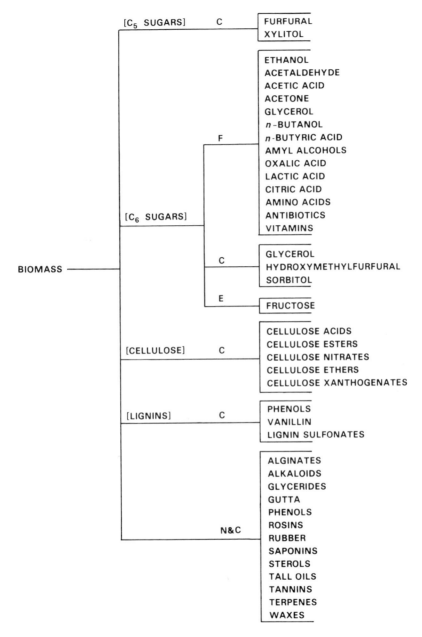

FIGURE 13.2 Primary building blocks and processing methods for production of biomass-derived chemicals.

were generally poor for most of the products. Consideration of the building blocks for most organic chemical commodities—synthesis gas, the light olefins ethylene, propylene, and butadiene, and BTX—and the chemical structures of the principal biomass components suggest some of the routes that might be developed to provide improved thermochemical processes for biomass feedstocks. With the exception of synthesis gas, each of these building blocks is a hydrocarbon, whereas the principal components in biomass, the polysaccharides, are polymers in which each carbon atom is bound to an oxygen atom. Depolymerization and carbon-oxygen bond scission are therefore necessary on a large molecular scale to synthesize the same building blocks. Such "brute force" degradation can be achieved at elevated temperatures. But there are opportunities to employ thermally balanced processes and energy conservation and to transfer the carbon and chemical energy in the feedstock to the desired products at near-theoretical yields. Low-moisture-content feedstocks are clearly preferred for thermochemical processes. Also, the costs of producing the basic building blocks from biomass feedstocks by thermochemical methods are usually more than the costs from petroleum and natural gas. This situation is expected to change as the fossil fuel era approaches its end.

Thermochemical gasification of biomass to high yields of synthesis gas (H_2 and carbon oxides), or syngas, can be carried out with most biomass feedstocks without regard to the structures of the biomass components (Chapter 9). Biomass is simply a source of carbon and energy. All of the organic components are thermochemically gasified under conditions that yield the desired syngas composition. Syngas can in turn be converted to a large number of chemicals and synthetic fuels by established processes (Fig. 13.3). The gasification technology for conversion of biomass to syngas has been developed and scaled up from the laboratory to the pilot plant to full-scale systems. It has been commercialized for fuel gas production, but not for the manufacture of syngas-derived chemicals. But there is no technical reason why commercial syngas

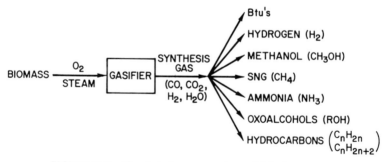

FIGURE 13.3   Chemicals from syngas by established processes.

processes cannot employ biomass feedstocks. The competitors for biomass feedstocks are natural gas and petroleum feedstocks, which are currently preferred in most countries for syngas production and conversion to methanol or high-hydrogen streams for other applications. Coal is used as a large-scale source of syngas for both chemical synthesis and synthetic liquid fuels by Sasol Limited in South Africa (Fourie, 1992; Ainsworth, 1996). Sasol's Synthol technology utilizes fluid-bed and slurry-bed reactors for conversion of syngas by Fischer–Tropsch chemistry to a broad product slate consisting of more than 100 fuels, solvents, and chemicals. The roster of commercial chemicals from syngas at Sasol includes aromatics, alcohols, ketones, oxidized and crystallized waxes, ammonia and a full spectrum of fertilizers, creosotes, cresylic acids, ethers, ethylene, α-olefins, phenol, propylene, and polyolefins. Sasol's program to convert coal to higher value products via syngas began in 1950. Experience showed that synthetic fuels from coal are unable to compete economically with petroleum-based fuels in isolation, so a limited coal-to-synfuels activity supported by oil refining and chemical production was undertaken by Sasol. The same rationale is applicable to the commercial production and conversion of syngas from biomass to chemicals. In addition, biomass can be utilized as a feedstock alone in gasifiers specifically designed for biomass feedstocks, or with coal in several types of coal gasifiers (Chapter 9).

The direct thermochemical production of the intermediate light olefin building blocks from biomass can be achieved by rapid pyrolysis at temperatures above those normally used to maximize liquid yields. Maximum liquid yields occur near temperatures of 500°C (Chapter 8). The volatile matter formed by biomass pyrolysis begins to participate in secondary, gas-phase reactions at temperatures exceeding about 600°C. This characteristic is similar to the cracking reactions employed by the petroleum industry in the manufacture of ethylene and propylene. Using gas-phase residence times of 0.5 to 1 s at temperatures above 500°C, the apparent rates of gas production from cellulose were measured for seven gas species: CO_2, H_2, CO, CH_4, C_2H_4, C_2H_6, and C_3H_6 (Antal, 1981). Figures 13.4 and 13.5 are graphs of log (k_j/m_i) vs the reciprocal of the temperature, where k_j is the experimentally determined rate of production of each gas species j. The slope of the lines connecting the values of log (k_j/m_i) provides the apparent activation energy E_j associated with the rate or production of each gas as shown in Table 13.3. The rate of conversion of volatile matter to ethylene is second only to those of CO and CH_4. In contrast to ethylene, the rate of production of propylene peaks at 675°C and declines at higher temperatures. These data suggest that short residence times at the appropriate pyrolysis temperature should afford good light olefin yields. Fast pyrolysis of biomass by the RTP® process, for example, provides reasonably good yields of light olefins at reactor temperatures and residence times of 700 to 900°C and 0.03 to 1.5 s (Graham, 1991). The gas yield can exceed

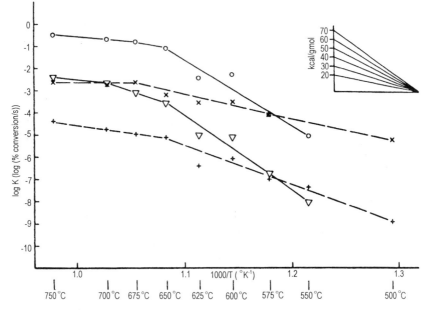

FIGURE 13.4 Arrhenius plot of gas-phase production rate for various gas species: (×) CO₂; (+) H₂; (○) CO; (∇) CH₄. From Antal (1981).

90 wt % of the biomass. Optimum yields of ethylene, total gaseous unsaturates, and total gaseous hydrocarbons from woody feedstocks were 8, 15, and 25% by weight of the feedstock, respectively. BTX and syngas are by-products.

Another method for production of ethylene and propylene from biomass feedstocks is via syngas and the synthesis of Fischer–Tropsch hydrocarbons. If this technology is commercialized on a large scale at some future time, the resulting hydrocarbons can be readily converted to these olefins. Almost any paraffinic or naphthenic hydrocarbon heavier than methane can be steam cracked to ethylene and propylene in good yields. The lower molecular weight paraffins normally afford higher yields of ethylene, whereas the higher molecular weight paraffins yield products that have higher propylene-to-ethylene ratios.

An approach to the production of ethylene from biomass that does not involve pyrolysis is ethanol dehydration. The catalytic conversion of syngas to ethanol from low-grade biomass (or fossil) feedstocks, and fermentation ethanol via advanced cellulose hydrolysis and fermentation methods, which make it possible to obtain high yields of ethanol from low-grade biomass feedstocks as well, are both expected to be commercialized in the United States (Chapter 11). Which technology becomes dominant in the market place has

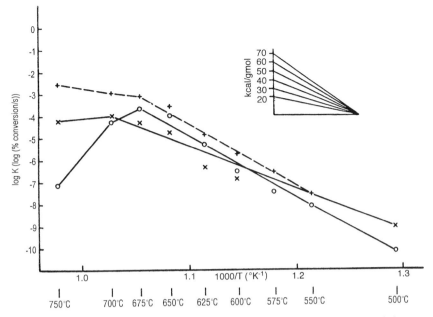

FIGURE 13.5 Arrhenius plot of gas-phase production rate for various gas species: (×) C_2H_6; (+) C_2H_4; (○) C_3H_6. From Antal (1981).

not yet been established. What is important is that these processes are expected to make large quantities of low-cost ethanol available for chemical uses that include ethanol conversion to ethylene as well as acetaldehyde and acetic acid. Once this technology is operational for the production of these chemicals, an

TABLE 13.3 Apparent Activation Energies of Product
Gases from Short-Residence-Time Cellulose Pyrolysis at
500 to 675°C[a]

Gas species	Activation energy (kJ/mol)
Carbon dioxide	88
Hydrogen	146
Ethane	159
Ethylene	230
Propylene	230
Carbon monoxide	251
Methane	280

[a]Antal (1981).

entire commodity chemicals industry founded on low-cost ethanol can follow (Fig. 13.6). Market forces will determine which low-cost ethanol technology is superior. Each is anticipated to be capable of providing ethanol at market prices in the same price range as methanol derived from natural gas.

It is noteworthy that the commodity organic chemicals acetic acid, acetaldehyde, ethyl chloride, and ethyl ether were manufactured directly from fermentation ethanol in Brazil in the early 1920s (Cunningham, 1993). Most of the fermentation ethanol produced in Brazil is derived from sugarcane. In 1958 and 1959, two fermentation ethanol-based ethylene units, as well as plants for converting the ethylene to acetaldehyde, butyl acetate, chloral, ethyl acetate, and vinyl acetate, began operations. When Brazil's fuel ethanol program, which is called PROALCOOL, was initiated in 1975, the production of commodity chemicals from fermentation ethanol increased rapidly. Table 13.4 lists the total capacities in 1993 of several Brazilian plants for conversion of ethanol and ethanol-based intermediates to commodity chemicals. Argentina has also established a commodity chemicals industry based on fermentation ethanol, although the production volumes are much less than those of Brazil.

Butadiene (1,3-butadiene) is manufactured in the petroleum industry by the catalytic dehydrogenation of the butanes and butenes, and by the direct cracking of naphthas and light oils. The overall butadiene yield by catalytic dehydrogenation, the most common industrial process, is as high as about 80% at selectivities of about 90%. The yields and selectivities of butadiene by

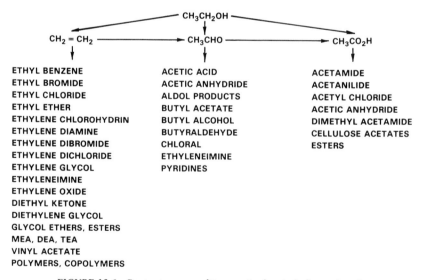

FIGURE 13.6   Routes to commodity organic chemicals from ethanol.

TABLE 13.4 Some Organic Commodity Chemicals from Fermentation Ethanol in Brazil[a]

Product	Production capacity (10^9 kg/year)	Product	Production capacity (10^9 kg/year)
Ethylene dichloride	1.011	Acrylonitrile	0.078
LD polyethylene	0.663	Ethyl acetate	0.060
Ethylbenzene	0.497	Ethylene glycol	0.030
Vinyl chloride	0.461	Acetic anhydride	0.026
HD polyethylene	0.397	Monochloroacetic acid	0.024
Acetic acid	0.182	Diethanolamine	0.012
Ethylene oxide	0.163	Triethanolamine	0.012
Diethylene glycol	0.147	Chloromethane	0.007
Monoethylene glycol	0.147	Pentaerithritol	0.007
Triethylene glycol	0.147	Chloral	0.004
Acetaldehyde	0.146	Acetylsalicyclic acid	0.003
Polyvinylacetate	0.143	Acetophenone	0.002
Ethylene	0.132	Ethyl ether	0.002
Monoethanolamine	0.122	Ethyl chloride	0.001
Vinyl acetate	0.080		

[a]Adapted from Cunningham (1993).

direct thermochemical conversion of biomass are nil, and the various routes to C_4 hydrocarbons from biomass are not yet practical. A few options exist for butadiene synthesis by alternative methods. Processes based on each of these routes to butadiene have been developed and commercialized. Butadiene is obtained by the following chemical reactions.

$$2\,CH_3CH_2OH \longrightarrow CH_2{=}CHCH{=}CH_2 + 2H_2O + H_2 \qquad (1)$$

$$CH_3CH_2OH + CH_3CHO \longrightarrow CH_2{=}CHCH{=}CH_2 + 2H_2O \qquad (2)$$

$$\underset{\underset{OH\ OH}{|\ \ |}}{CH_3CHCHCH_3} \longrightarrow CH_2{=}CHCH{=}CH_2 + 2H_2O \qquad (3)$$

$$\underset{\overset{|\qquad\quad|}{\underset{O}{CH_2\quad CH_2}}}{CH_2{-}CH_2} \longrightarrow CH_2{=}CHCH{=}CH_2 + H_2O \qquad (4)$$

Ethanol is the key reactant in Eq. (1), and also in Eq. (2) because it is readily converted to acetaldehyde. The process based on Eq. 1 was developed in Russia and the process based on Eq. 2 was developed in the United States. The yield of butadiene for the Russian process is about 30-35%. It is about 70% if mixtures of ethanol and acetaldehyde are employed as in the U.S. process. Equation (3) represents a process that involves 2,3-butylene glycol, a product from the microbial conversion of biomass. The process is carried out in two sequential steps via the glycol diacetate in overall yields to butadiene of about 80%. The process of Eq. (4) starts with a biomass derivative, the cyclic ether tetrahydrofuran, and can be carried out at high yields. When this process was first operated on a large scale in Germany, acetylene and formaldehyde were the raw materials for the synthesis of intermediate tetrahydrofuran. It is manufactured today from biomass feedstocks by thermochemical conversion, as will be discussed later.

In commercial practice, the aromatic building blocks for the synthesis of commodity organic chemicals, particularly benzene and p-xylene, are extracted from BTX. BTX is a mixture of the monocyclic aromatic hydrocarbons benzene, toluene, ethylbenzene, and the xylenes produced in the petroleum industry by dehydrogenation of naphthas using catalytic reforming and by the thermal reforming and catalytic and thermal cracking of middle distillates. Coal tars and coke oven gases also yield BTX. In modern coal coking processes, more that 90% of the benzene and toluene obtainable from coal is collected by scrubbing coke-oven gas. The direct thermochemical conversion of biomass to an equivalent BTX product, however, is not feasible under conventional conditions. Only small amounts of monocyclic aromatic hydrocarbons are formed on pyrolysis or the thermal cracking of biomass feedstocks. The pyrolysis of pure cellulose yields many different liquid products, including a few substituted phenolic compounds and phenol; only trace to small amounts of BTX are formed. The pyrolysis products of the carbohydrates in wood can be classified into three main groups: anhydrosugars and their dehydration products, low-molecular-weight oxygenates and other acetyl compounds, and alicyclic compounds and phenols (cf. Rejai et al., 1991). Thermolysis of the hemicelluloses and pentosans affords similar results. Pyrolysis at atmospheric or reduced pressures of the lignins, which are highly aromatic structures because of the abundance of substituted phenyl propane units, yields phenols instead of BTX-type products. The hardwoods generally yield phenols in the pyrocatechol and pyrogallol series, whereas softwoods yield phenols in the pyrocatechol series. A few of the aromatic compounds that have been identified in lignin pyrolysis products are phenol, eugenol, pyrocatechol, pyrogallol 1,3-dimethyl ether, o-cresol, guaiacol, vinylguaiacol, and propyl guaiacol. Most of the methanol formed on pyrolysis of woody biomass might be expected to be derived from the large number of methoxyl groups in the lignin fraction. But a substantial portion is also believed to be formed from the methoxyl groups present in the hemicelluloses.

Many research studies have been carried out to examine the possibilities for conversion of biomass feedstocks to liquid products by direct thermochemical treatment. These studies include the treatment of aqueous and nonaqueous slurries of wood particles with different reactants and catalysts at elevated temperatures and pressures, and the pyrolysis of wood under conditions that maximize liquid yields (Chapter 8). The prime objective was to develop processes for production of liquid fuels and not chemicals. The resulting products are generally acidic and contain high concentrations of carboxylic acids, phenolic compounds, and heterocyclic oxygen and alicyclic oxygenated compounds. With only a few exceptions, the products contain low concentrations of BTX.

Two basically different methods were found in this work, however, to maximize BTX formation during direct thermochemical conversion of biomass feedstocks. One is the hydrogenolysis of the substituted phenolic compounds that are formed from woody feedstocks, and the other is the aromatization of intermediates. An example of the results of the hydrogenolysis of wood is presented in Table 13.5. The noncatalytic hydropyrolysis of hardwood maple chips at 670°C and 1.02 MPa shows that aromatic compounds are formed in substantial yields on fast pyrolysis in hydrogen atmospheres. The liquid product yield is about 17 wt % of the feed, one-ring aromatic compounds make up about 39 wt % of the liquid product, and the yield of phenol is about 26 wt %. It is evident that to increase the yields of BTX, hydropyrolysis at greater severity or hydrocracking and hydrodealkylation should be used to aromatize more of the phenolic compounds. The organic vapors formed on fast pyrolysis of either woody biomass or RDF can be upgraded with shape-selective zeolite catalysts to selectively yield either BTX or C_2–C_5 olefins (Rejai et al., 1991). High yields of these products can be maintained at relatively high weight-hourly-space velocities (WHSVs) and low steam-to-biomass (S/ B) ratios. The BTX and C_2–C_5 olefin yields at 550°C, WHSVs of 10, and an S/ B ratio of 0.6 for RDF feedstocks ranged from about 10 to 19 wt % and 12 to 18 wt % of the feed, depending on the catalyst used. Assuming that 60% of the feed is converted to pyrolysis oil and that the empirical formulas of BTX and the olefins are $CH_{1.25}$ and CH_2, the yield limits of BTX and C_2–C_5 olefins based on the experimental stoichiometries for RDF were 25.6 wt % and 23.7 wt % of the feed. Using the same assumptions, the corresponding yield limits for wood were 25.5 and 23.6 wt % of the feed. It has also been found that biomass pyrolysis oils can be catalytically upgraded to high-BTX liquids after formation and collection (Bakhshi, Kaitikaneni, and Adjaye, 1995). Products containing 23 to 38 wt % BTX and 19 to 35 wt % aliphatic, gasoline-range hydrocarbons are obtained by thermal treatment at atmospheric pressure, 330–410°C, and a WHSV of 3.6. The selectivity for benzene is about 90% of

TABLE 13.5 The Non-catalytic Hydrogenolysis of
Hardwood Maple Chips[a]

Parameter	Condition or result
Feedstock	
Size	1.19 mm max.
Moisture	5.5 wt %
Volatile matter	81.2
Ash	0.21
Fixed carbon	13.1
Carbon	49.8
Hydrogen	6.05
Nitrogen	0.15
Higher heating value	19.52 MJ/kg
Conversion conditions	
Environment	H_2
Pressure	1.03 MPa
Temperature	670 °C
Free-fall reaction time	0.3 s (approx.)
Product yields	
Char	12.25 wt % of feed
Gases (C_1 to C_3)	54.00
Liquids	17.24
Composition of liquid products	
One-ring compounds	
Benzenes	34.0 wt %
Styrenes	0.2
Indans	1.4
Indenes	3.0
Two-ring compounds	
Naphthalenes	20.8
Biphenyl	0.7
Acenaphathene	1.3
Acenaphthalene	0.5
Three-ring compounds	
Fluorene	2.1
Phenanthrene and anthracene	2.5
Phenyl naphthalenes	0.2
Others	0.5
Four-ring compounds	2.2
Five-ring compounds	0.4
N-containing compounds	0.4
O-containing compounds	
Phenol	26.3
Benzofuran	1.3
Others	0.7
Unidentified	1.5

[a]Rose et al. (1981).

the BTX with aluminosilicate catalysts. Molecular sieve (HZSM-5) and similar shape-selective catalysts exhibited high selectivities for toluene, the xylenes, and other alkylated benzenes. An alternative approach to BTX from biomass discovered in the 1980s employs fast pyrolysis of woody biomass in methane atmospheres (methanolysis) rather than hydrogen (Steinberg, Fallon, and Sundaram, 1983). Preliminary studies of the methanolysis of dry fir wood particles in the presence of silica flour at 1 to 3-s residence times, temperatures of 1000°C, and pressures of 0.34 MPa converted 12% of the available carbon to BTX, 21% to ethylene, and 48% to CO without net methane consumption (Chapter 8). No products were formed in the absence of wood.

Thus, the building blocks for synthesis of commodity organic chemicals—synthesis gas, the light olefins ethylene, propylene, and butadiene, and BTX—can be obtained from biomass feedstocks by thermochemical conversion.

C. OTHER THERMOCHEMICAL SYNTHESES

Cursory examination of Table 13.1 shows that oxygenated compounds account for a large number of the commodity chemicals manufactured in the United States. Biomass might therefore be expected to be used as feedstock for many of the chemicals in this listing. Only a few are. If the list were extended, say, to include the top 500 or 1000 commercial organic chemicals, more biomass derivatives would begin to appear. But they would tend to be low-volume specialty and fine chemicals of complex structure that are more readily produced from biomass than from hydrocarbon feedstocks. Biomass often provides technical advantages in these cases. Sorbitol is manufactured by hydrogenation of dextrose, and although a thermochemical route to the same chemical can be devised for hydrocarbon feedstocks, it would be technically and economically impractical. When either a biomass or a hydrocarbon feedstock can be employed, economic factors usually, but not always, determine which is used. Methanol is a commodity chemical that can be manufactured from syngas by thermochemical methods using either biomass or fossil feedstocks, yet essentially all of it is made from natural gas because of the lower production cost in world-scale plants. In unleaded gasoline blends, synthetic ethanol is unable to compete with fermentation ethanol from biomass in the United States because of the motor fuel excise tax exemption. The exemption is also available for methanol made from biomass, but only small amounts of neat methanol can be used in unleaded gasolines compared to neat ethanol because of federal blending regulations. Unleaded gasoline-methanol blends are limited to 0.3 vol % methanol without cosolvent oxygenates, whereas up to 10 vol % anhydrous ethanol can be used. It is apparent that manufacturing costs alone may not determine whether a commodity chemical will be manufactured from biomass or a hydrocarbon feedstock. U.S. federal regulations can have a large impact, at least for oxygenated fuels.

Selected thermochemical syntheses are discussed here to show how the structural characteristics of biomass components can affect the methods chosen to manufacture a few specific commodity chemicals. The main components of biomass, the pentoses and hexoses from the polysaccharides and the lignins, lipids, and proteins, are examined as potential feedstocks.

Some Pentose-Derived Chemicals

Xylose-based chemicals analogous to some of the glucose derivatives discussed in the next section are commercially available. An example is xylitol, the pentahydric alcohol that is manufactured by hydrogenation of the aldehyde group in xylose. One of the unique reactions of the pentoses known since the nineteenth century is cyclodehydration to yield furfural. Xylose, for example, is dehydrated by acid treatment through one or two intermediates to form the furan-substituted aldehyde in quantitative yield, or 64.0 wt % of the xylose

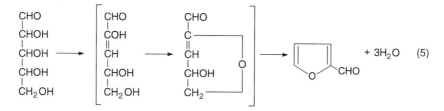

(*cf.* Wiley, 1953). Kinetic studies show that the rate of the reaction is directly proportional to both acid and xylose concentrations, and thus is at least a second-order process. The reaction is industrially significant because it made the furans available on a large scale from biomass feedstocks such as corn cobs, oat hulls, and rice hulls. The reactivity of the furan ring as a diene and in acid to form carbonium ions introduces a wide range of synthesis options. The two principal furans from which many derivatives are available are tetrahydrofuran and tetrahydrofurfuryl alcohol, both of which are manufactured from furfural. Furfural is converted by catalytic decarbonylation to furan (a), by catalytic hydrogenation of the aldehyde group to furfuryl alcohol (b), on further hydrogenation of the furan ring to tetrahydrofurfuryl alcohol (c), and by oxidation of the aldehyde group to furoic acid (d). Furan is converted to tetrahydrofuran (e) by catalytic hydrogenation, and undergoes a large number of substitution and addition reactions. Tetrahydrofurfuryl alcohol undergoes catalytic dehydration and rearrangement to dihydropyran (f), which can be hydrogenated to tetrahydropyran (g). Derivatives are formed from these cyclic ethers by standard reactions. Tetrahydrofurfuryl alcohol is also converted to levulinic acid (γ-ketovaleric acid) (h) on treatment with dilute acid. It is evident that the molecular structure of the pentoses facilitates the formation of many derivatives.

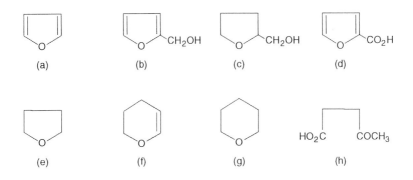

A large group of the furans and their derivatives have specialty and industrial uses. QO Chemicals, Inc., formerly the Chemicals Division of the Quaker Oats Company, has been the world's largest producer of furfural and its derivatives for many years. In addition to furfural, their commercial products include furfuryl alcohol, tetrahydrofuryl alcohol, tetrahydrofuran, furoic acid and its alkyl furoates, furfural amines, levulinic acid, methylfuran, acetyl furan, methyltetrahydrofuran, and various polyetherpolyols and resins.

The ring-cleavage chemistry of the furans has led to many suggestions for the manufacture of other established chemicals and polymers. Some of the commodity chemicals suggested for manufacture from the furans include *n*-butanol, 1,3-butadiene, styrene, adipic acid, maleic anhydride, 1,4-butanediol, and γ-butyrolactone (*cf.* Parker *et al.,* 1983). Tetrahydrofuran has been converted to adiponitrile, an adipic acid precursor, in a commercial two-step process via dichlorobutane (Cass, 1947). Du Pont used this process in their nylon operations in the 1950s. A single-step carbonylation method with nickel carbonyl catalysts to form adipic acid from tetrahydrofuran in high yield appears to offer advantages in similar applications. Tetrahydrofuran has also been converted to *n*-butanol in high selectivity, which simplifies and reduces the cost of purification (Smith and Fuzek, 1949), and to 1,3-butadiene via dehydration and dehydrogenation (Hasche, 1945). Both furan and furfural have been converted to maleic anhydride, with the former producing better yields and fewer by-products (Milas and Walsh, 1935).

Presuming the market for furfural and its derivatives ultimately grows to the point where commodity uses require the availability of large amounts of furfural, woody and waste biomass feedstocks can serve as a much more abundant and distributed source of supply than corn cobs, oat hulls, and rice hulls. For a plant supplied with woody feedstocks in which the hexoses are converted to fermentation ethanol, the net by-product credit for furfural made from the pentose fraction by cyclodehydration is projected to be able to reduce or subsidize the selling price of ethanol by about 50% (Parker *et al.*, 1983). The net by-product credit in this case is the difference between the cost of producing furfural and its selling price.

Four major carboxylic acids, acetic, formic, glyceric, and lactic, are formed under mild alkaline oxidation conditions from xylose (Rahardja *et al.*, 1994). At 40-60°C, lactic acid is formed from dilute aqueous solutions of xylose in the largest mass yield, about 50%, under continuous oxidation conditions. This is about 83% of the maximum theoretical yield of 60%. Under batch conditions, the yield of lactic acid from xylose is about 23%, while the yield from glucose is about 41%. These data indicate a commercial alkaline oxidation process for lactic acid from pentoses and hexoses is feasible, but it should be realized that the recovery of carboxylic acids having high water solubilities from dilute aqueous solutions can add significantly to production cost (Busche, 1984).

Some Hexose-Derived Chemicals

Several hexose sugars, sugar syrups, and starches are marketed as commodity chemicals, but only a few derivatives such as the disaccharide cellobiose, a glucose-β-glucoside, which is synthesized not from glucose but by careful hydrolysis of cellulose, are marketed. Despite the fact that the chemistry of the monosaccharides, oligosaccharides, and polysaccharides that contain repeating hexose units has been developed over several hundred years (*cf.* Raymond, 1943; Wolfrom, 1943; Nikitin, 1962), and the fact that these materials are very abundant, relatively few commodity chemicals are manufactured by thermochemical conversion of the hexoses. A federal U.S. survey of the synthetic organic chemicals manufactured in the United States includes only a few hexose derivatives out of thousands of commercial commodity compounds (U.S. International Trade Commission, 1995).

The hexoses yield a wide range of esters, ethers, and anhydro derivatives, and they undergo numerous rearrangement, substitution, isomerization, cyclodehydration, and reduction-oxidation reactions. So there are just as many options available for producing derivatives, if not more, from the hexoses as from the pentoses. Glucose is converted to gluconic acid (i) by oxidation of the aldehyde group to a carboxyl group, to the hexahydric alcohol sorbitol (j) by hydrogenation of the aldehyde group (or to sorbitol and mannitol when invert sugar is used in place of glucose), and to a large number of ethers by elimination of water such as 1,6-anhydroglucose (k), 2,4-anhydroglucose (l), and 2-hydroxymethyl-furfural (m).

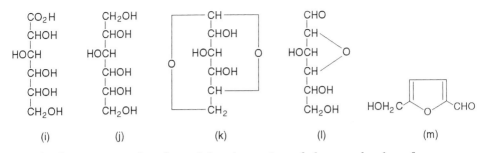

(i)   (j)   (k)   (l)   (m)

The last compound is formed by abstraction of three molecules of water from the hexose and is transformed into levulinic acid and formic acid on acid treatment:

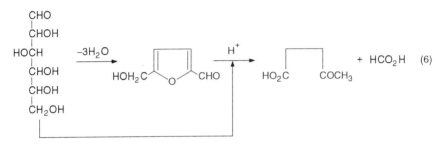

The hexoses can also be converted directly to levulinic acid and formic acid on acid treatment without isolation of 2-hydroxymethylfurfural. Note that levulinic acid can be synthesized from both pentoses and hexoses, but is usually manufactured from cane sugar or starch by boiling with hydrochloric acid. A process for the manufacture of levulinic acid has also been developed for paper-mill sludge feedstocks (Fitzpatrick and Jarnefeld, 1996).

Levulinic acid is a versatile chemical intermediate that can be converted through proven routes to higher value chemicals such as diphenolic acid, succinic acid, pyrrolidines, pyrrolidones, and the agricultural chemical δ-amino levulinic acid. When the price of levulinic acid drops to levels of \$0.09 to \$0.11/kg that are expected from large-scale production from sludge, it is believed that the acid will be used to manufacture 2-methyl-tetrahydrofuran, which contains 18.6 wt % oxygen, for use as an oxygenate in unleaded gasolines:

$$
\underset{\text{HO}_2\text{C} \qquad \text{COCH}_3}{\boxed{}} + 3\text{H}_2 \longrightarrow \underset{\text{O} \quad \text{CH}_3}{\boxed{}} + 2\text{H}_2\text{O} \qquad (7)
$$

However, a more direct synthesis of this ether is the catalytic hydrogenation of furfural at moderate temperatures. Angelica lactone, which is formed on distillation of levulinic acid has also been suggested as an oxygenated fuel extender (Thomas and Barile, 1953); it contains 32.6 wt % oxygen.

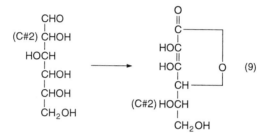

$$+ \ H_2O \qquad (8)$$

These are just a few of the many chemical reactions of the hexoses. Gluconic acid, glucono-δ-lactone from this acid, sorbitol, sorbitol esters, sorbitans (tetrahydric anhydrosorbitol), mannitol, mannitol esters, copolymers of fatty acid sorbitol esters and alkylene oxides, and a few other derivatives are currently marketed as commodity chemicals derived from the hexoses. Vitamin C (L-ascorbic acid) is perhaps one of the best examples of a commercial derivative of D-glucose that is manufactured primarily by chemical synthesis:

$$
\begin{array}{ccc}
\text{CHO} & & \text{O} \\
| & & \| \\
\text{(C\#2) CHOH} & & \text{C}{-}{-}{-}{-} \\
| & & \text{HOC}\qquad\quad| \\
\text{HOCH} & & \| \qquad\quad| \\
| & \longrightarrow & \text{HOC}\qquad \text{O}\quad(9) \\
\text{CHOH} & & | \qquad\quad| \\
| & & \text{CH}{-}{-}{-} \\
\text{CHOH} & & | \\
| & & \text{(C\#2) HOCH} \\
\text{CH}_2\text{OH} & & | \\
& & \text{CH}_2\text{OH}
\end{array}
$$

Although Vitamin C is not a commodity chemical and is manufactured by a sequential process that involves hydrogenation, oxidation, acetonide formation, oxidation, hydrolysis, and formation of the vitamin, this example illustrates that complex multistep processes are not excluded from commercial use and can yield products competitive with the natural product. Natural vitamin C for the commercial market is also extracted from certain plants.

The chemicals that are produced from polysaccharides by direct pyrolysis were discussed in Chapter 8. A few of them are expected to be marketed in the future, particularly some of the oxygenates from fast pyrolysis processes (Scott, Piskorz, and Radlein, 1993; Piskorz et al., 1995).

Since an extremely large number of synthetic carbohydrate derivatives are known and have been studied for decades, it is doubtful that many new additions to commodity status will be made from this group in the near future. It is more likely that some of the new derivatives such as synthetic oligosaccharides that are found to have unique properties in certain applications will become commodity chemicals. Sucrose was not synthesized until 1953, and general methods for the synthesis of oligosaccharides have not been perfected yet because of the difficulty encountered in replicating the repeating glycosidic bond between monosaccharide units and in controlling the bonding positions in molecules that offer several sites. Recent advances appear to provide solutions to the problem of synthesizing oligosaccharides (*cf.* Rouhi, 1996). New oligomers and biopolymers are expected to result from this effort.

Some Lignin-Derived Chemicals

Commodity lignin chemicals and their salts and derivatives are complex mixtures of compounds obtained from the spent liquors of the wood pulping industry. As previously described, the three basic pulping processes are the sulfite, sulfate (kraft), and caustic soda processes. The by-product sulfonate salts are made from the sulfite liquors, usually from the pulping of softwoods. The so-called black liquors from the other two processes yield by-product alkali lignins, which are also called kraft and sulfate lignins. Lignins can be obtained by direct wood hydrolysis, but the process is not used as a lignin resource in the United States. The lignins are complex natural polymers of condensed methoxypropylphenols. There is evidence of unsaturation and hydroxyl, carboxyl, and carbonyl groups in some of the lignins. The lignins are sold for a variety of industrial and specialty applications, including adhesives, binders, dispersing agents, emulsifiers, cement and gypsum board manufacture, oil well drilling, and road stabilization formulations.

Few pure commodity chemicals are synthesized from the lignins. Synthetic vanillin (3-methoxy-4-hydroxybenzaldehyde), a member of the catechol (1,2-dihydroxybenzene) series, is one example. This compound, a pharmaceutical intermediate and flavoring and perfume agent, is commercially produced by extraction of natural resources such as the vanilla bean, and by several synthetic methods using lignin or lignin-based chemicals. Among the synthetic methods are the alkaline air oxidation of spent sulfite liquor from pulping operations, the oxidation of coniferyl alcohol (3-(4-hydroxy-3-methoxy-phenyl)-2-propene-1-ol) from pine trees, and the conversion of guaiacol (*o*-methoxyphenol) from catechol or wood pyrolysis oils. Vanillin is obtained from guaiacol by several different synthetic methods, the simplest of which is probably the introduction of the aldehyde group by direct one-step reaction with alkaline chloroform (Reimer–Tiemann reaction).

The structures of the lignins are amenable to conversion to several classes of substituted phenols by thermochemical and thermal degradation methods. The displacement of pyroligneous tars by coal tars eliminated much of the demand that existed for the lignin-based products. Creosote oil or cresylic acid, a mixture of *o*-, *m*-, and *p*-cresols, is now manufactured mainly from coal tars, while only small amounts of cresols are made from wood tars. The use of wood tars and other biomass-derived tars as substitutes for a major portion of the phenol and formaldehyde in phenol-formaldehyde resins could reverse this trend (Himmelblau, 1995). The key to this process seems to be that the optimum phenolic character of the tar is not obtained at maximum tar yield, but at yields of about 25-30 wt % at reactor temperatures of about 600°C and short gas-phase residence times. A wide range of phenols, aldehydes, and other compounds capable of polymerization are formed under these conditions. If this effort and some of the others in progress to develop new applications for lignins are commercially successful, they will provide a great impetus to develop carbohydrate-based chemicals as commodity chemicals because the lignins are such a large fraction of woody biomass. The lignin fraction cannot be ignored. Indeed, the simultaneous occurrence of lignins and polysaccharides made up of C_5 and C_6 units in potential woody biomass feedstocks requires that the silvichemical facility of the future produce multiple products from the total feedstock, much like a petroleum refinery. A plant in which only one salable chemical is manufactured will not be competitive.

Some Lipid-Derived Chemicals

Lipids (fats) are found in seed and vegetable oils and fruit pulps (and animal tallows and greases) (Chapters 3 and 10). The vast majority are triglycerides, or C_6-C_{24} fatty acid esters of the trihydric alcohol glycerol (1,2,3-trihydroxypropane). Smaller amounts of mono- and diacylglycerides and unesterified fatty acids are present. The only exception to the glyceride structure is oil from the jojoba plant. Jojoba oil contains almost entirely a mixture of wax esters with 40-44 carbon atoms composed of long-chain alcohols esterified with long-chain fatty acids (Åppelqvist, 1989). There are about 50 fatty acids that occur as natural glycerides, a surprising characteristic of which is that they are almost invariably all composed of straight-chain acids containing an even number of carbon atoms and that they are all mixed glycerides containing two or three fatty acids. The odd-numbered, five-carbon isovaleric acid, which contains the same carbon skeleton as isoprene, is the exception; it has been found in glycerides from dolphin and porpoise blubber in relatively large amounts (Fieser and Fieser, 1950). The majority of the acids are C_{16} or greater. Alkaline hydrolysis of the glycerides provides large amounts of glycerol and the fatty acids as commodity chemicals. Other synthetic methods based on propylene feedstocks are also used commercially for glycerol production. But the largest source in the United States is the hydrolysis of fatty feedstocks. The fatty acids of most industrial significance as commodity chemicals from fat hydrolysis or as derivatives, usually esters, include the saturated lauric (C_{12}), myristic (C_{14}), palmitic (C_{16}), stearic (C_{18}), arachidic (C_{20}) and behenic (C_{24}) acids, the C_{18} monounsaturated oleic and ricinoleic acids, and the polyunsaturated C_{18} linoleic, linolenic, and eleostearic acids. Glycerol forms acetals, aldehydes, amines, ethers, esters, and halides and can also be converted to acrolein and other derivatives. Glycerol is thus a versatile chemical that is useful in hundreds of applications, the most important of which are in alkyd resins and for the manufacture of the explosive nitroglycerin (glyceroltrinitrate).

A new use for fatty acids in the United States is in the manufacture of synthetic sucrose esters for use as fat replacements in foods (Kirschner, 1997). This market is projected to be more than 200 million kilograms per annum in the United States because of the growing demand for diet foods. The ester is synthesized from oils such as cottonseed or soybean oil by transesterifying the glycerides in the oil with methanol to yield the methyl esters of the fatty acids, in the same manner that methyl biodiesel fuel is manufactured (Chapter 10). The methyl esters are isolated and then transesterified with sucrose to yield a sucrose polyester in which six to eight fatty acid molecules are bonded through ester linkages to the hydroxyl groups in the sucrose molecule. The liberated methanol is removed. The synthetic ester has many of the characteristics of and is a low-calorie replacement for natural fats. Unlike the glycerides in natural fats, digestive enzymes are unable to metabolize the synthetic ester; it passes through the body unchanged.

Some Protein-Derived Chemicals

Proteins are widely distributed in plants (and animals) and consist of one or more polymeric chains of amino acids that are bound in peptide linkages. Usually more that 100 amino acids are linked in the natural polypeptide. In dry biomass, the approximate protein content is the product of the percentage of Kjeldahl nitrogen in the biomass and the factor 6.25 (Chapter 3). Woody biomass normally contains very small amounts of protein, nil to a few percent, whereas herbaceous, aquatic, and marine biomass can contain up to 15 to 20 wt %. The polypeptide chains are converted to the amino acids by acid, alkali, and enzyme-catalyzed hydrolysis. More than 30 amino acids have been isolated from protein hydrolysates. The most common amino acids are the neutral acids glycine, alanine, serine, cysteine, cystine, threonine, valine, methionine, leucine, isoleucine, phenylalanine, tyrosine, proline, hydroxyproline, tryptophane, and asparagine; the basic acids are arginine, lysine, hydroxylysine, glutamine, histidine, and ornithine; and the acidic acids are aspartic and glutamic acids. All natural amino acids belong to the L-series, and in theory, each can be obtained by hydrolysis of separated protein and extraction from the hydrolysate. Amino acids are a class of organic chemicals that have wide use as foods and feed supplements. In the past, most were obtained by extraction from protein hydrolysates. But the technical difficulties of isolating and purifying the amino acids led to the development of other methods. Most are commercially produced by fermentation or enzymatic treatment of biomass-based substrates (next section). Glycine, alanine, and methionine are chemically synthesized; proline, hydroxyproline, cysteine, and cystine are obtained from protein hydrolysates—proline and hydroxyproline from gelatin normally of animal origin (collagen), and cysteine and cystine from keratin protein of animal origin.

It is evident that natural proteins are not a primary source of large amounts of amino acids, despite the fact that many of the acids are commercially significant chemicals and a few are commodity chemicals. The technical difficulties just alluded to include undesirable distributions of the amino acids in natural proteins, the sensitivity of proteins and amino acids to chemical hydrolysis conditions, racemization, the multiplicity of the product acids and the often low concentration of the desired acid or acids in the hydrolysate, and the consequent separation problems. Microbial synthesis of specific amino acids from biomass substrates or biomass-derived intermediates often has substantial advantages over thermochemical processing methods and is used for the commercial production of several of the amino acids. This is discussed in more detail in the next section.

Biopolymers

Excluding the polyisoprenes (i.e., natural rubbers) found in several biomass species, the main biomass constituents, the natural celluloses, hemicelluloses, and lignins, as well as the proteins and nucleic acids, are biopolymers. Some are chemically modified and marketed; some are marketed as natural products. The major classes of biopolymers include the cellulose esters and ethers, chitin and chitosan, hyaluronic acid, polyhydroxyalkanoates, silk and other natural polypeptides, starch esters, xanthan, dextran and other polysaccharides, synthetic polypeptides, and polymers synthesized from primary biomass derivatives such as glycerol, other polyols, and fatty acids. Several of these polymers are biodegradable plastics useful in packaging, molding, and extrusion applications, yield biodegradable detergents, furnish films and coatings that are selectively permeable to specific gases, are highly water-absorbing materials useful for moisture removal, or lend themselves to medical uses. New applications of many of these polymers are under development and should result in several additions to the roster of commodity chemicals from biomass.

D. MICROBIAL SYNTHESIS

Microbial processing by direct fermentation of primary biomass derivatives such as glucose can be used to synthesize a large number of organic chemicals. The cellular components that facilitate these processes are enzymes, the protein catalysts produced by the microorganisms. Some examples of direct fermentation with primary biomass derivatives as substrates are shown in Table 13.6. It is evident that most of the common chemicals listed are commercially available from nonbiomass sources. The molecular structures of the products range from simple compounds, such as ethanol, to complex compounds, such as the penicillins, to polymeric products, such as the polyhydroxybutyrates. Suitable substrates are generally monosaccharides and disaccharides and their original sources such as molasses and starch and cellulose hydrolysates. A wide range of biomass feedstocks are used for commercial fermentation systems. Examples are glucose for many different chemicals and products; beet sugar molasses for citric acid; hydrolyzed starch for citric acid, itaconic acid, and xanthan gum; thinned starch for ionophores and alkaline proteases; vegetable oil for terramycin; corn steep liquor for penicillin; and soybean meal for vitamin B_{12}. The microorganism is grown in a culture medium that contains the carbon source (substrate), a nitrogen source (usually ammonia, urea, or ammonium salts), and minor and trace nutrients and growth factors such as vitamins and amino acids, if necessary. The majority of the excreted chemicals are oxygenated compounds that contain carbonyl, carboxyl, or hydroxyl groups. During the fermentation process, part of the substrate is converted to cellular biomass, usually a small amount for anaerobic processes and a larger amount for aerobic processes. By-products such as CO_2 from aerobic processes and oligosaccharides and other water-soluble products are also formed. However, it appears from the broad range of compounds listed in Table 13.6 that the possibilities are almost unlimited as long as the appropriate organism and substrate are allowed to interact under suitable environmental conditions.

Many common organic chemicals can be manufactured by employing live microorganisms; fermentation ethanol is the best example. Certain microorganisms are also capable of performing syntheses that are very difficult to carry out by conventional thermochemical synthetic methods. The compounds produced in these cases are usually characterized by complex, chiral structures such as those of the antibiotics. Combinations of microbial and conventional thermochemical methods are sometimes employed for multistep syntheses when neither method alone is satisfactory. Recombinant DNA techniques are expected to open new routes for the efficient microbial production of complex compounds in high yields and make them available in large volumes at lower costs. Two examples of the application of recombinant DNA methods to chemical synthesis are cited in Table 13.6, indigo (Berry *et al.*, 1995) and polyhydroxyalkanoates (PHAs) (Williams, Gerngross, and Peoples, 1995). One PHA, poly-β-hydroxybutyric acid, a natural, biodegradable microbial polymer that has several desirable properties (*cf.* Kulprecha, Phonprapai, and Chanchaichaovivat, 1995), is expected to be manufactured with recombinant *Escherichia coli*. Projected prices of PHAs using genetically engineered *E. coli* are in the

TABLE 13.6 Examples of Organic Chemicals Produced by Direct Fermentation of a Primary Biomass Derivative[a]

Chemical	Substrate(s)	Microorganism(s)
Acetic acid	Various sugars	*Acetobacter aceti* *Clostridium thermoaceticum* *Pachysolen tannophilus*
Acetic acid (Ca/Mg salt)[b]	Cheese whey, hemicelluloses, others	Halophilic anaerobes Homolactic/homoacetic bacteria
Acetoin	Various sugars	Many species
Acetone	Various sugars	*Clostridium* sp.
Aureomycin (chlortetracycline hydrochloride)[c]	Sugars	*Streptomyces aureofaciens*
L-Alloisocitric acid	Glucose	*Penicillium purpurogenum*
2,3-Butanediol	Various sugars and acids	*Aerobacter aerogenes* *Bacillus polymyxa* *Klebsiella oxytoca* *K. pneuminiae*
n-Butanol	Various sugars and organics	*Clostridium* sp.
Butyraldehyde	Glucose	*Clostridium acetobutylicum*
Butyric acid	Various sugars	*Clostridium* sp.
Citraconic acid	Glucose	See itaconic acid
Cellulose	Glucose	*Acetobacter* sp.
Chitosan	Fungi, some algae	Literature cited
Citramalic acid	Glucose	See itaconic acid
Citric acid	Various sugars	*Aspergillus niger* *Saccharomycopsis lipolytica*
Comenic acid	Not known	Not known, literature cited
Curdlans	Various biomass	*Agrobacterium* sp. *Alcaligenes faecalis*
Dextrans	Glucose	*Lactobacillus brevis* *Lactobacteriacaeace dextranicum* *Leuconostoc mesenteroides*
Ethane	Various biomass	Literature cited
Ethanol	Various sugars	*Kluyveromyces* sp. *Candida utilis* *Saccharomyces cerevisiae* *Zymomonas mobilis*
C_{12}–C_{20} fatty acids	Sucrose	Bacteria, mold, yeast
Fumaric acid	Glucose	*Rhizopus* sp.
Gallic acid	Gallotannins	*Aspergillus wentii* *Rhus coriaria*

TABLE 13.6 (*Continued*)

Chemical	Substrate(s)	Microorganism(s)
Gibberellic acid	Glucose, starch, sucrose	*Gibberella fujikuroi*
Gluconic acid	Glucose	*Aspergillus niger* *Gluconobacter suboxydans*
Gluconolactone	Glucose	*Acetobacter suboxydans* *Gluconobacter suboxydans*
L-Glutamic acid[d]	Acetic acid, sugars, molasses	Glutamic acid bacteria *Brevibacterium flavum* *B. Lactofermentum* *Corynebacterium glutamicum*
Glycerol	Glucose, lactose	*Kluyveromyces fragilis* *Pichia farinosa* *Saccharomyces cerevisiae*
Hyaluronic acid	Glucose	Literature cited
Hydrobutyric acid	Fructose, glucose, acids	*Alcaligenes eutrophus* *Aspergillus griseus*
Indigo[c]	Glucose	Recombinant *Escherichia coli*
Isocitric acid	Glucose	*Clostridium brumptii*
Isopropyl alcohol	Various sugars	*Clostridium* sp.
Itaconic acid	Glucose, sucrose	*Aspergillus itaconicus* *A. terreus* *Ustilago zeae*
Itatartaric acid	Glucose	*Aspergillus terreus*
2- and 5-ketogluconic acids	Glucose	*Gluconobacter suboxydans* *Pseudomonas fluorescans*
α-Ketoglutaric acid	Glucose (?)	*Pseudomonas* sp.
Kojic acid	Glucose	*Aspergillus oryzae*
Lactic acid	Various sugars	*Bacillus dextrolacticus* *Lactobacillus delbrueckii*
L-Leucine[d]	Glucose	*Brevibacterium lactofermentum* *Serratia marcescens*
Levan	Sucrose	*Bacillus* sp. *Pseudomonas* sp. *Erwinia herbicola* *Leuconostoc mesenteroides* *Microbacterium laevaniformans* *Serratia marcescens* *Zymonas mobilis*
Linoleic acid	Glucose, lactose	*C. curvata*
Linolenic acid	Various sugars	*Mortierella ramammiana*

(continues)

TABLE 13.6 (*Continued*)

Chemical	Substrate(s)	Microorganism(s)
L-Lysine[c,d]	Glucose	*Brevibacterium flavum* *Corynebacterium glutamicum* *Micrococcus glutamicus* *Serratia marcesens*
Malic acid	Acetate, ethanol, fumarate, glucose, propionate	*Aspergillus flavus* *Brevibacterium flavum* *Lactobacillus brevis* *Paecilomyces varioti* *Pichia membranaefaciens*
Oleic acid	Glucose, lactose	*C. curvata*
Oxalic acid[c]	Various sugars	*Penicillium* and *Aspergillus* sp.
Palmitic acid	Glucose, lactose	*C. curvata*
Penicillins (4-thia-1-aza-bicyclo [3.2.0] heptanes)[c]	Corn steep liquor, lactose	*Penicillium chrysogenum* *Penicillium notatum*
Phosphomannans	Glucose, sucrose	*Hansenula capsulata*
Phosphoenolpyruvic acid	Glucose	*Debaryomyces couderti*
Polyhydroxyalkanoates (polyhydroxybutyrates)	Glucose, sucrose, xylose, others	Recombinant *Escherichia coli*[h] *Alcaligenes eutrophus* *Azotobacter* *Methylobacterium* *Methylocystis* *Pseudomonas cepacia* *Pseudomonas oleovorans* *Rhodospirillum rubrum*
Poly-β-hydroxybutyric acid[f]	Fructose	*Alcaligenes* sp. A-04
Propanediol	Algal biomass, glucose	*Clostridium pasteurianum* *C. thermosaccharolyticum*
n-Propanol	Glucose	*Clostridium fallax*
Propionic acid	Various sugars	*Clostridium* sp. *Propionibacterium shermanii*
Pullulans	Sucrose, other sugars	*Aureobasidium pullulans*
Pyruvic acid	Glucose	*Pseudomonas aeruginosa*
Scleroglucans	Glucose, sucrose	*Sclerotium* sp.
Sorbitol	Sucrose	*Zymomonas* sp.
Stearic acid	Glucose, lactose	*C. curvata*
Streptomycin[c]	Sugars	*Streptomyces* sp.
Succinic acid[c]	Various sugars	Many species
Tartaric acid	Glucose	*Gluc. suboxydans*
L-Tryptophan[d]	Glucose	*Corynebacterium glutamicum* *Brevibacterium flavum* *Brevibacterium subtilis* *Serratia marcescens*

TABLE 13.6 (*Continued*)

Chemical	Substrate(s)	Microorganism(s)
L-Valine[d]	Glucose	*Corynebacterium glutamicum* *Escherichia coli*
Vitamin B$_2$ (riboflavin)[c]	Sugars, whey	*Ashbya gossypii* *Eremothecium ashbyi*
Vitamin B$_{12}$ (cyanocobalamin)[c]	Sugars, whey	*Pseudomonas* sp.
Xanthan gum	Glucose, sucrose, others	*Xanthomonas* sp.
Xylitol[g]	Hemicelluloses	*Candida guilliermondii*
Xylonic acid	Xylose	*Pseudomonas fragi*
Zanflo gums	Lactose, others	Mutated soil bacterium

[a]Unless noted otherwise, the information is adapted from Leeper, Ward, and Andrews (1991), which contains specific references for each chemical listed.
[b]Hudson (1988); Yang *et al.* (1993).
[c]Fieser and Fieser (1950).
[d]Araki and Ozeki (1992).
[e]Berry *et al.* (1995).
[f]Kulprecha, Phonaprapai, and Chanchaichaovovat (1994).
[g]Sugai *et al.* (1994).
[h]Williams, Gerngross, and Peoples (1995).

range of \$4.40-\$5.50/kg, depending on grade, at an initial production capacity of only about 1000 t/year.

A few of the amino acids that are produced by microorganisms are listed in Table 13.6. Under normal growth conditions, the regulatory mechanisms in a given microorganism will ordinarily control the biosynthesis of amino acids in quantities that are only sufficient to meet growth requirements. However, some naturally occurring and mutant strains of Arthrobacter, Corynebacterium, Brevibacterium, and Nocardia species have been known to excrete large amounts of certain amino acids into the medium. The three groups of microorganisms that excrete amino acids in sufficient amounts to facilitate their manufacture are the glutamic acid bacteria, such as *Brevibacterium flavum*, *B. lactofermentum*, and *Corynebacterium glutamicum*; the enteric bacteria, such as *Escherichia coli* and *Serratia marcescens*; and certain Bacillus species, such as *Bacillus subtilis* (*cf.* Araki and Ozeki, 1992). The acids are accumulated by auxotrophic mutant strains, which are modified to require the addition to the medium of certain growth factors such as vitamins and amino acids. In these mutants, the formation of regulator effector(s) on amino acid biosynthesis is genetically blocked and the concentration of the effector(s) is kept low enough to release the regulatory mechanism and induce the overproduction of the corresponding amino acid and its accumulation outside the cells. The result is that the microbial process is more economical than natural protein hydrolysis or chemical synthesis. Glutamic acid, lysine, ornithine, tryptophan, and valine are manufactured by means of this technology.

Simple chemicals can often be converted to other chemicals with live microorganisms or cell-free enzyme catalysts. All processes that use microorganisms to produce chemicals involve the catalysis of chemical reactions by enzymes generated by the microorganisms. Enzymes have the unique capability of catalyzing specific chemical reactions at very high selectivities, particularly for the synthesis of stereospecific products. Enzymes are generally classified as hydrolases, which catalyze hydrolysis reactions; isomerases, which catalyze isomerization reactions; transferases, which catalyze the transfer of groups; oxidoreductases, which promote oxidation-reduction reactions; lyases, which catalyze the removal or addition of groups to double bonds; and ligases, which join molecules at the expense of high-energy bonds. An intensive research effort has been carried out over the past several decades to isolate and characterize enzymes generated by microorganisms, and to produce and market cell-free enzymes for use as catalysts in the synthesis of commodity and specialty chemicals. Recombinant DNA methods have made it possible to greatly expand the number of available cell-free enzymes. Among the commercial preparations are amylases, proteases, cellulases, glucose isomerases and oxidases, pectinases, invertases, and cofactors. The synthesis of nutritionally important amino acids and medically important pharmaceuticals and chemotherapeutic agents are examples of the commercial applications of enzyme catalysts.

Some examples of organic chemicals that are produced from simple organic intermediates or secondary biomass derivatives by live microorganisms and cell-free enzymes are listed in Table 13.7. Here, unlike the tabulation of microbial conversions listed in Table 13.6 with live organisms, the majority of the examples cited use as reactants simple chemicals that are not derived from biomass. The cell-free enzymes cited function more like traditional heterogeneous catalysts in thermochemical conversion processes for fossil-based feedstocks. Note that several of the transformations listed cannot be carried out as single-step processes by thermochemical conversion. Two or more steps may be necessary. This suggests that certain combinations of conventional catalysts, microorganisms, and cell-free enzyme catalysts might offer chemical synthesis possibilities that are not otherwise feasible. One of the attractive features of enzyme catalysis alone is that it is possible to design reactors in which immobilized layers of several enzymes in the proper sequence perform the equivalent of a complex, multistep, thermochemical reaction sequence in a single, small-volume reactor. A few techniques have been developed in which immobilized live microorganisms can function in the same manner.

Despite the great versatility and capabilities of microorganisms and cell-free enzymes, few microbial processes have been commercialized because of their inherent characteristics. Fermentation processes are carried out with live organisms in liquid water at low rates. Microorganisms are sensitive to inhibitors and operating conditions, especially temperature and pH. The process must often be performed under aseptic conditions to avoid contamination of the culture. Steady-state operations can be difficult to sustain because of the sensitivity of the microorganisms and the tendency for mutations to occur, especially under continuous processing conditions. Inocula of viable microorganisms must be available and on standby for a number of commercial processes in case of plant upsets. The recovery of the desired end product can be costly when it is water soluble and is formed at low concentration. Product recovery costs are often related to feedstock purity and can exceed the conversion costs, particularly for low end-product concentrations in aqueous media. Useful by-products can sometimes be extracted or separated from fermenter broths, but the effluents are in most cases still high-BOD (biological oxygen demand) waste streams that must be treated by other microbial processes, which create additional biosolids for disposal, before the water is discharged. Cell-free enzyme systems have some of the same characteristics as fermentation systems, among them being the maintenance of catalytic activity and the sensitivity to operating conditions.

These are the main reasons why relatively few microbial processes have been commercialized for the synthesis of commodity organic chemicals from biomass feedstocks. Numerically, only about 200 chemicals are commercially manufactured by microbial conversion of biomass, compared to the many thousands that are in commercial use. And well over half of the chemicals manufactured by microbial processes are high-value, low-volume antibiotics, enzymes, and amino acids. Many of these compounds have complex structures and multiple functional groups that make microbial synthesis technically and economically preferable to thermochemical synthesis. A few chemicals manufactured by fermentation of biomass feedstocks have clearly reached commodity status. They are ethanol, monosodium L-glutamate, citric acid (2-hydroxy-1,2,3-propanetricarboxylic acid), L-lysine (α,ε-diaminocaproic acid), and gluconic acid, as indicated by the data shown in Table 13.8 (Hinman, 1993). Together, these chemicals had annual worldwide production of about 17 million tonnes and annual sales of about \$17 billion in the late 1980s and early 1990s. The usual feedstocks for manufacture of the five commodity chemicals in Table 13.8 are the common carbohydrates. Note that the annual U.S. production of fermentation ethanol, citric acid, and especially lysine is a relatively large percentage of worldwide production, but the production of monosodium glutamate is a small fraction of the world's total annual production (cf. Table 13.1 and Table 13.8). This is due to the large market for lysine as an animal

TABLE 13.7 Examples of Organic Chemicals Produced by Fermentation or Enzyme-Catalyzed Conversion of a Biomass Component, Primary or Secondary Biomass Derivative, or Other Organic Chemical[a]

Chemical	Reactant(s)	Organism, enzyme, or enzyme source
Acetaldehyde	Ethanol	Alcohol dehydrogenase
	Ethanol	Candida utilis
	Glucose + O_2	Alcohol oxidase
Acetone	Isopropyl alcohol	Methylotrophic bacteria
Acrolein	Glycerol	Glycerol dehydrogenase
		Bacillus amaracrylus
		Bacillus welchii
		Candida boidinii
Acrylamide	Acrylonitrile	Brevibacterium sp.
Acrylic acid	β-Alanine	Megasphaera elsdenii
	Propionate	Clostridium propionicum
	Lactic acid	Literature cited
Adipic acid	Alkanes	Mutated organisms
	Cyclohexane	A five-enzyme sequence
L-Alanine[b]	L-Aspartic acid	Pseudomonas dacunhae
D-Aspartic acid[b]	DL-Aspartic acid	Aspartic-β-decarboxylase
L-Aspartic acid[b]	Fumaric acid	Aspartase
Benzoic acid	Phenylacetic acid	Literature cited
	Toluene	Methylotrophic bacteria
	Xanthine	Xanthine oxidase
Benzoin	Benzil	Xanthomonas oryzae
Benzyl alcohol	m-Chlorotoluene, toluene	Methylotrophic bacteria
Catechol	Benzene	Pseudomonas sp.
p-Cresol	Toluene	Methylotrophic bacteria
Diacetyl	Acetoin	Literature cited
1,4-Diaminobutane	Ornithine	Ornithine decarboxylase
1,5-Diaminopentane	Lysine	Lysine decarboxylase
Dihydroxyacetone	Glycerol	Acetobacter sp.
Dihydroxybiphenyl	Biphenyl	Fungal laccase
1,2-Epoxybutane	2,3-Butadiene	Methane monooxygenase
Ethylene	Ethanol	Penicillium cyclopium
Ethylene oxide	Tehylene	Literature cited
Formaldehyde + H_2O_2	Methanol + O_2	Alcohol oxidase
D-Fructose	D-Glucose	Glucose isomerase
Gluconic acid	Glucose	Glucose oxidase
Glucoconolactone	Glucose	Glucose oxidase

TABLE 13.7 (*Continued*)

Chemical	Reactant(s)	Organism, enzyme, or enzyme source
Glucosone + H₂O₂	Glucose	Pyranose-2-oxidase
Glycerol	Fats	Lipase
2-Heptanone	Octanoic acid	*Penicillium roquefortii*
Hydrolyzed lactose syrup	Lactose	Lactase
Hydroquinone	Benzene	Methylotrophic bacteria
	Benzoquinone	Glucose oxidase
m-, *p*-Hydroxybenzaldehydes	*m*-Cresol	Methylotrophic bacteria
o-, *p*-Hydroxyethylbenzenes	Ethylbenzene	Methylotrophic bacteria
p-Hydroxymethylstyrene	α-Methylstyrene	Methylotrophic bacteria
p-Hydroxypropylbenzene	Propylbenzene	Methylotrophic bacteria
p-Hydroxystyrene	Styrene	Methylotrophic bacteria
Isobutylene	Isovaleric acid	*Rhodotorula minuta*
2-Ketogluconic acid + H₂O₂	Glucosone	Glucose-1-oxidase
L-Lysine[b]	DL-α-Aminocaprolactam	Hydrolase + Racemase
	α,ε-Diaminopimelic acid	DAP decarboxylase
Malic acid	Fumaric acid	*Brevibacterium flavum*
Maltose	Starch	α-Amylase
Mandelic acid	Benzoylformic acid	Hydroxyisocaproate dehydrogenase
Methanol	Methane	Methane monooxygenase
Methacrylate	Isobutyric acid	*Clostridium propionicum*
5-Methyl-1,3-butanediol	*o*-Cresol	Methylotrophic bacteria
α-Naphthol	Naphthalene	Methylotrophic bacteria
β-Naphthol	Naphthalene	Methylotrophic bacteria
Phenol	Benzene	Methane monooxygenase
Phenylethanol	Ethylbenzene	Methylotrophic bacteria
Propylene oxide	Chlorohydrin	*Nocardia corallina*
	Propylene + H₂O₂	Haloperoxidase
	Propylene	Methane monooxygenase
Pyridine-*N*-oxide	Pyridine	Methylotrophic bacteria
Sebacic acid	Long-chain alkanes	Mutated organisms
Succinic acid	Malic acid	*Bacterium succinicum*
Styrene oxide	Styrene	Methylotrophic bacteria
n-Tetradecanol	*n*-Tetradecane	*Candida parapsilosis*
Vinyl acetate	Acetic acid, ethylene oxide	Literature cited

[a] Unless noted otherwise, the information is adapted from Leeper, Ward, and Andrews (1991), which contains specific references for each chemical listed.
[b] Araki and Ozeki (1992).

TABLE 13.8   Worldwide Production of Selected Fermentation Products[a]

Chemical	Approximate Production (t/year)	Sales (10^9)	For year
Ethanol	15,000,000	15.000	1990[b]
Monosodium glutamate	1,000,000	1.436	1992[b]
Citric acid	400,000	0.560	1991[b]
Lysine	115,000	0.288	1989[b]
Gluconic acid	50,000		1986
Ionophores (e.g., salinomycin)	3000	0.300	1992[c]
Ivermectins	15	0.600	1992[c]
Human insulin	6	2.000	1992[d]
Tissue plasminogen activator	0.01	0.230	1992[d]
Human growth hormone		0.375	1992[d]
Erythropoietin		0.500	1992[d]
Neupogen		0.550	1992[d]

[a]Hinman (1993).
[b]Data from SRI Consulting.
[c]Data from Pfizer, Inc.
[d]Data from Decision Resources and Pfizer, Inc.

feed supplement in the Americas and the large U.S. demand for fermentation ethanol as an oxygenate in motor gasolines. There is only a small U.S. demand for monosodium glutamate. Gluconic acid is not included in Table 13.1 because annual U.S. production is smaller than the production of the lower ranked chemicals in this table.

The ionophores and several other specialty products are included in Table 13.8 for comparison purposes. Products of mammalian cell culture such as plasminogen activator and erythropoietin are included as fermentation products in this listing because they are normally manufactured by cellular processes in bioreactors. Aside from the five commodity chemicals in this table, the most dramatic change in the commercial chemicals produced by fermentation results from the impact of genetic engineering and recombinant DNA methods on the specialty products. Antibiotics and biopolymers (hormones, enzymes, etc.) with molecular structures too complex for conventional chemical synthesis will continue to be manufactured by microbial processes (Hinman, 1993).

The potential of microbial processes has yet to be realized for the commercial syntheses of a large number of commodity organic chemicals. Biomass feedstocks, the availability of appropriate organisms, and also, therefore, the availability of appropriate enzymes do not limit the applications of microbial processes. Indeed, the capabilities of microorganisms as efficient synthesizers of organic chemicals led to the somewhat assonant Organic Chemist's Prayer (of unknown origin):

> *Dear God:*
> *I pray on bended knees,*
> *That all of my syntheses,*
> *Will never be inferior,*
> *To those conducted by bacteria.*

The questions of how and when the microbial synthesis of commodity organic chemicals will play a larger role in the world's chemical markets remain to be answered. The subject has appeared, disappeared, and then reappeared many times since 1950 when petroleum and natural gas prices increased to unacceptable levels in the United States. It is highly probable, however, that most of the specialty chemicals that are manufactured today by microbial conversion of biomass feedstocks will continue to be manufactured the same way.

In 1978, Irving S. Shapiro, chairman at that time of one of the world's largest chemical companies, E. I. du Pont de Nemours & Company, stated: "Energy requirements beyond the year 2000 indicate a need to turn to fusion and to extensive use of solar energy and biomass, to reduce demands on fossil resources. It is assumed that by then the chemical industry will be drawing more on coal as the initial source of its hydrocarbons, and less on what remains of the oil and natural gas." The key word here is "hydrocarbons." As shown in this chapter, hydrocarbon feedstocks are not essential for the manufacture of organic chemicals.

REFERENCES

Ainsworth, S. (1996). *Chem. and Eng. News* 74 (21), 28.

American Chemical Society (1996). *Chem. and Eng. News* 74 (26), 38.

American Chemical Society, American Institute of Chemical Engineers, Chemical Manufacturers Association, Council for Chemical Research, and Synthetic Organic Chemical Manufacturers Association (1996). "Technology Vision 2000." American Chemical Society, Washington, D.C., December.

Antal, M. J., Jr. (1981). *In* "Biomass as a Nonfossil Fuel Source," (D. L. Klass, ed.), ACS Symposium Series 144, p. 313. American Chemical Society, Washington, D.C.

Åppelqvist, L.-A. (1989). *In* "Oil Crops of the World," (Röbbelen, G., Downey, R.K., and Ashri, A., eds.), p. 22. McGraw-Hill, New York.

Araki, K., and Ozeki, T. (1992). *In* "Kirk-Othmer Encyclopedia of Chemical Technology" (J. I. Kroschwitz and M. Howe-Grant, eds.), 4th Ed., Vol. 2, p. 504. John Wiley, New York.

Bakhshi, N. N., Kaitikaneni, P. R., and Adjaye, J.D. (1995). *In* "Second Biomass Conference of the Americas: Energy, Environment, Agriculture, Industry," NREL/CP-200-8098, DE95009230, p. 1089. National Renewable Energy Laboratory, Golden, CO.

Berry, A., Battist, S., Chotani, G., Dodge, T., Peck, S., Power, S., and Weyler, W. (1995). *In* "Second Biomass Conference of the Americas: Energy, Environment, Agriculture, Industry" NREL/CP-200-8098, DE95009230, p. 1121. National Renewable Energy Laboratory, Golden, CO.

Busche, R. M. (1984). *In* "Energy From Biomass and Wastes VIII," (D. L. Klass and H. H. Elliott, eds.), p. 1295. Institute of Gas Technology, Chicago.

Cass, O. W. (1947). *Chem. Industries* 60, 612.

Chem. Mkt. Reporter (1997). 251 (12), March 24.

Cunningham, R. E. (1993). *In* "First Biomass Conference of the Americas: Energy, Environment, Agriculture, Industry," NREL/CP-200-5768, DE93010050, p. 1906. National Renewable Energy Laboratory, Golden, CO.

Fieser, L. F., and Fieser, M. (1950). "Organic Chemistry." D.C. Heath and Company, New York.

Fitzpatrick, S., and Jarnefeld, J. (1996). *In* "Bioenergy '96, Proceedings of the Seventh National Bioenergy Conference," Vol. II, p. 1083. Tennessee Valley Authority, The Southeastern Regional Biomass Energy Program, Muscle Shoals, AL.

Fourie, J. H. (1992). *In* "Asian Natural Gas—New Markets and Distribution Methods," (D. L. Klass, T. Ohashi, and A. Kutsumi, eds.), p. 583. Institute of Gas Technology, Chicago.

Goldstein, I. S. (1976). "Wood as a Source of Chemical Feedstocks." Paper presented at 69th Annual meeting. American Institute of Chemical Engineers, Chicago, December.

Graham, R. G. (1991). *In* "IEA Bioenergy Agreement, Task VII. Biomass Conversion, Activity 4: Thermal Gasification," (Compiled by Institute of Gas Technology), p. 1. Institute of Gas Technology, Chicago, December.

Hasche, R. L. (1945). *Chem. and Eng. News* 23 (20), 1840.

Himmelblau, D. A. (1995). *In* "Second Biomass Conference of the Americas: Energy, Environment, Agriculture, Industry," NREL/CP-200-8098, DE95009230, p. 1141, National Renewable Energy Laboratory, Golden, Colorado.

Hinman, R.L. (1993). *In* "First Biomass Conference of the Americas: Energy, Environment, Agriculture, Industry" NREL/CP-200-5768, DE93010050, p. 1441. National Renewable Energy Laboratory, Golden, CO.

Hudson, L. R. (1988). *In* "Energy From Biomass and Wastes XI," (D. L. Klass, ed.), p. 893. Institute of Gas Technology, Chicago.

Hunt, P. (1993). *In* "Asian Natural Gas—Development of the Domestic Industry," (D. L. Klass, H. Kataoka, and K. Ueda, eds.), p. 549. Institute of Gas Technology, Chicago.

Kirschner, E. (1996). *Chem. and Eng. News* 74 (15), 16.

Kirschner, E. (1997). *Chem. and Eng. News* 75 (16), 19.

Kulprecha, S., Phonprapai, and Chanchaichaovivat, A. (1994). *In* "Biomass for Energy, Environment, Agriculture, and Industry, Proceedings of the 8th European Biomass Conference, October 1994," (Ph. Chartier, A.A.C.M. Beenackers, and G. Grassi, eds.), Vol. 2, p. 1377. Elsevier Science, New York.

Leeper, S. A., Ward, T. E., and Andrews, G. F. (1991). "Production of Organic Chemicals via Bioconversion: A Review of the Potential," Report EGG-BG-9033. Idaho National Engineering Laboratory, Idaho Falls, ID

Lipinsky, E. S., and Ingham, J. D. (1994). "Brief Characterizations of the Top 50 U.S. Commodity Chemicals," Final Task Report. Pacific Northwest Laboratory, Richland, WA, September 1.

Lipinsky, E. S., and Wesson, R. (1995). "Characterization of the Top 12 U.S. Commodity Polymers," Draft Task Report, Battelle, Columbus, OH, July 6.

Milas, N. A., and Walsh, W. L. (1935). *J. Am. Chem. Soc.* 57, 1389.

Miller, K. D., Jr. (1993). *In* "First Biomass Conference of the Americas: Energy, Environment, Agriculture, Industry," NREL/CP-200-5768, DE93010050, p. 1159. National Renewable Energy Laboratory, Golden, CO.

Nikitin, N. I. (1962). "The Chemistry of Cellulose and Wood," (translated from the Russian by J. Schmorak, 1966). Israel Program for Scientific Translations Ltd., Jerusalem.

Parker, S., Calnon, M., Feinberg, D., Power, A., and Weiss, L. (1983). "The Value of Furfural/Ethanol Coproduction from Acid Hydrolysis Processes," SERI/TR-231-2000. National Renewable Energy Laboratory, Golden CO.

Peaff, G. (1994). Chem. and Eng. News 72 (43), 13.

Piskorz, J., Radlein, D., Majerski, P., and Scott, D. (1995). In "Second Biomass Conference of the Americas: Energy, Environment, Agriculture, Industry," NREL/CP-200-8098, DE95009230, p. 1151. National Renewable Energy Laboratory, Golden, CO.

Plastics Tech. (1997). "Pricing Update," 43 (3) 55, March.

Rahardja, S., Rigal, L., Gaset, A., Barre, L., Chornet, E., and Vidal, P. (1994). In "Biomass for Energy, Environment, Agriculture, and Industry, Proceedings of the 8th European Biomass Conference, October 1994," (Ph. Chartier, A. A. C. M. Beenackers, and G. Grassi, eds.), Vol. 2, p. 1420. Elsevier Science, New York.

Raymond, A. L. (1943). In "Organic Chemistry, An Advanced Treatise," (H. Gilman et al., eds.), Vol. II, p. 1605. John Wiley, New York.

Rejai, B., Evans, R. J., Milne, T. A., Diebold, J. P., and Scahill, J. W. (1991). In "Energy From Biomass and Wastes XV," (D. L. Klass, ed.), p. 855. Institute of Gas Technology, Chicago.

Rose, G. R., Singh, S. P., Onischak, M., and Babu, S. P. (1981). In "Energy From Biomass and Wastes V," (D. L. Klass and J. W. Weatherly, III, eds.), p. 613. Institute of Gas Technology, Chicago.

Rouhi, A. M. (1996). Chem. and Eng. News 74 (39), 62.

Scott, D. S., Piskorz, J., and Radlein, D. (1993). In "Energy From Biomass and Wastes XVI," (D. L. Klass, ed.), p. 797. Institute of Gas Technology, Chicago.

Shapiro, I. S. (1978). Science 202 (4365), 287.

Shapouri, H., Duffield, J. A., and Graboski, M. S. (1996). "Liquid Fuels and Industrial Products from Renewable Resources, Proceedings of the Third Liquid Fuel Conference," (J. S. Cundiff et al., eds.), p. 253. The American Society of Agricultural Engineers, St. Joseph, MI.

Smith, H. A., and Fuzek, J. F. (1949). J. Chem. Soc. 71, 415.

Steinberg, M., Fallon, P. T., and Sundaram, M. S. (1983). In "Energy from Biomass and Wastes VII," (D. L. Klass and H. H. Elliott, eds.), p. 1171. Institute of Gas Technology, Chicago.

Sugai, J. K., Delgenes, J. P., Preziosi-Belloy, L., Moletta, R., and Navarro, J. M. (1994). In "Biomass for Energy, Environment, Agriculture, and Industry, Proceedings of the 8th European Biomass Conference, October 1994," (Ph. Chartier, A. A. C. M. Beenackers, and G. Grassi, eds.), Vol. 2, p. 1270. Elsevier Science, New York.

Thomas, J. J., and Barile, R. G. (1984). In "Energy From Biomass and Wastes VIII," (D. L. Klass and H.H. Elliott, eds.), p. 1461. Institute of Gas Technology, Chicago.

Tonkovich, A. L. Y., and Gerber, M. A. (1995). "The Top 50 Commodity Chemicals: Impact of Catalytic Process Limitations on Energy, Environment, and Economics," Draft Report. Pacific Northwest Laboratory, Richland, WA, July 17.

U.S. Dept. of Energy (1995). "Technology Partnerships," DOE/GO-10095-170, DE95004086, Washington, D.C., April.

U.S. International Trade Commission (1995). "Synthetic Organic Chemicals, United States Production and Sales, 1994," ISITC Publication 2933, Washington, D.C., November.

Wiley, R. H. (1953). In "Organic Chemistry, An Advanced Treatise," (H. Gilman et al., eds.), Vol. IV, p. 723. John Wiley, New York.

Williams, S. F., Gerngross, U. T., and Peoples, O. P. (1995). In "Second Biomass Conference of the Americas: Energy, Environment, Agriculture, Industry," NREL/CP-200-8098, DE95009230, p. 1130. National Renewable Energy Laboratory, Golden, CO.

Wolfrom, M. L. (1943). *In* "Organic Chemistry, An Advanced Treatise," (H. Gilman *et al.*, eds.), Vol. II, p. 1532. John Wiley, New York.

Yang, S.-T., Zhu, H., Li, Y., and Tang, I-C. (1993). *In* "First Biomass Conference of the Americas: Energy, Environment, Agriculture, Industry," NREL/CP-200-5768, DE93010050, p. 1305. National Renewable Energy Laboratory, Golden, CO.

Integrated Biomass Production–Conversion Systems and Net Energy Production

I. INTRODUCTION

The costs of competitive fuels; fossil and biomass energy availability; environmental issues such as global warming, greenhouse gases, air and water quality, and waste treatment and disposal; the infrastructure needed to distribute biomass energy and biofuels to end users; government policies, energy demand, and national security; the development of new energy technologies; and new energy resources and reserves determine the role of biomass energy in the marketplace. Intensive as well as extensive parameters, many of which are "foggy" and difficult to assess, are among the influential factors. One of the greatest barriers to overcome is the fallibility of predictions. The problem is perhaps best illustrated by the difficulty of analyzing energy markets. Innumerable studies have been performed to predict future energy consumption patterns, oil imports, and oil prices. Unfortunately, many of these studies have been far off the mark, especially those that concern future oil prices. If the predictions of $40/bbl of crude oil and more in the United States had been accurate in the 1970s and 1980s and oil had stabilized at that price level, or even at its peak in 1980, the biomass energy industry would have exhibited much greater growth than it has. It is evident that the costs of refined petroleum products are relatively low in the United States and have been for decades except in times of international conflict and supply disruptions (Chapter 1).

Despite the apparent difficulty of large-scale entry of biomass energy into U.S. markets because of the availability of low-cost fossil energy, this situation is not expected to continue *ad infinitum*. There are also many areas of the world where fossil fuels are much more expensive than they are in the United States, and biomass energy costs are competitive. Two of the factors that will ultimately determine the role of biomass in future energy markets are the capability of biomass energy systems to make large, sustainable contributions to energy demand at competitive costs, and the capability of these systems to yield net salable energy. Basically, the goal is to design and operate environmentally acceptable systems that furnish new supplies of salable energy from biomass, whether they are low-energy gas, substitute natural gas (SNG), synthetic crude oil, methanol, ethanol, hydrogen, other fuels, and heat, steam, or electricity, at the lowest possible cost with a minimum consumption of energy inputs. It is necessary to quantify how much energy is consumed and how much salable energy is produced because the capital and operating costs are insufficient alone to determine the best systems. Economic data do not always correlate with net energy production. Comparative analyses of similar systems for production of synthetic liquid or gaseous fuels from the same feedstock or of different systems that yield the same fuels from different biomass feedstocks should be performed by consideration of economics *and* net energy production.

A complete net energy analysis (NEA) of an integrated synfuel-production system is an accounting of all of the energy inputs and salable energy products in such a way that even the energy required to build the integrated system is included as a capital energy investment, just as with a capital dollar investment. In other words, a complete NEA is not simply an input–output energy balance. It is important to know the energy cost of the salable products because energy and monetary costs are not necessarily related. The question "How many capital and operating energy units must be invested to produce an energy unit of salable product?" is not equivalent to the corresponding question regarding monetary units. Most NEAs, however, ignore capital energy investments and focus on the gross amount of salable energy produced minus the amount of energy needed to produce it.

The subject of NEAs was quite controversial in the early 1970s when first introduced, probably because different groups used different methodologies to perform them, and the results were inconsistent and not comparable. Today, NEAs can provide important data that are not easily obtained by other means because of system complexity and the interactions between system components. A detailed NEA of a complex integrated system can help pinpoint specific operating functions that may be amenable to improvement or that need to be modified. For a biomass energy system, the amount of nonrenewable fossil energy used to operate the system is obviously an important parameter. It is possible that less net salable energy will reach market than that withdrawn from the market to operate the system. For fuel ethanol from biomass, for example, an integrated biomass production-conversion system (IBPCS) can consume more fossil energy than the net ethanol output. Such a result would defeat the purpose of manufacturing fermentation ethanol if the objective is to displace and conserve petroleum fuels. This statement is not quite correct, as will be shown later, because of the energy content of co-products, and the fact that the chemical energy content of biomass is essentially derived from solar energy, which does not affect fossil energy reserves. The quality of salable energy products from an IBPCS is also a factor. If an IBPCS that produces hydrogen and SNG by biomass gasification uses diesel fuel to grow, manage, harvest, and transport the feedstock to the conversion plants, and No. 2 fuel oil is used to generate electricity to operate the conversion plants, how should the inherent differences in energy quality of the operating fuels and salable energy products be reconciled?

In the United States, the subject of NEAs has even been written into federal laws—the Federal Nonnuclear Energy Research and Development Act of 1974 (P.L. 93-577 as amended by P.L. 94-187) and Title II of the Energy Security Act of 1980 (P.L. 96-294). They require the U.S. Department of Energy to analyze the potential net energy yields of new energy technologies proposed under the Acts before funding their development. Although these particular laws are old, their provisions do not appear to have ended, and other federal laws occasionally include NEA requirements. The U.S. Comptroller General has stated that the U.S. Department of Energy has spent hundreds of millions of dollars on projects without performing NEAs. Indeed, NEAs have been largely ignored in the United States by developers of alternative energy systems.

As stated in Chapter 2, IBPCSs that are capable of producing quad blocks of energy at competitive prices are essential for biomass energy to have a large role in displacing fossil fuels. If such systems are not developed and commercialized, biomass energy usage will forever be limited to small-scale systems and niche markets. Some of the biomass energy systems of the future are therefore expected to be large IBPCSs that are able to deliver net salable energy in quantity at market prices. This chapter is devoted to an examination of IBPCSs and NEAs.

II. INTEGRATED SYSTEMS

A. DEFINITION AND FUNDAMENTALS

An IBPCS is usually defined to be a system in which all operations concerned with the production of virgin biomass feedstocks as dedicated energy crops and feedstock conversion are integrated to provide a balanced operating system. Multiple feedstocks, including combined biomass–fossil feedstocks and waste biomass, may be employed. Feedstock supply, or supplies in the case of a system that converts two or more feedstocks, is coordinated with the availability factor (operating time) of the conversion plants. Since growing seasons vary with geographic location and biomass species, provision is made for feedstock storage to maintain sufficient supplies to sustain plant operating schedules.

The proper design of an IBPCS requires the coordination of numerous operations such as biomass planting, growth management, harvesting, storage, transport to conversion plants, retrieval, drying, conversion to products, emissions control, product separation, recycling, wastewater and waste solids treatment and disposal, maintenance, and transport or transmission of salable products to market. The design details depend on the biomass species involved and the type, size, number, and location of biomass growth and processing areas. An example of a framework proposed for assessment and design of IBPCSs for electricity production is shown in Table 14.1. It is obvious that a multitude of parameters are involved. In the idealized case, the synfuel production plants are located in or near the biomass growth areas to minimize the cost of transporting the harvested biomass to the plants, all the nonfuel effluents of which are recycled to the growth areas as shown in Fig. 14.1. If this kind of integrated synfuel plantation can be implemented in the field, it would be equivalent to an isolated system with inputs of solar radiation, air, CO_2, and minimal water, and one output, synfuel. The nutrients are kept within the ideal system so that the addition of external fertilizers and chemicals is not necessary. Also, the environmental controls and waste disposal problems are minimized.

Various modifications of the idealized system can be conceptualized for large-scale usage. A few examples are presented here using the United States as the growth area because of its different climatic regions. One modification might consist of the controlled flow of wastewater effluent from several municipalities in the southern United States into an aquatic biomass growth area and the growth of water hyacinth for two purposes: the treatment of wastewater by luxuriant uptake of nutrients by water hyacinth, and the simultaneous growth, harvesting, and conversion of hyacinth to synfuel. In this case, inorganic material builds up in the system. So residual material from the conversion plant is partially removed or "bled" from the system as synfuel is produced. If it has useful applications, the residuum might be considered to be a co-product. Depending on its physical and chemical properties, useful applications developed for such co-products in the past include road surfacing coatings, concrete and asphalt additives, construction materials, and nutrients for fertilizers and land amendment.

TABLE 14.1 Framework for Assessment and Design of Integrated Biomass Production-Conversion System for Electric Power Production[a]

Biomass production	Logistics/transport	Conversion
Land base/uses	Plantation access	Local and regional demand (need for power, productive uses, seasonality)
Existing supplies	Road construction	
Assessment of site quality (rainfall, soils, elevation, slope)	Hauling distances	
	Storage (seasonality considerations)	Choice of technology (efficiency, scale, cost/kW, availability of equipment, repairs)
Plantation practices (species and sites, nursery, establishment)	Fuel preparation and quality	
	Co- and by-products	
Productivity	Plantation practice adjustments with hauling distance	Operational aspects (capacity factor, assurance of supply, local control and assurance of supply, local control and management)
Environmental benefits (deforestation, soil erosion, CO_2 and warming, biodiversity)	Alternative supplies	
	Cost/supply dynamics	Grid integration
Species improvement	Methods/equipment	

Energy/economic	Institutional/social	Environmental
Minimum economic returns	Local adaptations due to interactions among policies, resources, and markets	Soil stabilization
Food and fuel competition		Use of multiple clones (ecological diversity and resiliency, capacity of the system to respond to environmental stresses)
Power sector/energy resource planning issues	Identification of lead institutions	
Integration of biomass-fired power into local grid	Assisting institutional development	
Constraints to deployment	Examination of regional and local developmental and environmental programs	Protection of existing forests
Arrangements (financial, land tenure, and extension) to produce biomass		Land restoration
	Obtaining support at local, regional, and national levels	Effects of non-point-source chemicals
Integrating based on total fuel cycle to achieve maximum net benefits		Waste treatment and disposal
	Role of donor organizations	
Financing		

[a]Perlack and Ranney (1993).

Alternatively, short-rotation hybrid poplar and selected grasses could be multicropped on an energy plantation in forested areas of the United States and harvested for conversion to transportation fuels and cogenerated power for use on-site in centrally located conversion plants. The salable energy products are liquid biofuels and surplus electricity. This configuration might be

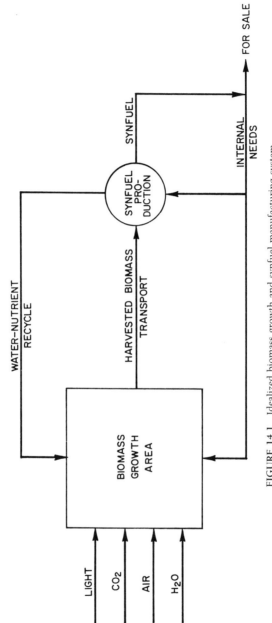

FIGURE 14.1 Idealized biomass growth and synfuel manufacturing system.

especially useful for the larger land-based systems. The system would be somewhat analogous to wood production on tree plantations and conversion by the pulp and paper industry.

Another possibility, especially for groups of small farms, is the integration of agricultural crop, farm animal, and fuel production into one system. For example, farmers in an appropriate region of the U.S. Midwest might grow corn as feedstock for conversion in a cooperative fuel ethanol plant. The equivalent of this is already done on a large scale in the U.S. Corn Belt, except most of the conversion plants are non-coop, commercial plants. A further variation is the return of the residual distillers' dry grains by the cooperative to the farmers for use as animal feed. The resulting animal manures are converted to medium-energy fuel gas in farm-scale anaerobic digestion units. The fuel gas is used on-site to generate heat, steam, and power, and the residual ungasified solids, which are high in nitrogen, potassium, and phosphorus, are recycled to the fields as fertilizer to grow more corn. The salable products are ethanol and co-products from the cooperative conversion plants and farm animals; the residuals are kept within the system.

Still another possibility might be a marine energy farm for the production and harvesting of *Sargassum* in Hawaii's tropical seas, where five species of this particular macroalga are indigenous to the state's coastal waters, and conversion to methane in a system similar to that described in Chapter 12. The SNG plant could be located either on a floating platform near the growth area or on shore. The biomass and fuel transport requirements would be different for each case.

It is apparent that many different integrated system configurations are conceptually possible. As the technology is refined and developed to the point where commercialization activities are well under way, it is expected that optimum designs will evolve. Many of them are referred to in previous chapters. In fact, several small-scale integrated biomass energy systems that can be considered to be modules of large-scale IBPCSs similar to those discussed here have already been designed, built, and tested.

B. IBPCS Characteristics

It is important to realize the general characteristics of IBPCSs and what is required to sustain their operation. As an example, consider an IBPCS that produces salable energy products at a rate of 10,000 BOE/day from virgin biomass. This is a small output relative to most petroleum refineries, but it is not small for an IBPCS. Assume that the conversion plant operates at an availability of 330 day/year and an overall thermal efficiency of feedstock to salable energy products of 60%, a reasonable value for established thermochemical conversion technologies. Equivalent biomass feedstock of average energy content 18.60 GJ/dry t would have to be provided at the plant gate to sustain conversion operations at a rate of 5291 dry t/day, or a total of 1,746,000 dry t/year. This amount of feedstock requires, at an average biomass yield of 25 dry t/ha-year, a biomass growth area of 69,840 ha (270 mi^2). For purposes of estimation, assume the product is methanol and no co-products are formed. The total annual methanol production is then approximately 1,237,000 L/year (327 million gal/year). If the product is methane, the total annual production is 496.4 million m^3(n)/year (18.48 billion SCF/year). If the conversion plant is centrally located within the IBPCS, a radial distance of 14.9 km (9.3 mi) from the plant circumscribes the growth area needed for biomass production exclusive of infrastructure. Fifty-four IBPCSs of this size (i.e., 10,000 BOE/day) are required to yield 1.0 quad of salable energy products per year. The total growth area required is 3,771,400 ha (14,561 mi^2). Again, exclusive of infrastructure and assuming the conversion facilities are all centrally located, this amount of growth area is circumscribed by a radial distance of 101.4 km (68.1 mi) from the plants. This basic analysis shows that the growth areas required for 10,000-BOE/day IBPCSs are large when compared with conventional agricultural practice, but not quite so large when compared with traditional wood harvesting operations in the forest products industry. A single one-quad IBPCS, however, would dwarf most IBPCSs that have been proposed or built to date.

Regarding the energy potential of IBPCSs, a comprehensive analysis of the manufacture of methanol and ethanol in the United States from dedicated agricultural biomass grown on suitable lands indicates that production of 8 quad/year of methanol or 9.4 quad/year of ethanol will be economically feasible by the year 2030 (Reese et al., 1993). It was concluded that such an industry can become commercially viable in the United States and that the agricultural community would benefit. Producers of traditional and biomass energy crops would benefit most. This conclusion was supported by what was termed a Low Biomass Yield scenario and required 45 million ha of growth area. The basic analysis just shown requires about 30 million ha of growth area for production of 8 quad/year of methanol.

These analyses allude to the potential impacts of fluctuating biomass productivities and of feedstock storage instabilities on sustaining IBPCS operations. Backup provisions in case of system upsets, which include those caused by upsets in the conversion plants and the biomass growth and harvesting operations, are desirable. Feedstock storage sites and conditions and the methods of transporting the feedstocks to the plant gate because of the distances and costs involved are significant as well. Just a cursory examination of feedstock storage illustrates how several several different options must be examined. Should feedstock be stored where harvested and shipped as needed, or shipped as harvested and stored in or near the conversion facility? If drying is involved, should the feedstock be shipped before or after drying? It was concluded from one comprehensive analysis of an IBPCS for electricity generation from woody biomass that under certain conditions, the lowest-cost feed supply strategy is to store harvested wood in the field as chips in or near the plantation site until required by the central conversion facility (Toft et al., 1995). Yet several operating wood-fueled IBPCSs for power generation ship logs and sometimes whole trees to the power plant for on-site conversion to chips. Some options even include whole-tree combustion (Ostlie and Drennan, 1989). Thus, the interactions between different components make it difficult to select the optimum pathways and system designs without detailed analysis.

C. EARLY IBPCS ASSESSMENTS

Detailed assessments of conceptual, large-scale IBPCSs were first carried out in the United States in the 1970s. Most of the early studies focused on electricity and SNG production. In one study, a conceptualized biomass plantation consisted of the multicropping of several biomass species, which entailed the integration of planting and harvesting schedules of annuals and perennials to achieve the greatest yields possible while affording a continuous, year-round supply of harvested biomass (Alich and Inman, 1974). Three crops of annuals such as sunflower or kenaf could be produced each year, and perennials, excluding tree species, might be harvested from two to five times each year. Solar drying of the harvested biomass to a moisture content of 10-15 wt % was employed to supply feedstock for direct firing or to an integrated gasification-combined cycle (IGCC) facility. The most favorable region of the United States for the year-round production of biomass according to the study is the Southwest, assuming water is available for irrigation. Irrigation is essential to maximize biomass energy yields in this region. A 1000-MW power plant supplied with this feedstock for direct firing, at a heat rate of about 10.54 MJ/kWh and a load factor of 80%, required 63,500 ha (245 mi^2) of growth area. The cost of electricity from biomass at that time (1973-1974) was estimated to be 1.31 cents/kWh, whereas the cost from a similar coal-fired plant was estimated to be 1.18 cents/kWh. The average retail cost of electricity supplied by privately owned utilities to industrial customers at that time was 1.25 to 1.69 cents/kWh (Energy Information Administration, 1982). A 1000-MW IGCC plant supplied with the same biomass, at a heat rate of about 12.02 MJ/kWh and a load factor of 80%, was projected to require 72,500 ha (280 mi^2) of growth area and to provide electricity at a cost of 1.55 cents/kWh. The cost from a similar coal-fueled plant was estimated to be 1.22 cents/kWh. Improvements in IGCC efficiency were expected to eventually afford lower-cost electricity than direct firing. That in fact has occurred since this study was carried out. The study also included an analysis of biomass gasification at a feedstock conversion efficiency of 60% in amounts equivalent to a 1000-MW power plant, or 2.04 million m^3(n)/day of SNG at an availability of 90%. The required growth area was 36,300 ha (140 mi^2) and the estimated cost of the SNG was $2.20/GJ, whereas the cost of a similar coal gasification facility was $1.70/GJ. The average cost of natural gas delivered to a conventional steam-electric plant at that time was $0.356 to $0.507/GJ.

Other early assessments of IBPCSs supported the position that large-scale systems are technically feasible, and under suitable conditions can be economically attractive (InterTechnology/Solar Corporation, 1977, 1978). An exemplary study was an extension of this initial work to an energy plantation model, the principal design parameters of which are shown in Table 14.2, for short-rotation hardwood production as a dedicated energy crop (Fraser et al., 1981). Although this study is not recent, the methods used for the assessment are still applicable today. The study also illustrates the complexity of IBPCSs and the interactions that can occur between system components. Fifteen sites were chosen for detailed analysis. The selection criteria included resource availability (land, wastewater, labor, climate) and markets for the energy products and services (electricity, SNG, and need for waste treatment and jobs). The biomass

TABLE 14.2   Design Parameters Used for Dedicated Energy Plantation[a]

Parameter	Description
Production	Variable, generally of the order of 200,000 dry t/year (approx. 3.72 TJ/year).
Crop	Fast-growing hardwoods with coppice regrowth.
Productivity	11.2 to 22.4 dry t/ha-year (5 to 10 dry ton/ac-year)
Planting density	0.37 to 1.49 m^2 (4 to 16 ft^2) per tree; i.e., approx. 24,700 to 6200 trees/ha (approx. 10,000 to 2500/ac).
Lifetime	One first-growth rotation followed by five coppice rotations.
Management	Mechanized weed control, fertilization, irrigation (in some modes of operation).
Harvesting	Self-propelled harvester-chipper.
Transportation	Green wood chips transported to conversion facility in center of integrated biomass production-conversion system.
Support	Nursery operation, equipment maintenance, repair, supervision.
Land	Plantation consisting of lots of the size on an average farm in the region distributed at random within a larger geographic area.

[a]Adapted from Fraser et al. (1981).

species included hybrid poplar, Eastern cottonwood, plains cottonwood, silver maple, American sycamore (southern and mid-latitude sites), and *Eucalyptus* (Florida sites only). Growth management options were carefully selected and included irrigation where necessary. The energy plantation is integrated with the conversion plant for production of electricity by direct combustion of hardwood chips, a plant for wastewater treatment and microalgae production, anaerobic digestion of the microalgae and CO_2 removal from the resulting biogas for SNG production, and recycling of the digested solids to the biomass growth area for use as fertilizer (Fig. 14.2). The average transport distance from the growing sites to the conversion facilities is between 11.3 and 32.2 km. Interactions between the plantation and the conversion system components were examined. They may range from the simple recycle of ash from the power plant, to recycle of residual biosolids from the anaerobic digestion plant, to application of wastewater into the plantation. Appropriately sized and specific types of equipment were used for each plantation and conversion plant. For the power plant, a "low-performance" steam cycle and a "high-performance" steam cycle were analyzed. The low-performance cycle represented state-of-the-art, wood-burning technology at the time of the analyses: a simple Rankine cycle with steam supplied at 4480 kPa and 400°C, and a turbine having an efficiency of 79%. The high-performance cycle was comparable to that used in utility systems, with steam supplied at 1650 kPa and 540°C, and a turbine having an efficiency of 90%.

The locations of the IBPCSs and their characteristics and projected economics (1977 dollars) under some of the conditions examined are shown in Tables 14.3 and 14.4. The microalgal pond was designed to handle the wastewater flow from each population center, and the energy plantation was sized to produce solid fuel at an optimally low cost. The ratio of wastewater flow to biomass production is an important quantity that determined the influence of the microalgal pond on the energy plantation. The amount of the SNG produced at each site from microalgae is small compared to the total demand for natural gas at each site. The optimum size of the energy plantation in terms of the cheapest fuel produced was found to be from 9700 to 14,600 ha for the sites analyzed. Differences in the cost of biomass from each energy plantation were related to a number of factors, including the investment, land rental, transportation cost, yield, and amount of nutrients recycled to the plantation. The biomass cost at the 15 sites ranged from \$18.63/dry t to \$26.01/dry t (municipal financing), which was comparable to the cost range estimated on the basis of other tree farm designs for similar locations at that time. Note that these are 1977 dollars. The cost calculated by the municipal method of financing for electricity generated by the high-performance cycle at the 15 sites ranged from 3.3 to 5.2 cents/kWh, as shown in Table 14.4. The cost range of electricity generated by a new coal-fired power plant owned and

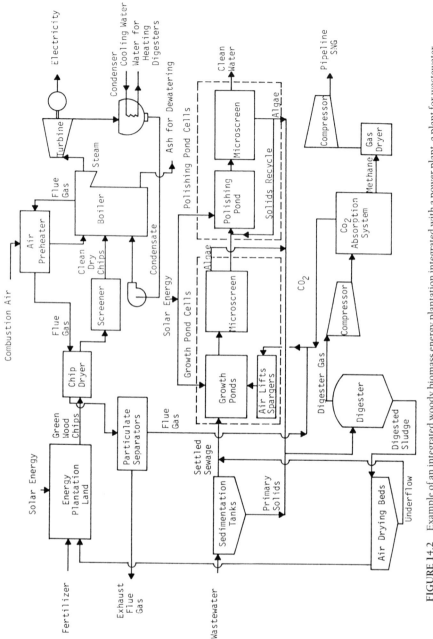

FIGURE 14.2 Example of an integrated woody biomass energy plantation integrated with a power plant, a plant for wastewater treatment and microalgae production, an anaerobic digestion plant for fermentation of the microalgae and methane recovery, and provision for recycling the digested solids to the biomass growth area. From Fraser *et al.* (1981).

TABLE 14.3 Characteristics of Integrated Biomass Production-Conversion System Consisting of Energy Plantation, Power Plant, SNG Plant, and Wastewater Treatment Plant[a]

Site	Energy plantation			Power plant		SNG produced $(10^3 \text{ m}^3 \text{ (n)/year})$	Wastewater plant	
	Size (ha)	Yield (dry t/ha-year)	Productivity $(10^3$ dry t/year)	Capacity (MW)	Production $(10^6$ kWh/year)		Size (ha)	Flow $(10^6$ L/day)
Pensacola, FL	9700	17.9	180	76	341	513	40.9	22.7
Kissimmee, FL	14,600	15.8	238	260	456	111	8.1	3.8
Chanute, KS	14,600	16.7	251	120	419	71	8.9	4.2
Maysville, KY	14,600	9.2	139	40	264	45	5.3	2.6
Natchitoches, LA	14,600	16.9	253	162	469	154	16.2	6.8
Minden, LA	14,600	15.8	237	150	446	129	13.8	5.7
Traverse City, MI	14,600	14.2	210	66	360	77	15.4	6.8
Bemidji, MN	10,900	20.3	227	91	381	55	8.5	4.2
Yazoo City, MS	14,600	12.6	190	105	360	94	10.9	4.2
Jamestown, NY	13,400	14.8	204	65	389	228	58.3	15.1
Hammond, NY	14,600	14.3	215	102	411	85	21.9	5.7
Greenwood, SC	12,100	14.6	182	88	362	180	18.6	7.9
Knoxville, TN	13,400	9.9	136	73	267	1460	197.1	64.3
Caldwell, TX	14,600	16.5	249	110	415	31	2.4	1.5
Martinsville, VA	12,100	15.8	198	88	394	214	25.1	9.5

[a]Adapted from Fraser et al. (1981). The plantation yield is the average above-ground biomass. The plantation productivity is the average sustained production including the contribution of root mass produced upon replanting. The power plant operates in the high-performance steam cycle. The wastewater plant is a carbon-limited design and the influent is wastewater from a primary treatment plant. The microalgae produced in the wastewater treatment plant are converted to SNG by anaerobic digestion.

TABLE 14.4 Economics of Energy Plantation, Power Plant, and Wastewater Treatment Plant in Integrated System[a]

Site	Energy plantation			Power plant				Wastewater plant	
	Capital cost ($10^6)	Annual cost ($10^6)	Fuel cost ($/dry t)	Capital cost ($10^6)	Installed cost ($10^3/kW)	Annual cost ($10^6)	Power cost (cents/kWh)	Capital cost ($10^6)	Cash flow ($10^6)
Pensacola, FL	5.27	4.06	22.09	80.16	1055	14.14	3.6	2.279	1.602
Kissimmee, FL	25.13	4.73	19.44	169.78	653	24.36	5.2	0.814	0.300
Chanute, KS	31.39	5.10	20.32	103.84	865	17.46	4.2	0.786	0.150
Maysville, KY	23.29	3.75	26.01	50.67	1267	10.84	4.1	0.592	0.027
Natchitoches, LA	7.34	5.28	20.48	123.90	765	20.00	4.2	1.100	0.421
Minden, LA	6.87	5.15	21.13	117.94	786	19.16	4.2	0.981	0.343
Traverse City, MI	8.81	5.32	25.31	72.09	1092	14.49	4.0	1.087	0.131
Bemidji, MN	7.47	4.22	18.63	90.31	992	15.13	4.0	0.782	0.021
Yazoo City, MS	6.53	4.05	20.96	97.20	926	15.91	4.3	0.834	0.235
Jamestown, NY	7.10	4.48	21.44	71.37	1098	13.72	3.4	2.155	0.374
Hammond, NY	7.55	4.54	20.89	95.63	938	16.23	3.9	1.143	0.165
Greenwood, SC	6.07	3.89	20.73	89.11	1013	14.97	4.0	1.203	0.504
Knoxville, TN	5.76	3.72	25.94	77.21	1058	13.44	3.3	5.883	4.619
Caldwell, TX	27.10	5.49	22.04	99.49	904	17.39	4.2	0.389	0.056
Martinsville, VA	6.18	4.21	20.56	89.20	1014	15.31	3.7	1.395	0.612

[a]Adapted from Fraser et al. (1981). The data in this table are calculated in 1977 dollars using municipal financing. The wastewater plant is a carbon-limited design for which a positive cash flow rather than an annualized cost is tabulated because of the wastewater treatment credit.

operated by a municipality or some other government entity was comparable to this range. It is not indicated in this table, but the analyses showed a significant correlation between electricity cost and load factor, with the cost decreasing as the load factor increases. Sensitivity analyses to ascertain the effects of scale also showed a marked decrease in electricity cost as generating capacity increased. With municipal financing, the annualized cash flow from the wastewater plant provided a treatment credit far more significant than the credit for the SNG produced. The credit applied to the annualized cost of the power plant enabled a reduction of about 0.16 to 0.25 cents/kWh generated.

It was concluded from this detailed analysis of an IBPCS consisting of a short-rotation hardwood energy plantation, a wood chip-fueled power plant, a microalgal wastewater treatment pond, and an anaerobic digestion unit for SNG production that the effluents and by-products from one system component can be beneficially used by other components and lead to increased energy conversion efficiencies and energy recovery at lower costs.

D. Examples of IBPCSs

It is apparent from these early assessments that large-scale IBPCSs present numerous problems to system designers because of the complexity of the system components and their interactions. Solutions to these problems are often not obvious. An early project to design and build a large-scale IBPCS illustrates the scope of the problem. One of the largest IBPCSs in the world was conceived and implemented by Donald K. Ludwig, a U.S. financier, for the production of energy, fuel, pulp, and chemicals in Brazil (Klass, 1983, 1984, 1985, 1987). The facility is located in the Rio Jari region of the Amazon and was originally designed around the growth of the deciduous tree *Gmelina arborea*. This species coppices well, exhibits rapid growth, provides average yields of paper with properties superior to those obtainable from most hardwood pulps, and is a good fuel (National Academy of Sciences, 1980). The original size of the plantation was reported to be 51,400 ha (198 mi^2) and the cost was more than $1 billion. The conversion plants were largely preassembled outside of Brazil and shipped by ocean-going barge to the plantation site. The main problem encountered in operating the system was that the yields of gmelina were considerably less than anticipated, except on about one-third of the growth areas that had the best soil conditions.

Technical and economic analyses indicated that since tree production was only about one-half the projected yield, the application of fertilizer was required at an added cost of about $7.2 million/year. The Brazilian government and a consortium of Brazilian companies finally took over ownership and operation of the integrated system (Hoge, 1982). It was concluded that the outlook for the system was "clearly bleak" without additional wood fuel and pulp plantations (Anonymous, 1983). Caribbean pine and eucalyptus have been grown successfully in the growth areas consisting of the sandy and transition soils that did not satisfactorily support gmelina. The project's managers estimated that ultimately, there would be 25,000 ha of gmelina, 42,000 ha of pine, and 32,000 ha of eucalyptus, or a total of 99,000 ha (382 mi^2) of growth area (Hornick, Zerbe, and Whitmore, 1984). The rotations are about 5 to 7 years for gmelina and 10 to 12 years for pine, and all species are planted at 2-m by 3-m spacings. Average annual biomass yields for properly managed sites were estimated to be about 13 m^3/ha-year.

The historical development of this project shows that large-scale biomass energy plantations must be planned extremely carefully and installed in a logical scale-up sequence. Otherwise, design errors and operating problems can result in immense losses and can be difficult and costly to correct after construction of the system is completed and operations have begun. It is also evident that even if the system is designed properly, the operation of IBPCSs can have a relatively long lag phase, particularly for large tree plantations, before returns on investment are realized. The financial arrangements are obviously critical and must take these factors into consideration.

Most of the other large IBPCSs that have been announced are either conceptual in nature or have not been fully implemented yet. Included are LEBEN, for Large European Bioenergy Project, in which integrated collection of woody biomass, agricultural residues, and dedicated energy crops are converted to fuel oil, low-energy gas, charcoal, electricity, and ethanol (Grassi *et al.*, 1987). This project was planned for the Abruzzo region of Italy and was supported by the Commission of the European Community. The total investment, excluding land costs, for implementing the project in the mid-1980s was 227 million ECU (1 ECU, or European Economic Community monetary unit, per $0.904 U.S. in February 1986). The growth areas supplying feedstock during the first 12 years of the project are estimated to be 95,000 ha for woody biomass and 20,000 ha for agricultural residues from vineyards and olive and fruit orchards. A network of 30 to 40 pyrolysis plants at strategic sites is planned for biomass processing. Later, sweet sorghum grown on 10,000 to 20,000 ha may be added to the system for conversion to ethanol. The costs of the salable products were estimated to be 130 ECU/t of oil equivalent for pyrolytic fuel, 0.28–0.30 ECU/ L for ethanol, and 0.04 ECU/kWh for electricity.

Some of the IBPCSs that have been proposed for power production in the United States are listed in Table 14.5. Most of them are considered to be modules of full-scale IBPCSs. Note that the power capacities are small relative to the capacities of fossil-fueled power plants. One of the largest commercial biomass-fueled power plants is the 60-MW plant in Williams Lake, British Columbia, Canada. It is fueled with waste biomass from nearby lumber mills

(Chapter 7), and is the equivalent of an IBPCS. The conversion technology is a proven, conventional steam cycle for the wood-waste-fired generating plant and consists of a steam generator and turbine. The largest IBPCS in Table 14.5 is the 75-MW baseload system proposed for Minnesota (Campbell, 1996; DeLong, 1995; DeLong *et al.*, 1995). This system is an innovative one in which an agricultural crop, the legume alfalfa, which is already grown in southwestern Minnesota, will be used as a dedicated energy crop. Electric power is generated from the plant stems in an IGCC plant, and the leaves are converted to co-product animal feed (Fig. 14.3). The virgin biomass growth area is within 80 km of an existing coal-fired power plant near Granite Falls, Minnesota, the site chosen for the IGCC plant. Established technology is used to dry and separate round hay bales of alfalfa into stems (about 53 wt % of the plant) and pelletized, protein-rich leaves (about 47 wt % of the plant). The hay bales arrive at the power plant with an average of 15 wt % moisture content. The stems are converted to power via low-energy gas from an air-blown, fluid-bed gasifier. Electricity is produced by two separately powered turbine generators. In the first cycle, the turbine is powered by combustion of the low-energy gas from the gasifier. The heat in the exhaust gas from the combustion turbine is reclaimed from the heat-recovery steam generator as steam, which powers the turbine in the second cycle. A total of 79.4 MW is generated in the IGCC plant at an overall efficiency of 40%, of which 4.3 MW is used for on-site needs. The feed rate to the IGCC plant is 41.41 t/h at 9.4 wt % moisture. The net electrical output is supplied to the grid at a net heat rate of 9.394 MJ/kWh and a net efficiency of 38.3%. Full-scale operation of the system will process alfalfa from within the 72,800-ha growth area at a rate of about 635,000 t/year, which will be supplied by 2000 farmers. About 291,000 t/year of pelletized co-product animal feed are produced. The total capital investment in the power and processing plants is $144.65 million, or $1,929/kW of installed net capacity, and the total fixed and variable costs are $3,466,000/year and $4,735,000/year, respectively. In 1994 dollars, the cost of electricity for a base case (one-third cost sharing) was estimated to be 6.52 cents/kWh; the no-cost-sharing case is 8.40 cents/kWh. The average retail price of electricity sold by utilities in 1994 in the United States was 6.91 cents/kWh (Energy Information Administration, 1996). The revenue streams from the combined power and animal feed co-products are expected to provide a suitable return on investment while simultaneously allowing payment to farmers for producing alfalfa at a level that is competitive with alternative crops.

Note that these economic results are not readily transferable to other regions of the country; most data of this type are site-specific. Biomass feedstock production costs are highly variable in different regions and depend on the climate, production technology, on-farm costs, and capital costs (*cf.* Table 4.15). Electricity costs are also highly variable in different regions of the

TABLE 14.5 Selected Integrated Biomass Production–Conversion System Power Modules Proposed for Development in the United States

Location	Feedstock	Conversion technology	Product(s)	Planned area (ha)	output (MW)	Remarks
California	Forest thinnings, mill wastes, rice straw, energy crops, hybrid popular	Modified boilers, fermentation	Power and possibly ethanol	809–1214	800	Waste biomass collected from several locations not included in growth area
Florida[a]	Sugarcane, leucaena, eucalyptus, elephant grass	Fermentation	Ethanol and on-site power	29,500		124 million L/year + stillage by-product and lignin fuel
Georgia	Agricultural and forest wastes and thinnings	Circulating fluidized-bed boilers	Power and steam		25	Waste biomass from several locations
Hawaii	Napier grass, sorghum	Circulating fluid-bed boilers, fermentation	Power and ethanol	4856	23	
Iowa[b]	Switchgrass	Cofiring in coal boiler or small gasifier	Power	16,188	35 and 5	Initially at 1619 ha, >10 dry t/ha-year
Minnesota[c]	Alfalfa	Integrated gasification combined cycle (IGCC)	Power from stems, feed from leaves	72,845	75	2000 farms within 80 km of plant

564

Minnesota[d]	Hybrid poplar	Whole-tree combustion	Power	19,425	50	
Minnesota	Plantation wood, agricultural waste	Fluidized bed or conventional boilers	Power	8094	50	
New Mexico	Gamma grass, hybrid poplar	Coal-fired boilers for pelletized fuel	Power	40		
New York[e]	Hybrid willow	Cofiring in pulverized coal boilers	Power	1073	25	3 to 4-year cycles, yields of 16.8 dry t/ha-year
North Carolina[f]	Wood harvesting and pulp mill residues (black liquor)	Direct and indirect gasification in IGCC	Power		43	More than 853,000 dry t/yr of residues available within 97 km of plant
Oregon	Forest thinnings	Stoker boilers	Power	4047	6	
Wisconsin	Willow and poplar, wood residues	Hydrograte spreader stoker boilers	Power		50	Waste biomass from several locations

Source: Adapted from Spaeth and Pierce (1996) and indicated references.
[a]Stricker et al. (1995).
[b]Cooper (1995).
[c]DeLong et al. (1995), DeLong (1995).
[d]Ostlie and Drennan (1989), Appel Consultants (1993).
[e]White, Abrahamson, and Robison (1995).
[f]Raymond and Kieffer (1995).

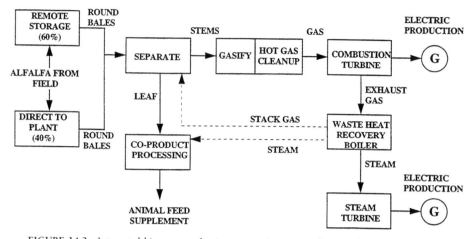

FIGURE 14.3 Integrated biomass production-conversion system for growth and processing of alfalfa to yield high-protein animal feed supplement and electricity for the grid. From DeLong (1995).

country. However, when the estimated cost of electricity in an IBPCS that converts wood chips from short-rotation hardwoods in a 91-MW plant in Minnesota (4.0 cents/kWh in 1977 dollars in Table 14.4) is converted to 1994 dollars by use of implicit price deflators, the result, 9.0 cents/kWh, is comparable to the no-cost-sharing price of 8.4 cents/kWh for the IBPCS fueled with alfalfa.

The combustion of whole trees by the Whole Tree Energy system (Ostlie and Drennan, 1989) in 50- and 100-MW plants in Minnesota (Table 14.5) is projected to supply electricity at a cost of 6.0 and 5.1 cents/kWh in 1991 dollars (Appel Consultants, Inc., 1993). Adjustment to 1994 dollars provides costs of 6.6 and 5.5 cents/kWh. The capital requirements in 1991 dollars for the 50- and 100-MW plants were estimated to be $1723 and $1342/kW of installed net capacity. Whole-tree feedstock production, consisting of careful matching of appropriate clones to sites, continuous harvesting year-round using high-speed continuous cutting, and replacement of cut trees with improved clones rather than by coppice regrowth, was estimated to be capable of supplying wood at delivered costs of $1.80/GJ compared to wood chip costs in the range $2.20 to $3.20/GJ from short-rotation forestry (Perlack *et al.*, 1996). The Whole Tree Energy system is therefore expected to supply electricity at lower costs than other IBPCs because of the reduction in harvesting and handling costs, and lower biomass losses from handling and storage.

III. NET ENERGY PRODUCTION

A. DEFINITION AND FUNDAMENTALS

The question now arises as to just how much net energy an IBPCS can deliver to market. As discussed in the Introduction, NEAs were quite controversial when first introduced. It was perceived by many energy specialists that since energy units are specifically defined and, unlike monetary units, are not a function of such variables as time, markets, living standards, and public policies, NEAs are more scientific, precise, and indicative of the real value and energy-producing capabilities of a system. In contrast, some economists were particularly concerned with the role of NEAs in energy assessments. Some believed that traditional economic analyses were being relegated to a secondary role in the assessment of energy resources by political decisions and legal mandates (cf. Huettner, 1976; U.S. Comptroller General, 1982). Some analysts went as far as to call NEAs a "red herring" and relatively insignificant (Leach, 1975). As will be shown here, NEAs supply a useful perspective that economic analyses *alone* are often unable to provide. More importantly, NEAs can help identify processing functions and system operations where improvements may be possible, how much of an improvement can be made, and sometimes how to make them.

A simple example should suffice for the moment to illustrate the value of NEAs. When fermentation ethanol manufactured from corn sugars was marketed in the United States as an octane enhancer and lead substitute for motor gasolines in the late 1970s and 1980s, some oil companies took the position that it takes more fossil energy to make ethanol from corn than the energy content of the resulting fuel, while the ethanol producers generally took the opposite position. Detailed NEAs carried out by different groups showed that more *or* less energy can be consumed than that contained in the ethanol, depending on whether energy credits are taken for salable energy co-product chemicals and cattle feed, whether the energy for drying the stillage is included as an input to the system, and whether potential fuels generated within the system are used to replace fossil fuel inputs (Klass, 1980a, 1980b). In addition, NEAs delineated specific unit operations where improvements made in the fermentation process can significantly reduce energy consumption. Most of these improvements have since been incorporated into the fermentation process and have resulted in processes that supply more net energy to market in the form of ethanol than that consumed in its production (Chapter 11). But the mere fact that more energy may be consumed in a process than that residing in the product fuel does not invalidate the utility of the process. As mentioned earlier, it depends on the qualities of the feedstock, energy inputs, and energy products. Thus, if a petroleum resid drives a process for conversion of wood wastes to a high-grade liquid transportation fuel, it does not necessarily follow that the consumption of more energy in the form of resid than that contained in the product precludes the manufacture of the transportation fuel.

Net energy has been defined in different ways. When based on the first law of thermodynamics, the definition ranges from the simple arithmetic difference between the gross amount of energy produced for consumer use by a system and the amount of energy needed to produce it (cf. U.S. Comptroller General, 1982), to the amount of salable energy that remains after the energy costs of finding, producing, upgrading, and delivering the energy to the consumer have been paid (cf. Odum, 1973). First-law treatments generally involve energy and material balances to relate the energy inputs and outputs, including heat losses and other undesirable outputs. They are used to calculate simple ratios of energy produced to energy consumed, or net energy production ratios (NEPRs), which are more useful than simple efficiencies. Arithmetically, the NEPRs can be defined by (Klass, 1976):

$$N_1 = [E_P - (nE_F + mE_X)]/(nE_F + mE_X),$$

where N_1 = net energy production ratio
$\quad E_P$ = sum of energy contents of salable energy products
$\quad E_F$ = energy content of feedstock or combined feedstocks
$\quad E_X$ = sum of energy values of all external energy inputs   (1)
$\quad\quad$ except feedstock
$\quad\quad n$ = fraction of primary energy source content (E_F) diverted to
$\quad\quad\quad$ other than salable energy products
$\quad\quad m$ = fraction of external energy inputs (E_X) diverted to other
$\quad\quad\quad$ than salable energy products.

In most cases, m is 1.0 because none of the external energy inputs contribute to E_P. Calculations can also be made in which nE_F is zero to delineate the net energy production delivered to market as a function of only the external energy inputs exclusive of feedstock. Presuming m is 1.0, the calculation for N_2 is then made by

$$N_2 = (E_P - E_X)/E_X.$$ (2)

Equation (1) yields an NEPR (N_1) that indicates the overall net energy production of a system and accounts for all energy inputs, whereas Eq. (2) yields an NEPR (N_2) that indicates how much more or less energy is produced as salable products than the external, nonfeed, energy inputs. Addition of 1.0 to the NEPR calculated by either Eq. (1) or (2) yields the number of multiples of ($nE_F + mE_X$) or E_X that appear in E_P. In either calculation, a positive N means that the system replaces the specified energy inputs, ($nE_F + mE_X$) or E_X, with the equivalent in salable energy products and also provides an additional amount of salable energy. Optionally, and depending on the system boundaries, E_X can include the capital energy investment needed for system construction or specific system units amortized over the life of the system, and the energy consumed in producing the materials and equipment that make up the operating system. This energy consumed within the integrated system consists of energy losses and the energy diverted to other than salable energy products. Diagramatically, the system is represented by

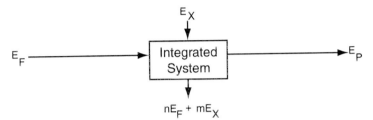

The first-law energy balance is simply:

$$E_F + E_X = E_P + nE_F + mE_X$$

To illustrate the utility of NEA using this model, the question has often been raised as to whether large-scale coal gasification processes to manufacture SNG can deliver as much net energy to the consumer as a natural gas system. Intuitively, it appears reasonable that an integrated coal-to-SNG system made up of the many required energy-consuming unit operations can not deliver net incremental energy supplies to the consumer compared to a similarly sized and relatively simple natural gas system. It might therefore appear that coal-to-SNG systems cannot compete with natural gas from the standpoint of net additions to energy supplies. To address these issues, a NEA was carried out using Eqs. (1) and (2) of comparably sized coal-to-SNG and natural gas systems starting with all energy resources and materials in the ground, to permit comparison of the NEPRs and to make the various energy inputs and outputs as additive as possible (Klass and Chambers, 1976). All capital energy investments, including those required to discover and develop the resources, to manufacture the materials of construction, and to build production, conversion, and delivery facilities, were amortized over a projected 20-year project life. Each system was sized to deliver approximately 6.7 million $m^3(n)$/day of natural gas or SNG via a model 1328-km pipeline to a local distribution utility. Figures 14.4 and 14.5 show some of the details of the basic systems that were analyzed; Table 14.6 is asummary of the first-law efficiencies and NEPRs that were calculated. As expected, the overall thermal efficiencies of the coal-to-SNG systems are lower than those of the natural gas systems. This is caused

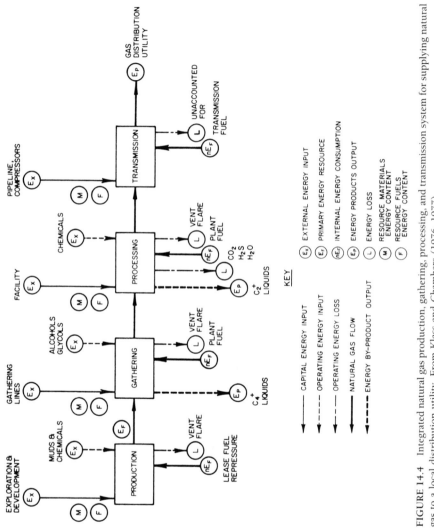

FIGURE 14.4 Integrated natural gas production, gathering, processing, and transmission system for supplying natural gas to a local distribution utility. From Klass and Chambers (1976, 1977).

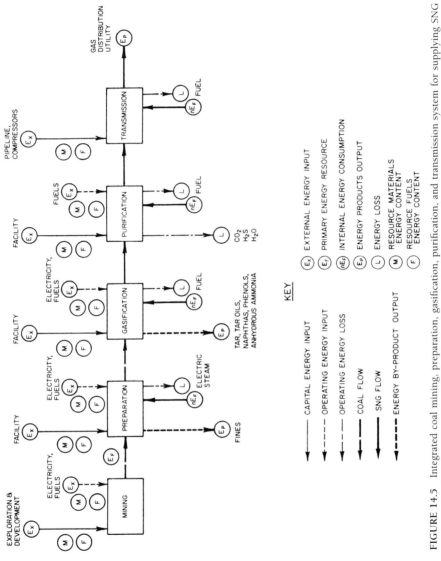

FIGURE 14.5 Integrated coal mining, preparation, gasification, purification, and transmission system for supplying SNG to a local distribution utility. From Klass and Chambers (1976, 1977).

571

TABLE 14.6 Comparison of Overall Thermal Efficiencies and Net Energy Production Ratios of Integrated Systems Supplying Substitute Natural Gas from Coal and Natural Gas to Local Distribution Company

| | Natural gas | | | | Substitute natural gas production by Lurgi gasification and methanation | | |
| | Large reservoir | | Small reservoir | | | | |
Parameter	50% Gas recovery (TJ/year)	90% Gas recovery (TJ/year)	50% Gas recovery (TJ/year)	90% Gas recovery (TJ/year)	North Dakota coal (TJ/year)	New Mexico coal (TJ/year)	Appalachia coal (TJ/year)
Primary energy source (E_F)	127.88	127.88	127.88	127.88	175.67	171.42	156.46
Primary energy consumed (nE_F)	26.41	26.41	26.41	26.41	59.04	62.57	57.69
External energy inputs (mE_X)							
Annualized capital	5.28	3.06	19.73	11.13	2.20	2.12	2.16
Operating					7.98	0.42	1.13
Total	5.28	30.6	19.73	11.13	10.18	2.54	3.29
Salable energy products (E_P)							
Pipeline gas	91.12	91.12	91.12	91.12	90.37	92.42	83.90
Other energy products	10.34	10.34	10.34	10.34	26.25	16.43	14.87
Total	101.46	101.46	101.46	101.46	116.62	108.85	98.77
Overall thermal efficiency, %	76.2	77.5	68.7	73.0	62.7	62.6	61.8
Net energy production ratio							
$(E_P - E_X)/E_X$	18.2	32.2	4.1	8.1	10.5	41.8	29.0
$[E_P - (nE_F + mE_X)]/(nE_F + mE_X)$	2.2	1.2	2.4	1.7	0.7	0.7	0.6

[a]Adapted from Klass and Chambers (1976, 1977). The systems supply approximately 6.7 million m³/day of pipeline gas. "Other energy products" include natural gas liquids for the natural gas systems, and tar oils, naphthas, crude phenols, and anhydrous ammonia for the SNG-from-coal systems. The overall thermal efficiency on a percentage basis is the energy content of salable energy products divided by the sum of all feedstock and external nonfeedstock energy inputs to the integrated system; i.e., $100E_P/(E_F + E_X)$.

mainly by the energy losses incurred on conversion of coal to SNG. However, the net energy production ratio, N_2, for each of the systems evaluated indicates that each is capable of supplying salable energy products that are a significant multiple of the external, nonfeed energy inputs, E_X, consumed. For N_1, the coal systems have ratios slightly lower than the N_1s of the natural gas systems, but both sets of NEPRs are positive values. The critical functional parameters that tend to equalize these NEPRs are the extensive exploration and development energy investments required to discover producing natural gas deposits, and the larger amounts of salable energy co-products from the coal systems. It was concluded from this application of the net energy model that integrated natural gas and integrated coal-to-SNG systems are comparable in terms of delivering net energy to the consumer. But as shown in Table 14.6, the percentage recovery of natural gas from a gas reservoir and the reservoir size can have a pronounced effect on net energy production. If either of these parameters is too small, the coal-to-SNG system evaluated in this study can deliver considerably more net energy to the consumer.

One of the problems perceived by some thermodynamicists who perform first-law NEAs is that when an intrinsically high-quality energy form, such as No. 2 fuel oil, is converted to a lower quality energy form, such as may be encountered in low-temperature heat transfer processes in which no other energy product is produced, the loss in energy quality is not accounted for by a first-law analysis because energy is always conserved (*cf*. Reistad, 1975). This argument is almost philosophical in nature because for most practical energy-producing systems, high-quality energy is not converted to low-quality energy except after consumer use. In the early days of NEAs, some energy analysts carried this rationale to the extreme. For example, it was reported that "The results of Second Law analyses are much more enlightening than First Law (energy) analyses, because the dissipations and efficiencies measured with availability are the *true* ones, whereas those measured with energy [i.e., first-law efficiencies] are erroneous and misleading" (Gaggioli and Petit, 1976). This statement itself is quite misleading because in the vast majority of cases, the energy losses in a NEA using first-law methods are readily accounted for and are usually not erroneous or misleading. But there are exceptions.

In an energy-producing process, the differences in energy quality can be accounted for by the concept of available energy (availability) using the second law of thermodynamics. To perform a second-law NEA, availability is defined as

$$A = (H - H_o) - T_o(S - S_0),$$

where A = availability of stream

H = enthalpy of stream at flowing condition

H_o = enthalpy of stream with all components in equilibrium with environment   (3)

T_o = temperature of environment
S = entropy of stream at flowing conditions
S_o = entropy of stream with all components in equilibrium with environment.

Availability is a thermodynamic property that is a measure of a system's ability to do work when restricted by the surroundings at a given temperature and pressure. It is not conserved, but is reduced by any irreversible degradation via friction or heat transfer. Availability is the maximum amount of work or thermal energy obtainable from a system as it attains equilibrium with its surroundings. It is the same as the change in Gibbs free energy at constant pressure and temperature, and is the same as the change in enthalpy if the system is isentropic.

Performance parameters in second-law NEAs have been defined as effectiveness (ε), which is the increase in availability of the desired product divided by the decrease in the availability required (Reistad, 1975). For example, in a mechanical-compression refrigeration system, refrigeration is the desired output, so the numerator is the availability of the refrigeration and the denominator is the decrease in availability required in the power input to the refrigeration system. The corresponding first-law efficiency (N_3) is the energy output in the desired product divided by the required energy input. Various devices can have the same or almost the same values for ε and N_3 such as electric generators, hydraulic turbines, and electric motors, each of which has energy inputs and outputs that are entirely available. For the fuel cell power plant, ε is slightly greater than N_3, but equal to N_3 for diesel and automobile engines, and is slightly less than N_3 for a steam-electric generating plant. The outputs of these devices are either electric or mechanical energy that is entirely available, so ε and N_3 are the same or similar. Effectiveness is dramatically less than N_3 for fossil-fired steam boilers, gas and oil furnaces, electric resistance and hot water heaters, gas heaters, and electric cooking and clothes-drying appliances. Detailed analysis often shows large variations in ε and N_3. The data in Table 14.7 compare the first-law energy and second-law availability losses for a coal-fired, steam-electric plant. The largest availability losses occur in the steam generator, because of the irreversibility of combustion and high temperatures, while the largest energy losses occur in the condenser. Analysis shows that if the availability losses in the steam generator are decreased, the energy losses in the condenser will also decrease. Thus, the irreversibility of the process taking place in the steam generator is a major cause of the limited efficiency of steam-electric power plants. First-law analysis alone suggests that improvements in condenser operations offer a better chance of increasing generating efficiency, whereas the steam generator is actually the largest source of inefficiency.

TABLE 14.7   Energy and Availability Losses in a Coal-Fired
Steam-Electric Generating Plant[a]

Component	Energy losses (% of plant input)	Availability losses (% of plant input)
Steam generator	9	49
Combustion		(29.7)
Heat exchangers		(14.9)
Thermal stack loss		(0.6)
Diffusional stack loss		(3.8)
Turbines	~0	4
Condenser	47	1.5
Heaters	~0	1.0
Miscellaneous	3	5.5
Total	59	61.0

[a]Reistad (1975).

Net energy production ratios can be calculated by second-law availability analyses for various types of hardware and chemical reactions to help pinpoint where improvements potentially reside and the magnitude of the improvements. But the calculations are often difficult to carry out for integrated systems, including IBPCSs. The availabilities of each component in a complex, integrated energy production-conversion system are not readily accessible or perhaps even conceptually feasible, such as those that must be calculated for goods and services, and in the case of an IBPCS fueled with virgin biomass, those for planting and managing the growth and harvesting of feedstock. The best approach to NEAs appears to be a combination of first- and second-law methodologies where possible. NEAs using first-law efficiencies can be performed first. A second-law treatment can then be given to devices and chemical processes amenable to availability calculations.

B. Net Energy Analysis of Terrestrial Biomass and Electricity Production

For most analyses of IBPCSs, the net energy production ratios defined by Eq. (2) are preferable. The ratio $(E_P - E_X)/E_X$ indicates how much more or less salable energy products are produced than the energy consumed by the integrated system if the external, nonbiomass energy consumed is replaced, and it is assumed that the biomass feedstock energy content is zero. This is a reasonable assumption because essentially 100% of the energy content of biomass is derived from solar radiation via photosynthesis. In a practical sense, biomass energy is also renewable over a relatively short span of time, unlike fossil energy. NEPRs of IBPCSs greater than zero indicate that an amount of energy equivalent to the sum of the external, nonbiomass energy inputs and an additional energy increment of salable fuel are produced; the larger the ratio, the larger the increment. The number of multiples of E_X produced by the IBPCS is $[(E_P-E_X)/E_X] + 1$. Thus, an IBPCS with a NEPR of 9 produces 10 times more energy than the nonbiomass energy consumed by the system. The overall thermal efficiency for the production of salable energy products is $100E_P/(E_F + E_X)$. Equation (2), of course, suggests that if the nonbiomass energy inputs for an IBPCS are reduced to zero, the NEPR becomes infinite. But for the next several decades, it is expected that fossil energy inputs will be necessary to operate IBPCSs. NEPRs as calculated by Eq. (2) will provide valuable information to help evaluate these systems.

A wide range of net energy analyses of IBPCSs were carried out in the United States at about the time of the First Oil Shock in the early 1970s. One of these studies targeted the production of terrestrial biomass in hypothetical 4047-ha (10,000-ac) growth areas to obtain feedstocks for processing in centrally located conversion facilities (Alich and Inman, 1974). Different species, including various types of woody and herbaceous biomass, and several conversion processes were considered. An updated version of the methodology used is discussed here because it is generally applicable to most terrestrial IBPCSs. The original analysis assumed higher-than-normal biomass yields (67.2 dry t/ha-year, 30 dry ton/ac-year) and projected they could be obtained at relatively high nitrogen fertilization rates (1121 kg NH₃/ha-year) and, for the U.S. Southwest, high annual irrigation rates (152.4 ha-cm, 5 ac-ft). These rates are more than adequate for most U.S. regions. The yields and nitrogen applications are reduced here to 30 dry t/ha-year and 225 kg NH₃/ha-year so that the model is more typical of what is projected for certain candidate energy species today. A 10% loss of biomass on handling was also added to bring the net biomass yield to 27 dry t/ha-year. The irrigation strategy was not changed to illustrate its effect on the energy consumption pattern. It could be substantially reduced for more temperate climates. But at a yield of 30 dry t/ha-year, irrigation is needed in many areas. Note that the irrigation rate of 152.4 ha-cm is exclusive of any natural precipitation. The energy consumed to manufacture the chemicals needed and the machinery used, amortized over the life of the equipment, are included in the analysis. An additional annual energy investment equivalent to the amortized energy investment for the manufacture of the machinery is included in the revised model for maintenance and repair services.

The design parameters for the hypothetical IBPCS module are summarized in Table 14.8. Three harvests of fast-growing perennials grown by no-till

TABLE 14.8   Design Parameters Assumed for a Hypothetical Terrestrial Biomass
Production System for Perennial Species[a]

Parameter	Description
Plantation size	4047 ha (10,000 ac).
Crop	Fast-growing herbaceous perennials. Higher yielding species located in center of plantation.
Operating schedule	7 day/week, 12 h/day. Irrigation performed 24 h/day.
Management	Fields cleared of weeds by application of herbicide. No-till planting and fertilization combined in one application. Sidedressing of fertilizer applied at appropriate times. Crop harvested with self-propelled combines, which also chop it into small pieces to facilitate drying. Replanted once every 3 to 5 years. Aircraft apply insecticides and fungicides when and where needed twice annually.
Irrigation	Irrigation water applied at 2-week intervals by an automatic, movable, center-pivot, overhead, sprinkling system capable of watering two 65-ha plots/day. About 3 h are needed to move the irrigation system.
Harvest frequency	Three times per year.
Productivity	The total yield per harvest is 30 dry t/ha, or a net of 27 dry t/ha-year after a 10% handling loss. At an energy content of 18.598 GJ/dry t (16 million Btu/dry ton), the energy yield is 502 GJ/ha-year.
Transportation	Chopped biomass trucked to solar drying area and dumped. Turned after sufficient drying has occurred, loaded into suitable conveyances, and transported to conversion plant gate.
Conversion plant	Located in central portion of plantation.

[a]Adapted with modifications from Alich and Inman (1974).

methods, and possibly multicropping, are projected in this updated model to supply feedstock at the conversion plant gate at an equivalent rate of about 300 dry t/day, or 344,200 BOE/year. The details of the net energy analysis are shown in Table 14.9. The NEPR is a strongly positive 13.5 at an overall thermal efficiency of 85%. This means that for this model, about 14.5 times more biomass energy in the form of feedstock is produced than the total amount of fossil energy consumed to operate the system, and that its integration with a reasonably efficient conversion process should result in a high-efficiency IBPCS. This result is somewhat unexpected because of relatively small biomass production levels, equivalent to about 110,000 dry t/year. The external energy investments in order of decreasing amounts of energy are the manufacture of chemicals, irrigation, other field operations, and the manufacture of the required machinery. The sum of these energy investments, however, is still small compared to the net biomass energy yield. Biomass yield is clearly a key factor that determines the net energy production capability. At a biomass yield of

TABLE 14.9 Net Energy Analysis of a Hypothetical, 4047-ha, Biomass Production System for Supplying Feedstock to a Centrally Located Conversion Plant[a]

Parameter	Energy consumption	
	Rate	MJ/ha-year
Manufacture of chemicals		
Nitrogen (at 225 kg NH₃/ha-year)	48.65 MJ/kg	10,946
Phosphorus (at 56.0 kg P₂O₅/ha-year	13.99	783
Potassium (at 112.1 kg K₂O/ha-year)	5.11	573
Herbicide (at 6.72 kg/ha-year)	101.3	681
Insecticide (at 3.36 kg/ha-year)	101.3	340
Fungicide (at 2.24 kg/ha-year)	101.3	227
Total for chemicals		13,550
Manufacture of machinery (for 4047 ha)		
6 tractors at 4.2 t each, 6-year life	4.2 t steel/year	
3 planters at 1.8 t each, 2-year life	2.7	
3 fertilizers at 1.8 t each, 2-year life	2.7	
3 herbicide rigs at 0.9 t each, 5-year life	0.5	
6 combines at 7.3 t each, 6-year life	7.3	
32 fresh haul units at 4.5 t each, 10-year life	14.4	
10 dry haul units at 5.4 t each, 10-year life	5.4	
1 turner at 4.5 t, 10-year life	0.5	
8 irrigation pumps at 0.9 t each, 20-year life	0.4	
33.2-km feeder lines at 12.0 t/km, 20-year life	19.9	
Sprinkler system at 30.9 t, 20-year life	1.5	
Total steel required	59.2 t steel/year	
at 20,799 kWh/t	1,231,301 kWh/year	
at 3.5983 MJ/kWh	4,430,590 MJ/year	1095
Total for machinery, maintenance, and repair, 2×		2190
System operations		
Field tasks per harvest		
Herbicide applications	2.853 L diesel/ha-year	
Plant and fertilize	7.080	
Fertilize (side dressing)	2.853	
Cut and chop	15.217	
Fresh haul	23.757	
Turn and dry	0.262	
Dry haul	20.371	
Pesticide application	0.159	
Total field tasks per harvest	72.551L diesel/ha-year	
Total field tasks, 3 harvests at 38.632 MJ/L		8408
Irrigation	6.240 kWh/ha-cm	
At 152.4 ha-cm	951 kWh/ha-year	
At 3.5983 MJ/kWh	3,423 MJ/ha-year	
At 33.3% thermal efficiency		10,276
Miscellaneous electricity use	12.36 kWh/ha-year	
At 3.5983 MJ/kWh	44.47 MJ/ha-year	
At 33.3% thermal efficiency		134
Total farming operations		18,818

(continues)

TABLE 14.9 (Continued)

Parameter	Energy consumption	
	Rate	MJ/ha-year
Seed production (0.3% of total)		104
Total external energy consumption, E_X		34,662
Total biomass energy production, E_F		557,940
Total net biomass energy production, E_P		502,146
Overall thermal efficiency, $100E_P/(E_F + E_X)$, %		85
Net energy production ratio, $(E_P - E_X)/E_X$		13.5

[a]Adapted with modifications from Alich and Inman (1974). Table 14.8 is a summary of the operating parameters of the plantation.

15 dry t/ha-year (6.7 dry ton/ac-year), which is 50% of that used for the analysis detailed in Table 14.9, the NEPR is still high, about 6.8. This yield level can be obtained for selected biomass species in many regions of the world without intensive growth management. For biomass species that do not require extensive chemical treatments, and where irrigation is not needed, the NEPR would be expected to exhibit considerably higher values. Also, operation at a larger scale would be expected to provide further improvements.

If it is assumed that the net biomass production from the hypothetical terrestrial system depicted in Table 14.9 is converted to electricity by conventional technology (Chapter 7), the overall thermal efficiency and NEPR for an IBPCS can be calculated as shown in Table 14.10 for Case 1. Case 2 in this table corresponds to a net biomass yield that is one-half that of Case 1. With a conventional generating system having a heat rate of 12 MJ/kWh, the equivalent thermal energy produced as electric power for Cases 1 and 2 corresponds to NEPRs of 3.3 and 1.2, and overall thermal efficiencies of 30.0% and 26.4%. If the power-generating system were an advanced combined-cycle process, such as IGCC systems that have thermal conversion efficiencies in the 40% range (Chapter 9), the corresponding figures would be higher as shown in Table 14.10—NEPRs of 4.8 and 1.9, and overall thermal efficiencies of 37.4 and 35.1%.

The alfalfa-based IBPCS proposed for power generation by an IGCC system that uses an IGCC system in Minnesota has an NEPR of 1.28 (energy output/input ratio of 2.28) without co-product credits (DeLong, 1995). There are several methods for inclusion of co-product credits in net energy analyses. The most direct method is to determine the energy value of any salable co-products, even if they are not marketed as fuels or energy, and to include it

TABLE 14.10   Net Energy Analysis of Integrated Biomass Production of Terrestrial Feedstock and Conversion to Electricity[a]

Parameter	Case 1	Case 2
Net biomass yield, dry t/ha-year	27	13.5
Total external energy consumption, MJ/ha-year (E_X)	34,662	34,662
Total net biomass energy production, MJ/ha-year (E_F)	502,146	251,073
Electricity generated by conventional steam technology:		
At heat rate of 12 MJ/kWh, kWh/ha-year	41,850	20,920
Energy equivalent, MJ/ha-year (E_P)	150,660	75,310
Overall thermal efficiency, $100E_P/(E_F + E_X)$, %	30.0	26.4
Net energy production ratio, $(E_P - E_X)/E_X$	3.3	1.2
Electricity generated by advanced IGCC technology:		
At heat rate of 9 MJ/kWh, kWh/ha-year	55,790	27,900
Energy equivalent, MJ/ha-year (E_P)	200,840	100,400
Overall thermal efficiency, $100E_P/(E_F + E_X)$, %	37.4	35.1
Net energy production ratio, $(E_P - E_X)/E_X$	4.8	1.9

[a]The biomass production data for Case 1 are from Table 14.9. The corresponding data for Case 2 were calculated from Case 1 assuming a lower net yield, but that E_X is the same for both cases. E_X does not include energy investments in the power plant.

in E_P. Another method is to assume a co-product can replace a specified amount of a product that is already commercially marketed, and then to calculate the energy content of that amount of commercial product for inclusion in E_P. An alternative to this approach is to include in E_P the amount of energy necessary to manufacture that amount of commercial product. This method was used for the Minnesota project. When credit is taken for the energy required to manufacture commercial soybean protein by the same amount of protein co-product from alfalfa leaves, the NEPR of the IBPCS rises to 2.04. Note that the average alfalfa yield, including stems and leaves, for the IBPCS proposed for Minnesota is in the range of 9 dry t/ha-year.

C. NET ENERGY ANALYSIS OF ETHANOL PRODUCTION

As mentioned earlier, a controversial issue that developed when fermentation ethanol was first marketed as a motor gasoline component in modern times concerned the amount of energy consumed in the fermentation process. The basic problem with conventional yeast fermentation of corn sugars and purification was that more energy can be consumed to manufacture a unit of ethanol than the energy contained in that unit. The problem was compounded in the

TABLE 14.11 Net Energy Production Ratios of Fermentation Ethanol Production from Corn[a]

Original source of energy data	External fuels used	Energy consumed (E_X)	Energy as ethanol (E_P)	Net energy ratio $(E_P - E_X)/E_X$
U.S. Office of Technology Assessment	Fuel oil	320 MJ	128 MJ	−0.60
Amoco Oil Company	Gas, fuel oil	46.29 MJ/L	21.29 MJ/L	−0.54
U.S. Department of Energy	Fossil fuels	0.612 MJ/h	0.643 MJ/h	0.05
Archer Daniels Midland Company	Gas, coal	6.44 TJ	8.97 TJ	0.39

[a]Klass (1980b).

United States by variations in the reported amounts of fossil energy consumed per ethanol unit produced. Conversion of these data to NEPRs showed that some were negative and some were positive (Table 14.11). A net energy analysis based on published energy consumption data for an integrated corn production-ethanol fermentation system at that time was performed to help clarify this apparent anomaly (Klass, 1980a, 1980b). The energy distributions to manufacture 1.00 L of anhydrous ethanol are shown in Fig. 14.6. The boundary of the integrated system shown in this figure circumscribes all the operations necessary to grow and harvest the corn, collect the residual cobs and stalks if they are used as process fuel, operate the fermentation plant for the production of anhydrous ethanol and by-product chemicals, and dry the

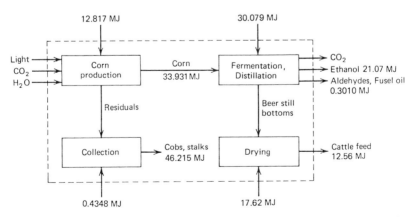

FIGURE 14.6 Energy inputs and outputs to manufacture 1 liter of anhydrous ethanol from corn. The ethanol yield is 385.9 liter/t (2.6 gal/bu) of corn. − − − − denotes system boundary. All figures are lower heating values. From Klass (1980a, 1980b).

stillage to produce distillers' dry grains plus solubles for sale as cattle feed. The capital energy investment is not included within the boundary. Various NEPRs were calculated as shown in Table 14.12. The ratio can be either positive or negative, depending on whether credit is taken for the by-product chemicals and cattle feed, whether the energy for drying the stillage is included as an input to the system, and whether a portion of the residual corn cobs and stalks is collected and used as fuel within the system to replace the fossil-fuel inputs. It is evident that a net energy analysis of fermentation ethanol production must be clearly specified in all details, including the boundary of the system, in order for NEPRs to be meaningful.

Since the early 1980s, the controversey regarding net energy production in IBPCSs for the manufacture of fuel ethanol by fermentation of corn sugars has continued, although the issue has abated somewhat (cf. Wang, 1996, 1997; Shapouri, Duffield, and Graboski, 1995, 1996; Tyson, 1992, 1993). Corn remains the primary feedstock for fuel ethanol in the United States, but is ultimately expected to be replaced by lower cost, cellulosic feedstocks (Chapter 11). The emphasis given to net energy production over the years has resulted, at least in part, in enhancement of various unit operations in modern fuel ethanol production systems. Several operations subject to improvement have been made more energy-efficient. Agricultural practices for corn production

TABLE 14.12 Net Energy Production Ratios for Fermentation Ethanol Production from Corn in an Integrated System[a]

Salable energy products, E_P	Non-feed energy inputs, E_X	$(E_P - E_X)/E_X$
Ethanol	Corn production, fermentation, bottoms drying	−0.65
Ethanol	Corn production, fermentation	−0.51
Ethanol, chemicals	Corn production, fermentation	−0.50
Ethanol, chemicals, cattle feed	Corn production, fermentation, bottoms drying	−0.44
Ethanol, chemicals, cattle feed	Corn production, fermentation, bottoms drying, 50% residuals	−0.10
Ethanol, chemicals, cattle feed	Corn production, fermentation, bottoms drying, 75% residuals	0.29
Ethanol	Corn production, fermentation, 75% residuals	1.43
Ethanol, chemicals	Corn production, fermentation, 75% residuals	1.47

[a]Klass (1980). The percentage figures are the amounts of cobs and stalks collected and used as fuel within the system to replace fossil fuel inputs.

and processes for manufacturing the chemicals used as nutrients have been improved. Average corn yields have risen 20 to 30% in the United States since the 1970s, and the energy inputs required per unit of corn grown have been significantly reduced. In 1975, energy consumption was about 5.052 MJ/kg (122,200 Btu/bu) of corn produced in the United States (Scheller, 1981). By 1991, the energy consumption averaged over five states (Illinois, Indiana, Iowa, Minnesota, and Nebraska) had been reduced to 1.972 MJ/kg (47,689 Btu/bu) (Shapouri, Duffield, and Graboski, 1995, 1996). Energy consumption has also been reduced in the fermentation process by incorporating energy-conserving improvements, such as by replacing azeotropic distillation with molecular sieve distillation. In 1979, the total energy consumption including electricity in a dry milling plant for fuel ethanol production was 19.4 MJ/L (69,600 Btu/ gal) of ethanol produced (Scheller, 1981). Modern wet milling plants now yield 1 liter of ethanol with inputs of 8.955 MJ of thermal energy and 0.5638 kWh of electrical energy, while the corresponding inputs for new dry milling plants per liter of ethanol are 9.471 MJ and 0.3170 kWh (Shapouri, Duffield, and Graboski, 1995, 1996). The ethanol yields for wet and dry milling plants are 0.371 L/kg (2.5 gal/bu) and 0.386 L/kg (2.6 gal/bu) of corn, respectively. The co-products are distillers' dry grains from the dry milling process, and gluten meal, gluten feed, corn germ meal, and corn oil from the wet milling process (Chapter 11); CO_2 is a co-product of each process. Conversion of the energy inputs for each modern process to equivalent thermal energy units yields values of 15.2 MJ/L and 13.0 MJ/L of ethanol for the wet and dry milling processes, respectively, assuming the electrical energy is generated at a heat rate of 11.0 MJ/kWh (32.8% efficiency). These energy requirements are about 60% of the energy content of ethanol, so intuitively, it would appear that the NEPRs of ethanol from modern fermentation plants are still relatively low. It is therefore worthwhile to examine the NEPRs of IBPCSs that utilize modern corn production and fermentation processes in more detail.

Table 14.13 is a summary of the energy consumption patterns for IBPCSs that yield a total of 378.5 million L/year (100 million gal/year) of fuel ethanol from a wet milling plant, and 113.6 L/year (30 million gal/year) from a dry milling plant (Shapouri, Duffield, and Graboski, 1996). The energy consumption calculations are based on ethanol yields of 371.1 L/t (2.5 gal/bu) for the wet milling plant and 385.9 L/t (2.6 gal/bu) for the dry milling plant, and an ethanol HHV of 23.387 MJ/L. The 1994 weighted average of corn yield from five states (Illinois, Iowa, Nebraska, Minnesota, and Indiana) used for this study is 9.338 t/ha-year (148.2 bu/ac-year). These states account for about 65 and 93% of U.S. corn and ethanol production, respectively. The fermentation plant uses molecular sieve distillation and coal-fired cogeneration for steam and electric power.

TABLE 14.13   Net Energy Analysis of Ethanol Manufacture in a Modem Integrated Corn Production–Fermentation System with and without Co-product Energy Content Credit[a]

	Energy consumption per ethanol unit	
Operation	Wet milling (MJ/L)	Dry milling (MJ/L)
Corn production	5.313	5.109
Corn transport to plant	0.351	0.338
Fermentation	13.611	13.058
Ethanol distribution to retailer	0.884	0.884
Energy losses	1.349	1.301
Total external energy inputs, E_X	21.508	20.689
Total product energy without co-products, E_P	23.387	23.387
Net energy production ratio, $(E_P - E_X)/E_X$	0.09	0.13
Total product energy with co-products, E_P	32.207	31.067
Net energy production ratio, $(E_P - E_X)/E_X$	0.50	0.50

[a]Energy consumption data from Shapouri, Duffield, and Graboski (1995, 1996).

Calculation of the NEPRs without co-product energy credits results in small but positive values of 0.09 and 0.13 for the wet and dry milling cases, respectively, or slightly more than replacement by ethanol of the total fossil energy consumed. When co-product energy credits are taken, the NEPRs for each case are 0.50. This corresponds to replacement by ethanol of the total fossil energy consumed and an additional 50% increment of energy in the form of co-products. Note that no credit is taken for co-product CO_2, which can have a market value depending upon the location of the conversion plants. The net amount of new transportation fuel made available as ethanol according to this NEA is not as much as might be expected for modern, integrated corn production-fermentation systems, despite the improvements that have been made over the years in corn yields and fermentation technology. The energy consumption patterns presented in Table 14.13 suggest that additional opportunities for enhancing IBPCSs based on corn sugar fermentation reside in further improvements in corn yield and energy consumption in the fermentation plants. The wet and dry milling fermentation plants each require about 63% of the total fossil energy inputs. As pointed out in Chapter 13, any effort made to improve the energy conservation characteristics of the fermentation process for ethanol from corn sugars should be directed to reducing process energy consumption rather than to increasing fermentation ethanol yield. The yield of fermentation ethanol per unit of corn is already near the stoichiometric maximum for sugar fermentation. It is evident, however, that since considerable improvements have been made in the IBPCS for corn-derived ethanol over a period of many years, a point of diminishing returns may have been reached. Nonfermentative methods coupled with lower grade biomass feedstocks may offer a more practical approach to the development of advanced fuel ethanol IBPCSs.

To examine this possibility, an NEA was performed for each of three IBPCSs, two based on biomass feedstocks and one based on natural gas feedstock, in which the thermochemical synthesis of ethanol is substituted for ethanol synthesis by fermentation. The following discussion illustrates how NEAs can help guide the development of advanced technologies. The thermochemical conversion process—steam reforming and synthesis gas conversion to ethanol via methanol intermediate (Chapter 11) and purification—yields 199-proof ethanol. The first biomass case represents an integrated biomass production system and a conversion plant operating at an availability of 330 day/year and a thermal efficiency of 54.8%; 797 t/day of feedstock (dry equivalent) is required at the plant gate. Ethanol production is 114 million L/year (30 million gal/year). The second biomass case is similar in that ethanol is produced in a conversion plant of the same plant capacity and availability but is operated at a thermal efficiency of 81.5%. The biomass feedstock requirement is 536 t/day (dry equivalent) at the plant gate. The second biomass case approximates the maximum thermal yield possible including energy losses in a sugar fermentation plant, whereas a thermal efficiency of 54.8% used in the first biomass case is the approximate thermal efficiency of natural gas conversion to ethanol by steam reforming via methanol intermediate. Since the thermochemical synthesis of ethanol from biomass-derived synthesis gas may permit ethanol to be manufactured in small-scale plants at costs similar to the cost of methanol from natural gas, a third case using natural gas feedstock was included. For this case, a world-scale plant of capacity 836 million L/year (220 million gal/year) operating at 54.8% thermal efficiency and an availability of 330 d/year was examined; 2000 t/day of feedstock is required at the plant gate. Reformer fuel for each of the three cases is in-plant waste gases.

The natural gas case starts with all resources and materials in the ground and is based on a large reservoir with 50% gas recovery (Table 14.6). All operating fuels are natural gas or waste gases except those for transporting product to the retailer; the latter fuel is included in mE_X. The energy co-products are produced in the gas processing plant (Table 14.6). The two biomass cases are based on the NEA for the management-intensive, terrestrial biomass production system described in Table 14.9, and include the capital energy investments for machinery. The capital energy investment in the ethanol plant is excluded for each of the three cases. A small amount of by-product is recycled for the biomass cases, and no credits are available for energy co-products. The coefficient m for E_X is 1.0 for each of the three cases, which means that none of the nonfeed, fossil-energy inputs contributes to E_P. The results of the NEAs projected for each integrated system are presented in Table 14.14.

TABLE 14.14 Net Energy Analysis of the Thermochemical Conversion of Natural Gas and Biomass to Ethanol in an Integrated Feedstock Production–Conversion System[a]

Parameter	Natural gas feedstock		Biomass feedstock			
Ethanol production, t/day	2000		273			
t/year	660,000		90,000			
L/year	836 million		114 million			
PJ/year	19.67		2.68			
Energy co-products, PJ/year	4.09		0.00			
Total energy products production, PJ/year (E_P)	23.76		2.68			
Conversion plant availability, day/year	330		330			
Conversion plant thermal efficiency, %	54.8		81.5			
Total feedstock production, t/year	928,000		196,536			
PJ/year (E_F)	50.60		3.656			
Feedstock at conversion plant gate, t/year	658,200		176,882			
PJ/year	35.89		3.290			
Energy consumption	PJ/year	MJ/L	PJ/year	MJ/L	PJ/year	MJ/L
Feedstock						
Handling losses	N/A	N/A	0.543	4.76	0.366	3.21
Production	8.71	10.42	0.316[d]		0.213[d]	
Transmission or transport to conversion plant	2.04	2.44	0.003[d]		0.002[d]	

586

Conversion						
Processing losses	10.95	13.10	2.211	19.39	0.610	5.35
In-plant fuel	5.14	6.15	0.697[d]		0.469[d]	
Product distribution						
Transport to retailer	0.74[c]		0.100[c]		0.100[c]	
Total feedstock energy consumed (nE_F)	26.84	32.11	2.754	24.16	0.976	8.56
External energy						
Annualized capital energy investment	2.09	2.50	0.021	0.18	0.014	0.12
Fossil fuels used for system	0.00	0.00	1.016	8.91	0.684	6.00
Transport to retailer[b]	0.74	0.89	0.100	0.88	0.100	0.88
Total external energy consumed (mE_X)	2.83	3.39	1.137	9.97	0.798	7.00
Total energy consumed/ethanol unit		35.50		34.13		15.56
Overall thermal efficiency, $100E_P/(E_F + E_X)$, %		44.5		40.8		60.2
Net energy production ratio, $(E_P - E_X)/E_X$		7.40		1.36		2.36
$[E_P - (nE_F + mE_X)]/(nE_F + mE_X)$		−0.20		−0.31		0.51

[a]Figures may not add because of rounding.
[b]The energy cost per liter to transport ethanol to the retailer is the same as that used in Table 14.13
[c]Included in external energy, transport to retailer.
[d]Included in external energy, fossil fuels used for system.

The overall thermal efficiencies of the IBPCS are about 41 and 60% and compare favorably with that of the integrated, world-scale natural gas system, which is projected to operate at an overall thermal efficiency of 45%. The natural gas case operates at a slightly negative NEPR when all energy inputs, about 90% of which are natural gas, are included. Modern, world-scale, methanol plants based on the steam reforming of natural gas and the reduction of carbon oxides operate at about 70% conversion efficiencies, excluding the efficiencies of finding, producing, and delivering natural gas feedstock to the plant gate. It is believed that there is little opportunity for further significant improvement in conversion efficiencies for methanol synthesis (Mansfield and Woodward, 1993). However, the NEA of the natural gas system for thermochemical ethanol production indicates there are several opportunities for improving operating efficiencies and NEPRs, particularly in the conversion process, which exhibits high energy consumption. Development of a direct synthesis process for converting carbon oxides to ethanol without proceeding through a methanol intermediate would be expected to significantly increase the net ethanol production capability. Thermochemical conversion efficiencies approaching those of methanol may be feasible.

The IBPCS for the lower efficiency biomass case also has a slightly negative NEPR when all energy inputs are accounted for. But when it is assumed that the feedstock energy content is zero, the NEPR as calculated by $(E_P - E_X)/E_X$, which is the approach used for biomass-based systems, is 1.4. This is about three times higher than the NEPRs for fermentation ethanol produced from corn feedstock by an IBPCS of the same ethanol capacity including co-product energy credits, and about 10 to 15 times higher than the NEPRs without co-product credits (Table 14.13).

The higher efficiency biomass case, again with biomass feedstock produced by the management-intensive system and no credit for energy co-products, is projected to be capable of delivering about 3.4 times as much energy in the form of ethanol than was in the fossil-fuel inputs needed to operate the system. The NEPRs of each of the thermochemical biomass cases would be higher still if low-grade feedstocks were used instead of those produced by the system described in Tables 14.8 and 14.9.

It was concluded from these NEAs that the thermochemical synthesis of ethanol from biomass feedstocks can potentially offer significantly more opportunity to deliver net ethanol energy to the consumer than fermentation technology. The NEAs described here also suggest that relatively small-scale thermochemical ethanol plants for biomass feedstocks can compete from a net energy production standpoint with world-scale ethanol plants supplied with fossil feedstocks. The transport problems encountered in the distribution of fuel ethanol-gasoline blends (Chapter 11) might also be alleviated by the wider geographic distribution possible for smaller thermochemical ethanol plants.

The key to the future application of thermochemical ethanol technology depends on completion of its development and commercialization.

From the somewhat limited discussion presented in this chapter on net energy analysis, the value of conducting NEAs for energy- and synfuel-producing systems is apparent. The effort needed to complete detailed NEAs is a time-consuming task, but can provide the basic information needed to select the critical pathway to higher efficiency systems, and to help determine whether the desired results can be achieved. Without the perspective provided by an NEA, the effort to perfect a biomass energy system can easily be misdirected.

It is evident that the design of IBPCSs requires detailed analyses to minimize the problems inherent in large-scale, energy-producing systems in which the various components interact and affect each other. The use of net energy methodologies in these analyses is an invaluable aid, along with economic analyses, for the optimization of the system components and their integrated operation.

REFERENCES

Alich, J. A., Jr., and Inman, R. E. (1974). "Effective Utilization of Solar Energy to Produce Clean Fuel," Final Report, June 1974, under Grant 38723 Initiated May 1, 1973. National Science Foundation, Washington, D.C.

Anonymous, (1983). "The New Jari Project in Brazil: Is It Technically Viable?", "The New Jari: Risks and Prospects of a Major Amazonian Development," *Olsen's Agribusiness Report* 4 (8), 1,4, February.

Appel Consultants, Inc. (1993). "Strategic Analysis of Biomass and Waste Fuels for Electric Power Generation," Final Report, EPRI TR-102773, Project 3295-02. Electric Power Research Institute, Palo Alto, CA, December.

Campbell, K. (1996). *In* "Bioenergy '96," Vol. I, p. 68. Southeastern Regional Biomass Energy Program, Tennessee Valley Authority, Muscle Shoals, AL.

Cooper, J. T. (1995). *In* "Second Biomass Conference of the Americas: Energy, Environment, Agriculture, and Industry," NREL/CP-200-8098, DE95009230, p. 1592. National Renewable Energy Laboratory, Golden, CO; *idem.* "Economic Development Through Biomass Systems Integration, Final Report," Chariton Valley Resource Conservation & Development, Inc., Subcontract No. AAW-4-13326-09. National Renewable Energy Laboratory, Golden, CO.

DeLong, M. M. (1995). "Economic Development Through Biomass Systems Integration," Vols. 1-4, Northern States Power Company, Minneapolis, MN (NREL/TP-430-20517). National Renewable Energy Laboratory, Golden, CO, December.

DeLong, M. M., Swanberg, D. R., Oelke, E. A., Onischak, M., Schmid, M. R., and Wiant, B. C. (1995). *In* "Second Biomass Conference of the Americas: Energy, Environment, Agriculture, and Industry," NREL/CP-200-8098, DE95009230, p. 1582. National Renewable Energy Laboratory, Golden, CO.

Energy Information Administration (1982). *Monthly Energy Review*, DOE/EIA-0035(82/05). U.S. Department of Energy, Washington, D.C., May.

Energy Information Administration (1996). *Monthly Energy Review,* DOE/EIA-0035(96/03). U.S. Department of Energy, Washington, D.C., March.

Fraser, M. D., Henry, J. F., Borghi, L. C., and Barbara, N. J. (1981). *In* "Biomass as a Nonfossil Fuel Source," (D. L. Klass, ed), ACS Symposium Series 144, p. 495. American Chemical Society, Washington, D.C.

Gaggloli, R. A., and Petit, P. J. (1976). *In* "Symposium on Net Energetics of Integrated Synfuel Systems," preprints of papers, **21** (2), 56. 171st National Meeting, American Chemical Society, Division of Fuel Chemistry, New York, April 5-9.

Grassi, G., Ciammaichella, P., Pace, N., and Gheri, F. (1987). *In* "Energy from Biomass and Wastes X," (D. L. Klass, ed.), p. 1545. Institute of Gas Technology, Chicago.

Hoge, W. (1982). "An Amazon Buyout by Brazil." *New York Times,* Business Section, pp. 1,4, January 15.

Hornick, J. R., Zerbe, J. I., and Whitmore, J. L. (1984). *J. Forestry* **82** (11), 663.

Huettner, D. A. (1976). *Science* **192** (4235), 101.

InterTechnology/Solar Corporation (1977). Final Report, U.S. Energy Research and Development Administration, ERDA Contract EX-76-C-01-2548. Washington, D.C., June; *idem.* (1978). Final Report, U.S. Environmental Protection Administration, EPA Contract 68-01-4688. Washington, D.C., July; *idem.* U.S. Defense Advanced Research Projects Report 260675. Washington, D.C.; *idem.* "Solar SNG: The Estimated Availability of Resources for Large-Scale Production of SNG by Anaerobic Digestion of Specially Grown Plant Matter," Report 011075, Project No. IU 114-1. American Gas Association, Arlington, VA.

Klass, D. L. (1976). *In* "Symposium on Net Energetics of Integrated Synfuel Systems," Preprints of papers, **21** (2), 1. 171st National Meeting, American Chemical Society, Division of Fuel Chemistry, New York, April 5-9.

Klass, D. L. (1980a). *In* "Energy from Biomass and Wastes IV," (D. L. Klass and J. W. Weatherly III, eds.), p. 32. Institute of Gas Technology, Chicago.

Klass, D. L. (1980b). *Energy Topics* **1**, April 14.

Klass, D. L. (1983). *In* "Energy from Biomass and Wastes VII," (D. L. Klass, ed.), p. 26. Institute of Gas Technology, Chicago.

Klass, D. L. (1984). *In* "Energy from Biomass and Wastes VIII," (D. L. Klass, ed.), pp. 22, 25. Institute of Gas Technology, Chicago.

Klass, D. L. (1985). *In* "Energy from Biomass and Wastes IX," (D. L. Klass, ed.), p. 27, Institute of Gas Technology, Chicago.

Klass, D. L. (1987). *In* "Energy from Biomass and Wastes X," (D. L. Klass, ed.), p. 55. Institute of Gas Technology, Chicago.

Klass, D. L., and Chambers, W. C. (1976). "A Comparison of the Net Energy Production Ratios of Integrated Systems Supplying Natural Gas, and SNG from Coal." Paper presented at the 171st National Meeting, American Chemical Society, Division of Fuel Chemistry, New York, April 5-9. American Chemical Society, Washington, D.C.

Klass, D. L., and Chambers, W. C. (1977). *Hydrocarbon Processing* **135**, April.

Leach, G. (1975). *Energy Policy* **3** (4), 332.

Mansfield, K., and Woodward, K. (1993). *In* "Asian Natural Gas—Development of the Domestic Industry," (D. L. Klass, H. Kataoka, and Ueda, K., eds.), p. 533. Institute of Gas Technology, Chicago.

National Academy of Sciences (1980). *In* "Firewood Crops, Shrub and Tree Species for Energy Production," pp. 46-47. National Academy of Sciences, Washington, D.C.

Odum, H. T. (1973). *Ambio* **2**, 220.

Ostlie, L. D., and Drennan, T. E. (1989). *In* "Energy from Biomass and Wastes XII," (D. L. Klass, ed.), p. 621. Institute of Gas Technology, Chicago

Perlack, R. D., and Ranney, J. W. (1993). *In* "First Biomass Conference of the Americas: Energy, Environment, Agriculture, and Industry," NREL/CP-200-5768, DE93010050, Vol. III, p. 1855. National Renewable Energy Laboratory, Golden, CO.

Perlack, R. D., Walsh, M. E., Wright, L. L., and Ostlie, L. D. (1996). *Bioresource Technology* 55, 223.

Raymond, D. R., and Kiefer, J. A. (1995). *In* "Second Biomass Conference of the Americas: Energy, Environment, Agriculture, and Industry," NREL/CP-200-8098, DE95009230, p. 1557. National Renewable Energy Laboratory, Golden, CO.

Reese, R. A., Aradhyula, S. V., Tyson, K. S., and Shogren, J. F. (1993). *In* "Energy from Biomass and Wastes XVI," (D. L. Klass, ed.), p. 331. Institute of Gas Technology, Chicago.

Reistad, G. M. (1975). *J. Engineering Power Transactions ASME,* 429, July.

Scheller, W. A. (1981). *In* "Biomass as a Nonfossil Fuel Source," (D. L. Klass, ed.), ACS Symposium Series 144, p.419. American Chemical Society, Washington, D.C.

Shapouri, H., Duffield, J. A., and Graboski, M. S. (1995). "Estimating the Net Energy Balance of Corn Ethanol," Agricultural Economic Report No. 721. U.S. Department of Agriculture, Economic Research Service, Washington, D.C., July.

Shapouri, H., Duffield, J. A., and Graboski, M. S. (1996). "Liquid Fuels and Industrial Products from Renewable Resources, Proceedings of the Third Liquid Fuel Conference," (J. S. Cundiff *et al.,* eds.), p. 253. The American Society of Agricultural Engineers, St. Joseph, MI.

Spaeth, J. J., and Pierce, L. K. (1996). *In* "Bioenergy '96," Vol. I, p. 52. Southeastern Regional Biomass Energy Program, Tennessee Valley Authority, Muscle Shoals, AL.

Stricker, J. A., Rahmani, M., Hodges, A. W., Mishoe, J. W., Prine, G. M., Rockwood, D. L., and Vincent, A. (1995). *In* "Second Biomass Conference of the Americas: Energy, Environment, Agriculture, and Industry," NREL/CP-200-8098, DE95009230, p. 1608. National Renewable Energy Laboratory, Golden, CO.

Toft, A., Bridgwater, T., Mitchell, P., Watters, M., and Stevens, D. (1995). *In* "Second Biomass Conference of the Americas: Energy, Environment, Agriculture, and Industry," NREL/CP-200-8098, DE95009230, p. 1524. National Renewable Energy Laboratory, Golden, CO.

Tyson, K. S. (1993). "Fuel Cycle Evaluations of Biomass-Ethanol and Reformulated Gasoline," Vol. 1, NREL/TP-463-4950, DE4000227, November; Vol. II, Appendices A to I, October. National Renewable Energy Laboratory, Golden, CO.

U.S. Comptroller General (1982). "DOE Funds New Energy Technologies Without Estimating Potential Net Energy Yields," Report to the Congress of the United States, GAO/IPE-82-1, July 26. Washington, D.C.

Wang, M. Q. (1996). "GREET 1.0—Transportation Fuel Cycles Model: Methodology and Use," ANL/ESD-33. Argonne National Laboratory, Argonne, IL, June.

Wang, M. Q. (1997). "Energy Use and Greenhouse Gas Emissions of Biomass Ethanol Calculated with GREET 1.2," memorandum. Argonne National Laboratory, Argonne, IL, February 24.

White, E. H., Abrahamson, L. P., and Robison, D. J. (1995). *In* "Second Biomass Conference of the Americas: Energy, Environment, Agriculture, and Industry," NREL/CP-200-8098, DE95009230, p. 1534. National Renewable Energy Laboratory, Golden, CO.

EPILOGUE

As long as liquid and gaseous organic fuels remain the mainstay of the energy market place, biomass will gradually displace petroleum and natural gas as energy resources. It is inevitable that this will happen because as stated in the Preface, the first derivative of the law of supply and demand is the law of energy availability and cost. And this most assuredly applies to natural gas and petroleum depletion and the accompanying rise in energy prices. In addition, environmental problems are unavoidable. I am not certain that global warming is a reality or that continued fossil fuel consumption over many years is the primary culprit. The data I have examined indicate that they are not, and that the *accumulated loss* of biomass growth areas and the *annual reduction* in photosynthetic fixation of carbon because of changes in land usage have already had and will continue to have a serious impact on the build-up of atmospheric CO_2. However, I am reasonably certain that localized air-quality problems, especially in urban areas, are caused by excessive fossil fuel consumption. This is the other driving force that will stimulate the transformation of biomass into one of the dominant energy resources.

When will this happen? Actually, the transition has already started, albeit slowly. I do not believe, however, that environmental issues alone will lead to rapid growth of biomass energy consumption. The transition from the fossil fuel era to the renewable energy era will gather more momentum when natural gas and petroleum depletion are measurable, new reserves are not found at a sufficient rate to replace what is being consumed, and fossil fuel prices begin to increase disproportionately and irreversibly. My projections indicate these events will begin in earnest between 2030 and 2040, and that the business of energy production and distribution will have changed forever by the end of the twenty-first century.

Units and Conversion Factors

INTRODUCTION

The SI is used throughout this book, as explained in the Preface. The system is reviewed in this appendix. Definitions and sufficient conversion factors are presented to enable the reader to understand the system, and to convert SI units to other common energy units used in the United States. Additional information is presented on the equivalencies of a few common U.S. energy units.

SI has not been uniformly adopted by the United States, even though it is responsible for about one-quarter of the world's annual energy consumption. After a laborious multiyear effort, the Metrication Task Committee of the American Gas Association published its application guide for SI in 1980 for use by its members, but there has been little or no acceptance of the system. Indeed, many more years will pass before SI is integrated into the U.S. energy industry despite the benefits of conducting international energy transactions in SI units that are accepted by most of the world, and the U.S. federal actions that have been taken to encourage the adoption of SI, such as the Metric Conversion Act of 1975. This act declared conversion to SI to be a national policy of the United States.

A.1 DIMENSIONALLY INDEPENDENT BASE UNITS

Unit name	Quantity	Symbol
ampere	electric current	A
candela	luminous intensity	cd
kelvin	temperature	K
kilogram	mass	kg
meter	length	m
mole	amount of substance	mol
radian*	plane angle	rad
second	time	s
sterradian*	solid angle	sr

*Note: Supplementary units.

A.2 DERIVED UNITS WITH SPECIAL NAMES

Name	Quantity	Symbol	Expression in other units
coulomb	electric charge	C	A·s
degree Celsius	Celsius temperature	°C	K
farad	capacitance	F	C/V
henry	inductance	H	Wb/A
hertz	frequency	Hz	s^{-1}
joule	energy, work, heat	J	N·m
newton	force	N	kg · m/s^2
ohm	electric resistance	Ω	V/A
pascal	pressure, stress	Pa	N/m^2
power	watt	W	J/s
volt	electric potential	V	W/A

A.3 MULTIPLES AND SUBMULTIPLES

Factor	Prefix	Symbol
10^{18}	exa	E
10^{15}	peta	P
10^{12}	tera	T
10^{9}	giga	G
10^{6}	mega	M
10^{3}	kilo	k
10^{2}	hecto*	h
10^{1}	deka*	da
10^{-1}	deci*	d
10^{-2}	centi*	c

Factor	Prefix	Symbol
10^{-3}	milli	m
10^{-6}	micro	μ
10^{-9}	nano	n
10^{-12}	pico	p
10^{-15}	femto	f
10^{-18}	atto	a

*Note: Use of these prefixes should generally be avoided except for SI unit multiples raised to powers, such as those for area, volume, moment, etc., and for nontechnical use of centimeter as for body and clothing measurement.

A.4 SPECIAL NAMES FOR SOME UNITS EXPRESSED IN TERMS OF TWO OR MORE BASE UNITS OR UNIT MULTIPLES

Quantity	Unit name	Symbol	Formula
acceleration	meter per second squared	m/s^2	
area	hectare	ha	1 ha = 10^4 m^2
	square hectometer	hm^2	1 hm^2 = 10^4 m^2
	square kilometer	km^2	1 km^2 = 10^6 m^2
	square meter	m^2	
density	gram per liter	g/L	
	kilogram per cubic meter	kg/m^3	
energy and power	kilojoule	kJ	1 kJ = 10^3 J
	kilowatt	kW	1 kW = 10^3 J/s
	kilowatthour	kWh	1 kWh = 3.6 MJ
	megajoule	MJ	1 MJ = 10^6 J
enthalpy	joule per kilogram	J/kg	1 J/kg = 1 m^2/s^2
entropy	joule per kelvin	J/K	1 J/K = 1 m^2·kg/s^2·K
force	kilonewton	kN	1 kN = 10^3 kg·m/s^2
frequency	hertz	Hz	1 Hz = 1 s^{-1}
	megahertz	MHz	1 MHz = 10^6 Hz
heat capacity	joule per kelvin	J/K	1 J/K = 1 m^2·kg/s^2·K
heat flux density	watt per square meter	W/m^2	1 W/m^2 = 1 kg/s^3
heat transfer coefficient	watt per square meter kelvin	W/(m^2·K)	1 W/m^2·K = 1 kg/s^3·K
heating value	joule per cubic meter	J/m^3	1 J/m^3 = 1 kg/m·s^2
	joule per kilogram	J/kg	1 J/kg = 1 m^2/s^2
length	micrometer	μm	1 μm = 10^{-6} m
mass	tonne	t	1 t = 10^3 kg
	tonne	Mg	1 Mg = 10^6 g

Quantity	Unit name	Symbol	Formula
plane angle	degree	°	$1° = (\pi/180)/rad$
	minute	′	$1′ = (\pi/10.8)mrad$
	second	″	$1″ = (\pi/648)mrad$
pressure	bar	bar	$1\ bar = 10^2\ kPa$
	kilopascal	kPa	$1\ kPa = 10^3\ Pa$
	newton per square meter	N/m²	$1\ N/m^2 = 1\ kg/m·s$
	pascal	Pa	$1\ Pa = 1\ n/m^2$
speed of rotation	revolution per minute	r/min	$1\ r/min = (\Pi/30)rad\ /s$
surface tension	millinewton per meter	mN/m	$1\ mN/m = 10^{-3}\ kg/s^2$
	newton per meter	N/m	$1\ N/m = 1\ kg/s^2$
tensile strength	megapascal	MPa	$1\ MPa = 10^6\ Pa$
time	minute	min	$1\ min = 60\ s$
	hour	h	$1\ h = 3.6\ ks$
	day	day	$1\ day = 86.4\ ks$
	year	a	$1\ a = 31.536\ Ms$
torque	newton meter	N·m	$1\ N·m = 1\ kg/s^2$
viscosity, dynamic	millipascal second	mPa·s	$1\ mPa·s = 10^{-3}\ kg/m·s^2$
	pascal second	Pa·s	$1\ Pa·s = 1\ kg/m·s^2$
velocity	meter per second	m/s	
	kilometer per hour	km/h	$1\ km/h = 10^3\ m/h$
viscosity, kinematic	square meter per second	m²/s	
	square millimeter per second	mm²/s	$1\ mm^2/s = 10^{-3}\ m^2/s$
volume	cubic centimeter	cm³	$1\ cm^3 = 10^{-3}\ L$
	cubic decimeter	dm³	$dm^3 = 1\ L$
	cubic meter	m³	$1\ m^3 = 10^3\ L$
	liter	L	$1\ L = 1\ dm^3$
	milliliter	mL	$1\ mL = 10^{-3}\ L$
	microliter	μL	$1\ \mu L = 10^{-6}\ L$

A.5 STANDARD REFERENCE CONDITIONS

The standard cubic meter [m³ (st)], according to the International Gas Union's (IGU) reference conditions, is measured at a temperature of 288.15 K (15°C) and a pressure of 101.325 kPa, dry. The normal cubic meter [m³ (n)], according to the IGU's reference conditions, is measured at a temperature of 273.15 K (0°C) and a pressure of 101.325 kPa, dry. The standard cubic foot (SCF), according to conventional U.S. practice, is measured at a temperature of 60°F (15.6°C) and a pressure of 14.73 lbf/in.², dry. A standard volume of gas (STP), according to conventional scientific practice, is measured at a temperature of 0°C (273.15 K) and 1.00 atmosphere.

The international standard ISO 5024—*Petroleum Liquids and Gases— Measurement—Standard Reference Conditions* published by the International

Organization for Standardization reads: "The standard reference conditions of pressure and temperature for use in measurements on crude petroleum and its products, both liquids and gaseous, shall be 101.325 kPa and 15°C, with the exception of liquid hydrocarbons having a vapour pressure greater than atmospheric at 15°C, in which case the standard pressure shall be equilibrium pressure at 15°C."

A.6 CONVERSION OF SI UNITS TO UNITS COMMONLY USED IN THE UNITED STATES

Quantity	Multiply	by	to Obtain
length	cm	0.3937	inches (in.)
	m	39.37	in.
	m	3.281	feet (ft)
	km	0.6214	miles (mi)
area	cm^2	0.1550	square inches (in.^{2})
	m^2	10.76	square feet (ft^2)
	m^2	0.0002471	acres (ac)
	km^2	247.1	ac
	km^2	0.3861	square miles (mi^2)
	ha	2.471	ac
	ha	1.0764×10^5	ft^2
	hm^2	2.471	ac
	hm^2	1.0764×10^5	ft^2
volume	L	0.2642	gallons (gal)
	L	0.035315	cubic feet (ft^3)
	L	0.006289	barrels (bbl)
	m^3	6.289	bbl
	m^3	264.2	gal
	m^3	35.315	cubic feet (ft^3)
	m^3	2.838	bushel (bu)
	m^3 (st)	35.290	standard cubic feet (SCF)
	m^3 (n)	37.226	SCF
mass	kg	2.205	pounds (lb)
	kg	0.001102	short tons (ton)
	Mg	0.002205	lb
	t	2205	lb
	t	1.1025	ton
pressure	bar	9.869×10^{-7}	atmospheres
	kPa	0.01	bar
	kPa	7.5	millimeters of mercury (mm Hg)
	kPa	0.009869	atmospheres
	kPa	0.1450	pounds per square inch (psi)
	kPa	20.89	pounds per square foot (lb/ft^2)
	MPa	145.0	psi
	kg/m^2	0.2048	lb/ft^2

Quantity	Multiply	by	to Obtain
energy and power	MJ	239.01	kcal*
	GJ	0.9485	10^6 British thermal units (MBtu)*
	EJ	0.9485	10^{15} Btu
	EJ	448,200	barrels oil equivalent/day (BOE/day)
	kW	1.3410	electrical horse power (HP)
	kW	0.1019	boiler horse power (HP)
	kWh	3412	Btu (heating value at 100% efficiency)
	kWh	10,409	Btu (at conversion efficiency of 32.8%)
	MJ	948.5	Btu
	MJ/h	0.3724	horse power (HP)
velocity	km/h	0.6214	miles per hour (mph)
	m/s	2.237	mph
dynamic viscosity	mPa·s	1.000	centipoise (cP)
	Pa·s	1000	cP
heating value	kJ/kg	0.4302	Btu/lb
	MJ/kg	239.0	kcal/kg
	MJ/kg	430.2	Btu/lb
	GJ/t	0.8603	10^6 Btu/ton
	MJ/L	3590	Btu/gal
	MJ/m³ (n)	25.47	Btu/SCF
	MJ/m³ (st)	26.88	Btu/SCF
density	kg/m³	0.06244	lb/ft³
	kg/m³	0.008346	lb/gal
	kg/m³	0.3505	lb/bbl
	kg/m³	0.001	g/cm³
	L/t	0.2396	gal/ton
yield	MJ/ha·year	383.9	Btu/ac·year
	MJ/hm²·year	383.9	Btu/ac·year
	kg/ha·year	0.8924	lb/ac·year
	t/km²·year	2.855	ton/mi²·year
	t/ha·year	0.4461	ton/ac·year
	t/hm²·year	0.4461	ton/ac·year
	L/ha·year	0.0169	gal/ac·year
insolation	W/m²	7.616	Btu/ft²·day
	W/m²	2.065	cal/cm²·day

*Note: Thermochemical values of kcal and Btu used throughout. MBtu is 10^6 Btu.

A.7 SOME EQUIVALENCIES COMMONLY USED IN THE UNITED STATES

	Quantity	Equivalency
molar units	1.0 kgmol	1000 gmol
	1.0 MJ/kgmol	430.2 Btu/lb·mol
	1.0 kJ/gmol	430.2 Btu/lb·mol
energy	1.0 quad	10^{15} Btu
		1.05435 EJ

Quantity		Equivalency
		62.5×10^6 ton of dry biomass
		170×10^6 bbl of crude oil
		470,000 BOE/day for one year
		45×10^6 ton of coal
		10^{12} ft^3 of dry natural gas
biomass	1.0 dry ton	16×10^6 Btu (16 MBtu)
		0.762 ton of coal
		2.857 bbl of crude oil
		16×10^3 ft^3 of dry natural gas
	1.0 cord of wood	128 ft^3 (4 ft $\times$ 4 ft $\times$ 8 ft)
		3.63 m^3
		3000 lb (approx.)
		1361 kg (approx.)
	1.0 bushel (bu)	8 gal
coal	1.0 ton	21×10^6 Btu (21 MBtu)
		1.31 ton of dry biomass
		3.75 bbl of crude oil
		21×10^3 ft^3 of dry natural gas
crude oil*	1.0 bbl	5.6×10^6 Btu (5.6 MBtu)
		5.904 GJ
	1.0 barrel oil equivalent (BOE)	5.6×10^6 Btu (5.6 MBtu)
	1.0 bbl	42 gal
		159 L
		0.159 m^3
		0.35 ton of dry biomass
		0.27 ton (540 lb) of coal
		5.6×10^3 ft^3 of dry natural gas
		0.150 ton
		0.136 t
natural gas	1.0 standard cubic foot (SCF)	1.0 ft^3 (at 60°F, 30.00 in. Hg, dry)
		0.028328 m^3 (st)
		0.026853 m^3 (n)
	1.0 SCF	1000 Btu
	1.0 ft^3	1000 Btu
	10^3 ft^3	0.0625 ton of dry biomass
		0.047 ton (92 lb) of coal
		0.18 bbl (7.4 gal) of crude oil
	1.0×10^6 ft^3	10^9 Btu
	1.0 Mcf	10^9 Btu
	1.0 lbmol	378 SCF (15.6°C, 30 in. Hg, dry)
		359 ft^3 (0°C, 760 mm Hg, dry)
electricity†	1.0 kWh	3412 Btu
		3.6 MJ
		2.655×10^6 ft·lb
		1.341 HPh
	10^3 kWh	0.213 ton of dry biomass
		0.61 bbl of crude oil
		0.163 ton of coal
		3412 ft^3 of dry natural gas

Quantity		Equivalency
power	1.0 kW	3412 Btu/h
		56.87 Btu/min
		44,250 ft·lb/min
		1.341 HP (electrical)
		0.101942 HP (boiler)
	1.0 HP(electrical)	2546 Btu/h
		33,013 ft·lb/min
		0.7457 kW
	1.0 HP (boiler)	33,520 Btu/h
		9.8095 kW

*Note: The heating values of common crude oils are quite variable, as are their chemical compositions and physical properties, and are generally in the range 39.68 to 43.00 MBtu/t. This corresponds to 5.40 to 5.85 MBtu/bbl. Europe commonly uses 10,000cal/g or 39.68 MBtu/t to define the tonne of oil or petroleum equivalent. An average of the range 5.40 to 5.85 MBtu/bbl, or 5.6 MBtu/bbl, is used here.

†Note: Because of the net energy losses associated with electricity generation, it takes about 0.6 ton of dry biomass, 1.8 bbl of crude oil, 0.47 ton of coal, or 10^4 ft³ of natural gas to generate 10^3 kWh of electricity in conventional generation systems.

Table A.1 presents a comparison of approximate fuel equivalents for three energy quantities, 1.0 GJ, 1.0 MBtu, and 1.0 MWh, in SI and U.S. units for dry biomass, bituminous coal, crude oil, and natural gas, assuming the nominal heating values are 16 MBtu/ton, 21 MBtu/ton, 5.6 MBtu/bbl, and 1000 Btu/ft³, respectively.

TABLE A.1   Approximate Fuel Equivalents

Energy quantity	Dry biomass		Coal		Crude oil		Natural gas	
	t	ton	t	ton	m³	bbl	m³	ft³
1.0×10^9 J	0.054	0.059	0.041	0.045	0.027	0.169	26.9	948
1.0×10^6 Btu	0.057	0.063	0.043	0.048	0.028	0.179	28.3	1000
1.0×10^3 kWh	0.193	0.213	0.148	0.163	0.097	0.610	96.7	3412

A.8   PERCENTAGE MOISTURE CONTENT

Throughout this book, a moisture content that is stated in "wt %" units refers to the percentage by weight of moisture in the material of interest containing both dry matter equivalent and moisture. In other words, 200 kg of a material that contains 60 wt % moisture consists of 120 kg of water and 80 kg of dry matter equivalent. Such compositional data are often designated by units of "wt % (wet basis)" in the literature. Another method of designating moisture content that is used in the literature is "wt % (dry basis)." These units are sometimes confusing because the percentage amounts of moisture in a material can then be more than 100 wt %. For example, on a dry basis, the 200-kg sample above contains 150 wt % moisture because dry basis calculations are made by dividing the weight of moisture in the sample by the dry weight of the sample, that is, 120/80. These units are not commonly used in this book. When they are, the designation is "wt % (dry basis)."

Calculation of Fossil Fuel Reserves Depletion Times

The model used for these calculations assumes that the annual growth rate of fuel consumption, P ($\% \times 10^{-2}$), is constant. The reduction in the amount of reserves, R, then corresponds to a compounded fuel consumption. If the baseline annual consumption is C:

Then for year 1:	$C_1 = C + PC = C(1 + P)$
And for year 2:	$C_2 = C(1 + P) + PC(1 + P) = C(1 + P)^2$
And for year 3:	$C_3 = C(1 + P)^2 + PC(1 + P)^2 = C(1 + P)^3$
And for year n:	$C_n = C(1 + P)^n$
Summing for all years:	$\Sigma C = C(1 + P) + C(1 + P)^2 + C(1 + P)^3 + \cdots + C(1 + P)^n$
Let:	$S = \Sigma C/C$
Then:	$S = 1 + (1 + P) + (1 + P)^2 + (1 + P)^3 + \cdots + (1 + P)^n$
Multiplying by $(1 + P)$:	$S(1 + P) = (1 + P) + (1 + P)^2 + (1 + P)^3 + \cdots + (1 + P)^{n+1}$
And subtracting S:	$S(1 + P) - S = (1 + P)^{n+1} - 1$
Or:	$1 + PS = (1 + P)^{n+1}$
And:	$n = [\ln(1 + PS)/\ln(1 + P)] - 1$
Substituting for S:	$n = [\ln(1 + P\Sigma C/C)/\ln(1 + P)] - 1$
At 100% depletion of reserves:	$R = \Sigma C$
Then:	$n = [\ln(1 + PR/C)/\ln(1 + P)] - 1.$

A few examples of the use of this formula are cited here. In their base-case projection, the Energy Information Administration projects the annual growth rate in consumption of petroleum products to be 1.2% from 1992 to 2010. The world's proved and currently recoverable reserves of crude oil as of January 1, 1993, was estimated to be 6448 EJ, or 1092.4 billion barrels of oil [*World Oil* **214**(8), August 1993], and the world's consumption of crude oil was 144 EJ in 1992 (Energy Information Administration). Presuming the growth rate in consumption does not change, the depletion time is then calculated by

$$n = [\ln(1 + 0.012 \times 6448/144)/\ln(1 + 0.012)] - 1$$
$$n = 35.1 \text{ years, or the year 2027.}$$

If the total estimated remaining recoverable reserves are five times the proved and currently recoverable reserves as of January 1, 1993, the depletion time is calculated by

$$n = [\ln(1 + 0.012 \times 32,240/144)/\ln(1 + 0.012)] - 1$$
$$n = 108.4 \text{ years or the year 2100.}$$

For the conditions cited earlier, the year, Y, when the reserves-to-consumption ratio drops to 10 years, for example, is given by

$$Y = 1992 + [\ln(1 + 0.012 \times 6448/144)/\ln(1 + 0.012)] - 11$$
$$Y = 2017.$$

Carbon Dioxide Emissions From Fossil Fuel Combustion, and Human and Animal Respiration

The factors for converting fossil fuel consumption to carbon dioxide emissions were derived from the emissions data in Table C.1. The energy consumption in energy units is multiplied by the factor to obtain the mass of carbon dioxide emitted.

The carbon dioxide emitted by human respiration was estimated as follows. The population was assumed to be 5.64 billion (Table 1.3). The normal rate of respiration of an adult male is 13 to 18 breaths/min (Zoethout, 1948); it increases on muscular exertion. In females, the rate is 2 to 4 breaths/min greater. The rate is 40 to 70 at birth; at 15 years of age, it is about 20. Sleep decreases the rate by up to 25%. For this estimate, the average human respiration rate was assumed to be 15 breaths/min. Expired air is about 4.00 mol % carbon dioxide (Zoethout, 1948). Inspired air contains about 0.03 mol % carbon dioxide; it was ignored for this estimate since it is only 0.75% of that in the expired air. The average volume of air per breath is about 0.5 L (Zoethout, 1948). At standard pressure and temperature conditions, the expired air is assumed to have a density of 1.873×10^{-6} t/L. The calculation of global carbon dioxide emissions from human respiration is then

TABLE C.1   Carbon Dioxide Emissions per Energy
Consumption Unit[a]

Fuel[b]	(lb/MBtu)	(kg/GJ)
Methane	115	49.5
Natural gas	118	50.8
Propane	139	59.8
Petroleum liquids	165	71.0
Bituminous coal	202	86.9
Coal-based liquids	320	138
Coal-based electricity	536	230

[a]Data adapted from Klass (1990).
[b]The emissions for methane and propane were calculated for direct combustion of the pure chemical. The emissions for coal-fired electricity production are for a generic process having 38% thermal efficiency. The range reported for coal-fired electricity production was 536 to 900 lb/MBtu. The other emissions are averages from numerous references.

$$525,600 \text{ min/year} \times 15 \text{ breaths/min} \times 0.5 \text{ L/breath} \times 0.04 \times 5.64 \times 10^9 \text{ people} \times 1.873 \times 10^{-6} \text{t/L} = 1.67 \times 10^9 \text{ t/year}$$

From consideration of the populations and species of air-breathing animals and their projected respiration characteristics, the total global carbon dioxide emission from animal respiration is estimated to be about a factor of 2 times that of human respiration.

REFERENCES

Klass, D. L. (1990). In "Asian Natural Gas—for a Brighter '90s," (D. L. Klass and T. Ohashi, eds.), p. 91. Institute of Gas Technology, Chicago.
Zoethout, W. D. (1948). In "Introduction to Human Physiology," Chapt. 14, p. 186. The C. V. Mosby Company, St. Louis, Missouri.

INDEX